INTRODUCTION TO STATISTICS AND COMPUTER PROGRAMMING
PILOT EDITION
by

Carl F. Kossack and Claudia I. Henschke

HOLDEN-DAY, INC., San Francisco
Düsseldorf   Johannesburg   London   Mexico   Panama
São Paulo   Singapore   Sydney

This book is a PILOT EDITION published in a format that can
be revised and updated quickly in order to incorporate new
developments in both subject matter and computer technology.
Both the publishers and authors are anxious to receive your
comments and suggestions for improvements. These should be
forwarded directly to Holden-Day, Inc., 500 Sansome Street,
San Francisco, CA. 94111

INTRODUCTION TO STATISTICS AND COMPUTER PROGRAMMING

PILOT EDITION

INTRODUCTION TO STATISTICS AND COMPUTER PROGRAMMING

Preface

As our modern society becomes more oriented to the computer age it is expected that the use of statistical analyses will significantly increase. The availability of remote console computer input devices and the existence of integrated information files within an organization will lead to the use of the computer as a computational tool by many individuals.

It is expected that the statistical analyses that will be useful to individuals in the future will differ from the classical techniques, since the increased computer capability will enable one to develop techniques which recognize the particular characteristics of the problem being studied instead of using some standard technique which makes assumptions about the problem that cannot be met. In anticipation of such an approach, this textbook has been written with an emphasis on the role of data analysis in statistics, and has coupled such analyses with the use of decision theory so as to assist individuals in resolving practical problems.

Since individuals will need a programming capability that will enable them to evolve their own computer program to solve particular problems, instruction and experience in computer programming has also been integrated into the text. However, the basic instruction in FORTRAN programming has been concentrated in Chaps. 4, 5, and 6 and is written so that students who are already familiar with programming can skip or skim over these chapters without adverse effect on their use of the rest of the text.

It is felt that the availability of illustrations and problems encountered in a statistical course such as this one will greatly facilitate the instruction in FORTRAN programming. On the other hand, the emphasis on statistical data analysis, both in the text and in the assignments, provides a recognition of the need of the computer as an analytical tool. The approach followed stresses learning-by-doing and thus students are expected to write their own computer programs and to use these programs to solve problems during the course. The assignments are restricted in number, but since each contributes to the appreciation of the concepts being considered, it is expected that every student will work every problem in the exercises. Where needed, more practice problems are also given.

The text book is written for the advanced lower division or beginning upper division student in the social, biological, management and behavioral sciences. Illustrative examples include those from business and personnel management, public health and medicine, manufacturing and production control, and scientific experimentation. The combination of statistical data analyses with decision techniques should provide the student with an insight into the characteristics of different types of problems encountered in applying statistics to real problems.

The textbook is written for either a two semester, three hour per week sequence or a two quarter, five hour per week sequence. Variation could be accomodated since the chapters beyond Chap. 10 do not need to be taken in sequence. If some of these chapters are omitted, it is hoped that students could then study the omitted ones at their own convenience.

Although no mathematical prerequisite other than high school algebra is required for the use of the text, we assume a mature approach on the part of the student. The availability of the computer allows us to introduce such mathematical concepts as integration, matrices, and the use of the more complicated mathematical functions from a numerical evaluation point of view. The use of existing computer routines including the computer simulation techniques enable students to appreciate these mathematical concepts and their usefulness in providing solutions to practical problems.

Since the text utilizes the FORTRAN IV language without specializing it to any computer system, instructors will find it necessary to supplement the programming instructions by introducing their own system constraints, i.e. system cards. In addition, the mechanism by which the programs are made machine readable is left to the instructor, although experience indicates that having students keypunch their own program deck and, in particular, their corrections is a worthwhile approach. If a time sharing, remote console system is to be used, changes will be needed in the input/output concepts presented. It is felt, however, that the instructor can easily introduce such changes.

TABLE OF CONTENTS

PAGE

LIST OF FIGURES

LIST OF PROGRAMS

CHAPTER 1

INTRODUCTION

Most problems encountered in the real world involve an
element of uncertainty. Uncertainty becomes critically important when man tries to solve the more complex problems of
our modern society. In attempting to make more intelligent
decisions, he studies the world about him. More specifically, he makes observations, studies them, makes inferences
based on these observations, and then makes his decisions,
taking into account the uncertainty inherent in this approach. For example, to effectively plan for an adequate
future transportation system, we must take into consideration many elements of uncertainty, including the future
growth and past and present geographic location of the population to be served, and the need. The results of such
studies can then be projected into an estimate of future
needs. Thus, based on the present situation and its projec-
tion, alternative solutions are generally formulated and
then evaluated so that a decision can be made as to which
alternative to follow in making the final plan.

Statistics is involved in each decision-making step.
Statistical sampling and experimental design consider methods of efficiently obtaining representative observations;
descriptive statistics considers methods of summarizing the
observations to more readily appreciate the information contained in the observed data, and statistical inference and
decision theory are used to reach final conclusions.

To illustrate how statistics is involved, let us consider the problem that a college has in determining whether
or not to admit an applicant. Because high school records
often do not provide sufficient information relative to an

1

applicant's potential future success in college, the applicant must also take college entrance examinations and submit
a personal history. These observations are then summarized
into a set of descriptive statistics to facilitate the evaluation of his potential. Next, the relationship between the
observations and the applicant's chance of succeeding in
college must be determined. This can be done by compiling
data about students who have already demonstrated their success in college and whose admission observations are on
record. This experience of both successful and unsuccessful
college students is compared with the data of the applicant
and then used to formulate a decision rule for the new college applicant. However, before the decision rule is used,
the chances of rejecting some successful students and accepting some unsuccessful students should be calculated to
see if the chances of making these errors are sufficiently
small. Statistical decision theory assists in this type of
activity. Thus, from the initial data-acquisition step to
the final decision, statistics is being used.

In considering statistical techniques such as statistical inferences and related decision procedures, three fundamental concepts must be defined: population of interest,
sample, and random variable.

A population is the collection of units about which information is desired. Information is usually desired about
certain variables that can be observed or measured on each
unit in the population, such as age, sex, test results,
weight, and height. For example, the population of interest
may be the candidates for admission to a particular college
for a given year, and information about their high school
grades and college entrance test scores are the variables to
be studied. This population is certainly finite (limited)
because there are only a finite set of grades and scores to
be considered. The population of interest could be the set
of results that would be obtained if a particular student
was given the battery of college entrance tests over and
over. In this case the variable of interest is the college
entrance score, and the possible set of scores is considered
to be infinite (unlimited) because the number of times the

test could be given could always be increased.

Thus populations can be finite or infinite. When the population of interest is infinite, it is not possible to observe all the units in the population. Even for a large finite population, it is usually not economical to study all the units in the population. In such cases, a subset of the observations in the population can be selected; this subset is called a sample. Because the sampled units are chosen by using a selection rule in which chance plays a role, the variable being observed is called the random variable.

Problems of statistical inference arise whenever the sample results are used to make statements about the characteristics of an entire population. Statistics considers how such inferences can best be made and how the uncertainty involved in the inference can be measured.

Statistical techniques have been developed to study problems encountered in many areas, such as quality control, medical diagnosis, personnel assignments, inventory management, manufacturing scheduling, product development, and systems planning and design. Because many statistical techniques involve a large amount of data processing and/or computing, analysts have found the digital computer a useful tool. For this reason, we shall introduce techniques of computer programming and their use in statistical analyses throughout the text. Thus a student will learn how to use the computer to help him solve problems that require statistical analysis. Since the computer can also be used to study how changing inputs or assumptions affect the answers obtained, the student will also find that the computer is a useful tool in gaining insight into how such analyses work. With such an insight an individual should be better able to select the appropriate analysis and also be able to integrate the results of such analyses into meaningful and valid conclusions.

To illustrate how statistical techniques are used to solve problems encountered in the real world, we shall use examples selected from the following four problem areas throughout the text:

(1) In the health and medical science area, the pre-

3

vention and treatment of illnesses is the major problem. General considerations of many problems encountered in this area involve decisions that relate to the lives and health of many people over a long period of time, and as such they should necessarily include consideration of such complex factors as the social interaction of people, nutrition, health practices, and disease susceptibility. All these factors contribute in one way or another to the overall health picture of a community as well as to the effectiveness of any medical treatment given to a sick person. The evaluation of the effectiveness of any health plan as it relates to the general health problem surely involves many different elements of uncertainty. It is easy to recognize that such a general approach involves difficulties of such a magnitude that no satisfactory approach has yet been developed. However, statistics has made a significant contribution to the health and medical science area in more specialized problem areas. We shall consider some of these areas in our illustrations and related exercises.

(2) The research scientist in his study of the laws of nature through controlled experimentation encounters many problems whose solutions can be obtained most effectively by the use of statistical techniques. Although the nature of many of the general research problems being considered by a scientist is usually complex and involves numerous factors, he generally restricts his current considerations in order to be able to study a particular part of the general problem through a sample survey or through some laboratory experiment. In this restricted setting, the scientist must still develop satisfactory response criteria, determine the size of the experiment, isolate the variables that should be controlled, and then properly analyze and interpret the results of the experiment. The data he obtains from his studies are subject to experimental or survey error, and thus there is always the problem of discriminating between those characteristics that represent real effects and those that are due simply to chance. A significant number of statistical techniques have been evolved to assist the scientist in his research design and analysis efforts; we shall introduce some

4

of these in our consideration of different statistical methods.

(3) The development, testing, manufacturing, and operation of physical systems such as automobiles, radios, or even Geiger counters provide still another area where statistical analyses are required. The space-exploration program has brought the problems associated with research and developmental programs in this area to public attention, and we all appreciate the magnitude of the systems research and development effort that was required to land the first man on the moon. Again, each individual decision, such as how long to test a given rocket component or what types of astronomy training should be given to each crew member, had to be coordinated with other decisions encountered within the system to ensure the timely success of the entire man-on-the-moon program. Although the complexity of the NASA space program is fascinating, simpler systems also reflect such problems. Statistics provides a systematic approach to these types of problems, and some of the techniques that we shall introduce have direct application to many such problems.

(4) A major concern of many organizations is personnel utilization because all organizations wish to use their manpower to increase the effectiveness of their organization. This area of study presents many challenging and difficult problems, from the initial development of appropriate measures of individual and organizational effectiveness to the determination of when an individual should be transferred to a new position. There are also simpler problems that require solution in this area, such as whether an individual should be hired for a given job or whether he should be sent to a new training program. Surely uncertainty is present even in these simpler problems because whenever one is dealing with problems directly related to individuals, the actual outcome of any approach is always in doubt. Several examples related to such problems will be used as illustrations or exercises throughout the text.

The primary goals of this textbook are to familiarize students with various statistical techniques and have them gain insight into the fundamental properties of each technique. An important feature in obtaining this insight is the study of the behavior of a technique under different conditions. The use of the computer in this regard will be stressed. Because it is useful for an individual be be able to write his own computer program to perform such studies, an ever-increasing progamming capability will be provided as he works his way through the text. As a long-range goal we hope that this combination of statistics and programming will enable the student to critically evaluate any future problems involving uncertainty that he may encounter and to better determine the appropriate technique to be used in its solution.

Chapters 1 to 3 present descriptive statistic concepts and techniques, emphasizing data tabulation and summarization procedures. The basic principles of computer programming are then introduced in Chaps. 4 to 6 so that the student can take advantage of the computer to perform statistical analyses. The problems of statistical inferences and decision making are presented in Chaps. 7 to 17, with more advanced computer applications being introduced whenever appropriate.

CHAPTER 2

TABULATIONS, GRAPHICAL REPRESENTATIONS, AND
PROPERTIES OF OBSERVED DISTRIBUTIONS

2.1   Introduction

Descriptive statistics are techniques that can be used
to process a set of observations on a variable of interest
so as to convey to the reader the information he requires.
The approaches used can be simple tabulations, graphical
representations, or numerical summarizations. Often the information needed involves some description of how the variable is distributed over the population of interest, so a
model of the distribution must be developed. Descriptive
statistics are also useful for providing the necessary information when such a model is being formulated.

The appropriate descriptive statistical technique for a
particular set of observations depends to a great extent
upon the type of information required and the audience who
will view the results. To avoid getting lost in general
considerations, let us concentrate on a particular problem.
Suppose that the personnel manager of a large organization
is interested in learning about the reading capability of
his employees. He wants to obtain a quantitative measure of
their reading capability, and thus he decides to use a
standard text of a certain length and to measure the time
needed to read the material. He recognizes that comprehension should also be checked and so prepares a true-false
test with 16 questions about the material in the text so he
can determine the number of incorrect responses made by an
individual. Such quantitative measures as the reading time
and number of incorrect responses on the test are called

random variables and are usually designated by capital letters such as Y or X.

In our example, the two variables of interest are of different types. The reading time is obtained by using a measurement device, and the response on the true-false test is obtained by a count. When the observations on a variable are obtained by measurement, the variable is called a continuous random variable. When, however, the observations on the variable are obtained by a counting procedure, the variable is called a discrete random variable.

The observations obtained for a discrete random variable must be from a set of possible values that are known exactly. Observations for a continuous random variable are from a set of values that cannot be known exactly because of the limitation in the accuracy of all measurement devices. Because the values can be measured with only some degree of accuracy, we deal with "recorded" rather than true values. For example, when we say an individual has a reading time of 4.3 min, we know that this is a recorded value involving some error due to the measurement-device limitation, the use of a round-off rule, or both. Since we shall be using such recorded values in much of our future considerations, the student should learn the rules that govern arithmetic operations involving these values; rules are given in Appendix A.

In the following sections we shall show how observations can be summarized and how the sample properties can be discerned.

2.2 Tabulation of Observed Distributions

As the observations on the variable of interest are obtained, each value is recorded. Figure 2.1 illustrates how such data, often called ungrouped data, can be represented in a general form.

General Representation of
Ungrouped Data

Observation, i	Value, Y
1	y_1
2	y_2
3	y_3
.	.
.	.
.	.
i	y_i
.	.
.	.
.	.
n	y_n

Fig. 2.1 Ungrouped data.

Following are examples of ungrouped listings for a continuous random variable and a discrete random variable. These are observations that were obtained when 12 individuals took a reading test.

Continuous Random Variable
Reading Time
(measured to nearest tenth
of a minute)

Discrete Random Variable
Number of Incorrect
Responses on True-False Test
(16 questions)

i	Y	i	Y
1	4.2	1	3
2	5.1	2	0
3	3.3	3	7
4	4.3	4	1
5	5.2	5	3
6	3.1	6	0
7	4.2	7	5
8	5.3	8	4
9	6.1	9	2
10	2.2	10	0
11	4.2	11	1
12	5.2	12	2

Even for this small set of observations it is not easy to assimilate the information contained in the data. And as the number of observations increases, such ungrouped listings become even more difficult to assimilate. If the number of observations is large, but we still wish to present the data, we can tabulate them into a frequency distribution.

The simplest of such a tabulation can be made when the number of different observed values is small, such as when the variable is discrete and the range of observed counts is small. In this case an ordered listing of all the possible different values of the variable can be made. The frequency distribution then simply gives the number of observations that have each value. In such a tabulation we still have a two-column listing, but now the left column is reserved for the ordered list of different possible observed values of the variables, and the right column gives the corresponding frequencies of occurrence.

Figure 2.2 shows the general representation of such a simple frequency distribution; it gives the tabulation of the observations obtained by administering the 16-items reading comprehension test to 390 employees. In the representation, y_j denotes the observation value, and f_j denotes the frequency with which the value occurred.

By using this simple frequency distribution we are able to condense the total number of observations given by $n = 35 + 45 + 120 + \cdots + 0 = 390$ into $m = 17$ entries. We can also readily observe from the simple frequency distribution facts that would be difficult to obtain if all 390 observations were presented in a long ungrouped listing. Thus we can note, for example, that only 1 employee made as many as eight errors, 35 employees had perfect papers ($Y = 0$), and the most frequently occurring score was $Y = 2$ because 120 employees made two errors. We have, however, lost the identity of the individual observations and thus are unable from the frequency distribution to determine such details as the identity of the employee making the largest number of errors.

When it is not practical to list all possible different values of the variable because they are too numerous, we

	Simply Frequency Distribution		Frequency of Incorrect Response on True-False Test (16 questions)	
	Y	F	Y	F
	y_1	f_1	0	35
	y_2	f_2	1	45
	y_3	f_3	2	120
	y_4	f_4	3	112
	.	.	4	42
	.	.	5	23
	.	.	6	10
	y_j	f_j	7	2
	.	.	8	1
	.	.	9	0
	.	.	10	0
	y_m	f_m	.	.
			.	.
			.	.
			16	0

Fig. 2.2 Simple frequency distribution.

tabulate the observations into nonoverlapping intervals that also cover the entire observed range of values. A general representation is shown in Fig. 2.3, along with a numerical illustration. (The approach will be discussed in greater detail.) In such a tabulation, the left column indicates the end points of each interval; the corresponding frequency is given in the right column.

Note that in the numerical example the end points of the intervals are designated in tenths in sympathy with the recording precision. However, to determine the true end points (class boundaries) and the midpoint of each interval, one must undo the process used in recording the observations. In our illustration the reading time was recorded to the nearest tenth; appropriate class boundaries are given after Fig. 2.3.

11

| | Frequency Distribution of Reading |
| | Time for a Group of 390 New |

Regular			Employees	

Frequency Distribution			(recorded to the nearest tenth)	

Interval	F		Interval	F
$b_1 - c_1$	f_1		2.0 - 2.4	21
$b_2 - c_2$	f_2		2.5 - 2.9	43
$b_3 - c_3$	f_3		3.0 - 3.4	68
.	.		3.5 - 3.9	97
.	.		4.0 - 4.4	72
.	.		4.5 - 4.9	53
$b_j - c_j$	f_j		5.0 - 5.4	21
.	.		5.5 - 5.9	11
.	.		6.0 - 6.4	4
.	.			
$b_m - c_m$	f_m			

Fig. 2.3 Regular frequency distribution.

Interval	F	Class Boundaries	Midpoint
2.0 - 2.4	21	1.95 - 2.45	2.2
2.5 - 2.9	43	2.45 - 2.95	2.7
3.0 - 3.4	68	2.95 - 3.45	3.2
.	.	. .	.
.	.	. .	.
.	.	. .	.

One can appreciate the mechanism used to evolve the
class boundaries by realizing that an individual whose reading time is recorded as 3.1 could have a score anywhere between 3.05 and 3.15; likewise, a recorded time of 2.5 could
fall anywhere in the range 2.45 to 2.55. We might be concerned about where we would place a person whose score was
exactly 2.45 because this value occurs as both the upper
class boundary of the first class and the lower class
boundary of the second class. The placement of each indivdual into the interval has already been determined by the recording mechanism used at the data-acquisition stage. The

midpoint of each interval is obtained by averaging the two true class boundaries of each interval.

For a discrete random variable the interval designation and the class boundaries would be the same as in the following illustration:

Frequency Distribution
Number of Incorrect Responses

Interval	F	Class Boundaries	Midpoint
0 - 1	80	0 - 1	0.5
2 - 3	232	2 - 3	2.5
4 - 5	65	4 - 5	4.5
6 - 7	12	6 - 7	6.5
8 - 9	1	8 - 9	8.5

In the discrete variable case, the true range of the first interval is still 0 to 1 because the number of incorrect responses on the true-false test is actually counted and recorded exactly.

The arbitrary round-off rules that can be used in recording data are (1) round to the nearest unit, (2) round to the next larger unit, and (3) round to the next smaller unit. To determine the true class boundaries in each of these cases, we must first determine the recording accuracy a, which is given by

$$a = b_{j+1} - c_j$$

where b_{j+1} is the lower interval value for the j+1st interval and c_j is the upper interval value for the jth interval. In our illustration we find that a = 2.5 - 2.4 = 0.1. We denote the lower class boundary by ℓ_j and the upper class boundary by u_j. These class boundaries can then be obtained for each round-off rule as follows:

(1) Recorded to nearest unit

$$\ell_j = b_j - \frac{a}{2}$$

$$u_j = c_j + \frac{a}{2}$$

(2) Recorded to next higher unit

$$\ell_j = b_j - a$$

$$u_j = c_j$$

(3) Recorded to next lower unit

$$\ell_j = b_j$$

$$u_j = c_j + a$$

For a discrete variable,

$$\ell_j = b_j$$

$$u_j = c_j$$

In all cases, the midpoint of each interval is found by averaging the upper and lower class boundary

$$y_j = \frac{u_j + \ell_j}{2}$$

We have already illustrated how round-off rule (1) is applied; we shall now illustrate the other two rules for the same continuous variable.

(2) Recorded to the next largest tenth

Interval	Class Boundary	Class Midpoint
2.0 - 2.4	1.9 - 2.4	2.15
2.5 - 2.9	2.4 - 2.9	2.65

(3) Recorded to the nearest smaller tenth

Interval	Class Boundary	Class Midpoint
2.0 - 2.4	2.0 - 2.5	2.25
2.5 - 2.9	2.5 - 3.0	2.75

The rules used in making up the intervals for a frequency distribution are flexible, but usually the following criteria should be satisfied:

(1) The number of intervals should be between 8 and 15.

(2) The intervals should cover the range, so the first interval must include the smallest (or largest) observation, and the last interval must include the largest (or smallest) observation.

(3) The widths of the intervals should all be equal.

(4) The use of round-numbers (multiples of 10, 5, or 2) is desirable in designating midpoints or intervals.

(5) Each observation must fall into one and only one interval.

Some of the above criteria may not always be desirable. For example, sometimes unequal class intervals are used to display particular characteristics of observed data.

2.3 Graphical Representations of Distributions

There are many ways of making a graphical representation of observed data, but we shall restrict our attention to bar graphs, histograms, frequency polygons, and ogives. We shall use the two distributions whose tabulations are given in Figs. 2.4 and 2.5 to illustrate such graphical representations.

Employees

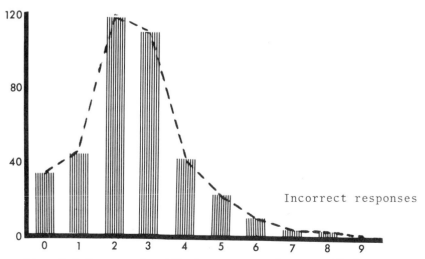

Fig. 2.4 Bar graph with frequency polygon. Number of incorrect responses on true-false test.

Employees

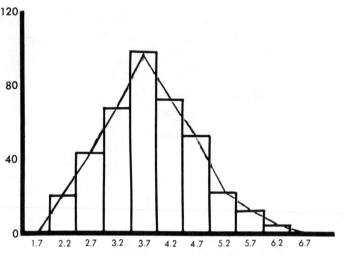

Reading time

Fig. 2.5 Histogram with frequency polygon. Reading
time of a group of 390 new employees.

Discrete distributions are usually represented by bar
graphs, and continuous distributions are usually graphically
displayed by histograms. In both cases, the midpoints of
each interval can be connected to form the frequency polygon.
Discrete polygons are indicated by dotted lines; continuous
polygons are denoted by continuous lines.

In making such graphs, relative frequencies can be used
instead of the absolute frequencies by converting each observed frequency to its corresponding fraction of the total
number of observations. The use of these relative frequencies makes it easier to compare the data with theoretical
models that consider probabilities instead of frequencies.

Another graphical representation is based on the cumulative frequencies. In this case, we associate with each

16

interval the number of observations whose value is less than (greater than) or equal to the upper (lower) interval of the class. If we are using relative frequencies, we use the cumulative fraction of the individuals in the study instead of the cumulative frequency. We graph such cumulative frequencies by using a frequency polygon, which in this case is also called an ogive.

We demonstrate these ideas by using the frequency distribution of reading times considered above and obtaining the cumulative (less than) frequencies and relative frequencies (Fig. 2.6). The corresponding ogive is given in Fig. 2.7.

Reading Time

(measured to nearest tenth)

Interval	F	Less than cum F	rel F	cum(rel F)
2.0 - 2.4	21	21	0.0538	0.0538
2.5 - 2.9	43	64	0.1103	0.1641
3.0 - 3.4	68	132	0.1744	0.3385
3.5 - 3.9	97	229	0.2487	0.5872
4.0 - 4.4	72	301	0.1846	0.7718
4.5 - 4.9	53	354	0.1359	0.9077
5.0 5.4	21	375	0.0538	0.9615
5.5 - 5.9	11	386	0.0282	0.9897
6.0 - 6.4	4	390	0.0103	1.0000

Fig. 2.6 Relative and cumulative (less than) relative frequency distribution.

Students should note that the less than ogive is obtained by connecting the upper end points of the intervals rather than the midpoints because the cumulative frequencies are less than or equal to the end point of each interval.

In observing the graphs students should appreciate that in many cases only one of the possible graphical representations is needed. In addition, the following points may be mentioned:

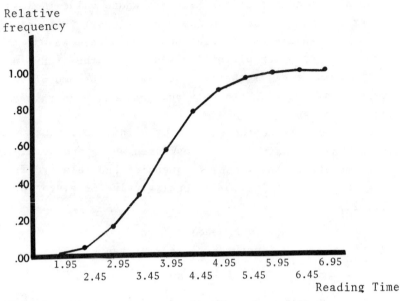

Fig. 2.7 Less than ogive for the reading
time of 390 new employees.

(1) Graphs need a title, both axes labeled, and the
    scale indicated.

(2) Bar graphs are used to represent the distribution
    of a discrete variable because the gaps between the
    bars may serve to emphasize the discrete nature of
    the variable being graphed.

(3) Dotted lines are used for discrete data; solid
    lines are used for continuous data.

(4) The frequency polygon should be "anchored" by in-
    troducing zero-frequency intervals.

(5) Areas are proportional to observed frequencies.
    This is important when the intervals have unequal
    widths.

2.4 The Characteristics of Observed Distributions

We have considered how to tabulate data into a frequency distribution and how to display the distribution graphi-

cally. Since one reason behind these activities is to display the characteristics exhibited by the distribution, we
shall consider some of the more meaningful characteristics
in this section and illustrate how changes in these characteristics affect the distribution.

To describe these characteristics we shall use a smooth
curve rather than the frequency polygon to represent the
graph of the distribution. Now the smooth curve is used as
a matter of display convenience, but later in our considerations of mathematical models for distributions we shall actually introduce such smooth representations more formally.
We then think of the graphical representation of an observed
distribution as represented in Fig. 2.8:

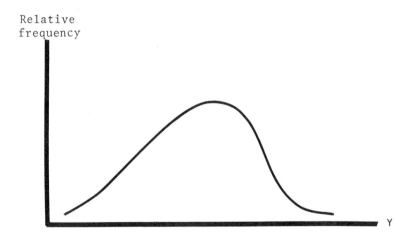

Fig. 2.8 A smooth frequency polygon of a distribution
of an observable random variable.

There are four characteristics generally associated
with the distribution of a random variable. The first relates to the fact that observations generally have a tendency to center or group themselves around a central value.
Thus, we are interested in obtaining a measure that reflects
the value about which this central tendency occurs. One
measure of central tendency is sometimes called an average.
For example, we may make a study of the reading time of

19

males under 20 years of age who have recently joined our organization and another study of the reading time of 60-year old males who have been in the organization for 10 or more years. The results obtained may show differences, like those formally represented in the curves of Fig. 2.9 when they are plotted.

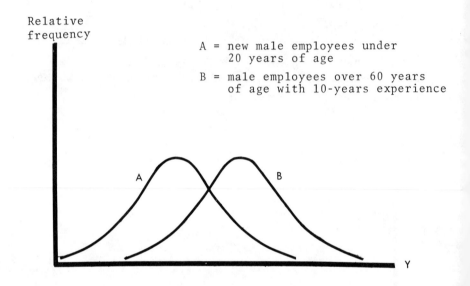

Fig. 2.9 Distributions with different central tendencies.

By looking at these two distributions we see that they differ as to the point about which they center. Thus, any measure of central tendency should indicate the location of the point about which the distribution is centered.

Next let us consider the distributions in Fig. 2.10. Although the two distributions exhibit essentially the same central tendency, distribution A shows less variation about the central point than does distribution B. Thus the observations in distribution A have values that cluster closer around the center of the distribution than the observations from distribution B. We refer to this characteristic as dispersion or variability. One of our tasks is to evolve a numerical measure that indicates the amount of dispersion

Relative
frequency

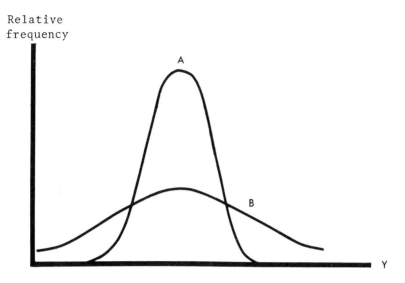

Fig. 2.10 Distributions with different dispersions.

exhibited by a distribution.

Let us continue to explore how distributions may vary.
Consider Fig. 2.11. Figure 2.11 shows the graphs of three
different distributions. Distribution A is symmetrical in
the manner in which the observations distribute themselves
about the point of central tendency; distribution B (though
having about the same central tendency and the same amount
of dispersion as distribution A) shows a lack of symmetry by
there being some extremely large observations for which there
are no compensating extremely small ones. Distribution C
reflects the opposite type of asymmetry, with many observations having small values and only a few observations having
large values. We are interested in being able to measure
the amount and direction of the asymmetry in a distribution.

The distribution of heights of adult males is an example of a symmetrical distribution, and the distribution of
weights of the same group will have a "positive" asymmetry
because there will be some extremely heavy men whose over-
weight is not matched by men with the same amount of under-
weight. For the "negative" asymmetry case we can consider
the distribution of test scores of students. With the tests
averaging (centering) about 70, the largest possible score

21

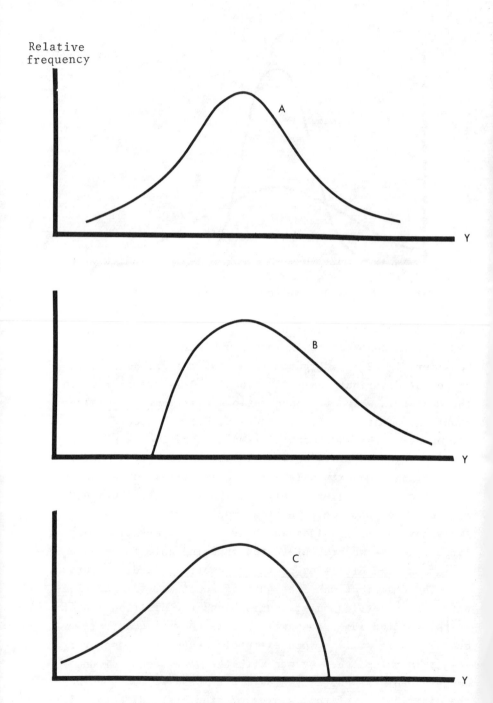

Fig. 2.11 Distributions with different asymmetry.

of 100 is 30 points above the average; the zero score is 70
points below the average.

The final characteristic is exhibited by Fig. 2.12. As
the standard we use the bell-shaped symmetrical distribution
labeled A. This distribution is called the normal distribution, which will be considered in detail later in the text.

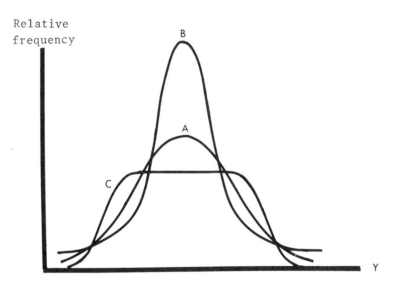

Fig. 2.12 Distributions with different peakedness.

Distribution B, although it has the same central tendency, the same amount (aggregate) of dispersion, and the
same symmetry, is quite peaked, whereas distribution C is
quite flat. We refer to this characteristic as peakedness.
We would like a measure that would tell us if a distribution
is similar to A, B, or C. We are therefore looking for numerical measures that will indicate the (1) central tendency
of a distribution, (2) amount of dispersion, (3) amount and
direction of asymmetry, and (4) amount of peakedness. Furthermore, we need to be able to calculate these measures for
ungrouped observed data, for both simple and regular frequency distributions, and for the mathematical models we
shall use to represent the observed distributions. We con-

23

sider the problems of finding these numerical measures for observed data in the next chapter and for mathematical models in subsequent chapters.

2.5 Summary

In the preceding discussion we introduced the concept of using data to obtain information about the distribution of a variable. The need for a model to represent the distribution is present whenever we attempt to reduce extensive observed data to a few characteristic measures or to perform any type of statistical inference. Thus in both situations, (1) using observed data to obtain information about the distribution of a random variable and (2) reducing a large number of observed data to a few measures, a model is required.

In this introductory chapter on descriptive statistics we have noted that if one is studying the distribution of a variable through the acquisition of actual observed data, such data are ungrouped when initially recorded. Because one of our problems is to use the observed data to determine the essential characteristics of the distribution, we can often simply tabulate or group the data so we can "see the forest." In addition, such tabulation can then be graphed to present a picture of the observed distribution. Finally, for any real future analysis we must determine a mathematical model that approximates the form of the observed data; the model could then be used in future analyses in place of the actual data.

EXERCISES

1. The following observations represent measures of DDT in parts per million (ppm) reported in 40 soil samples:

1.2	1.1	1.6	4.3	0.9	3.7	5.1	2.8
0.8	2.2	2.5	5.4	2.8	4.1	3.9	1.5
1.4	2.9	1.6	5.5	3.2	1.8	2.4	3.1
1.9	3.5	1.8	2.9	2.1	3.7	3.3	4.6
1.5	3.2	2.2	2.4	2.1	2.1	1.2	1.6

(a) Make a frequency table for the observations.
(b) Draw the corresponding histogram with the frequency polygon.
(c) Plot the less than ogive.

2. The following data represent measurements of the error in 50 radar observations of the horizontal component of the position vector of a stationary communications satellite. The unit of measurement is one nautical mile.

0.32	0.24	1.47	1.57	1.53
0.35	1.35	0.77	1.20	1.03
1.64	0.94	0.84	1.15	0.94
0.83	1.05	0.32	1.38	1.81
1.24	1.12	1.11	1.62	1.43
1.92	1.05	0.53	1.34	1.55
1.43	1.14	0.65	1.95	1.12
0.65	1.78	1.23	1.44	0.99
0.93	1.45	1.35	0.82	1.19
1.64	1.35	0.64	0.79	0.83

(a) Set up a frequency table, and plot the corresponding histogram.
(b) Find the cumulative (more than) frequencies, and plot the more than ogive using relative frequencies.
(c) Using the plotted ogive, estimate the value of the observation (percentile) such that (i) 50 percent (fifty percentile), (ii) 75 percent (seventy-five percentile), and (iii) 25 percent (twenty-five percentile) of the observations are below the value

obtained.

3. Adequate facilities for patient care is an important facet of a total community health plan. The available personnel, sociomedical pattern of the community, and many other considerations are involved in determining such a plan. A first step is to investigate the present status of the facilities. Then a pertinent question is to determine the average number of days of hospitalization per patient in each hospital in the community. Because of the multiplicity of causes of hospitalization, it is decided to restrict the current study to measuring the distribution of the number of days that patients remain hospitalized subsequent to minor surgery. Two hospitals in the community are sampled. The number of days of hospitalization of a sample of 25 patients in each hospital is

Hospital A: 1, 3, 1, 2, 3, 3, 3, 4, 2, 4, 6, 3, 6,
              5, 4, 3, 3, 5, 3, 4, 5, 3, 4, 4, 2

Hospital B: 1, 3, 4, 1, 2, 6, 8, 1, 3, 4, 1, 2, 7,
              2, 1, 2, 3, 4, 1, 2, 3, 5, 1, 3, 2

(a) Make a simple frequency table for the days of hospitalization for each hospital, and plot the two bar graphs using relative frequencies.

(b) Using the bar graph representation, compare the two distributions in terms of the four characteristics discussed in the chapter (central tendency, dispersion, asymmetry, and peakedness).

(c) Another method of comparing two distributions is to plot the cumulative less than frequency polygon of each one on the same graph. Then certain percentile points such as the 25, 50, and 75 percent ones can be compared. Plot the polygons on the same graph, and make these three comparisons.

4. A newly automated manufacturing process has an output of 1,000 programmed circuit cards per hour. The manager is

concerned about the quality of the production method, so as a check on the quality he picks five cards at random from each hour's production. He submits these five cards to a destructive test to determine whether they are defective. The results of the 24 samples studied during 1 day are

No. of defectives observed (in random samples of size 5)
$$\begin{cases} 1,\ 0,\ 2,\ 0,\ 3,\ 1,\ 1,\ 2,\ 4, \\ 2,\ 1,\ 2,\ 3,\ 1,\ 1,\ 2,\ 1,\ 3, \\ 0,\ 1,\ 2,\ 3,\ 1,\ 3 \end{cases}$$

(a) Assuming that there are no changes in the production process during the day, the 24 samples of five cards can each be considered random samples of the production. Set up a simple frequency table, and plot the corresponding bar graph for the observed number of defective items.

(b) If no defects are observed in a sample of five cards, then the estimated probability of obtaining a defective item is zero ($p = 0$); when one defective item is observed, then the estimated probability of obtaining a defective item is $p = 0.20$; and so forth. If management considers the manufacturing process superior when $p \leq 0.05$ (5 defectives in 100 items), the process satisfactory when $p = 0.10$ (10 defectives in 100), and the process unsatisfactory when $p = 0.25$ (25 defectives in 100), how would you rate the process using the results of the 24 samples? State your reason.

CHAPTER 3

THE NUMERICAL CHARACTERISTICS OF OBSERVED DISTRIBUTIONS

3.1 Introduction

In this chapter we shall develop numerical measures of the characteristics of the distribution of a variable using observed data. The characteristics that we are interested in measuring are central tendency, dispersion, asymmetry, and peakedness. We need to define these numerical measures and to compute their values from ungrouped data, simple and regular frequency distributions. In addition, we need to interpret these measures relative to the charactistics of interest. The satisfactory solution of this latter requirement, however, must be delayed until we have considered different mathematical models that are available for representing the distribution of the observed variable.

3.2 The Algebra of Summations

Before developing numerical measures of the four characteristics, let us introduce the summation notation that we shall use in subsequent formulas. We are often interested in summations such as

$$SUM = y_1 + y_2 + \cdots + y_n$$

where the three dots indicate that we continue the summation process until we have stepped through all the values of index i from 1 to n. It is convenient to use a shorthand notation for a summation process, such as the one on the right-hand side of the above equation: the Greek capital letter Σ (sigma) will represent the process. We can thus write

28

$$\text{SUM} = \sum_{i=1}^{n} y_i = y_1 + y_2 + \cdots + y_n$$

When we use the symbol $\sum_{i=1}^{n}$ ( ), the expression within the parentheses that involves the index i (parentheses may not always be needed) is "incremented" by letting i take on successively the values 1, 2, 3, and so forth, up to n (or to whatever limit is indicated). These n values are then added to obtain the value of the summation. The index can be any alphabetical character, but usually i, j, or k are used. Thus

$$\sum_{i=1}^{7} y_i = y_1 + y_2 + y_3 + y_4 + y_5 + y_6 + y_7$$

$$\sum_{j=3}^{m} (y_j^2 - 2y_j) = (y_3^2 - 2y_3) + (y_4^2 - 2y_4) + \cdots + (y_m^2 - 2y_m)$$

and

$$\sum_{y=1}^{5} \frac{1}{y} = \frac{1}{1} + \frac{1}{2} + \frac{1}{3} + \frac{1}{4} + \frac{1}{5}$$

In addition to the summation notation, we shall frequently use the concept of a numerical function. A numerical function defines the relationship between two sets of numbers; for example,

$$G(Y) = 10 + 2Y \quad Y = 1,2,3, \ldots , n$$

where for each possible value of Y a unique value G(Y) is obtained. Thus when Y = 1, G(Y) = 12; when Y = 2, G(Y) = 14; and so on. For G(Y) to be a function, there can be only a single value G(Y) for each value of Y.

Suppose we wish to sum the values G(Y). We can write

$$\sum_{j=1}^{n} G(y_j) = G(y_1) + G(y_2) + \cdots + (G(y_n)$$

For example, when G(Y) = 10 + 2Y,

29

$$\sum_{j=1}^{4} G(y_j) = \sum_{j=1}^{4} (10 + 2y_j)$$

$$= (10 + 2y_1) + (10 + 2y_2) + (10 + 2y_3) + (10 + 2y_4)$$

$$= 40 + 2y_1 + 2y_2 + 2y_3 + 2y_4$$

The Σ symbol will be most useful when writing statistical formulas and equations. As we become more familiar with this notation we may at times (when there is no ambiguity of reference) omit the index and its limits and write $\Sigma G(y_i)$ instead of $\sum_{i=1}^{n} G(y_i)$.

The summation symbol has only three simple algebraic properties that we need to consider:

(1) We can separate the sum (difference) of two (or more) terms into separate summations.

$$\sum_{i=1}^{n} [G(y_i) \pm H(y_i)] = \sum_{i=1}^{n} G(y_i) \pm \sum_{i=1}^{n} H(y_i)$$

(2) We can factor any constant (a quantity that is unaffected by the index) to the outside of the summation sign.

$$\sum_{i=1}^{n} cG(y_i) = c \sum_{i=1}^{n} G(y_i)$$

(3) The sum of a constant is the constant multiplied by the number of terms in the summation.

$$\sum_{i=1}^{n} c = nc$$

We leave the verification of the three properties to the student. Now we shall use this summation notation in developing the numerical measures of the characteristics of interest.

3.3 Numerical Measures of the Characteristics
of Observed Data

Numerical measures of the four characteristics can be obtained by first introducing the concept of moments of a numerical function of a random variable. We define the hth moment of a numerical function of a random variable $G(Y)$ by:

$$\text{hth moment of } G(Y) = \sum_{i=1}^{n} \frac{G(y_i)^h}{n}$$

$$= \frac{G(y_1)^h + G(y_2)^h + \cdots + G(y_n)^h}{n}$$

We shall show that we can obtain the required numerical measures of the four characteristics by specializing the function $G(Y)$ and the value of h. We shall discuss each characteristic in turn in this section and present the techniques of obtaining its measure for both ungrouped and grouped data in the following sections.

A measure of central tendency

We obtain a measure of central tendency by letting $G(y_i) = y_i$ and $h = 1$. Thus, for ungrouped data,

$$m_1 = \frac{\sum_{i=1}^{n} y_i}{n} = \frac{y_1 + y_2 + \cdots + y_n}{n} = \bar{y}$$

which is called the first moment about the origin or the arithmetic mean of the observed y_i's.

To support the fact that m_1 is a measure of central tendency, consider what would happen to m_1 if each observation were increased by 50; that is, for each i, we define $y_i^* = y_i + 50$. It is apparent that the distribution of the new variable Y* has been shifted over 50 units, and thus the central tendency of Y* should be the central tendency of Y plus 50. To establish that the m_1's are so related we use the algebra developed for the summation notation. To distinguish between the central tendency of Y and Y*, we let $m_{1:y}$ represent the central tendency of Y, and $m_{1:y*}$ represent the central tendency of Y*.

31

$$m_{1:y^*} = \frac{\sum_{i=1}^{n} y_i^*}{n} = \frac{\sum_{i=1}^{n} (y_i + 50)}{n} = \frac{\sum_{i=1}^{n} y_i + \sum_{i=1}^{n} 50}{n}$$

$$= \frac{\sum_{i=1}^{n} y_i}{n} + \frac{n \cdot 50}{n} = \frac{\sum_{i=1}^{n} y_i}{n} + 50 = m_{1:y} + 50$$

showing that m_1 reflects the 50-unit shift of the central tendency.

A measure of dispersion

To obtain a measure of dispersion, we let $G(y_i) = (y_i - \bar{y})$, and take $h = 2$. We thus obtain

$$v_2 = \frac{\sum_{i=1}^{n} (y_i - \bar{y})^2}{n}$$

which is called the second moment about the mean. By using the deviation of each observation from its mean $(y_i - \bar{y})$, we have removed the effect of the mean and thus are averaging the square of the deviations about the mean.

To note that v_2 is a measure of dispersion, consider the following three distributions of ages:

	Distribution A			Distribution B			Distribution C		
i	Y	$Y-\bar{Y}$	$(Y-\bar{Y})^2$	Y	$Y-\bar{Y}$	$(Y-\bar{Y})^2$	Y	$Y-\bar{Y}$	$(Y-\bar{Y})^2$
1	15	0	0	3	-12	144	3	-12	+144
2	15	0	0	20	5	25	10	-5	+25
3	15	0	0	15	0	0	15	0	0
4	15	0	0	17	2	4	20	+5	+25
5	15	0	0	20	5	25	27	+12	+144
Sum	75	0	0	75	0	198	75	0	338

$$\bar{y} = \frac{\Sigma y_i}{n} = \frac{75}{5} = 15 \qquad \bar{y} = \frac{75}{5} = 15 \qquad \bar{y} = \frac{75}{15} = 15$$

$$v_2 = \frac{\Sigma (y_i - \bar{y})^2}{n} = \frac{0}{5} = 0 \qquad v_2 = \frac{198}{5} = 39.6 \qquad v_2 = \frac{338}{5} = 67.6$$

Distribution A has no dispersion because all observations have the same value, which is reflected by $v_2 = 0$. Distribution B has a limited amount of dispersion, which results in $v_2 = 39.6$, and distribution C has a much larger dispersion, which is reflected in the large value of $v_2 = 67.6$. This demonstrates that v_2 can be considered as a measure of dispersion. Note also that $\Sigma(y_i - \bar{y}) = 0$ for all three distributions. It can be shown that $v_1 = \Sigma(y_i - \bar{y})/n = 0$ for any distribution and thus is not useful in describing the distribution. We call v_2 the variance and use the symbol s_y^2 or s^2 to represent it. The use of a square in s_y^2 recognizes that the variance has square units as its dimension. Since a measure involving square units is often awkward to interpret, we can reduce the dimension to that of the original measurements by taking the square root of the variance. We call this measure the standard deviation; its definition[1] is

$$s = s_y = \sqrt{v_2} = \sqrt{\frac{\Sigma(y_i - \bar{y})^2}{n}}$$

At present we use the standard deviation to compare the dispersion of one distribution with that of another distribution, as we did for the previous distributions A, B, and C.

A measure of asymmetry

To obtain a measure of asymmetry we let $G(y_i) = (y_i - \bar{y})/s$ and $h = 3$, which yields

$$a_3 = \frac{\sum_{i=1}^{n}\left[(y_i - \bar{y})/s\right]^3}{n} = \frac{1}{s^3}\frac{\sum_{i=1}^{n}(y_i - \bar{y})^3}{n}$$

This measure is called the third standard moment because it is dimensionless. The loss of the dimension is due to the fraction $(y_i - \bar{y})/s$ being used. This means that a_3 is independent of the choice of the "yard stick" used in measuring

---

[1] Students may encounter a similar formula with $n - 1$ as the divisor. We shall introduce this variation in Chap. 10.

the variable as long as an arithmetic scale of measurement is used (not a nonlinear scale such as the logarithmic scale). Under this restriction we are free to choose whatever measurement unit we please (for weight, we could use either pounds, ounces, grams, kilograms, or stones), and the resulting a_3 would have the same value. This property is surely not true for m_1 or v_2.

Let us examine why a_3 is a measure of asymmetry by comparing the symmetrical distribution B with two asymmetrical distributions A and C:

	Distribution A			Distribution B			Distribution C		
i	Y	Y-Ȳ	$(Y-\bar{Y})^3$	Y	Y-Ȳ	$(Y-\bar{Y})^3$	Y	Y-Ȳ	$(Y-\bar{Y})^3$
1	3	-12	-1728	3	-12	-1728	3	-12	-1728
2	5	-10	-1000	10	-5	-125	20	5	125
3	6	-9	-729	15	0	0	15	0	0
4	9	-6	-216	20	+5	+125	17	2	8
5	52	+37	+50653	27	+12	+1728	20	0	125
Sum	75	0	46980	75	0	0	75	0	-1470

$$\bar{y} = \frac{\Sigma y_i}{n} = \frac{75}{15} = 15 \qquad \bar{y} = \frac{75}{5} = 15 \qquad \bar{y} = \frac{75}{5} = 15$$

$$s = \sqrt{\frac{\Sigma(y_i - \bar{y})^2}{n}}$$

$$= \sqrt{\frac{1730}{15}} = 18.6 \qquad s = \sqrt{67.6} = 8.22 \qquad s = \sqrt{\frac{198}{5}} = 6.29$$

$$a_3 = \frac{\Sigma(y_i - \bar{y})^3}{ns^3}$$

$$= \frac{46980}{(5)(18.6)^3} \qquad a_3 = \frac{0}{5(8.22)^3} \qquad a_3 = \frac{-1470}{5(6.29)^3}$$

$$= 1.46 \qquad\qquad = 0 \qquad\qquad = -1.18$$

We note that in the symmetrical distribution B, a_3 equals zero because for every positive deviation from the mean there is a compensating negative deviation. Distribution A has a positive value for a_3 because the observation of 52 has no compensating value below the mean, and thus

34

the large positive deviation (52 - 15) = 37, which is then
cubed, dominates the sums of the cubed deviations. A similar reasoning applies to distribution C, except that there
is a large negative deviation from the mean without any compensating positive deviation. Thus asymmetrical distributions with a "positive tail" have values for a_3 that are
greater than zero, and asymmetrical distributions with a
"negative tail" have values for a_3 that are less than zero.
The numerical value of a_3 reflects the extent of the asymmetry or <u>skewness</u> of the distribution, with the sign of a_3 indicating the direction of the asymmetry. Right now we cannot give any mathematically based interpretation of how the
value of a_3 reflects the lack of symmetry of a distribution.
However, distributions having an a_3 numerically less than
0.2 hardly show their lack of symmetry graphically, whereas
distributions having an a_3 numerically greater than 0.6 are
obviously asymmetrical. The following diagram illustrates
the relationship between a_3 and the shape of the distribution:

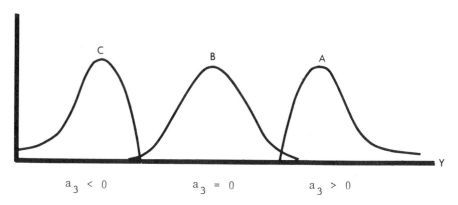

$$a_3 < 0 \qquad\qquad a_3 = 0 \qquad\qquad a_3 > 0$$

<u>A measure of peakedness</u>

To obtain a measure of peakedness we let $G(y_i) =$
$(y_i - \bar{y})/s$ and $h = 4$, which yields the fourth standard moment

$$a_4 = \frac{\sum_{i=1}^{n} \left[(y_i - \bar{y})/s \right]^4}{n} = \frac{1}{s^4} \frac{\sum_{i=1}^{n} (y_i - y)^4}{n}$$

This measure is also dimensionless because the same fraction
$(y_i - \bar{y})/s$ used in finding a_3 is used for a_4. To illustrate
the relationship between a_4 and the peakedness exhibited by
distributions, we use the three representative distributions
introduced in Fig. 2.12 in Chap. 2. Note that these are all
symmetric distributions having the same dispersion because
the illustration becomes too involved if these characteristics are allowed to vary.

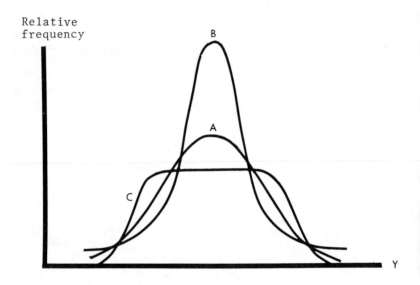

Let us discuss the a_4 values for distributions B and C
by comparing them with the a_4 value for distribution A since
A has a bell-shaped ("normal") distribution that is used as
a standard for purposes of comparison and an a_4 of 3. We
are using fourth powers of the deviations, so a_4 is always
greater than or equal to zero, with a "nonnormal" distribution having a value of a_4 either less than or greater than
3. Looking at the distribution of B as compared with A, we
note the existence of an excessive number of numerically
large deviations: both tails of the B distribution stretch
out higher than those of A. The $\Sigma(y_i - \bar{y})^2$ or variances of
the two distributions appear to be about the same because
the excessively large deviations of B are compensated for by
the excessive number of numerically small deviations and the

lack of moderately sized deviations. Although the excessive number of small deviations of B produce its "peak," the large value of a_4 is determined by the numerically large deviations that dominate the summation due to the fourth power. Thus $\sum\limits_{B}(y_i - \bar{y})^4 > \sum\limits_{A}(y_i - \bar{y})^4$, and accordingly, a_4 for distribution B will be greater than the standard value of 3 that we associate with the distribution A. Distribution C has a deficiency of large deviations and will therefore have a smaller $\sum\limits_{C}(y_i - \bar{y})^4$ than distribution A; thus its a_4 will be less than 3. We therefore conclude that if a set of observed data yields an a_4 significantly less than 3, the distribution will generally be flatter than the normal distribution; but if a_4 is greater than 3, the distribution will be more peaked. The value of a_4, however, really depends upon the behavior at the tails of the distribution rather than upon the peakedness in the middle. If asymmetry exists also, this simple interpretation of a_4 will require some modification because of the influence of the skewness.

The different classes of moments

We have generated four numerical measures of the characteristics of interest. The central tendency of a distribution can be measured using the arithmetic mean $\bar{y}$; its dispersion about this mean can be measured by the standard deviation s or the variance s^2. The skewness a_3 is a measure of the asymmetry of the distribution, and the kurtosis a_4 is a measure of the peakedness of the distribution. The latter two measures are sometimes referred to as shape parameters.

To obtain these numerical measures, we specialized the definition of the hth moment of G(Y) by using different functions and values of h. We introduced three general classes of moments, although we only used one or two moments from each class.

The moments in the first class are called moments about the origin. These moments are defined by

$$m_h = \frac{\sum\limits_{i=1}^{n} y_i^h}{n} = \frac{y_1^h + y_2^h + \cdots + y_n^h}{n}, \quad h = 1,2,3, \ldots$$

A measure of central tendency is given by m_1, but we shall see that the higher-order moments m_2, m_3, and m_4 are useful computationally in finding the remaining numerical measures.

The moments in the second class are called moments about the mean; these moments are defined by

$$v_h = \frac{\sum\limits_{i=1}^{n} (y_i - \bar{y})^h}{n}, \quad h = 1, 2, 3, \ldots$$

We use v_2 and $\sqrt{v_2}$ as the measures of dispersion, but again v_3 and v_4 will be useful computationally, and v_1 will always equal zero.

The moments in the third class are called standard moments and are defined by

$$a_h = \frac{\sum\limits_{i=1}^{n} \left[(y_i - \bar{y})/s \right]^h}{n} = \frac{1}{s^h} \sum\limits_{i=1}^{n} \frac{(y_i - \bar{y})^h}{n}, \quad h = 1, 2, 3, \ .$$

In this class we use a_3 as a measure of asymmetry and a_4 as a measure of peakedness. Note that $a_1 = 0$ and $a_2 = 1$ for any set of values.

We now need to develop the necessary computational methods for obtaining these numerical measures for ungrouped and grouped data. We shall also illustrate the relationship between the different classes of moments and how these relationships can be used to find the numerical measures.

3.4 Computation of the Numerical Characteristics for Ungrouped Data

We shall first consider the computational procedure for the simplest case, that of ungrouped data. We shall demonstrate the procedure for obtaining all four measures by using the moments about the origin rather than the defining formulas. Consider the ungrouped data given in Fig. 2.1, Chap. 2:

Reading Times of 12 New Employees
(minutes)

i	Y	Y^2	Y^3	Y^4
1	4.2	17.64	74.088	311.1696
2	5.1	26.01	132.651	676.5201
3	3.3	10.89	35.937	118.5921
4	4.3	18.49	79.507	341.8801
5	5.2	27.04	140.608	731.1616
6	3.1	9.61	29.791	92.3521
7	4.2	17.64	74.088	311.1696
8	5.3	28.09	148.877	789.0481
9	6.1	37.21	226.981	1384.5841
10	2.2	4.84	10.648	23.4256
11	4.2	17.64	74.088	311.1696
12	5.2	27.04	140.608	731.1616
Sum	52.4	242.14	1167.872	5822.2342

By finding the second, third, and fourth power of each value y_i and summing the appropriate columns, we obtain the following summations that are needed in the moment calculations:

$$n = 12$$
$$\Sigma y_i = 52.4$$
$$\Sigma y_i^2 = 242.14$$
$$\Sigma y_i^3 = 1167.872$$
$$\Sigma y_i^4 = 5822.2342$$

Using these summations, we can readily find the first four moments about the origin to be

$$m_1 = \frac{\Sigma y_i}{n} = \frac{52.4}{12} = 4.3667$$

$$m_2 = \frac{\Sigma y_i^2}{n} = \frac{242.14}{12} = 20.1783$$

$$m_3 = \frac{\Sigma y_i^3}{n} = \frac{1167.872}{12} = 97.3227$$

$$m_4 = \frac{\Sigma y_i^4}{n} = \frac{5822.2342}{12} = 485.1862$$

Thus the arithmetic mean is equal to

$$\bar{y} = m_1 = 4.3667 = 4.37.$$

To find the other characteristics we first make use of the relationship that exists between the moments about the origin and the moments about the mean. We have

$$v_2 = \frac{\Sigma(y_i - m_1)^2}{n} = \frac{\Sigma y_i^2 - 2m_1\Sigma y_i + nm_1^2}{n} = m_2 - 2m_1^2 + m_1^2$$

$$= m_2 - m_1^2 = 20.1783 - (4.3667)^2 = 1.1102$$

Using the same type of algebra,

$$v_3 = \frac{\Sigma(y_i - m_1)^3}{n} = m_3 - 3m_2 m_1 + 2m_1^3$$

$$= 97.3227 - 3(20.1783)(4.3667) + 2(4.3667)^3$$

$$= -0.4860$$

$$v_4 = m_4 - 4m_3 m_1 + 6m_2 m_1^2 - 3m_1^4$$

$$= 485.1862 - 4(97.3227)(4.3667) + 6(20.1783)(4.3667)$$

$$- 3(4.3667)^4 = 3.0636$$

The student should attempt to reproduce these results on a desk calculator. The next chapter will present metho by which these computations may be executed on a high-spee digital computer. Comparisons of the labor required by th two methods leaves most students favorably impressed with the efficiency of the digital computer.

From these moments about the mean we obtain

$$s^2 = v_2 = 1.1102$$

$$s = \sqrt{v_2} = \sqrt{1.1102} = 1.0537 = 1.05$$

$$a_3 = \frac{v_3}{s^3} = \frac{-0.4860}{(1.0537)^3} = -0.4154 = -0.42$$

$$a_4 = \frac{v_4}{s^4} = \frac{3.0636}{(1.0537)^4} = 2.4852 = 2.49$$

Using these numerical characteristics we can describe the distribution of reading times for the 12 new employees as centering around 4.37 min, with a standard deviation of 1.05 min. At this stage we interpret this measure by comparing it with the standard deviation of a different population, say, of older employees. If they have a standard deviation greater than 1.05 min, we would recognize that the distribution for the older employees had more dispersion than our new employees have. The distribution also has a negative skewness of -0.42, meaning that there were some relatively small test scores for which there were no comparable large scores. Finally, the peakedness is measured by a kurtosis of 2.49, and because this is somewhat less than our standard of 3 we feel that the distribution may be a little less peaked as compared with the normal one.

In many situations only measures of central tendency and dispersion are needed. We would therefore like a more straightforward procedure for computing the mean and standard deviation of a distribution. For this we need only n and the sum and the sum of squares of the observations. Thus from our observed ungrouped data we would obtain

$$n = 12$$

$$\Sigma y_i = 52.4$$

$$\Sigma y_i^2 = 242.14$$

The mean is computed as before by a simple division:

$$\bar{y} = \frac{\Sigma y}{n} = \frac{52.4}{12} = 4.37$$

For the standard deviation we can use an equivalent formula involving only the above summations since

$$s = \sqrt{v_2} = \sqrt{\frac{\Sigma(y_i - \bar{y})^2}{n}} = \sqrt{\frac{\Sigma y_i^2 - (\Sigma y_i)^2/n}{n}}$$

41

Thus we obtain

$$s = \sqrt{\frac{242.14 - (52.4)^2/12}{12}}$$

$$s = \sqrt{\frac{242.14 - 228.8133}{12}}$$

$$s = \sqrt{1.1106} = 1.0539 = 1.05$$

There are many different equivalent formulas that can be used to compute these four numerical measures. Each formula has its own advantages, depending upon the computational facilities being used. The preceeding formulas are most appropriate for desk-calculator evaluations. When using the computer, however, different formulas are often more appropriate.

3.5 Computation of Numerical Characteristics for Grouped Data

We can use the mechanics introduced in Sec. 3.4 to develop the procedures for computing the numerical characteristics for grouped data. We need to find the appropriate moments or summations for the grouped data and then follow the same mechanics as used for the ungrouped data.

We first illustrate this concept by using the following distribution given in Fig. 2.2. There are $m = 9$ values of y_j, where each value has an associated frequency of occurrence f_j greater than zero.

Y	F
0	35
1	45
2	120
3	112
4	42
5	23
6	10
7	2
8	1
	390

The moments about the origin are given by

$$m_h = \frac{\sum\limits_{j=1}^{m} y_j^h f_j}{\sum\limits_{j=1}^{m} f_j}$$

where the powers of Y are weighted by the frequency, and the total number of observations is given by $n = \sum\limits_{j=1}^{m} f_j$. Thus

$$m_1 = \frac{\Sigma y_j f_j}{\Sigma f_j} = \frac{986.0}{390} = 2.5282 = 2.53$$

$$m_2 = \frac{\Sigma y_j^2 f_j}{\Sigma f_j} = \frac{3302}{390} = 8.4667$$

$$m_3 = \frac{\Sigma y_j^3 f_j}{\Sigma f_j} = \frac{12953}{390} = 33.2051$$

$$m_4 = \frac{\Sigma y_j^4 f_j}{\Sigma f_j} = \frac{58022}{390} = 148.7744$$

The general rule is to replace $\sum\limits_{i=1}^{n} G(y_i)$ in the un-grouped computation by

$$\sum\limits_{j=1}^{m} G(y_j) f_j \quad \text{and n by} \quad \sum\limits_{j=1}^{m} f_j$$

to find the corresponding summation for the grouped data. Thus

$$v_h = \frac{\sum\limits_{j=1}^{m} (y_j - \bar{y})^h f_j}{\sum\limits_{j=1}^{m} f_j}, \quad h = 1, 2, \ldots$$

and

$$a_h = \frac{\sum\limits_{j=1}^{m} (y_j - \bar{y})^h f_j}{ns^h}, \quad h = 1, 2, \ldots$$

43

are the definitions of the moments about the mean and the standard moments for grouped data.

Once the moments about the origin are calculated, the same relationships between the different classes of moments hold. For example,

$$s^2 = v_2 = m_2 - m_1^2$$

$$= 8.4667 - (2.5282)^2 = 8.4667 - 6.3918$$

$$= 2.0749$$

Likewise, the third and fourth moments about the mean are given by

$$v_3 = m_3 - 3m_2 m_1 + 2m_1^3$$

$$v_4 = m_4 - 4m_3 m_1 + 6m_2 m_1^2 - 3m_1^4$$

and the skewness and kurtosis are given by

$$a_3 = \frac{v_3}{s^3} \qquad a_4 = \frac{v_4}{s^4}$$

When only the mean and standard deviation of grouped data are needed, the following computational formulas can be used:

$$\bar{y} = \frac{\sum_{j=1}^{m} y_j f_j}{\Sigma f_j}$$

and

$$s = \sqrt{\frac{\sum_{j=1}^{m} y_j^2 f_j - \left(\sum_{j=1}^{m} y_j f_j\right)^2 / \sum_{j=1}^{m} f_j}{\Sigma f_j}}$$

We illustrate the use of these computational formulas by applying them to the following set of grouped data:

Frequency of Incorrect Responses to True-False Test

Y	F	YF	Y^2F
0	35	0	0
1	45	45	45
2	120	240	480
3	112	336	1008
4	42	168	672
5	23	115	575
6	10	60	360
7	2	14	98
8	1	8	64
Sum	390	986	3302

$$\Sigma f_j = 390$$

$$\Sigma y_j f_j = 986$$

$$\Sigma y_j^2 f_j = 3302$$

$$\bar{y} = \frac{\Sigma y_j f_j}{\Sigma f_j} = \frac{986}{390} = 2.5282 = 2.53$$

$$s = \sqrt{\frac{\Sigma y_j^2 f_j - \left(\Sigma y_j f_j\right)^2 / \Sigma f_j}{\Sigma f_j}}$$

$$= \sqrt{\frac{3302 - (986)^2/390}{390}} = \sqrt{2.0748} = 1.4404 = 1.44$$

To determine the characteristics for frequency distributions where the class is designated by an interval rather than by a single representative value of the variable, we simply use the midpoint of each class and assume that all observations in the class have that value.

Consider the frequency distribution given in Fig. 2.3:

Reading Times of a Group of 390 New Employees
(measured to nearest tenth of a minute)

Interval	F	Class Boundaries	Midpoint(Y)	YF	Y^2F
2.0 - 2.4	21	1.95 - 2.45	2.20	46.20	101.6400
2.5 - 2.9	43	2.45 - 2.95	2.70	116.10	313.4700
3.0 - 3.4	68		3.20	217.60	696.3200
3.5 - 3.9	97		3.70	358.90	1327.9300
4.0 - 4.4	72	.	4.20	302.40	1270.0800
4.5 - 4.9	53	.	4.70	249.10	1170.7700
5.0 - 5.4	21	.	5.20	109.20	567.8400
5.5 - 5.9	11		5.70	62.70	357.3900
6.0 - 6.4	4	5.95 - 6.45	6.20	24.80	153.7600
Sum	390			1487.00	5959.2000

Using the midpoint y_j, of each class, we can proceed to compute the mean and standard deviation using the computational formulas just given:

$$\Sigma f_j = 390$$

$$\Sigma y_j f_j = 1487.00$$

$$\Sigma y_j^2 f_j = 5959.2000$$

$$\bar{y} = \frac{\Sigma y_j f_j}{\Sigma f_j} = \frac{1487.00}{390} = 3.8128 = 3.81$$

$$s_y = \sqrt{\frac{\Sigma y_j^2 f_j - \left(\Sigma y_j f_j\right)^2 / \Sigma f_j}{\Sigma f_j}} = \sqrt{\frac{5959.2000 - (1487.00)^2 / 390}{390}}$$

$$= \sqrt{\frac{5959.2000 - 5669.6641}{390}} = \sqrt{0.7423} = 0.8615 = 0.86$$

Although these computational formulas are straightforward, there are more efficient techniques that can be used. One alternative technique is given in the following section.

3.6 Transformations

When computing the numerical characteristics of observations, transformations of the observations are often useful. A linear transformation is particularly useful because it allows us to change the origin or scale of the original data for ease of calculation or, more importantly, for comparison purposes without causing changes in the higher-order characteristics of the distribution. Most national scores such as college board or graduate record examinations are obtained by transforming the actual scores into scores which have a predetermined mean and standard deviation so that the results of different examinations can be more easily compared.

To transform any set of observations y_i, a function G of the observations is defined, say

$$u_i = G(y_i)$$

In particular, a linear transformation has the general form

$$u_i = G(y_i) = \frac{y_i - A}{B}$$

For this general linear transformation, the mean of the u_i's is related to the mean of the y_i's by the relationship

$$\bar{u} = \frac{\Sigma u_i}{n} = \frac{\Sigma (y_i - A)/B}{n} = \frac{1}{B} \frac{\Sigma y_i - nA}{n} = \frac{1}{B} \left(\frac{\Sigma y_i}{n} - A \right) = \frac{\bar{y} - A}{B}$$

Thus the mean of the original Y's can be found by the formula

$$\bar{y} = B\bar{u} + A$$

Likewise, the variance of the Y's can be calculated from the variance of U using the relationship

$$s_u^2 = \frac{\Sigma (u_i - \bar{u})^2}{n} = \frac{\Sigma \left[(y_i - A)/B - (\bar{y} - A)/B \right]^2}{n}$$

$$= \frac{1}{B^2} \frac{\Sigma (y_i - \bar{y})^2}{n} = \frac{s_y^2}{B^2}$$

The standard deviation of Y is given by

$$s_y = Bs_u$$

A linear transformation does not change the basic characteristics of the observations other than for the location of the origin and the scale. The shape of the graphs for the original and transformed values are thus identical except for the unit of measurement. Therefore a_3 and a_4 remain the same for the transformed and original observations.

Linear transformations can be used to "code" the data to facilitate the computations necessary to determine $\bar{y}$, s^2, a_3, and a_4, particularly when working with frequency distributions. If we use the transformation $u = (y - y_o)/w$, where y_o is the class midpoint of a specific interval and w is the class width for each interval, then the class midpoints expressed in the u scale always take on the integer values ... -4, -3, -2, -1, 0, $+1$, $+2$, $+3$, ... and thus the summations, $\Sigma u_j^h \cdot f_j$ are easily accomplished.

We illustrate this coding method by the following example, in which we compute the mean and standard deviation for the frequency distribution of the 390 reading times given in Fig. 2.3.

Reading Times of a Group of 390 New Employees

Interval	F	True Class Limits	Class Midpoint	U	UF	U^2F
2.0 - 2.4	21	1.95 - 2.45	2.20	-3	-63	198
2.5 - 2.9	43	2.45 - 2.95	2.70	-2	-86	172
3.0 - 3.4	68	.	3.20	-1	-68	68
3.5 - 3.9	97	.	3.70	0	0	0
4.0 - 4.4	72	.	4.20	+1	+72	72
4.5 - 4.9	53	.	4.70	+2	+106	212
5.0 - 5.4	21	.	5.20	+3	+63	189
5.5 - 5.9	11	.	5.70	+4	+44	176
6.0 - 6.4	4	.	6.20	+5	+20	100
Sum	390				+88	1178

$$\Sigma f_j = 390 = n$$

$$\Sigma u_j f_j = 88$$

$$\Sigma u_j^2 f_j = 1178$$

$$\bar{u} = \frac{\Sigma u_j f_j}{\Sigma f_j} = \frac{88}{390} = 0.2256$$

$$s_u = \sqrt{\frac{\Sigma u_j^2 f_j - \left(\Sigma u_j f_j\right)^2 / n}{n}} = \sqrt{\frac{1178 - (88)^2/390}{390}}$$

$$= \sqrt{2.9695} = 1.7232$$

We decode the results to find

$$\bar{y} = w\bar{u} + y_o = 0.5(0.2256) + 3.70 = 3.8128 = 3.81$$

$$s_y = ws_u = (0.5)(1.7232) = 0.8616 = 0.86$$

If the higher-order characteristics a_3 and a_4 had been calculated for the u_i's, they would not need to be decoded.

An important special case of the general linear transformation is

$$z_i = G(y_i) = \frac{y_i - \bar{y}}{s_y}$$

where the mean $\bar{y}$ is subtracted from each observation and the difference is divided by the standard deviation s_y. The u_i are called standard scores because the mean of z_i will always be zero and the standard deviation will always be 1. In this case, the higher-order characteristics can be calculated by

$$a_3 = \frac{\Sigma z_i^3}{n} = \frac{\Sigma\left[(y_i - \bar{y})/s_y\right]^3}{n} = \frac{\Sigma(y_i - \bar{y})^3}{ns_y^3}$$

$$a_4 = \frac{\Sigma z_i^4}{n} = \frac{\Sigma\left[(y_i - \bar{y})/s_y\right]^4}{n} = \frac{\Sigma(y_i - \bar{y})^4}{ns_y^4}$$

49

We demonstrate the use of this special linear transformation in the computation of a_3 and a_4 below using the ungrouped times of 12 new employees. We shall use the rounded values for the mean and standard deviation of $\bar{y} = 4.37$ and $s_y = 1.05$.

Reading Times of 12 New Employees

i	Y	$Y' = Y - \bar{Y}$	Z	Z^2	Z^3	Z^4
1	4.2	-0.17	-0.16	.0256	-0.0041	0.0007
2	5.1	+0.73	+0.70	.4900	+0.3430	0.2401
3	3.3	-1.07	-1.02	1.0404	-1.0612	1.0824
4	4.3	-0.07	-0.07	0.0049	-0.0003	0.0000
5	5.2	+0.83	+0.79	0.6241	+0.4930	0.3895
6	3.1	-1.27	-1.21	1.4641	-1.7716	2.1436
7	4.2	-0.17	-0.16	0.0256	-0.0041	0.0007
8	5.3	+0.93	+0.89	0.7921	+0.7050	0.6274
9	6.1	+1.73	+1.65	2.7225	+4.4921	7.4120
10	2.2	-2.17	-2.07	4.2849	-8.8697	18.3604
11	4.2	-0.17	-0.16	0.0256	-0.0041	0.0007
12	5.2	+0.83	+0.79	0.6241	+0.4930	0.3895
Sum		-0.04	-0.03	12.1239	-5.1890	30.6470

From the deviations $y'_i = y_i - \bar{y}$ we can compute the variance and the standard deviation using the sum of squares. Thus,

$$\Sigma(y_i - \bar{y})^2 = \Sigma y'^2_i = 13.3268$$

$$s^2_y = \frac{\Sigma(y_i - \bar{y})^2}{n} = \frac{13.3268}{12} = 1.1106 = 1.11$$

and

$$s_y = \sqrt{1.1106} = 1.0539 = 1.05$$

In our computations the sum of Z's does not exactly equal zero and the sum of squares of the Z's equals 12.1239 rather than its theoretical value of n = 12. This is not due to a numerical error but to round-off error. We used s = 1.05 in our computation rather than s = $\sqrt{1.1106}$... =

1.05385008.... If we had used more significant digits in the values for $\bar{y}$ and s and had carried more decimal places in all the intermediate computations. we would find these discrepencies from the theoreti‹

We have from the above

$$a_3 = \frac{\Sigma z_i^3}{n} = \frac{-5.1890}{12} = -0.4324 = -0.43$$

$$a_4 = \frac{\Sigma z_i^4}{n} = \frac{30.6470}{12} = 2.5539 = 2.55$$

which agrees except for round-off error with the previous results obtained in Sec. 3.4. Thus we see how the choice of an appropriate linear transformation facilitates the manual computations. In computer computations transformations are important because truncation errors can give meaningless results. In statistics, nonlinear transformations are also important in meeting assumptions that are needed for statistical inference, as the student will see in the later chapters.

3.7 Other Measures of the Characteristics of Observed Distributions

In the development of numerical measures of the four principal characteristics of a distribution, we have so far restricted ourselves to using the moments of the distribution because these will be used in most of our statistical studies. However, there are many other numerical measures that can be used to characterize a distribution; in this section we shall introduce a selected number of them.

To illustrate the use of these measures, we shall use the ungrouped reading-time data in Fig. 2.1 and the grouped reading-time data in Fig. 2.3. The data is:

(a) ungrouped data

Reading Time for 12 New Employees

Id	1	2	3	4	5	6	7	8	9	10	11	12
Y	4.2	5.1	3.3	4.3	5.2	3.1	4.2	5.3	6.1	2.2	4.2	5.2

(b) grouped data

Time Class Limits	Frequency	Less than Cum. Freq.	Relative Frequency	Less than Cum. Rel. Freq.
1.95 - 2.45	21	21	0.0539	0.0538
2.45 - 2.95	43	64	0.1103	0.1641
2.95 - 3.45	68	132	0.1744	0.3385
3.45 - 3.95	97	229	0.2487	0.5872
3.95 - 4.45	72	301	0.1846	0.7718
4.45 - 4.95	53	354	0.1359	0.9077
4.95 - 5.45	21	375	0.0538	0.9615
5.45 - 5.95	11	386	0.0282	0.9897
5.95 - 6.45	4	390	0.0103	1.0000

One class of measures is called <u>percentile measures</u>; these are defined in terms of the cumulative frequency or relative frequency of the observed distribution. Because they are independent of the scale used in making the observations, they are somewhat similar to the z transformation used in obtaining standard moments. Since the cumulative (relative) frequencies characterize a distribution, the percentile measures are not only a measure of location but can also be converted into measures of central tendency, dispersion, and asymmetry. We shall therefore first introduce percentiles and then consider how they, along with other types of measures, can be used to obtain measures of the characteristics of a distribution.

<u>Percentiles</u>

A percentile P_k is the value of the variable such that k percent of all observations are below P_k and (100 - k) percent are above P_k. Although percentiles can be obtained for ungrouped data, they are usually obtained for grouped data.

For grouped data, we first determine the k percentile class. This is the class whose less than cumulative relative frequency first exceeds k/100, that is,

$$\left[\frac{\sum\limits_{j=1}^{c} f_j}{n} \right] > \frac{k}{100}$$

where the k percentile class is the cth class.

If we are willing to assume that the observations in the k percentile class are uniformly distributed over the class, the k percentile can be obtained by the following formula:

$$P_k = l_c + w \left[\frac{n \cdot k/100 - \sum\limits_{j=1}^{c-1} f_j}{f_c} \right]$$

where l_c = lower class boundary of k percentile class

f_c = frequency of k percentile class

w = width of class

$\sum\limits_{j=1}^{c-1} f_j$ = cumulative frequency up to but not including k percentile class

$n = \sum\limits_{j=1}^{m} f_j$ = total number of observations

We demonstrate the technique by finding P_{10} for the grouped reading-time data. Since the cumulative relative frequency for the second class is $0.1641 > 0.10$, $c = 2$, we obtain

$$P_{10} = 2.45 + (0.5) \left[\frac{390 \cdot (0.10) - 21}{43} \right] = 2.6593 = 2.66$$

Thus 10 percent of all the observations have values less than 2.66.

In characterizing a distribution, the percentiles that are of particular interest are P_{25}, P_{50}, and P_{75}. These percentiles are also called the first, second, and third quartile, respectively. Using the same formula given above, we find that for the grouped reading-time data,

$$P_{25} = 2.95 + (0.5) \left[\frac{390 \cdot (0.25) - 64}{68} \right] = 3.1963 = 3.20$$

$$P_{50} = 3.45 + (0.5) \left[\frac{390 \cdot (0.5) - 132}{97} \right] = 3.7747 = 3.77$$

$$P_{75} = 3.95 + (0.5) \left[\frac{390 \cdot (0.75) - 229}{72} \right] = 4.3909 = 4.39$$

Percentile ranges are used to locate an observation relative to all the observations. Thus an observation is said to be in the 10th percentile range if its value falls between P_9 and P_{10}.

Measures of Central Tendency

1. The mode

In many observed distributions the frequency of occurrence increases as the center of the distribution is approached. Thus it is reasonable to use the value of the variable that occurs most frequently as a measure of central tendency. This measure is called the mode and is denoted by M_o.

To determine the mode for ungrouped data, the single value which occurs most frequently is picked, although for a particular set of ungrouped data it is possible that the mode does not exist or that several modes might exist. The illustrative ungrouped data, however, does have a mode, which is $M_o = 4.2$.

To determine the mode for grouped data, the modal class must first be identified; it is the class that has the highest frequency. Let us designate the modal class as the cth class. To determine the mode within this class, the frequency of the adjacent classes can be used as weights. Thus

$$M_o = 1_c + w \left[\frac{f_{c+1}}{f_{c-1} + f_{c+1}} \right]$$

where 1_c = lower class boundary of modal class
$\quad$ w = class width
$\quad f_{c-1}$ = frequency of (c-1)th class
$\quad f_{c+1}$ = frequency of (c+1)th class

For the grouped reading-time data, the modal class is c = 4 because this class has the highest frequency of 97.

Thus

$$M_o = 3.45 + (0.5) \left[\frac{72}{68 + 72} \right] = 3.7071 = 3.71$$

2. The median

The median is defined as the value such that 50 percent of the observations fall below the value and 50 percent fall above the value. The median will be denoted by M_d and is a measure of central tendency, which is particularly useful when the data are highly skewed.

The student should recognize that the median is also the 50th percentile, that is, $M_d = P_{50}$; we have already illustrated how it can be found for grouped data. For ungrouped data, the values of the observations are first ranked in either ascending or descending order. The median M_d is then found.

(a) If the number of observations n is odd, then M_d is equal to the middle value or

$$M_d = y_{(n+1)/2}$$

(b) If n is even, then the median is the average of the two middle values, or

$$M_d = \frac{y_{(n/2)} + y_{(n/2+1)}}{2}$$

For the ungrouped reading-time data, the ranked observations are

2.2 3.1 3.3 4.2 4.2 4.2 4.3 5.1 5.2 5.2 5.3 6.1

and because n = 12 is even,

$$M_d = \frac{y_6 + y_7}{2} = \frac{4.2 + 4.3}{2} = \frac{8.5}{2} = 4.25$$

Measures of dispersion

1. The range

The range R is defined as the difference between the largest and smallest observation. Thus for ungrouped data the range is

$$R = y_{max} - y_{min}$$

When the data is grouped, the range can be considered equal to the difference between the upper boundary of the last interval and the lower boundary of the first interval, or

$$R = u_{max} - \ell_{min}$$

2. The average absolute deviation

The average absolute deviation $\bar{D}$ is defined as the average of the absolute deviations from the mean. For ungrouped data, it is given by

$$\bar{D} = \frac{\sum\limits_{i=1}^{n} |y_i - \bar{y}|}{n}$$

and for grouped data it is given by

$$\bar{D} = \frac{\sum\limits_{j=1}^{m} |y_j - \bar{y}| \, f_j}{\sum\limits_{j=1}^{m} f_j}$$

3. The semi-interquartile range

The semi-interquartile range is given by

$$R_q = \frac{P_{75} - P_{25}}{2}$$

R_q is a measure of dispersion because it can be written as

$$R_q = \frac{(P_{75} - M_d) + (M_d - P_{25})}{2} = \frac{P_{75} - P_{25}}{2}$$

where the difference $P_{75} - M_d$ represents the distance one must go above the median to include 25 percent of the observations, and $M_d - P_{25}$ is the distance one must go below the median to obtain 25 percent of the data.

Asymmetry

The relationship between the arithmetic mean $\bar{y}$, the mode M_o, and the median M_d can be used to determine the asymmetry of an observed distribution. By looking at some skewed distributions, the student can observe that if the distribution has a positive skewness, then $M_o < M_d < \bar{y}$, but if it has a negative skewness, then $\bar{y} < M_d < M_o$. Using these measures of central tendency, a measure of asymmetry can be obtained by

$$b_1 = \frac{\bar{y} - M_o}{s}$$

or by

$$b_2 = \frac{3(\bar{y} - M_d)}{s}$$

The second formula should only be used for moderately skewed distributions.

Using the grouped reading-time data, for which $s = 0.86$, we find that

$$\bar{y} = 3.81 \qquad M_d = 3.77 \qquad M_o = 3.71$$

Thus $M_o < M_d < \bar{y}$, and we say that the skewness is positive. We can evaluate

$$b_1 = \frac{3.81 - 3.71}{0.86} = 0.1163 = 0.12$$

and

$$b_2 = \frac{3(3.81 - 3.77)}{0.86} = 0.1395 = 0.14$$

These measures compare with the a_3 of 0.25, which we can obtain by using the first three moments. Although the values do not agree, the sign is the same, and when the distribution is symmetric, $M_o = M_d = \bar{y}$ and all three measures are equal.

3.8 Summary

This chapter can best be summarized by listing the concepts that were introduced and the formulas that were developed.

1. The summation symbol was introduced, where

$$\sum_{i=1}^{n} G(y_i) = G(y_1) + G(y_2) + \cdots + G(y_n)$$

The symbol has the following algebraic properties:

(a) $\Sigma[G(y_i) \pm H(y_i)] = \Sigma G(y_i) \pm \Sigma H(y_i)$

(b) $\Sigma c G(y_i) = c \Sigma G(y_i)$

(c) $\sum_{i=1}^{n} c = nc$

2. Descriptive numerical characteristics of observed distributions were defined first generally:

	Ungrouped	Grouped
Moments about the origin	$m_h = \dfrac{\sum_{i}^{n} y_i^h}{n}$	$m_h = \dfrac{\sum_{j}^{m} y_j^h f_j}{\Sigma f_j}$
Moments about the mean	$v_h = \dfrac{\sum_{i}^{n} (y_i - \bar{y})^h}{n}$	$v_h = \dfrac{\sum_{j}^{m} (y_j - \bar{y})^h f_j}{\Sigma f_j}$
Standard moments	$a_h = \dfrac{\sum_{i}^{n} \left[(y_i - \bar{y})/s_y\right]^h}{n}$	$a_h = \dfrac{\sum_{j}^{m} \left[(y_j - \bar{y})/s_y\right]^h f_j}{\Sigma f_j}$

where $\bar{y} = m_1$, mean, a measure of central tendency

$s_y = \sqrt{v_2}$, standard deviation, a measure of dispersion

a_3 = skewness, a measure of symmetry

a_4 = kurtosis, a measure of peakedness

58

3. Computation techniques for computing these numerical characteristics were developed for ungrouped data, the simple frequency distribution, and the general frequency distribution. The role of the recording accuracy in the determination of the class boundaries of a frequency distribution was considered in determining the y_j or class midpoint for an interval.

4. The effect of coding data through the linear transformation $U = (Y - A)/B$ results in

$$\bar{y} = B\bar{u} + A$$

$$s_y = Bs_u$$

$$a_{3:y} = a_{3:u}$$

$$a_{4:y} = a_{4:u}$$

5. Other measures that numerically describe an observed distribution were also presented. These measures included percentiles, median, mode, range, average absolute deviations, and the semiinterquartile range.

The computations involved in finding the numerical characteristics of observed distributions are lengthy, and thus the use of the computer is indicated. Before introducing computer programming, however, it is useful to become familiar with the computations by using desk calculators, because the increased familiarity with the computational procedures, the use of desk calculators, and the associated difficulties are useful in learning programming. We must also have data to check the correctness of a program once it is written. Thus we can use the answers obtained for the exercises at the end of Chap. 3 as test cases for the programs that will be written.

1. Calculate the first four moments about the origin for the ungrouped data given in Exercise 1, Chap. 2. Then evaluate s^2, a_3, and a_4. From the value obtained for a_3, would you expect some extremely large or extremely small values of the variable without the compensating opposite values? Check your conclusion by examining the actual data. Does the distribution exhibit more peakedness than the normal distribution? Carry four decimal places in your computations.

2. For each hospital, calculate $\bar{y}$, s^2, a_3 and a_4 from the simple frequency tables set up in Exercise 3, Chap. 2. Compare the two distributions in terms of these measures.

3. The range is often used as a measure of dispersion. Find the range of the data given in Exercise 5, Chap. 2. The standard deviation can be estimated using the range. The following relationship usually holds: $R/6 \leq s \leq R/4$. The estimate of R/6 is used when there is a large amount of data; the estimate tends toward R/4 as the number of observations decreases. Check whether the above inequality holds true for the data in Exercise 2, Chap. 2.

4. Use the ungrouped data given in Exercise 2, Chap. 2, and determine the mean, variance, skewness and kurtosis (use moments about the origin). Change each observation into its corresponding z value given by the z transformation, and calculate the mean, variance, skewness, and kurtosis for the z values. Compare the four characteristics of the y and z values.

5. Find the mean, median, and mode for the data given in Exercise 2, Chap. 2. Use these measures to determine the two alternative measures of skewness, and compare them with the a_3 value found in Exercise 4 above.

CHAPTER 4

INTRODUCTION TO FORTRAN PROGRAMMING

4.1   Introduction

By now most students have probably seen a computer and
have an idea of what computers can generally do. We shall,
therefore, not take time in this course to give any description of computers in terms of the hardware involved — the
actual physical components that make up a computer system —
or the software — the language which can be used to harness
the power of the computer to solve a wide scope of problems.
We should appreciate that the computer can be a useful
tool in statistical analyses, especially after encountering
the computational problems in Chap. 3. We therefore wish to
introduce the computer as a tool that can be useful in solving problems. The approach we shall use stresses learning
by doing rather than from abstract and theoretical considerations. It is expected that each student will run a computer progam within 1 or 2 days after beginning this chapter.
He will not know all the fine details that are involved in
programming, but he will know the general principles. Starting from this base, we shall build up a more thorough understanding of the techniques involved in computer programming.
We shall restrict our consideration to the digital computer because it has very general capabilities. The digital
computer must be given the data that are to be processed together with the instructions as to how to process the data.
The set of instructions is called a program, and the digital
computer has the capacity to store both the program and data
in its memory. This memory can be recalled almost instantly
and repeatedly. Another important characteristic of the

digital computer is that it can perform or <u>execute</u> the instructions at incredible speeds, although it can only process one instruction at a time.

The computer can store only information (program and data) that is coded in terms of binary digits (digits which can only be either 0 or 1). Thus all program instructions and data must be converted into the 0-1 codes that are used by the particular computer. The set of 0-1 codes that a computer uses is called an <u>object code</u> or <u>absolute machine language</u>.

To avoid the tedious task of such coding, languages that are more familiar to humans have been devised. In fact, it was recognized that the computer could serve as its own translator, and thus programs which convert these "high level" languages (set of codes) into the appropriate machine language are available. The program that accomplishes the translation is called a <u>compiler</u>; many are available: FORTRAN, COBOL, ALGOL, and BASIC, for example.

We shall study the FORTRAN language (the word FORTRAN is derived from <u>For</u>mula <u>Trans</u>lation). FORTRAN is a computer language that uses expressions similar to those used in algebra. The computer cannot directly execute programs written in this high-level language but must use the FORTRAN compiler to translate or compile the program into an object (machine) language program. Once the program is stored in the computer memory in the object language code, the program can be executed. Fortunately for the FORTRAN programmer, the details of the compilation and execution processes of the digital computer are of little concern, and thus his major emphasis can be on how to use the FORTRAN language to solve his problems.

Before we can proceed with the discussion of how to write a FORTRAN program, we need to have a basic understanding of how to read the program instructions and data into the computer memory. FORTRAN programs and data can be punched into punch cards; the cards are then read by a card reader. A standard punch card is shown in Fig. 4.1. In any of the 80 columns, a digit, letter, or special character can be punched, using the appropriate code as illustrated in

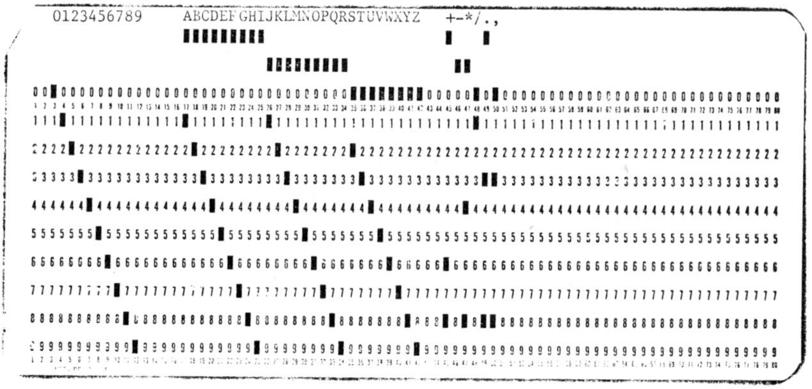

Fig. 4.1 An 80-column punch card.

the card. The required punches are inserted into the card
by a key punch which resembles a typewriter, except that
only capital letters can be punched.

There are many other ways that programs and data can be
read into the computer. For example, information can be
read directly into the computer by using a remote terminal.
The same information that is punched into a card is typed
and read directly by the computer.

To learn the basic concepts associated with the reading
of information into the computer and to avoid being too abstract we shall assume that a punch card reader is available.
Once the student is familiar with the punch card input mechanism, he can easily learn to utilize the other types of
reading machanisms if needed. Similarly, the results of
computer processing can be obtained in many forms, such as
on punch cards, magnetic tapes, remote terminals, and so
forth, but we shall assume that a printer is used and thus
that the output will appear on a printed page.

The student should visit his computer center to learn
how programs and data are to be prepared for input. In
fact, one of our first exercises will be to prepare a simple
FORTRAN program and to submit it to the computer center for
execution.

4.2 Steps in Computer Problem Solving

Let us now consider the mechanics involved in using a digital computer to solve a problem. To keep our discussion from becoming too abstract or wordy, we shall use a simple problem: We want to write a FORTRAN program to find the mean, variance, and standard deviation of five observations. Although this is like swatting a fly with a sledge hammer, we can observe the steps we should follow.

Step 1.

The first step is the selection of the method of solving the problem, including the algorithm to be used in determining the actual numerical solution to a specific problem. An algorithm is the actual step-by-step procedure to be followed for finding the solution. One algorithm to determine the mean, variance, and standard deviation involves finding the sum and sums of squares of the observations and then computing the mean, variance, and standard deviation using the formulas

$$\text{YMEAN} = \frac{\sum_{i=1}^{n} y_i}{n}$$

$$\text{YVAR} = \frac{\sum_{i=1}^{n} y_i^2 - \left(\sum_{i=1}^{n} y_i \right)^2 / n}{n}$$

$$\text{YSTD} = \sqrt{\text{YVAR}}$$

Step 2.

The steps to be followed in actually using the algorithm are outlined either formally or informally in a flow diagram. A flow diagram is used to show the successive stages or evaluations that are to be made during the actual computation; it may be very general and indicate only the major steps, or it may be very detailed. Often a flow diagram that shows only the major steps is developed, and then each major step is flow diagramed individually in greater detail. This hierarchical process may be extended through several stages.

There are several different conventions that may be used in preparing a flow diagram, but because the flow diagram serves mainly as an aid to the programmer, programmers often evolve their own variations of any set of conventions. However, we should try to maintain some regularity in our use of symbols and follow the American Standard Flowchart Symbols because other programmers may want to follow our logic and procedural steps. The symbols that we shall use include

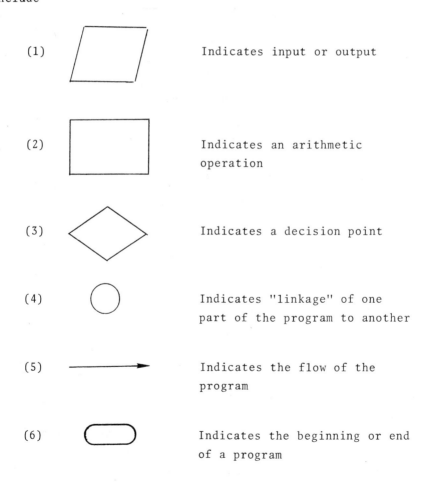

(1)   Indicates input or output

(2)   Indicates an arithmetic operation

(3)   Indicates a decision point

(4)   Indicates "linkage" of one part of the program to another

(5)   Indicates the flow of the program

(6)   Indicates the beginning or end of a program

The idea of flow diagraming will become more apparent
as we progress in our discussions. For now the concept is
illustrated only by the flow diagram for the program to find
the three numerical measures for five observations:

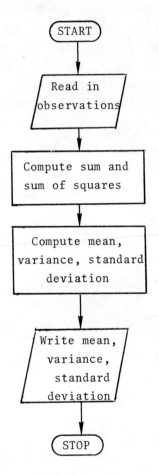

Step 3.

The computer program is written following the outline
given in the flowchart. We shall use the FORTRAN language
in writing such programs, but as mentioned, different languages such as COBOL (Common Business Oriented Language),
ALGOL (Algorithm Language), and BAL (Basic Assembly Language)
can also be used. The computer program for our particular
illustration will be introduced after we have completed our

discussion of the different steps involved in solving a problem.

Step 4.

After the computer program has been written, it must be put into a form readable by the computer. This is generally done by keypunching the program onto a set of punch cards called the source deck. We shall demonstrate this step by listing the deck of punch cards:

The deck of punch cards (source deck and data cards) are now submitted to the computer for processing. This processing usually involves at least three steps: (a) the FORTRAN program is compiled or translated into the object language. The resulting program is called the object program; this program can be saved and reused by punching it on cards, thus creating an object deck. (b) The object deck is executed, using the data cards that have been submitted with the program. (c) The results are printed out by the printer. During these steps the program deck may first be read onto a magnetic tape or disk to reduce the computer read-in time, the output may be printed on a remote printer, and so forth. As beginning programmers, all we need to know is where to submit our program decks and where to collect the resulting output sheets.

Step 5.

Once the computer results for the data are available, the results have to be interpreted in light of the problem being studied. In programming, unfortunately, the initial outputs from a computer run are often in the form of an "error message" telling us that something is wrong with the program we have written. Under these circumstances, our interpretation will be oriented in the direction of how we need to modify the program to get correct results. Fortunately, these error messages contain diagnostic statements to help us in isolating the programming errors. We shall learn more about this debugging activity later.

Because we plan to learn programming by doing, we intend to "get into the water" immediately. Each student will be expected to learn programming by actually writing programs and running his program on a computer.

4.3 The FORTRAN Program

The FORTRAN program for finding the mean, variance, and standard deviation of five observations is shown in Program 4.1, followed by a brief commentary. As we have already indicated, each student should keypunch the program, data cards, and the required system cards and submit the resulting program deck to be run on the computer.

Only a specified number of symbols can be "understood" by the computer. These symbols are the capital (uppercase) letters of the English alphabet, the digits 0, 1,2, ..., 9, and certain other symbols, such as ., ,, (), -, +, /, *, =. There may be confusion between the printed zero 0 and the alphabet character O so we shall follow the convention of slashing the numerical 0. Thus the 0 in 1∅.∅ is slashed, but alphabetical letter O will be unslashed. We may use the lowercase b with a slash, ∅, for indicating a space: X = ∅1; this would indicate a space between the equal sign and 1.

Before presenting and commenting on the FORTRAN program, it is important to point out that in processing a program the computer evaluates one FORTRAN statement at a time, and after taking the action indicated by the statement goes to the next statement unless told otherwise. Each FORTRAN statement is punched on separate cards according to a prescribed format, which is

columns 1 to 5   Blank or statement number
column 6   Blank
columns 7 to 72   FORTRAN statement
columns 73 to 80 Blank or identification or
                    sequence numbers

If a FORTRAN statement does not fit into columns 7 to 72, it can be continued on succeeding cards. Such a "continuation" card is recognized by a character (1 to 9 or letter) being punched in column 6. The format of a continuation card is therefore

68

columns 1 to 5   Blank or statement number

column 6   Any character (1 to 9 or letter)

columns 7 to 72   The continuation of the FORTRAN state-
                    ment, with column 7 being adjacent to
                    column 72 of the preceding card

columns 73 to 80   Blank or identifying characters

Although the FORTRAN statements must be punched in cards according to the above format, data cards do not have a pre-scribed format, so the values can be punched into any of the card's 80 columns.

When the FORTRAN statements are punched into cards, the deck of cards is called the source deck. This deck must be merged with system and data cards to form the program deck that is processed by the computer. Some system cards have to precede the source deck in order to indicate that the deck is written in FORTRAN. If data cards are to be read also, they must be placed behind the system cards that in-dicate the end of the source deck. The program deck arrange-ment is illustrated by Program 4.1. Note that the ∅'s are used simply to indicate blanks in writing the program prior to keypunching but are not keypunched.

```
        SYSTEM CARD(S)
    C      GSTAT-MEAN, VARIANCE, AND STANDARD DEVIATION OF 5 OBSERVATIONS
           READ(5.1) Y1,Y2,Y3,Y4,Y5
    1      FORMAT(5F1∅.2)            Y3*Y3 + Y4*Y4 + Y5*Y5
         ' SUMY= Y1+Y2+Y3+Y4+Y5
           SUMY2= Y1*Y1 + Y2*Y2 + Y3*Y3 + Y4*Y4 + Y5*Y5
Source     YMEAN= SUMY/5.
program    YVAR= (SUMY2 - SUMY*SUMY/5.)/5.
           YSTD= SQRT(YVAR)
           WRITE(6,2) YMEAN,YVAR,YSTD
    2      FORMAT(1H1,7HMEAN∅IS,1X,F1∅.2/1H∅,11HVARIANCE∅IS,1X,F1∅.3/1X,
           1∅∅∅21HSTANDARD∅DEVIATION∅IS,1X,F1∅.2)
           STOP
           END
        SYSTEM CARD(S)
Data       1∅.5       14.6       17.8       13.6       18.7
        SYSTEM CARD(S)
```

Program 4.1 GSTAT — mean, variance and standard
           deviation of five observations.

Because each computer facility has its own operational system, the system cards required to process a FORTRAN program will vary from installation to installation. On a particular system, these cards are the same for all FORTRAN programs, and thus the student should obtain the appropriate set of system cards and use them with each source deck that is submitted.

Comments on the FORTRAN Program

Line 1.

The appropriate system card(s) specifies the necessary accounting information, the language in which the program is written, and whether compilation and/or execution is requested. These cards depend upon the computer center and the computer model.

Line 2.

The C in column 1 indicates a comment card. The message following the C will be printed in the program listing, but otherwise the card is ignored by the computer. Thus comment cards are not really part of the FORTRAN program but are helpful in identifying the program steps and sections.

Line 3.

This is the first card of the FORTRAN program. It is a FORTRAN statement and therefore begins in column 7, as all FORTRAN statements must. This particular statement tells the computer to read the data card from unit 5 (standard unit number for card reader) according to format (the card layout) described in statement number 1. The names Y1, Y2, Y3, Y4, Y5 are the variables whose values will be read from the data cards.

Line 4.

The number 1 is the statement number assigned to this statement. The number can have one to five digits that can be punched anywhere in columns 1 to 5. This particular

21HST...IS   Print out the next 21 characters following H.

1X   Skip one position.

F10.2   Print out the third variable YSTD, with the same rules used for YMEAN and YVAR.

All format specifications are separated by commas or slashes; different ways of specifying the carriage-control character can also be used.

Line 13.

The STOP statement tells the computer to stop its calculations.

Line 14.

The END statement is always the last card in every FORTRAN program.

Line 15.

Necessary system cards to indicate end of program and beginning of data cards.

Line 16.

Data cards (only one needed):

Columns 1 to 10	10.5	(Y1)
Columns 11 to 20	14.6	(Y2)
Columns 21 to 30	17.8	(Y3)
Columns 31 to 40	13.6	(Y4)
Columns 41 to 50	18.7	(Y5)

Line 17.

Necessary system cards to indicate the end of data cards.

The FORTRAN statements and appropriate system cards in Program 4.1 are keypunched and submitted to the computer center. Included in the pages of computer output that are returned along with the original FORTRAN deck and system

cards are the following results. We note with satisfaction
that the program not only ran but gave us the correct
answers:

```
           C    GSTAT -MEAN, VARIANCE, AND STANDARD DEVIATION OF 5 OBSERVATIONS
  ØØØ1          READ(5,1) Y1,Y2,Y3,Y4,Y5
  ØØØ2        1 FORMAT(5F1Ø.2)
  ØØØ3          SUMY=Y1+Y2+Y3+Y4+Y5
  ØØØ4          SUMY2=Y1*Y1 + Y2*Y2 + Y3*Y3 + Y4*Y4 + Y5*Y5
  ØØØ5          YMEAN=SUMY/5.
  ØØØ6          YVAR=(SUMY2-SUMY*SUMY/5.)/5.
  ØØØ7          YSTD=SQRT(YVAR)
  ØØØ8          WRITE(6,2) YMEAN,YVAR,YSTD
  ØØØ9        2 FORMAT(1H1,7HMEAN IS,1X,F1Ø.2/1H , 11HVARIANCE IS,1X,F1Ø.2/1X,
               1    21HSTANDARD DEVIATION IS,1X,F1Ø.2)
  ØØ1Ø          STOP
  ØØ11          END


               ::   ::   ::   ::   ::


  MEAN IS        15.Ø4
  VARIANCE IS          8.78
  STANDARD DEVIATION IS         2.96
```

In addition to the program listing and the requested answers, the computer printout will generally contain accounting information and a storage map indicating the system routines and variables that were used and where they are
stored. These details, however, are not important to our
introductory effort. We are principally interested in the
program listing and the resulting output that we just
illustrated.

Let us for the sake of demonstration consider what type
of output would result if an error is made in writing the
program. There are unfortunately many types of errors that
might be made, some of which would not even permit any compilation or output. There are essentially three categories
of errors: (1) errors that do not permit compilation, (2)
errors that do not permit execution but are not checked at
the time of compilation, and (3) logic errors. The logic
errors are the most difficult to find because they will not
be detected by the computer; they are found ony by comparing
the answers obtained with known correct answers. We shall
illustrate first some compilation and execution errors. The
same program deck was submitted with six mistakes. The resulting printout with diagnostic messages is

```
         C     GSTAT -MEAN, VARIANCE, AND STANDARD DEVIATION OF 5 OBSERVATIONS
ØØØ1           REA (5,1)Y1,Y2,Y3,Y4,Y5
                   $ $
         Ø1)  IEYØØ1I ILLEGAL TYPE          Ø2)  IEYØ13I SYNTAX
ØØØ2         1 FORMAT(5F1Ø.2)
ØØØ3           SUMY=Y1+Y2+Y3+Y4+Y5
ØØØ4           SUMY2=Y1"Y1 + Y3"Y3 + Y4"Y4 + Y5"Y5
ØØØ5           YMEAN=SUMY/5.
ØØØ6           YVAR= SUMY2-SUMY"SUMY/5.)/5.
                                        $
         Ø1)  IEYØ13I SYNTAX
ØØØ7           YSTD=SQRT(VAR)
ØØØ8           WRITE(6,2) YMEAN, VAR,YSTD
ØØØ9         1 FORMAT(1H1,6HMEAN IS,1X,F1Ø.2/1H ,11HVARIANCE IS,1X,F1Ø.2/1X,
               $                        $
         Ø1)  IEYØØ6I DUPLICATE LABEL       Ø2)  IEYØ13I SYNTAX
             1 21HSTANDARD DEVIATION IS,1X,F1Ø.2)
ØØ1Ø           STOP
ØØ11           END

                      ::   ::   ::   ::   ::

                     1EYØ221    UNDEFINED LABEL
   2

::OPTIONS IN EFFECT::  NOID,EBCDIC,SOURCE,NOLIST,NODECK,LOAD,NOMAP
::OPTIONS IN EFFECT::  NAME = MAIN    , LINECNT =      5Ø
::STATISTICS::    SOURCE STATEMENTS =      11,PROGRAM SIZE =       5Ø6
::STATISTICS:: ØØ6 DIAGNOSTICS GENERATED, HIGHEST SEVERITY CODE IS 8
```

The first and second error messages were caused by the D being omitted from the READ; a parenthesis omitted in statement 0006 resulted in the third error message; the use of statement number 1, which had already been used, yielded the fourth error; and a 6H instead of 7H in the second FORMAT caused the next error. The final "undefined label" message is given because no FORTRAN statement 2 was found. Note that these error messages are often of a general nature and thus may not exactly pinpoint the error but rather indicate that a difficulty exists. Since these errors were detected in the compilation phase and the program was not executed, there were two further errors that were not detected. These errors exist because the variable VAR in the WRITE statement is not defined, and the first field on the data card, Y1, has two decimal points (see next printout).

Let us correct the first four errors that yielded the five error messages shown and resubmit the program. Since all compilation errors have been corrected, the program is also executed. The resulting printout is

```
         C     GSTAT -MEAN, VARIANCE, AND STANDARD DEVIATION OF 5 OBSERVATIONS
ØØØ1           READ(5,1) Y1,Y2,Y3,Y4,Y5
ØØØ2         1 FORMAT(5F1Ø.2)
ØØØ3           SUMY=Y1+Y2+Y3+Y4+Y5
ØØØ4           SUMY2=Y1"Y1 + Y2"Y2 + Y3"Y3 + Y4"Y4 + Y5"Y5
ØØØ5           YMEAN=SUMY/5.
ØØØ6           YVAR=(SUMY2-SUMY"SUMY/5.)/5.
ØØØ7           YSTD=SQRT(YVAR)
ØØØ8           WRITE(6,2) YMEAN, VAR,YSTD
ØØØ9         2 FORMAT(1H1,7HMEAN IS,1X,F1Ø.2/1H ,11HVARIANCE IS,1X,F1Ø.2/1X,
             1 21HSTANDARD DEVIATION IS,1X,F1Ø.2)
ØØ1Ø           STOP
ØØ11           END
```

```
IHC215I CCNVERT - ILLEGAL    DECIMAL    CHARACTER .

 .1Ø.5     14.6      17.8      13.6       18.7

TRACEBACK  ROUTINE  CALLED FROM ISN   REG.  14    REG.  15   REG.    Ø   REG.   1

        IBCOM                          ØØØ7Ø994   ØØØ7ØBCØ   FFØØØØØ8  ØØØ7F7F8

        MAIN                           4Ø1799AE   ØØØ7Ø8Ø8   FFØØØØØ8  ØØØ7F7F8

ENTRY POINT= ØØØ7Ø8Ø8
STANDARD FIXUP TAKEN , EXECUTION CONTINUING

              ::    ::    ::    ::    ::


MEAN IS      12.96
VARIANCE IS        Ø.ØØ
STANDARD DEVIATION IS       6.71
```

In spite of the two errors, answers are obtained, although they are incorrect because Y1 is not included and VAR
is not defined. The decimal error is diagnosed, but the
fact that VAR is undefined could be determined only from the
incorrect answer. When these two errors are corrected, the
program runs as when it was originally submitted without
errors.

Because the error messages obtained depend upon the
particular computer system and will often be either general
or cryptic, the student will frequently have to seek the ad-
vice of an "expert" to help interpret the diagnostic information obtained as output. This process of program checking
and modification is called debugging. In complex programs
debugging often takes many hours of effort and many computer
runs before a correct program is obtained. To help in this
debugging process, sample problems are used to test the program logic; however, the possibility exists that the program
will run properly and give the correct answers using the
sample data but will still encounter problems using new data
that cause other program sections or options to be used.
Another suggestion is to print all data being read in and
intermediate results to try to isolate the source of the
error. In spite of these precautions, we rarely can say
with certainty that a complex program is without bugs and
will always give the proper answer.

4.4 Summary

We can summarize the major steps to be followed in writing a computer program to solve a problem:

1. Identify the problem to be solved.
2. Select an algorithm to be used to obtain the numerical solution.
3. Flow diagram the algorithm.
4. Write the computer program.
5. Prepare the program deck (punched source deck, data cards, and system cards) for submittal to the computer center.
6. Debug the program until the correct answers are obtained.

Once an algorithm has been decided upon and the flow diagram of the steps has been outlined, the computer program can be written in any of the computer languages (that is, FORTRAN, COBOL, ALGOL). Punch the program on cards according to a prescribed format; the resulting punch cards are called the source deck.

The source deck is compiled by the computer. This compilation process translates the source program into an object language (absolute machine language). If no errors are found during the compilation, the execution process can be started (if requested) because the program instructions are stored in the computer memory. The execution process of processing each instruction in the sequence indicated by the program. It is only during execution of the program that the data cards are used because when a READ instruction is encountered, the computer reads the data card.

It is important to understand this two-step process of compilation and execution, although the student might never see the object program. In many computer systems, the source deck is read and then the object program is temporarily stored until the execution is completed. The object program, however, can always be punched on cards, and then the program can be executed without compilation by using the resulting object deck and data cards. This can greatly reduce the computer time required to process the data and so is a worthwhile cost consideration.

We presented a sample program to illustrate the steps to be followed in writing a computer program, but the detailed discussion of FORTRAN programming will be covered in the following chapters. Although we shall limit our consideration to the FORTRAN IV language, the student should be aware that there are many other computer languages and even FORTRAN versions. However, the basic ideas of FORTRAN are the same, and a computer program written in one FORTRAN version can be modified "easily" to run on a different system. In addition, once a student has mastered FORTRAN programming he will find that other programming languages can be readily acquired.

EXERCISES

The student should copy the FORTRAN program to calculate the mean, variance, and standard deviation (GSTAT) on his own programming form, inserting the system cards used by his computer center. Using his own copy of the program, he should punch the program, using the appropriate key punch and system symbols. Be sure that every symbol is correctly punched and in the right column. When making corrections, be sure that the cards are kept in their proper order.

Assemble the program deck and data cards with the required systems cards, and submit it to be run by the computer center.

1. Debug the program, using the data card given in Program 4.1. If you have made an error, rerun the program until you have a satisfactory output.

2. Use your GSTAT program deck to find the mean, variance, and standard deviation of your own set of five observations. To obtain these observations, take a sample of five individuals of the same sex from among your fellow students. Measure the height of each one to the nearest tenth of an inch. After the computer results are available, compare your results with others who took samples from the same population. Note the differences in the observations used and the mean, variance, and standard deviation obtained, although they are all from the same population. This type of random variation forms the basis for much of our statistical analyses and is why statistical inference is required when observed data are used in a study.

CHAPTER 5

THE FUNDAMENTALS OF FORTRAN

5.1 Introduction

The FORTRAN language has, as does any language, a structure or grammer. The advantages of FORTRAN are the simplicity of its structure and the lack of exceptions. This simplicity makes it possible to learn FORTRAN and to appreciate its logic and flexibility by studying the rules of the language. Having once learned the rules of FORTRAN, the student will appreciate that the major programming difficulties are finding numerical methods or algorithms to solve the applied problem.

The FORTRAN program GSTAT presented in Chap. 4 illustrates how descriptive words are used in FORTRAN and FORTRAN's similarity to algebra. The brief description given for each FORTRAN statement of the program should help the student in interpreting other programs in a general way. Before we attempt to write our own program, however, we need a more systematic discussion of FORTRAN.

5.2 FORTRAN Constants and Variables

In FORTRAN, as in algebra, we distinguish between two types of quantities: constants and variables. For example, the algebraic equation y = 3x + 4 involves two constants, 3 and 4, and two variables or unknown quantities, y and x. The equivalent FORTRAN statement Y = 3.*X+4. also has two fixed quantities or constants, 3. and 4., and two variables, Y and X. Other examples of FORTRAN constants are 3,4.5, .0013, 10013, -14, -14.; examples of FORTRAN variables are X, Y, IMAX, J1, and YBAR. FORTRAN variables can be represented by a single letter or by a word involving several letters or characters, which gives us greater flexibility than algebra, where only a single letter can be used. In contructing these FORTRAN words, two restrictions must be observed:

(1) A variable name can be formed using any of the capital alphabetical letters (A,B,C, ... , Z) or any digit (∅,1,2, ... , 9).

(2) The beginning character of a variable name must always be a letter.

(3) The maximum number of characters that can be used for a variable name depends upon the particular FORTRAN compiler being used but is usually limited to six characters.

When a variable name is first introduced, it is assigned a storage location in the computer, and the value of the variable is stored in this location for future use.

The observant student will already have noticed that some FORTRAN constants have decimal points, while some do not. There is no numerical difference between 4 and 4., but in FORTRAN the decimal point differentiates the constants as to their mode, which can be real or integer. Integer constants are numbers without any decimal point; real constants are those with a decimal point. Thus 27 is an integer constant, and 27. is a real constant.

Variables are also differentiated as to integer or real modes. Integer variables are represented by words or names starting with the letters I, J, K, L, M, and N. Real variables must have starting letters other than these and so must start with any letter from A to H or O to Z. Thus I and IMAX are integer variables, and Y and YMEAN are real variables. Integer variables can have only whole-number values, with no decimal point being allowed; real variables can have whole-number values (with a decimal point at the end) and/or fractional values. This distinction is important because the computer has two distinct computational modes, real and integer, and the variables and constants determine the appropriate mode.

81

5.3 FORTRAN Expressions

FORTRAN variables and constants can be combined with the FORTRAN operators to form algebraic exprssions. The operators used to connect variables and constants are

Operation	FORTRAN Operator
Addition	+
Subtraction	-
Multiplication	* (asterisk)
Division	/ (slash)
Exponentiation	**

FORTRAN expressions may be formed by using the following rules:

(1) There must be an operator between every two variables and/or constants.
(2) Two or more operators cannot occur contiguously.
(3) Real and integer modes may not be in the same expression (exception: exponentiation).

Examples of algebraic expressions and the equivalent FORTRAN expressions are

ALGEBRAIC Expression	FORTRAN Expression	MODE
$3x + \dfrac{5}{y}$	3.*X+5./Y	Real
$b^{3.5} + \dfrac{z}{a}$	B**3.5+Z/A	Real
$j^3 + 4j$	JMAX**3+4*JMAX	Integer
$y^3 + 4$	Y**3+4.	Real

A real variable can be raised to an integer or real power (variable or constant), but an integer variable can be raised only to an integer power (variable or constant).
Examples of invalid FORTRAN expressions are

3*X+5	Mixed types
A B+C	Adjacent variables
A+-B	Two adjacent operators

Suppose we wish to write the FORTRAN expression that is equivalent to the algebraic expression (3x + 5)/y. The student can appreciate that FORTRAN must have specific rules as to the order by which expressions are evaluated because the FORTRAN expression 3.*X+5./Y could just as readily represent the FORTRAN expression (3x + 5)/y as 3x + 5/y. Thus the order in which operations are performed or the priority of the operators must be specified. In FORTRAN this hierarchical ordering of the operations is the same as in ordinary algebra. That is, exponentiations are performed first, the multiplication and division operators arc on the priority level below exponentiation, and addition and subtraction operators are on the lowest priority level. If operators of equal priority exist in a FORTRAN expression, the operations are performed in the order they occur from left to right.

According to this priority ordering, the expression 3.*X+5./Y is evaluated in the following steps. There is no exponentiation, and therefore the expression is scanned for the next priority level operations from left to right. Because there are two operations on the same level, a multiplication and a division, the evaluation proceeds from left to right; thus 3. is multiplied by X first and then 5. is divided by Y. No further multiplication or division operations remain, so the scanning starts again from left to right for addition and subtraction operators. The intermediate results of 3.*X and 5./Y are added, and the evaluation is completed.

A more complicated FORTRAN expression Z/Y+X**3.+X/Y+Z*3./2. is evaluated by

(1) Raising X to the third power
(2) Dividing Z by Y, then X by Y, then multiplying Z by 3, then dividing that result by 2
(3) Adding the four resulting intermediate values

Now, if we wish to write the FORTRAN expression for the algebraic expression (3x + 5)/y, we must have the ability to change the prescribed order of the priorities. In algebra the same problem is solved by the use of parenthesis. Thus the algebraic order implied by 3x + 5/y is changed when we write (3x + 5)/y. FORTRAN uses the same device; the required FORTRAN expression can be written as (3.*X+5.)/Y. As in algebra, several sets of parentheses can be nested, although in FORTRAN the nested parentheses must all use the same symbol. For example, the algebraic expression

$$\{[(7x - 2)x + 2]x - 6\}8 - 9$$

can be written in FORTRAN as

$$(((7.*X-2.)*X+2.)*X-6.)*8.-9.$$

The expression contained in the innermost set of parenthesis is evaluated first; this process continues until the entire expression is evaluated.

Parentheses play an even more important role in FORTRAN since the slash in FORTRAN does not have the advantage of the division indication used in algebra. In algebra, we would write the expression a + b/cd, but if we wrote in FORTRAN A+B/C*D, the hierachical ordering would divide B/C first, then multiply this result by D, and finally add A to that intermediate result. So we need to use parentheses and write A+B/(C*D). In fact, a good rule to follow is "When in doubt as to the FORTRAN interpretation of an expression, use parentheses to make sure you evaluate the expression you want."

FORTRAN will not allow the use of double operators, so we cannot write 7. - -Y, but we can change the expression into a valid FORTRAN expression by adding praentheses and writing 7.-(-Y). In certain other situations, parentheses must be also used, as, for example, when an exponent consists of more than one constant or variable. The algebraic expression

$$x^{-\frac{1}{2}x^2}$$

is written in FORTRAN as X**(-∅.5*X**2).

In the remaining chapters the student will have ample opportunities to observe different ways of combining FORTRAN variables, constants, and operators into FORTRAN expressions and to explore such possibilities on his own.

5.4 The FORTRAN Arithmetic Statement

You observed in the FORTRAN program in Chap. 4 that each line or punch card contains a FORTRAN statement. The most basic FORTRAN statement is the arithmetic statement, which is similar to any algebraic statement. Examples of arithmetic statements are

$$Y = A * B$$
$$YBAR = Y1 + Y2$$
$$I = A + B$$
$$J = 4 * LMAX$$

Whenever an arithmetic statement is encountered, the expression on the right side of the equal sign is evaluated, and the result is stored in the location designated by the variable on the left side of the equal sign. Thus the symbol = means "execute (find the value of) the expression on the right side and assign the resulting value to the variable on the left side" and does not indicate the usual logical equality. Thus we can write the statement

$$X = X + 1.$$

which in algebra does not make sense. In programming, however, the value previously assigned to variable X is increased by 1, and this new value is then assigned to X. Note that the original value of X is lost.

Every arithmetic statement must have a single variable (real or integer) on the left side of the equal sign and any valid FORTRAN expression (real or integer) on the right side. FORTRAN cannot manipulate unknown quantities, so all variables appearing in the expression on the right side must have previously had a value assigned to them and the same mode. However, the mode of the variable on the left side does not have to agree with the mode of the expression on the right side.

The evaluation of an arithmetic statement depends upon the mode of the variables and constants that appear in the expression on the right side of the equal sign. For example, the expression

$$10./3.$$

has the computed value 3.333... with as many decimal places as the particular computer allows. If we have the expression

$$10/3$$

the integer mode is used, and the computed value would be the integer value 3 since any decimal fraction is dropped (not rounded).

Although the computational mode is determined by the FORTRAN expression, the mode in which the resulting value is stored depends upon the mode of the variable on the left side of the equal sign. If we have

$$I = 10./3.$$

the real expression 10./3. = 3.333... is evaluated and then the decimal places are truncated because only the integer part of the result will be stored in the location designated by I. On the other hand, if we have

$$X = 10/3$$

the integer value 3 obtained by evaluating 10/3 will be stored in the location specified by X as 3..

We can therefore use an arithmetic statement to change the mode of a variable. For example, if we want to compute the mean we cannot use the instruction

$$YMEAN = SUMY/N$$

since SUMY is the real variable name representing (we assume) the sum of the Y's, and N is an integer name representing the number of observations. Using the fact that the variable mode on the left-hand side of the equal sign does not have to agree with the mode of expression on the right, we assign the integer value of N to a real variable through the statement

$$AN = N$$

and then we can write the required FORTRAN statement

$$YMEAN = SUMY/AN.$$

We could have written also

$$YMEAN = SUMY/FLOAT(N)$$

which makes use of a special FORTRAN function FLOAT ( ),
which is described below.   In Program 4.1 GSTAT we encountered a FORTRAN function SQRT ( ) that was used to find the
square root of an expression.   Many other functions are
available in the FORTRAN language and can be used in any
FORTRAN expression:

SQRT( )   Takes the square root of the expression in
            the parentheses, which must be nonnegative
            and real.

ALOG( )   Takes the natural logarithm of whatever is in
            the parentheses, which must be nonnegative
            and real.

EXP( )   Raises e = 2.718... to the power of whatever
            is in the parentheses.

FLOAT( )   Changes the value of the integer expression
            in the parentheses to its equivalent real
            value.

IF1X( )   Changes the value of the real expression in
            the parentheses to its equivalent integer
            value.

Nested functions can also be used.   Thus

$$Y = SQRT(ALOG(X*X))$$

represents the algebraic expression

$$y = \sqrt{\log x^2}$$

There are other available functions, and the student should
find out which ones are available on the particular computer
system being used.

We can now write the FORTRAN expressions for any algebraic expression.   For example, the formula for the volume
of a fluid in a spherical container is

$$V = \frac{1}{6} \pi h(3r^2 + h^2)$$

where h = depth of container

     r = radius of top surface

     π = 3.1416

The FORTRAN statement is

$$VOL = 1./6.*PI*H*(3.*R**2+H**2)$$

and it is assumed that the values of PI = π and H and R have previously been read in or specified.

We can write the FORTRAN statements to calculate the variance, skewness, and kurtosis by using (1) deviations about the mean and (2) moments about the origin, if we know that the observations have already been read into the computer. Suppose we have three observations Y1, Y2, and Y3:

(1) The formulas using deviations about the mean are

$$\bar{y} = \frac{\sum\limits_{i=1}^{n} y_i}{n}$$

$$s^2 = \frac{\sum\limits_{i=1}^{n} d_i^2}{n} \qquad\qquad d_i = y_i - \bar{y}$$

$$a_3 = \frac{\sum\limits_{i=1}^{n} z_i^3}{n} \qquad a_4 = \frac{\sum\limits_{i=1}^{n} z_i^4}{n} \qquad z_i = \frac{y_i - \bar{y}}{s} = \frac{d_i}{s}$$

and the corresponding statements are

```
YMEAN = (Y1+Y2+Y3)/3.
D1 = Y1-YMEAN
D2 = Y2-YMEAN
D3 = Y3-YMEAN
YVAR = (D1**2+D2**2+D3**2)/3.
YSTD = SQRT(YVAR)
Z1 = D1/YSTD
Z2 = D2/YSTD
Z3 = D3/YSTD
A3 = (Z1**3+Z2**3+Z3**3)/3.
A4 = (Z1**4+Z2**4+Z4**4)/3.
```

(2) The formulas using moments about the origin are

$$\bar{y} = m_1 \qquad\qquad s^2 = m_2 - m_1^2$$

$$a_3 = \frac{m_3 - 3m_2 m_1 + 2m_1^3}{s^3} \qquad a_4 = \frac{m_4 - 4m_3 m_1 + 6m_2 m_1^2 - 3m_1^4}{s^4}$$

$$m_h = \frac{\sum\limits_{i=1}^{n} y_i^h}{n}$$

and the corresponding statements are

```
YM1 = (Y1+Y2+Y3)/3.
YM2 = (Y1**2+Y2**2+Y3**2)/3.
YM3 = (Y1**3+Y2**3+Y3**3)/3.
YM4 = (Y1**4+Y2**4+Y3**4)/3.
YVAR = YM2-YM1**2
YSTD = SQRT(YVAR)
A3 = (YM3-3.*YM2*YM1+2.*YM1**3)/YSTD**3
A4 = (YM4-4.*YM3*YM1+6.*YM2*YM1**2-3.*YM1**4)/YSTD**4
```

Knowing only these basic rules for writing FORTRAN arithmetic statements, we can prepare programs that would evaluate any given formula. Because most problems involve more than a simple formula evaluation, however, we will need additional programming concepts. In any case, we need to know how to read in values, how to write results, and how to indicate the end of a program.

5.5 The END and STOP Statement

The END statement is required in every FORTRAN program and must be the last card of the program. It is a nonexecutable instruction (as is the comment card) and therefore does not cause any computer action during execution; it serves to indicate the end of the program. Like any other FORTRAN statement, it starts in column 7 of the punch card.

The other FORTRAN statement, STOP, actually affects the computer execution of the program. It can be placed anywhere in the program, and when it is encountered all execution is discontinued. For example, it can be used to stop execution if certain conditions that are not permissable are encountered. A STOP statement is usually placed in front of the END statement that indicates the termination of the executable portion of the program, although it can be omitted in certain cases.

5.6 Input and Output Statements

We discussed a program GSTAT in Chap. 4, in which data cards were read and results printed out. Although data can be read from tape, disk, and other storage devices, we shall assume that the data have been punched into data cards according to some format. In fact, the required data may be only a part of the information that is punched into the cards, and thus only certain fields from the card are needed in the particular program. Since the card reader progresses from left to right as it reads each card, we must be able to skip columns, to select those fields that are to be read, and to assign the appropriate variable name to each field. The two FORTRAN statements that accomplish this are

$$\text{READ}(a,b) \text{ var1,var2,var3,}\ldots$$
$$b \text{ FORMAT(spec1, spec2,spec3,}\ldots)$$

In the READ statement, the number a refers to the input unit from which the data is being read, and in most computer centers the number 5 refers to the card reader. The b is the cross-reference statement number that links the READ statement to its associated FORMAT statement. Although any FORTRAN statement can have a statement number (a positive integer that can be punched anywhere in the columns 1 through 5), the FORMAT statement must have the statement number b because it does not need to follow the READ statement; it can be placed anywhere in the program.

The "list" var1, var2, var3,... represent the variables names that will sequentially be assigned the values that are read from the data card(s). These names may be integer or real and must satisfy the constraints listed earlier (i.e. one to six characters, and so on). The process is such that when the reader reaches the first data field, it reads the numbers that are punched and assigns their value to the variable whose name is var1, then goes to the second data field and assigns its value to var2, and so forth.

In the FORMAT statement, the specifications given by spec1, spec2, spec3,... are used to identify the data fields to be read from the card. This is done by certain formats that are defined in the actual FORMAT statement (lowercase letters must be replaced by integers) by

wX   Skip w columns.

nIw   Repeat the Iw format n times, and read each w columns as one integer field. No decimal point can be punched in the data field.

nFw.d   Repeat the Fw.d format n times, and read each w columns as one real field. If there is no decimal punched in the field, the number read should have d digits to the rights of the decimal point. However, a decimal point that is punched in the data card always overrides the format specification.

When n = 1 in the above format specifications, it can be omitted.

Using these conventions data can be made available to the program. The statements

$$READ(5,10)IMAX,JMAX,VAR1$$
$$10 \quad FORMAT(12X,I8,I6,I4X,F10.2)$$

indicate that the data is on cards since a = 5 (the standard designation for the card reader on most computer systems) and is to be read according to FORMAT statement 10 (b=10). Three variables are to be read: the first two are integer variables, and the third is a real variable. The data card containing the values for the variables is given on the next page. All columns of the card are considered until the final field is read; the format specifications must take this into consideration. Since in the example the first field starts in column 13, 12 columns must be skipped, which is denoted by the first specification, 12X. The value of IMAX is found in the next eight columns (columns 13 to 20), and

91

because IMAX is an integer variable this information is communicated by the next format specification, I8. The value must be punched in the rightmost columns of the field (right-justified; i.e., all blanks appear to the left of the first significant number). The value of JMAX is found in the adjacent six columns (columns 21 to 26), and because it is also an integer variable, the field is specified by I6; the value must be right-justified. The next 14 columns (columns 27 to 40) must be skipped before the value of VAR1 is read; this is done by the specification 14X. The last variable is found in columns 41 to 50 (10 columns). This is a real variable and is specified by F10.2. The value of 10 specifies the number of columns, and the 2 specifies the number of decimal places. If the decimal point is punched in the card, the value is read as it is punched on the card. If the decimal point is not punched, format decimal specification holds, and the number would be read with two decimal places, counting from the right of the field.

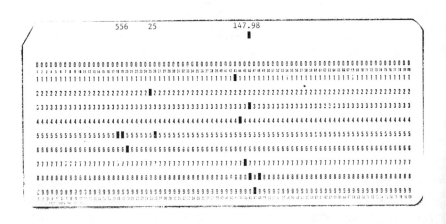

The effect of the format specification and punching the decimal point can best be illustrated by the following example. Consider the effect of using the following data cards to read the value of VAR1 specified by F10.2:

	1	2	3	.	.	.	41	42	43	44	45	46	47	48	49	50	.	.	.
(1)									1	4	7	.	9	8					
(2)												1	4	7	9	8			
(3)									1	4	7	9	8						

If the decimal point is punched in the data field, the value
can be placed anywhere within columns 41 to 50. If the same
value is to be read without punching the decimal point, the
digits must be in columns 46 to 50 since the 10.2 indicates
that the implied decimal point is between columns 48 and 49.
If the digits are placed in any other columns, as in the
third example, the value would be read in with the decimal
point still implied between columns 48 and 49. Thus the
value of 147980.00 would be read in if the third data card
is used because blank columns are read as zeros.

In reading real variables, we can always ensure that
the correct value is read by punching the decimal point.
Integer variables, however, do not have decimal points, and
thus the values must be punched in the rightmost columns of
the field. If the value IMAX, for example, had been punched
in columns 16 to 18, (instead of 18 to 20), then IMAX would
be assigned the value of 55600 instead of 556.

The student should recall that in GSTAT the READ statement is placed in the FORTRAN program before the values of
variables are needed by the program. For example, all variables found on the right-hand side of an arithmetic statement must have appeared either in a READ list or on the
left-hand side of an arithmetic statement. The data cards
actually containing the specific values for the variables
are placed behind the entire program in the same order as
the READ statements occur in the program.

Each READ statement causes a new card to be read. This
fact can be used to read data from several cards. There
are, however, several mechanisms that can be used, in addition to using a new READ statement for each card:

(1) The READ statement list var1, var2, var3, ... must be satisfied regardless of how many specifications are given in the FORMAT statement. Thus, if the last close parenthesis of the FORMAT statement is reached before the READ list is satisfied, a new card will be read according to the specifications given after the innermost open parenthesis. Therefore, the statements

READ(5,100)A,B,C,D,E
100 FORMAT(F8.2)

will cause _five_ cards to be read. The values found in columns 1 to 8 on each card will be assigned to the variables respectively. On the other hand,

READ(5,100) A,B,C,D,E
100 FORMAT(F8.4,(F6.2,F4.0))

will cause A, B, and C to be read from the first data card as per the three specifications. Since D and E must have values assigned to them, a new card is read according to F6.2, F4.0 so that the values of D and E must be in columns 1 to 6 and 7 to 10, respectively.

(2) Any slash (/) appearing between specifications will cause a new card to be read. Thus

READ(5,1) X,Y,Z,W
1 FORMAT(2F4.0/2F6.0)

will cause X and Y to be read from columns 1 to 4 and 5 to 8, respectively, and Z and W will be read from the next data card from columns 1 to 6 and 7 to 12.

Now consider the problem of writing results. We shall assume that the results will always be printed. Although different printers have a different number of print positions, the first position is always used for the carriage control; it is not available for printing. The student should determine the maximum number of print positions available at his computer center.

The FORTRAN statements used to write results are

$$WRITE(a,b) \ var1,var2,var3,...$$
$$b \ FORMAT(\ spec1,spec2,spec3,...)$$

In the WRITE statement, a is the number assigned to the output unit. The standard usage in most computer centers is
that a = 6 designates the printer. The letter b again represents the statement number of the corresponding FORMAT
statement. The list var1, var2, var3, ... are the names of
the variables whose values are to be printed. In the format
specifications, spec1 provides carriage-control information.
This specification is generally written in the form 1Hw, and
when w is equal to

   β or anything else single-space before printing the
                                results
   0   double-space before printing results
   1 skip to a new page before printing the results.

The specifications other than the first specification
indicate the format to be followed in printing the line and
can involve

   wX   Skip w print positions.
   nIw Write the values of the next n integer variables
          named in the WRITE list, using w positions for each
          value.
 nFw.d Write the value of the next n real variables named
          in the WRITE list, using w positions with d places
          to the right of the decimal.
   wH   Write the next w characters following H as a
          message.

We see that the write specifications are similar to
those of the read specifications except for the carriage
control and the H message printing format. A simple illustrative example demonstrates the approach:

```
    WRITE(6,276)J,Y
276 FORMAT(1H1,14X,5HIMAX=,I8,3X,18HSMALLEST VALUE IS=,F10.3)
```

These statements cause the printer to start on a new page;
then 14 spaces are skipped, the five characters after 5H are
printed, followed by the value of J; then three spaces are
skipped, the 18 characters after 18H are printed, and the
value of Y is printed. The resulting printout is:

IMAX=   556   SMALLEST VALUE IS=   23.765

Format specifications must always be separated from
each other by commas or slashes. The printing of variables
is always right-justified. The value of w for each integer
field specification must be sufficiently large to allow for
a sign and the entire integer value. The value of w for
each real field specification must allow for at least the
sign, the decimal point, the whole number, and the d decimal
places. If w is not sufficiently large, an error message is
given in place of the result. Several special features need
to be noted:

(1) The printer uses first character in the FORMAT
specification for carriage control whether the 1H form is
used or not, and thus the following two specifications are
equivalent:

1H߶,10X   or   11X

(2) A slash (/) introduced between specifications in
the FORMAT statement causes the printer to start on a new
line. The first character after the slash (/) is used for
carriage control.

(3) The variable list given in the WRITE statement must
be satisfied. If there are not enough specifications given
in the FORMAT statement, the computer returns to the innermost open parentheses in the FORMAT statement and prints the
remaining values on the next line. Thus, if we had

```
       WRITE(6,17)X,Y,Z
17  FORMAT(5X,F8.2)
```

The computer would single space to a new line, skip an additional four positions and print the value of X in the next
eight columns, go to a new line, skip four positions and
print the value of Y in the next eight columns, and so forth.

Finally, some mention should be made of the appearance
of the resulting printed page. We surely do not want all
the output to be located in the upper left-hand corner of
the paper, and generally we do not want only numbers to
appear without identifying context or labels. To help in
this output design special papers are available, but one can
do a good job with just a working sketch.

We should now be able to set up the appropriate READ,
WRITE, and FORMAT statements for any problem. For example,
if we wish to read in the five observations, each of which
is found on a separate card in columns 11 to 20 with the
decimal point punched, the appropriate statement is

```
        READ(5,100)Y1,Y2,Y3,Y4,Y5
100  FORMAT(10X,F10.2)
```

We can now write these values, one per line, by using the
statement

```
        WRITE(6,100)Y1,Y2,Y3,Y4,Y5
```

and specify the identical FORMAT statement used by the READ
statement.

Let us assume we want to print the value of the mean,
variance, skewness, and kurtosis that we calculated from
these observations (end of Sec. 5.4), and we also want to
identify each value which we print. These WRITE statements
would be placed after the arithmetic statements given in
Sec. 5.4. They are

```
      WRITE(6,101)YMEAN,YVAR,A3,A4
  101 FORMAT(1H1,7HMEAN IS,F12.4/1HØ,11HVARIANCE IS,F12.4/
     A1HØ,11HSKEWNESS IS,F12.4/1HØ,11HKURTOSIS IS,F12.4)
```

The FORMAT STATEMENT 101 extended past the 72 columns allowed for each FORTRAN statement punch card. This problem often occurs when punching the FORMAT statement connected with WRITE statement because most printers have at least 120 print positions and the message to be written might be long. In this case the letter A in column 6 on the next card is used to indicate the continuation. Using the arithmetic, READ, WRITE, FORMAT, STOP, and END statements, we are now in position to write a simple FORTRAN program. Such programs, however, need to be straightforward because each statement is executed sequentially starting with the first one, and we have so far not introduced the necessary statements to change this operation.

5.7 Transfer Statements

Often it is necessary to branch (transfer control) to a different section of the program under one set of conditions and to another section of the program for another set of conditions. This branching in FORTRAN can be accomplished by four types of statements: the GO TO, the computed GO TO, the arithmetic IF, and the logical IF. The simplest GO TO is

$$GO\ TO\ n$$

which is an unconditional transfer. When this statement is reached during execution, the control is transferred to the statement labeled n so that this statement is executed instead of the next sequential statement. A computed GO TO allows control to be transferred to different program sections depending upon the value of an integer variable. This statement is given by

$$GO\ TO(n_1,n_2,n_3,\ldots,n_k),\ i$$

where $n_1, n_2, n_3, \ldots, n_k$ are statement numbers, and i is an integer variable. When this statement is encountered during execution, the value of i must have been previously defined as having some integer value between 1 and k. According to this value of i, the control is transferred to statement numbered n_i. An example of this statement is

GO TO (11,11,31), ITALLY

If ITALLY has the value 1, control is transferred to statement 11; if ITALLY = 2, control is transferred to statement 11; and if ITALLY = 3, control is transferred to statement 31. If ITALLY has some other value, the computer action is unpredictable, but often an error message will result.

Now let us consider the arithmetic IF statement. This statement can be used to transfer control to three different program sections, depending upon the value of a FORTRAN arithmetic expression. The arithmetic IF is given by

IF(a) n_1, n_2, n_3

where a is any FORTRAN expression, integer or real, and n_1, n_2, n_3 are statement numbers. When this statement is encountered during execution, the expression a is evaluated. If a is less than zero, control is transferred to statement numbered n_1; if a is equal zero, control is transferred to n_2; and if a is greater than zero, control is transferred to n_3. For example, when the statement

IF(B**2-4.*A*C) 101,27,53

during execution is enountered, the expression $b^2 - 4ac$ is evaluated, and if it is less than zero, control is transferred to the statement 101; if it is equal to zero, control is transferred to statement 27; and if it is greater than zero, control is transferred to statement 53.

The logical IF statement has the form

$$IF(a)\ c$$

where a can be any logical expression and c can be any FOR-
TRAN statement other than another IF or DO (introduced in
Chap. 6) statement. If, during execution, the logical expression a is true, the computer then executes statement c
before proceeding to the next sequential statement after c.
If, however, the logical expression is not true, the computer simply goes to the next statement after the IF statement.

A logical expression a can be simply two FORTRAN arithmetic expressions connected by a _relational_ operator or several such expressions joined by a _logical_ operator. The
allowable _relational_ operators are

.EQ.	=	equal
.GT.	>	greater than
.LT.	<	less than
.GE.	≥	greater than or equal
.LE.	≤	less than or equal
.NE.	≠	not equal

For example, when the following set of statement is
executed,

$$J=1$$
$$IF\ (M.GE.10)\ J=2$$
$$N=J*M$$

the computer will first set J equal to 1. It then checks if
M is greater than or equal to (≥) 10. If M ≥ 10, J will be
set to 2, and the next sequential statement N=J*M will be
executed with J=2. If however M was less than (<) 10, then
N=J*M would be executed with J=1. The logical IF statement
can also be used to transfer control to a different program
section. The statement

$$IF\ (P.LT.\ 100.)\ GO\ TO\ 25$$

100

causes the value of P to be compared with 100. If P < 100, the statement on the same line is executed, which is a GO TO statement, and thus the next statement that will be executed is statement 25. If P $\geq$ 100, the computer goes to the next statement after the IF statement. Notice that the mode of the expressions on either side of the relational operator must be the same whether they are real or integer.

The logical expression can involve more than one operator by using the <u>logical</u> operators .AND., .OR., or .NOT. Thus we can write

IF(X .LT. 0. .OR. N .EQ. 100) ICODE=2

If either X < 0 or N = 100, the logical expression is true, and ICODE is set equal to 2. Otherwise ICODE retains its former value, and the statement following the IF is executed. Notice that the expressions on either side of a logical operator can have different modes.

We have discussed four different transfer statements. Each causes the sequential execution of the FORTRAN statements to be interrupted. When the appropriate branch to a different program section is made, the sequential execution is again continued until another transfer statement is enountered.

Because there are many ways in which the above transfer statements can be used to evaluate complex problems, the student needs to give careful consideration to their use in writing programs. In the next section, we illustrate some of these programming techniques.

5.8 Illustrative Programs

In many mathematical and statistical computer problems, a series must be evaluated. Because the programming techniques used to evaluate different series are very similar and use transfer statements, we shall illustrate the technique by evaluating

$$S = \frac{1}{1} + \frac{1}{2} + \frac{1}{3} + \cdots + \frac{1}{n}$$

The flowchart for the required programming steps is

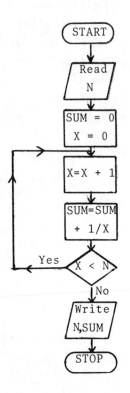

We write the FORTRAN program, SERIES as

```
C      SERIES - EVALUATION OF SERIES
       READ(5,1)N
     1 FORMAT(I3)
C      N PUNCHED IN THE FIRST THREE COLUMNS OF THE DATA CARD
       SUM = Ø
       X = Ø.
     2 X = X+1.
       SUM = SUM+1./X
       N1 = X
       IF(N1.LT.N) GO TO 2
       WRITE(6,4)N,SUM
     4 FORMAT(1H1,37HTHE SUM OF THE RECIPROCALS OF 1,2,...,I3
      1  /1H ,11HIS EQUAL TO,F8.5)
       STOP
       END
```

Program 5.1 SERIES — evaluation of a series.

The above program introduces the concept of a loop, which we shall consider in more detail in Chap. 6. However, the student should consider both the flow diagram and the program to see how the program "loops" back and then "steps" X by increasing its value by 1 and increments SUM until the required number of reciprocals have been added. The loop consists of the statements

```
2 X = X+1.
  SUM = SUM+1./X
  N1 = X
  IF(N1.LT.N) GO TO 2
```

Before entering the loop, SUM and X are both set equal to zero. The first time through the loop, X= 0.+1. and SUM = (0.)+1./1.. The second time through, X becomes 2. and SUM becomes (1.)+1./2., then X = 3. and SUM = (1.5)+1./3., and so forth. The sample input to the program is

Columns 1 to 3   ∅∅5

Then the corresponding output is

THE SUM OF THE RECIPROCALS OF 1,2,..., 5
IS EQUAL TO   2.28333

To further demonstrate the use of branching or transfer statements, let us modify the original program for finding the mean, variance, and standard deviation of observations. In the initial program we assumed that there were only five observations, but now we want to write a program that will handle any number of observations.

We shall assume that the value for each observation will be in separate cards and punched in columns 11 to 20, with the end of the data deck shown by a card with the digits of 99999999 in columns 13 to 20. In this case the value of each new observation must be added to the variable SUM, designated to represent the sum of the y's, and the sum of squares of each value must be added to the variable SUM2, which is used to accumulate the sum of squares. When all observations are read and accumulated, the mean, variance, and standard deviation must be found and the results printed out. The flowchart is

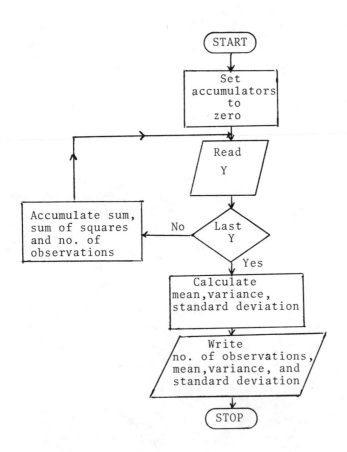

We have the following FORTRAN program, called GSTAT1:

```
C     GSTAT1 - MEAN, VAR, STD DEV FOR UNKNOWN NO OF OBS
      SUMY = Ø.
      SUMY2 = Ø.
      XN = Ø.
    2 READ(5,1)Y
    1 FORMAT(1ØX,F1Ø.2)
      IF (Y.GE. 9999999999.) GO TO 1Ø
      SUMY = SUMY + Y
      SUMY2 = SUMY2 + Y**2
      XN = XN+1.
      GO TO 2
   1Ø YMEAN = SUMY/XN
      YVAR = (SUMY2-SUMY**2/XN)/XN
      YSTD = SQRT(YVAR)
      WRITE(6,3) XN
    3 FORMAT(1H1,22HNUMBER OF OBSERVATIONS, F5.Ø)
      WRITE(6,4) YMEAN,YVAR,YSTD
    4 FORMAT(1H ,7HMEAN IS,1X,F1Ø.2/1H ,11HVARIANCE IS,1X,
     1 F1Ø.2/1X,21HSTANDARD DEVIATION IS,1X,F1Ø.2)
      STOP
      END
```

Program 5.2 GSTAT1 — mean, variance, and standard
deviation for unknown number of observations.

To test the program, we use the 12 reading times given in
Fig. 2.1, Chap. 2. The data deck consists of the 13 cards
(the values can be anywhere in the fields 11 to 20 since the
decimal point is punched):

```
                    4.2
                    5.1
                    3.3
                    4.3
                    5.2
                    3.1
                    4.2
                    5.3
                    6.1
                    2.2
                    4.2
                    5.2
                99999999
```

The output for these data cards is

```
        NUMBER OF OBSERVATIONS    12.
        MEAN IS          4.37
        VARIANCE IS          1.11
        STANDARD DEVIATION IS      1.Ø5
```

105

The concept of incrementing or indexing is useful in many types of calculations, for example, in the evaluation of factorials. The definition of n! (n factorial) is $1 \cdot 2 \cdot 3 \cdots$ $(n - 1)n$, with n being any positive integer. Thus $5! = 1 \cdot 2 \cdot 3 \cdot 4 \cdot 5$, and $2! = 1 \cdot 2$, with $0! = 1$ by definition. Because factorials occur in many mathematical representations, it is useful to write a program to evaluate n! for any value of n. The flowchart for such a program is

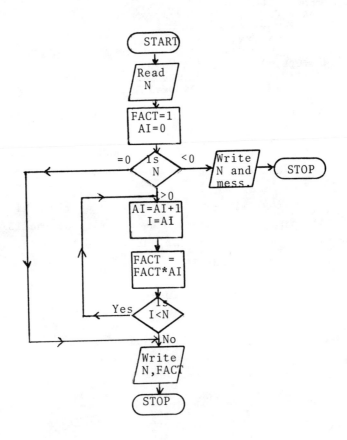

The corresponding program, called FACT, is

```
C     FACT - EVALUATION OF N FACTORIAL
      READ(5,1)N
    1 FORMAT(I3)
      FACT = 1.
      AI = Ø.
      IF(N) 2Ø,1Ø,2
    2 AI = AI+1.
      FACT = FACT*AI
      I = AI
      IF(I.LT.N) GO TO 2
   1Ø WRITE(6,4)N,FACT
    4 FORMAT(1H1,I4,2X,21HFACTORIAL IS EQUAL TO,F12.Ø)
      STOP
   2Ø WRITE(6,21)N
   21 FORMAT(1X,19HINVALID VALUE FOR N,I5)
      STOP
      END
```

Program 5.3 FACT — evaluation of n factorial.

The sample input is

$$\text{columns 1 to 3} \qquad \text{ØØ5}$$

The sample output is

$$\text{5 FACTORIAL IS EQUAL TO} \qquad \text{120.}$$

This program increments upward by 1 until the value of N is reached. Another approach would be to start with the value of N and decrease the multiplier by 1 until the value of 1 is reached.

The student might ask why the program does not use the integer variable IFACT to represent n! instead of the real variable FACT since the value will always be a whole number. To understand the reason for using a real variable, the student needs to know how numbers are stored in the computer and the restrictions on the magnitude of the numbers. We shall see in the next section that this problem is less critical when using real variables than when using integer values.

5.9 Additional FORTRAN Concepts

Since the available internal storage of digital computers is limited, all computers restrict the number of digits
that can be used to represent a number when storing the value of a constant or variable. The maximum number of digits
for an integer number can be as low as 1, although most dig-
itial computers can accommodate a maximum of 10 to 20 digits.
Since 15! = 1,307,674,386,000 is already a large number,
we appreciate that the use of an integer variable to represent factorial values would soon exceed the capacity of any
computer and result in an overflow message being printed.

Obviously the digital computer must have a wider range
of numbers that can be stored if it is to be truly useful as
a scientific tool. To overcome the restriction of the integer representation, scientific notation is used: The real
numbers are changed into a mantissa and exponent form when
they are stored in the computer. For example, A = 3.456789
can also be written and stored as $.3456789 \times 10^1$ — the decimal portion is called the mantissa and the 1 is called the
exponent. The general form used for storing real values is

$$\pm \underbrace{.xxxxxxxx}_{\text{mantissa}} E \underbrace{\pm xxx}_{\text{exponent}}$$

The number of places in the mantissa and the exponent depend
upon the computer model, with both the mantissa and exponent
requiring signs. If the exponent becomes numerically too
large with a positive sign, an overflow message is printed;
if it is numerically too large with a negative sign, an
underflow message is printed. Although a 10-digit number
such as 4567891234 might cause an overflow if stored as an
integer, the real variable $.4567891234 \times 10^{10}$ would not.
However, depending upon the number of digits allowed for
the mantissa, some of the right-hand digits may be lost
through round-off without receiving an error message. For
example, in the factorial problem where n = 15, FACT is
stored as .13076744 E+13 if only eight digits are allowed
for the mantissa, with the last two significant digits being
lost although no overflow occurs.

It is left to the student to determine the maximum number of digits and values for both integer and real variables allowed on the particular computer model being used by modifying Program 5.3 so as to be able to see when overflow occurs for either variable type. Alternate computational methods must therefore frequently be used, such as evaluating the logarithm of n! instead of n!.

The scientific representation of real variables can also be used in reading data and is particularly useful in printing results because in many scientific evaluations a particular variable can have a wide range of values. Although the Fw.d format is attractive for output since the actual value of the variable is given with the decimal point properly located, one would have to allow for the maximum number of digits both to the right and left of the decimal point to accommodate a wide range of values. The E format specification is given by

nEw.d   Read (print) n real variables in w columns (print positions), with d places to the right of the decimal.

Values that are printed using the E-format specification will always have the form

$$\pm \underbrace{x.\underbrace{xxxxx...xE\pm xx}_{\text{d places}}}_{\text{w places}}$$

and thus the total number of print positions must always exceed the number of decimal places by 6 (w>d+6) to accommodate the sign, decimal and the exponent E±xx. For example, the statement

WRITE(6,100) X
100 FORMAT(1H ,E13.6)

would result in

ʬ.274162Eʬ3

being printed, which means that X is .274162 × 10^3 or 274.162.

In reading data using an E-format specification, the values can be punched into cards in several different ways. Assume we wish to read X = 274.162 using the statements

READ(5,1) X
1 FORMAT(E16.8)

Then we can use any of the following data cards:

Columns

	1	2	3	4	5	6	7	8	9	10	11	12	13	14	15	16
(1)							2	7	4	.	1	6	2			
(2)									.	2	7	4	1	6	2	E 3
(3)									.	2	7	4	1	6	2	+ 3

Thus the values can be punched in the standard or scientific notation with the punched decimal point overriding the specification d given in the FORMAT. If the scientific notation is used, the values must be right-adjusted in the field since blanks are read as zeros. The letter E in the data card can be omitted [example (3)].

There are other special FORTRAN considerations, such as the explicit and implicit type statements, that can be used to change the mode of variable names, the assigned GO TO, and the additional format specifications such as Lw, and Dw.d. We have decided to omit them in this introductory text because they are not necessary for the programming problems we shall be considering.

5.10 Summary

The fundamentals of FORTRAN presented in this chapter included integer and real constants and variables, the construction of FORTRAN variable names, FORTRAN operations and expressions, order of operations, use of parentheses and how they can be used to construct a FORTRAN arithmetic statement. Other FORTRAN statements such as the STOP, END, READ, WRITE, and FORMAT were also introduced.

The READ statement with its corresponding FORMAT statement uses the format specifications wX,nFw.d, nEw.d and nIw and causes data to be read in from punched data cards. The WRITE statement with its corresponding FORMAT statement uses the format specifications wX, nFw.d, nEw.d, nIw and wH and causes messages and values of variables to be printed out. Special attention is required for the carriage-control character that is the first specification in the format associated with a WRITE statement. Special consideration of the role of the slash (/) and how the computer behaves when the format specifications do not correspond to the list was also given.

The transfer statements GO TO n; GO TO $(n_1, n_2, \ldots, n_k), i$; IF(a) n_1, n_2, n_3; and IF(a)c are used to transfer control to another portion of the program rather than to continue the sequential processing of each statement. The branching capability makes it possible to build loop's into computer programs. Since computer programs are generally written for large-scale problems that involve repetitious steps or accumulation of sums, this looping concept is most useful and should be understood before proceeding to the next chapter.

These FORTRAN concepts enable us to write programs for most statistical problems that can be formulated numerically. Three programs were presented to indicate the flexibility of these FORTRAN concepts to accomplish a specific purpose. Any problem can be programmed in different ways, some more ingenious than others, and it is up to the student to explore and study FORTRAN so he can write programs with a minimum number of statements and/or with the shortest execution time.

We do not have the indexing capability that enables us to indicate one element of a list Y, say y_i, which we found most useful in the algebraic formulation of statistical formulas. This indexing capability is closely connected with the summation capability and would be most useful in FORTRAN, particularly when we have large numbers of observations that

we want to retain. Presently we have only two ways of handling such cases: (1) We can use a different name for each
observation, which means we must always change the program
whenever we change the number of observations, or (2) we can
read an observation, manipulate it, and then read another,
destroying the previous value. These concepts were illustrated by Programs 4.1 and 5.2, but neither is very satisfactory for large problems.

The next chapter introduces the FORTRAN concepts that
give us the desired indexing capability so that we can retain numerous observations of a random variable for later
manipulations in a more general fashion than either of the
two alternatives just mentioned permit.

1. Modify the GSTAT program in Chap. 4 so that the variance is calculated by the formula

$$s^2 = \frac{\sum\limits_{i=1}^{n} d_i^2}{n} \qquad d_i = y_i - \bar{y}$$

and the printout is

```
THE FIVE OBSERVATIONS ARE
±XXXXXX.XX    ±XXXXXX.XX    ±XXXXXX.XX    ±XXXXXX.XX    ±XXXXXX.XX
THE MEAN IS ±XXXXXX.XX
THE VARIANCE IS ±XXXXXX.XX
THE STANDARD DEVIATION IS ±XXXXXX.XX
```

To calculate each d_i, the program must first read in the five observations and calculate their mean. Check the program by submitting the program to run with the same data card used for GSTAT in Chap. 4.

2. Write a program to calculate the mean, variance, standard deviation, skewness, and kurtosis for any number of un-grouped observations. Calculate the quantities using the moments about the origin. (Hint: Use the concept illustrated by program GSTAT1 in this chapter and the other portions given in different sections of this chapter.) The printout for the program should be

```
THE NUMBER OF OBS. IS XXXXXX
THE MEAN IS ±XXXXXX.XXX
THE VARIANCE IS ±XXXXXX.XXX
THE STD. DEV. IS ±XXXXXX.XXX
THE SKEWNESS IS ±X.XXX
THE KURTOSIS IS ±X.XXX
```

(a) Check the correctness of the program by using the un-grouped data on the reading times of 12 employees given in Fig. 2.1, Chap. 2.

(b) Run the program, using the data in Exercise 2, Chap. 2.

3. Write a program similar to the one for Exercise 2, which calculates the same five characteristics of grouped data. Read in the number of intervals and then the midpoint of each interval y_j and the frequency for that interval, f_j. Use the formulas with the moments about the origin for grouped data, and print out the same information as in Exercise 2.

   (a) Check the program, using the grouped data on the reading times given in Fig. 2.3, Chap. 2.

   (b) Run the program, using the data in the frequency table constructed for Exercise 4, Chap. 2.

4. Change program 5.3 so that an integer variable is used for the value of n!, and perform the operations in the integer mode. Determine what value of n causes an overflow by having the program compute first 1!, then 2!, 3!, ..., until the overflow occurs. The overflow condition is indicated by an error message that is printed.

5. Include a program section in the program written for Exercise 2, which calculates the range of the observation. (Hint: Find the minimum and maximum values and then subtract. The maximum value can be found by initializing YMAX to zero and then comparing each observation to the variable YMAX using

$$IF(Y.GT.YMAX)YMAX = Y$$

Thus, any observation greater than the value of YMAX is then called the maximum value. A similar technique can be used to find the minimum value.

PRACTICE PROBLEMS

1. In each of the following cases, state whether a constant is integer (I), real (r), or invalid (W):

-47	43	0	-17.78
1,385004	1%	2.	0.67

2. State whether the following variables are integers (I), real (R), or invalid (W):

SUM	DIFF	KKK	MEAN
DIVISOR	NIDOT	I1A	XII
IIA	LOVE	AMEAN	4N

3. Consider the following program section:

   ```
   SUM = 0.
   SUM = SUM+7.
   SUM = SUM+12.
   SUM = SUM+16.
   ```

   (a) What is the value of SUM after each statement is processed by the computer?

   (b) Are the intermediate values of SUM saved?

4. Suppose we want to evaluate $\sqrt{a^2 + 2b}$ when a = 7 and b = 3. We write the program

   ```
   A = 7.
   B = 3.
   X = SQRT(A**2+2.*B)
   ```

   so the answer is stored in the computer as X. If we want to know or use the value of $\sqrt{a^2 + 2b}$, we must use X as its representation.

(a) Write the FORTRAN statement(s) to calculate

$\sqrt{a^2 + 2b}$ + $\sqrt{b^2 + 2a}$, which would follow the three statements given above.

5. Correct the following arithmetic statements if they are wrong:

```
Y = 2X
Y = 2*X
MEAN = SUM/N
SSA = SUM*SUM+2
N = N-1
X = X-1
I = I+2
N = N+P+Q
RESULT = 4-5/X
X/2 = Y
CIRC = 3.1416*DIAMETER
PI = 3.1416
PIHA = 1/2*PI
VOL = LONG*WIDE*HIGH
J = 2*I+5
```

6. Write a FORTRAN arithmetic statement for each of the following mathematical formulas and the statements to print the calculated value with identifying information:

(a) $T = 2\pi\sqrt{\dfrac{e}{g}}$

(b) $C = \sqrt{2gh}$

(c) $Y = \sqrt{S + \sqrt{10X}}$

(d) $Y = \dfrac{a}{b} + \dfrac{c - d}{e} + \dfrac{x(a + b)}{u + v}$

(e) $r = \dfrac{e - ab/n}{(c - a^2/n)(d - b^2/n)}$

(f) $s = \sqrt{\dfrac{x^2 - (x)\,2/n}{n \cdot 1}}$

7. Write a mathematical formula corresponding to each of the following FORTRAN expressions:

(a) Y = SQRT(C-S**2/AN)
(b) Y = B/2.+SQRT(B**2/4.+C)
(c) Y = (((2.*X+4.)*X-7.)*X+8.)*X-5.
(d) Y = 1./(X+2.(X+3.1(X+4.)))

8. Give the FORTRAN READ and FORMAT statement for the following data found in a single data card:

    columns 14 to 16   An integer variable
    columns 20 to 25   A real variable (with punched decimal point)
    columns 77 to 80   An integer variable

9. Give the WRITE statement to print out on one line the variables as given on the data card in Prob. 8.

10. Three data cards are to be read to find the values for I, YVAR, and KL. The values are columns 1 to 10 (card 1), columns 20 to 30 (card 2), and columns 1 to 10 (card 3), respectively. Give the READ and FORMAT that can be used to read the three values. Give one WRITE and FORMAT statement that prints the values on separate lines with a blank line in between.

11. Write a program to determine the value of the series

$$\sum_{i=1}^{N} \frac{1}{i^2} = 1 + \frac{1}{2^2} + \frac{1}{3^2} + \cdots + \frac{1}{N^2}$$

for any positive integer value of N.

12. Write a program to determine the value of the series

$$\sum_{i=1}^{N} \frac{1}{i(i + 1)} = \frac{1}{1 \cdot 2} + \frac{1}{2 \cdot 3} + \frac{1}{3 \cdot 4} + \cdots + \frac{1}{N(N + 1)}$$

for any positive integer value of N.

CHAPTER 6

FORTRAN LISTS AND LOOPS

6.1 Introduction

In most data processing problems involving the computer
and statistics, large arrays or lists of numerical data are
read in. Thus, efficient methods of handling are needed.
In statistics, an algebra was developed that utilized an index (usually written as a subscript) and the summation symbol Σ, which enabled us to write general formulas in compact
notation. In FORTRAN we want to have the same indexing (or
subscripting) and summation capability. Without such a capability, we would have to, for example, invent a new name
for each element of a list that we wish to retain, and in
some cases we would even have to change the program every
time the number of observations changed, although the procedure would remain the same.

This chapter will introduce FORTRAN lists (arrays) and
the DO LOOP that give us the desired statistical data processing capability and will show their applicability to statistical problems.

6.2 FORTRAN Lists or Arrays

Since FORTRAN does not allow subscripts or super-
scripts, a list like $y_1, y_2, \cdots, y_i, \cdots, y_n$, must be
written as $Y(1), Y(2), \ldots, Y(I), \ldots, Y(N)$. By using this
notation, any variable, regardless of the name or mode, can
be indexed. The variable name then identifies the list, and
the index indicates the specific observation in the list.
For example, YDATA(7) is the value of the 7th observation in
the real list called YDATA, and ICT(J) is the value of the

119

jth observation in the <u>integer</u> list called ICT.

It can be seen from the above illustrations that a FORTRAN list consists of a name and an integer expression within parentheses. The value of this integer expression must be a nonnegative natural number <u>other than zero</u>. The form of the integer expression must be

$$i*name\pm j$$

where i and j are nonnegative integer constants, and name is any valid integer variable name. Thus we could write ALIST(2*I+3) but not ALIST(3+2*I) or ALIST(I*J). If we write COL(3*N-5), the integer expression is valid as long as $N \geq 2$.

The mode according to which the values of the elements in the list are stored is determined by the list name, which can be integer or real. The list name therefore has the same function as the variable name in determining whether or not fractional values are stored. For example, ICOL(J) is an integer list, so only integer values are stored; YD(J) is a real list, so real values are stored.

6.3 The DIMENSION Statement

We have already used the parentheses in such functional designations as ALOG(2.*XN+7.) and SQRT(X+7.). The question now arises as to how the computer can differentiate between lists and functions. This differentiation is achieved by the DIMENSION statement, which defines lists or arrays that will be used in the program. For example, the statement

DIMENSION X(50),Y(75),ICT(40),SUM(30)

indicates that the variable names X, Y, ICT, and SUM will be used <u>only</u> for lists, and the maximum number of elements in the lists will be 50,75, 40, and 30, respectively. In any particular use of the program the actual number of elements in the list can be less than or equal to the maximum given in the DIMENSION statement.

The DIMENSION statement is usually the first statement in the program, although in most FORTRAN compilers it can be placed anywhere in the program as long as it precedes the first statement using the list.

6.4 The Use of Lists

In working with lists, we want to be able to read and manipulate them and to output either the original list or modified lists. One way to do this is to use the "stepping up" or "looping" concept. Consider the preparation of a computer program that would compute the mean of a set of observations up to 200 in number. We may follow the flow diagram

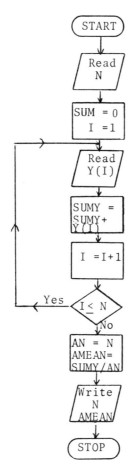

Our FORTRAN program would be

```
C     GSTAT2 - MEAN OF N OBSERVATIONS
      DIMENSION Y(2ØØ)
      READ(5,1) N
    1 FORMAT(I3)
      SUMY = Ø
      I = 1
    2 READ(5,3) Y(I)
    3 FORMAT(F1Ø.2)
      SUMY = SUMY+Y(I)
      I = I+1
      IF (I.LE.N) GO TO 2
      AN = N
      AMEAN = SUMY/AN
      WRITE(6.4) N, AMEAN
    4 FORMAT(1HØ,16HTHE MEAN OF THE ,I3, 16H OBSERVATIONS IS,
    1 F1Ø.2)
      STOP
      END
```

Program 6.1 GSTAT2 — mean of n observations.

We could utilize the above program to process a set of test scores obtained from a group of up to 200 subjects. If there is need to process a longer list, the number specified in the DIMENSION statement could be increased. However, a limit to the number used in the DIMENSION declaration exists because the the computer has a limited amount of internal storage. Thus, if you are too ambitious in the DIMENSION declaration, the program may exceed the storage capacity of the computer and could not be executed.

To illustrate the input and output of the program, we shall utilize the data on reading times that were presented in Fig. 2.1, Chap. 2. The input to the program is

```
Ø12
4.2
5.1
3.3
4.3
5.2
3.1
4.2
5.3
6.1
2.2
4.2
5.2
```

The output is:

THE MEAN OF THE 12 OBSERVATIONS IS   4.37

6.5 The DO Statement

The student of statistics should learn to appreciate
the correspondence between the summation symbol Σ and the
stepping or looping technique already illustrated in the
previous chapter. Whenever a summation is involved in algebra, a loop is used in FORTRAN programming. In fact, if we
have

$$\sum_{i=1}^{n} y_i$$

we program

```
          DIMENSION Y(100)
          SUM = 0
          I = 0                    initialize
        1 I = I+1                   step
Loop      SUM = SUM+Y(I)
          IF(I.LE.N) GO TO 1       check and loop
          .
          .
          .
```

Since this set of statements occurs so frequently, FOR-
TRAN has a single statement, the DO, which combines the index initialization, stepping, checking, and branching statements shown above. To illustrate, we rewrite the above
statements using the DO statement

```
          SUM = 0
          DO 25 I = 1,N,1     initialize, step, check
Loop   25 SUM = SUM+Y(I)      loop
```

When the computer sees the DO statement it automatically forms a loop starting at the DO statement and continuing through the statement having the statement number specified by the DO statement, and executes the loop for each value of the index specified in the DO statement.

The general form of the DO statement is

DO n index = lower, upper, increment

The DO statement initializes the index at the lower limit (first value after the equal sign) and executes all statements down to and including the statement given the statement number n for this value of the index. After executing all the statements in the loop for the lower value of the index, the computer increments the index by the increment (the third value after the equal sign). The new value of the index is checked to see if if is still <u>less than or equal to</u> the upper limit (the second value after the equal sign). If the index is less than the upper limit, the computer again executes all statements down to and including statement n for that value of the index and again increments the index and checks it against the upper limit.

The range of the loop statement is from the DO statement to statement n. During the execution of the loop, the specific value of the index assigned during each step of the loop is substituted in all statements where the index variable is found. Thus the statements

```
          SUM1 = 0
          SUM2 = 0
          SUM3 = 0
          DO 10 I=1,3
          SUM1 = SUM1+Y(I)
          SUM2 = SUM2+Y(I)**2
       10 SUM3 = SUM3+Y(I)**3
```

result in the following operations: The SUM's are set to zero, I is set to the lower value, which is 1, then SUM1 =

$0+Y(1)$, SUM2 $= 0+Y(1)**2$, and SUM3 $= 0+Y(1)**3$. The loop is completed for I = 1, and the index is now incremented by the third value. If this value is not specified, as in this case, the increment is assumed to be 1. The index I is now equal to 2, which is less than the upper limit of 3, so the loop is executed again, and SUM1 $= Y(1) + Y(2)$, SUM2 $= Y(1)**2+Y(2)**2$, SUM3 $= Y(1)**3+Y(2)**3$. The index is incremented again and is now 3, and because this is equal to the upper limit, the loop is again executed. Therefore SUM1 $= Y(1)+Y(2))+Y(3)$. After SUM2 and SUM3 are accumulated for I = 3, the index is again incremented by 1, and since I = 4 > 3, the computer has completed the loop and proceeds to the first statement after statement number 10. If the DO statement had been

DO 10 I=1,5,2

the loop would be executed only for values of I = 1, I = 3, and I = 5.

The index specified in the DO statement must be an integer variable. The lower and upper limit as well as the increment can be either integer variables or constants. If variables are used, they must be assigned specific values before coming to the DO statement. Thus the following DO statement is valid:

```
        SUM = 0
        LL = 3
        LU = 10
        IN = 2
        DO 15 JJ1 = LL,LU,IN
     15 SUM = SUM + Y(JJ1)
```

The rules for DO loop's are
(1) A DO loop must always be entered at the DO state-
    ment since the index must be initialized.
(2) The value of the index should not be changed by any
    statements within the range of the LOOP.

(3) It is permissable to exit from a DO LOOP at any point using any type of transfer statement.

(4) The last statement of a DO LOOP cannot be an IF, GO TO, or DO statement.

If it is necessary to end a loop with a transfer or DO statement, a CONTINUE statement must be used as the last DO statement in the loop, although it causes no computer action (a nonexecutable statement). An example of this is

```
SUM = 0
DO 10 K=1,10
SUM = SUM + Y(K)
IF(SUM.GE.200.) GO TO 12
10 CONTINUE
12    .
      .
      .
```

In this program the Y's are being accumulated into SUM. However, if SUM becomes greater than or equal to 200, the computer will discontinue the loop and transfer to statement 12. If SUM never becomes greater than or equal to 200, the entire loop is completed, and the computer than proceeds to the next statement after statement 10, which in this case is statement 12.

The CONTINUE statement can always be used as the last statement in the loop. The following program section performs the identical steps as the loop on page 124 except that the CONTINUE statement is utilized:

```
SUM1 = 0
SUM2 = 0
SUM3 = 0
DO 10 I=1,3
SUM1 = SUM1 + Y(I)
SUM2 = SUM2 + Y(I)**2
SUM3 = SUM3 + Y(I)**3
10 CONTINUE
```

6.6 The Use of the DO Loop

Let us use the power of the DO loop to write the FACT program given in Chap. 5 to calculate $n! = 1 \cdot 2 \cdot 3 \cdots n$.

```
C     FACT2 - CALCULATE N FACTORIAL
      READ(5,1) N
    1 FORMAT(I3)
      FACT = 1.
      IF (N .LE. 1) GO TO 2
      DO 10 I=1,N
      AI = I
   10 FACT= FACT * AI
    2 WRITE(6,3)N,FACT
    3 FORMAT(1H1,I6,22H FACTORIAL IS EQUAL TO,F12.0)
      STOP
      END
```

Program 6.2 FACT — a factorial program using DO loops.

Suppose we have an array of X values and another array of Y values. We wish to find

$$\sum_{i=1}^{n} x_i \quad \sum_{i=1}^{n} y_i \quad \sum_{i=1}^{n} x_i^2 \quad \sum_{i=1}^{n} y_i^2 \quad \sum_{i=1}^{n} x_i y_i$$

We shall assume that the lists of X's and Y's are already read into the computer, together with the value of n. The program statements for finding the summations are

```
      SX = 0.
      SY = 0.
      SSX = 0.
      SSY = 0.
      SP = 0.
      DO 136 J=1,N
      SX = SX+X(J)
      SY = SY+Y(J)
      SSX = SSX+X(J)**2
      SSY = SSY+Y(J)**2
  136 SP = SP+X(J)*Y(J)
```

The student should appreciate that the above program will compute Σx, Σy, Σx^2, Σy^2, and Σxy for a set of bivariate observations. J is set equal to 1, and the first time through the loop each sum is set equal to the appropriate value using the first observation. Then J is incremented by 1; thus J = 2. The value using the second observation is added to the appropriate sum, J is again incremented, until all N observations have been considered.

We can therefore appreciate that as soon as we see a summation sign in statistics, we should think of a DO loop in FORTRAN programming. The program statements should first set the variable name that will represent the sum equal to zero, followed by a DO statement which indicates the index, its upper and lower limit, the increment, and the range of the loop. The necessary accumulation statements must then be placed within the range of the loop. Thus, to find

$$v = \sum_{i=1}^{n} (y_i - m_1)^2$$

we program

```
      V = 0.
      DO 15 I=1,N
   15 V = V+(Y(I)-YM1)**2
```

and after completion of the loop, the desired value of v is obtained.

6.7 Read and Write Loops

We have shown how the DO loop and FORTRAN lists give us the summation capability we want. The next question is how to read in and write out FORTRAN lists. We can read and write lists within a DO loop as in the following example:

```
      DIMENSION Y(100)
      N = 100
      DO 1 I=1,N
    1 READ(5,2) Y(I)
    2 FORMAT(1X,F10.2)
```

128

```
        DO 3 I=1,N
    3 WRITE(6,2) Y(I)
        STOP
        END
```

This program reads Y(1) on the first data card in columns 2
to 11, Y(2) on the second data card in columns 2 to 11, and
so on.   Recall that every time a READ statement is encountered the next card (record) is read.   After all N values
are read, the READ loop is completed, and the next statement, the WRITE loop, is entered.   Each Y(I) is written on a
separate line according to the same format used by the READ
statement.   The 1X now serves as a carriage-control code,
and because it is a blank, the printing will be single
spaced.   In fact, we could have accomplished the same result
by using a single loop that included both the READ and WRITE
statement as given below:

```
        DO 1 I=1,N
        READ(5,2) Y(I)
    1 WRITE(6,2) Y(I)
    2 FORMAT(1X,F10.2)
```

Suppose however, our data is already punched in cards
with eight elements per card, with the format on each data
card being

First card		Second card	
Columns 1 to 10	Y(1)	Columns 1 to 10	Y(9)
Columns 11 to 20	Y(2)	Columns 11 to 20	Y(10)
.		.	
.		.	
.		.	
Columns 71 to 80	Y(8)	Columns 71 to 80	Y(16)

To help read in arrays or lists such as these, a condensed
version of the DO loop, an implied loop, is available.   This
condensed version can only be used with READ and WRITE
statements.   The READ statement, for example, would be

```
READ(5,1) (Y(I),I=1,N)
1 FORMAT(8F10.2)
```

and the corresponding WRITE statement would be

```
WRITE(6,2) (Y(I),I=1,N)
2 FORMAT(1X,8F10.2)
```

The implied loop (Y(I),I=1,N) can be interpreted as simply a shorthand notation for Y(1), Y(2), Y(3),..., Y(N), where the value of N must be defined before the READ statement. The lists to be read or written must always be enclosed by parentheses. Note that the 8F10.2 format is used repeatedly until all Y's are read or written. The return to the open parenthesis after every eight fields causes a new card to be read or a new line to be printed.

The standard READ and WRITE statement for unsubscripted variables can also be combined with reading lists using the implied DO loop. Suppose we wish to read the value of N in columns 4 to 5 on one card and then the list values in columns 11 to 20 on the succeeding cards. The appropriate READ statement is

```
READ(5,1) N,(Y(I),I=1,N)
1 FORMAT(3X,I2/(10X,F10.2))
```

The read and write lists are always satisfied so that the number of data cards read depend upon the value of N. Since in this case we wish only to read the value of N once we must enclose the format designation for the list in an extra pair of parenthesis.

There are many tricks that can be used in programming. In fact, the student will learn to appreciate their value as he pursues programming beyond this introductory level and will evolve his own set of tricks. We simply note an example of the printing of lists to indicate the flexibility of the FORMAT statement in coordination with the WRITE statement.

Suppose we have a list Y(I) with 400 entries, and we

```
```

wish to print the list of Y's in a table with 10 entries per row and to start a new page (1H1) before printing the list. We could program

```
        WRITE(6,90)(Y(I),I=1,400)
     90 FORMAT(1H1,5X,10F9.1/(6X,10F9.1))
```

The introduction of the slash (/) before the second parenthesis starts a new line, and, following the doubling back principle, the computer continues writing the values on each new line according to (6X,10F9.1) until it has finished writing all 400 entries on a total of 40 lines.

If, instead, we wanted Y(1) through Y(100) to be printed 10 per line on the first page, Y(101) through Y(200) on the second page, we could write

```
        K = 0
     20 K = K+1
        KX = 100*(K-1)+1
        KY = KX+99
        WRITE(6,90) (Y(I),I=KX,KY)
     90 FORMAT(1H1,5X10F9.1/(6X,10F9.1))
        IF(K.LT.4) GO TO 20
```

The execution of these statements would initialize K = 1, and then Y(1) to Y(100) would be printed. Since K = 1, the computer would return to statement 20, increment K and then Y(101) to Y(200) would be written. This process is continued until Y(301) to Y(400) are written. We thus have one implied loop and see that the stepping of K actually forms another loop. We would have therefore one loop within another; the conditions for such nested loops are discussed in the following section.

6.8 Nested DO Loops

The program flow in executing a DO loop given by a statement such as DO 10 K=1,N can be represented by

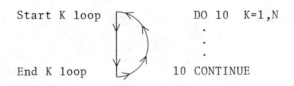

Start K loop   DO 10   K=1,N
.
.
.
End K loop   10 CONTINUE

The DO statement starts the loop and sets K = 1. The statements down through the end of the loop (statement 10) are
executed for K = 1. The index K is incremented and checked
against the upper limit. The computer then loops back to
the start, K is set equal to 2, and the computer goes down
to the end again. The looping continues until K is greater
than N.

The diagram for a nested loop can be represented by

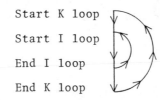

Start K loop
Start I loop
End I loop
End K loop

For each value of the index K, the inner loop (index I) must
be completed before K is incremented.

An example of such nested loops was shown at the end of
the last section, where 10 lines per page were written by an
inner implied loop, and the four pages were written by the
outer loop. The same result can also be achieved by using a
DO statement for the outer loop as the following diagram
shows:

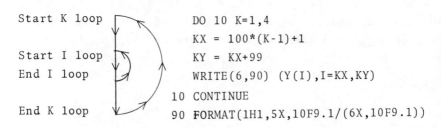

Start K loop   DO 10 K=1,4
                      KX = 100*(K-1)+1
Start I loop   KY = KX+99
End I loop   WRITE(6,90) (Y(I),I=KX,KY)
                   10 CONTINUE
End K loop   90 FORMAT(1H1,5X,10F9.1/(6X,10F9.1))

At the beginning of the outer loop, K = 1, and then Y(1) to
Y(100) is printed on the first page to satisfy the inner implied WRITE loop. When the inner loop is satisfied, K is

incremented to 2, and again the inner loop must be completed
for this value at K. For each value of K, the inner loop
writes the appropriate set of Y's on a new page and then
goes to the outer loop, until both loops have been processed
through all their values.

This example shows how an implied loop can be nested
within a standard DO loop. An example of nested DO loops is

Start I loop   DO 4 I=1,N
                        READ(5,3) Y(I)
Start J loop   3 FORMAT(F10.0)
                        DO 4 J=1,4
End I and J loop   4 SUM(J) = SUM(J)+Y(I)**J

These nested DO loops have a diagram similar to the one just
given, although both loops end with the same statement. In
this example, the Y's values are read in, one at a time, by
the outer loop. The inner loop accumulates each Y(I) into a
sum when J=1 (SUM(1)), a sum of squares when J=2 (SUM(2)), a
sum of cubes when J=3 (SUM(3)), and a sum of fourth powers
when J=4 (SUM(4)). At the completion of both the inner and
outer loops, all values of Y have been read, and both the
values of the list Y(I) and the sums SUM(J) are stored.

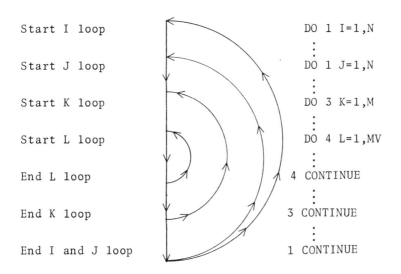

Start I loop   DO 1 I=1,N
                          ⋮
Start J loop   DO 1 J=1,N
                          ⋮
Start K loop   DO 3 K=1,M
                          ⋮
Start L loop   DO 4 L=1,MV
                          ⋮
End L loop   4 CONTINUE
                          ⋮
End K loop   3 CONTINUE
                          ⋮
End I and J loop   1 CONTINUE

The only rules governing nested loops are

    (1) The indices should not be changed within a loop (therefore nested loops cannot have the same indices).

    (2) Each loop in the nested set must be entered at the beginning of the loop, the DO statement, each time.

    (3) Intersecting loops are not permitted.

    (4) Inner loops are always completed before an outer loop is incremented to the next value.

Care must be taken in using nested loops to ensure that these rules are always followed. Examples of invalid nested loops are

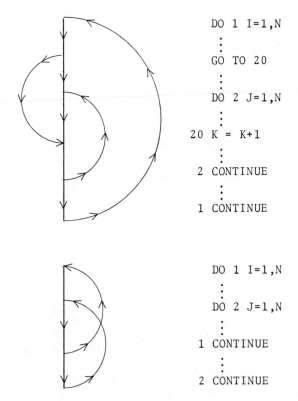

```
DO 1 I=1,N
    .
    .
    .
GO TO 20
    .
    .
    .
DO 2 J=1,N
    .
    .
    .
20 K = K+1
    .
    .
    .
2 CONTINUE
    .
    .
    .
1 CONTINUE
```

```
DO 1 I=1,N
    .
    .
    .
DO 2 J=1,N
    .
    .
    .
1 CONTINUE
    .
    .
    .
2 CONTINUE
```

The concept of nested loops provides a powerful tool for evaluating multiple summations. If all nested loops had to be written by using only the stepping and comparing techniques given in Chap. 5, it would soon become difficult to keep track of all the indices and necessary comparisons.

6.9 Computer Programs and Accuracy Problems

We can now write a computer program using lists and DO loops that will compute the four characteristics $\bar{y}$, s^2, a_3, and a_4 of a distribution of ungrouped data and retain the list of values. Let us follow the computational algorithm that uses the moments about the origin shown in Chap. 3. A flow diagram showing these steps would be

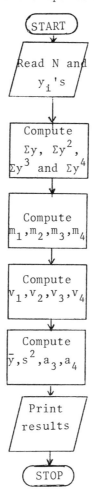

We have the following program:

```
C       UNGROUPCHAR1 - MEAN,S2,A3,A4,FOR UNGROUPED DATA
        DIMENSION Y(1ØØØ),SUM(4),YM(4)
        READ(5,1) N
      1 FORMAT (I3)
        AN = N
        DO 2 K=1,4
      2 SUM(K) = Ø.
        DO 4 I=1,N
        READ(5,3) Y(I)
      3 FORMAT(F1Ø.Ø)
        DO 5 K=1,4
      5 SUM(K) = SUM(K)+Y(I)**K
      4 CONTINUE
        DO 6 I=1,4
      6 YM(I) = SUM(I)/AN
        YVAR = YM(2)-YM(1)**2
        YSTD = SQRT(YVAR)
        A3 = (YM(3)-3.*YM(2)*YM(1)+2.*YM(1)**3)/YSTD**3
        A4 = (YM(4)-4.*YM(3)*YM(1)+6.*YM(2)*YM(1)**2-3.*YM(1)
      1 **4)/YVAR**2
        WRITE(6,1ØØ) (YM(K), K=1,4)
    1ØØ FORMAT(1H1,26HTHE FIRST FOUR MOMENTS ARE/1HØ,
        13HM1=,F12.4,3X,3HM2=,F12.4,3X,3HM3=,F12.4,3X,3HM4=,
        2F12.4)
        WRITE(6,1Ø1) YM(1),YVAR,A3,A4
    1Ø1 FORMAT(1HØ,4HMEAN,F12.4,3X,8HVARIANCE,F12.4,3X,
        18HSKEWNESS,F12.6,3X,8HKURTOSIS,F12.6)
        STOP
        END
```

Program 6.3 UNGROUPCHAR1 — ungrouped data characteristics.

Sample input to the program is

```
                          Ø12
                          4.2
                          5.1
                          3.3
                          4.3
                          5.2
                          3.1
                          4.2
                          5.3
                          6.1
                          2.2
                          4.2
                          5.2
```

The resulting output is

```
THE FIRST FOUR MOMENTS ARE
M1=    4.3667  M2=    20.1783  M3=    97.3225  M4=    485.1854

MEAN    4.3667  VARIANCE    1.1106  SKEWNESS   -0.417162  KURTOSIS    2.497248
```

136

One problem is writing a general computer program is that the program must accommodate data of all types. The sample data on reading times used to test Program 6.3 involved only 12 observations whose values were numerically small with only two significant digits, but even in this case the final results will usually differ from those obtained on a desk calculator. If we used the same program to calculate the four characteristics of many observations, all having large values, such difficulties may even be enlarged. It seems necessary to indicate the potential problem areas and possible solutions.

We saw in Chap. 5 that the use of an integer variable to represent the value of n! in a program would restrict one to a relatively small value of n to avoid overflow. This type of error, though it may be critical, is detected because a message is printed to indicate the overflow condition.

A more crucial problem is the loss of accuracy caused by round-off technique, which is used because only a fixed number of the significant digits can be stored by the computer. For example, when two 8-digit numbers are multiplied, the resulting 16-digit result cannot be stored if the computer allows for only eight significant digits. The last eight digits on the right would then be truncated (after round-off), and in further computations only the eight-digit result would be used. We mentioned some of the problems caused by the use of rounded or truncated values in Chap. 2 and 3 (also in Appendix A) and showed the different results that can be obtained as the number of digits carried throughout the computational process changes.

This round-off technique introduces difficulties even in determining whether a result is zero or, for that matter, equal to any exact number. Consider, for example, the FORTRAN statement A = 1./3. + 1./3. + 1./3., in which A is evaluated by the computer. We certainly know that the value of A should be equal to 1. The computer would evaluate 1./3. as .33333333..., but keep only the specified maximum number of digits, say eight. Thus in this case the value of

A that is stored is .99999999, which, though it is essentially 1, is not exactly equal to 1. Difficulties can then occur if comparisons are made in a subsequent program section such as

IF (A.EQ.1.) GO TO 10

This logical comparison will be true only if A is stored as 1.00000000 and will never be true if A is stored as .99999999. On the other hand, the evaluation of

$$B = 2./3. - 1./3. - 1./3.$$

would cause the value of .00000001 to be stored for B since 2./3. is evaluated .666666... and stored as .66666667; similar difficulties would now be encountered in checking if B is equal to zero. These two examples serve only to demonstrate the problems introduced by the round-off technique used by digital computers. These difficulties become even more pronounced when many arithmetic operations must be made in finding a particular result since all intermediate computations must be rounded off; thus the last digits of the result may no longer be correct.

A good programmer attempts to anticipate the problems of round-off and overflow by carefully choosing the computational algorithm or performing the evaluations in a certain optimum order. For example, if A, B, C are all to a magnitude of 10^{25}, the evaluation of Y = A*B/C might cause an overflow error when the intermediate result of A*B is evaluated (depending on the computer being used), but Y = A*(B/C) would not cause an overflow. The problem of comparing real variables to specific values should be avoided because a real number can be represented in two ways; for example, a 1 can be stored as 1.00000. or .99999.

Let us return to the specific problems that might occur if Program 6.3 is used. There may be an overflow message when very large observations are read because they are raised to the second, third, and fourth powers. In addition, the loss of significant digits can cause undetected accuracy

problems because the four characteristics are obtained by
first evaluating the sum; the sum of squares, cubes, and
fourth powers; and their subsequent subtraction. An example
of a problem that can arise from using this method is when
the first moment about the origin m_1 is 145.7844 and the
second moment about the origin m_2 is 21545.66042106. The
variance, obtained by evaluating the difference $m_2 - m_1^2$
equals 00000.00033770. However, if the computer stored only
eight digits, m_1^2 would be evaluated as 21545.660, and when
this value is subtracted from the value of m_2, which is
stored (21545.660), the result would be zero.

To avoid the particular problem, we can use the
approach of coding the data (Sec. 3.8, Chap. 3). Instead of
subtracting an arbitrary constant from each observed value,
however, we subtract the mean and perform the remaining calculations using deviations from the mean. We thus avoid
such worrisome subsequent subtractions as those just shown.
The revised program for the characteristics using this
algorithm is

```
C     UNGROUPEDCHAR2 - MEAN,S2,A3,A4 FOR UNGROUPED DATA
      DIMENSION Y (1000),SUM(4)
      READ(5,1) N
    1 FORMAT (I3)
      AN = N
      DO 2 K=1,4
    2 SUM(K) = 0.
      DO 4 I=1,N
      READ(5,3) Y(I)
    3 FORMAT(F10.2)
    4 SUM(1) = SUM(1)+Y(I)
      YMEAN = SUM(1)/AN
      SUM(1) = 0
      DO 5 I=1,N
      YDEV = Y(I)-YMEAN
      DO 5 K=1,4
    5 SUM(K) = SUM(K)+YDEV**K
      DO 6 K=1,4
    6 SUM(K) = SUM(K) /AN
      YVAR = SUM(2)
      YSTD = SQRT(YVAR)
      A3 = SUM(3)/YSTD**3
      A4 = SUM(4)/YVAR**2
      WRITE(6,100) (SUM(K),K=1,4)
  100 FORMAT(1H1,37HTHE FIRST FOUR MOMENTS ABOUT THE MEAN/
     11H0,3HV1=,F12.4,3X,3HV2=,F12.4,3X,3HV3=,F12.4,3X,3HV4=
     2,F12.4
      WRITE(6,101) YMEAN,YVAR,A3,A4
```

```
1Ø1 FORMAT(1HØ,4HMEAN,F12.4,3X,8HVARIANCE,F12.4,3X,
   18HSKEWNESS,F12.6,3X,8HKURTOSIS,F12.6)
    STOP
    END
```

Program 6.4 UNGROUPCHAR2 — ungrouped data characteristics.

The identical sample input that was used for Program 6.3 served as input for Program 6.4; the results are

THE FIRST FOUR MOMENTS ABOUT THE MEAN ARE

V1=	0.0000	V2=	1.1106	V3=	-.4882	V4=	3.0790

MEAN	4.3667	VARIANCE	1.1106	SKEWNESS	-0.417169	KURTOSIS	2.496489

Since the input values are small, the answer differs only in the last decimal places that are printed. If observations having large values had been used, however, a difference would have been detected in the more significant digits.

These two programs illustrate the use of nested loops and stored arrays. Of the two, Program 6.4 is the one that should be used because of the truncation problem.

6.10 Summary

In this chapter we introduced FORTRAN lists and the DO loop and discussed the rules to be followed in using them. FORTRAN lists give us the ability to store a variable number of observations easily, and the DO loop provides us with the desired ability to perform summations on a stored list of values. The DO loop can be used in conjunction with lists or simply as a counting mechanism to control the number of times certain calculations should be performed. The DO loop is also useful as a programming device because the index, the upper and lower limits, the increment, and the range of each loop can be easily recognized.

As the student has already noticed, programs can be written in many different ways. By using FORTRAN lists we have an additional capability. Thus many programs written in Chap. 4 and 5 could be written using lists. Before writing a program using lists, we should, however, evaluate the benefit of storing a list of values versus the cost of using computer storage that is required to save the list. The storage capacity of the computer and the size of the list determines to a large extent which approach might best be used.

Chapters 4 to 6 introduced the major FORTRAN statements and capabilities. We now have the capability to write general computer routines that will help us in understanding the statistical investigations in the next chapters and be useful in obtaining the answers. Since we have mastered the essential FORTRAN concepts and associated rules, our emphasis now returns to the statistical problems. Thus we shall use either the general computer routines or place programs that are useful to solve particular problems at the end of the chapter. There are still further powerful FORTRAN concepts that are useful in writing general purpose programs; these will be presented in subsequent chapters where they are needed. However, there are advanced FORTRAN techniques, such as implicit variable typing , that are restricted to particular computer systems; they will not be included in this text. The student interested in these techniques should acquire a programming manual for his particular computer system.

EXERCISES

1. Write a program that determines the median for a set of
   ungrouped data. Use SUBROUTINE SORT in Appendix B to
   sort the data in ascending order, and then apply the
   appropriate rule to find the median, depending upon
   whether n is odd or even. Use the truncation characteristic of FORTRAN integer division to determine whether n
   is even or odd.

2. Write a program that reads in a list of Y's, calculates $\bar{y}$
   and s, and then calculates the z, z_2, z_3, and z_4 for each
   value of y_i. Also find $\bar{z}$, s_z and a_3 and a_4 for the z values. Print out the following information for each observation:

   $$i \qquad y \qquad y - \bar{y} \qquad z \qquad z^2 \qquad z^3 \qquad z^4$$

   and after all observations have been printed, print the
   values

   $$\Sigma y_i \qquad \Sigma(y_i - \bar{y}) \qquad \Sigma z_i \qquad \Sigma z_i^2 \qquad \Sigma z_i^3 \qquad \Sigma z_i^4$$

   and the values of

   $$\bar{y} \qquad s_y \qquad \bar{z} \qquad s_z \qquad a_3 \qquad a_4$$

   (a) Use the ungrouped data on the reading time of 12 new
       employees given in Fig. 2.1, Chap. 2 to test the
       program.
   (b) Run the program using the data given in Exercise 1,
       Chap. 2. Use the program written in Exercise 2 with
       the same data used by Programs 6.3 and 6.4. Check to
       determine if there is any discrepancy in the answers,
       and, if so verify the correct answer using a calcu-
       lator.

3. Selection of outliers: In statistics, extreme observations are called outliers; they often require identification to determine whether they should be deleted from the
   analysis. One method of selection is to consider the
   size of absolute value of the deviation of the observa-

tion from the mean. If the absolute deviation is greater than some multiple, say 3, of s, the standard deviation, then we consider the observation to be an outlier. We want to write a computer program that reads a list of un-grouped observations and determines for each observation y_i whether

$$\left| \frac{y_i - \bar{y}}{s} \right| = \left| z \right| \geq 3$$

If the inequality is satisfied, the program should print out the identification i and the value of y_i. If no outlier exists, this fact should be printed out. To find the absolute value of z, using either ABS(Z) or SQRT(Z**2).

Use the following flow diagram as an aid in preparing the FORTRAN program. Use the data from Exercise 2.1, but also make up one set of data with no outlier and another with two outliers. (Note: a program test with data having only one outlier would not debug the program for no or more than one possibility.)

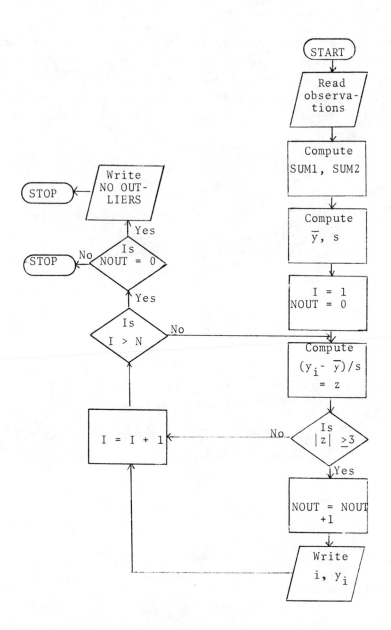

CHAPTER 7

MATHEMATICAL MODELS FOR DISCRETE RANDOM VARIABLES

7.1 Introduction

In this chapter we develop some mathematical models
that can be used to explain the occurrence or distribution
of an observable discrete random variable. The availability
of such a mathematical representation will not only provide
us with a means of mathematically summarizing the information contained in a set of observations, but it will also
serve as a basis for analyses, inferences, and decisions
that we may wish to make.

Observed distributions provide us with information
about the underlying population. We can use this information in selecting the appropriate mathematical model to represent the distribution of the population. In particular,
we can compute the numerical characteristics of the observed
data and the model, which will provide a means of linking
the particular set of data with the mathematical model.

In this chapter we shall also introduce the basic concepts of probability that are needed to be able to appreciate the use of mathematical models. We shall illustrate too
how the computer can be used to evaluate the probabilities
for different discrete mathematical models.

7.2 An Introduction to Probability

In statistics our interest centers around an observable
event that is the outcome of an experiment. An example of a
simple experiment is the tossing of a die, with the possible
outcomes of the experiment being the number of spots showing on the face of the die. Although many experiments exist

145

in which the number of possible outcomes is infinite, we shall restrict ourselves to the finite case in this introduction.

The <u>set</u> of all possible outcomes of an experiment is called the <u>sample space</u> and will be denoted by S. Individual outcomes are called elementary events and will be denoted by e_i, i = 1, 2, ... , m, where m represents the total number of possible outcomes. In the die example, the sample space S has only six elementary events: e_1, e_2, e_3, e_4, e_5, and e_6, which correspond to the number of spots on the face of the die. We can represent the sample space and its elementary events by the following Venn diagram:

Elementary events are defined as being mutually exclusive. This implies that one and only one event can occur as the outcome of a single experiment. The set of elementary events is also exhaustive, i.e., one event must occur. With each elementary event in S we can associate a probability. This probability is denoted by $Pr(e_i)$ and must be a value between zero and 1. Since the set of the elementary events in S is exhaustive, the probability of the set S, $Pr(S)$, must be equal to 1. For completeness we also consider the null set Φ and define $Pr(\Phi) = 0$. In the die experiment, for example, each elementary event is equally likely; thus $Pr(e_i) = 1/6$.

We can define subsets of the sample space S. For example, the subset A would consist of all elementary events having some property A in common. Similarly, we could define other sets, say B and C. Having defined certain sets, we might be interested in such sets as

(1) The union of two sets A U B. This is the set of all elementary events that belong either to A, to B, or to both.

(2) The intersection of two sets A ∩ B. This is the

set of all elementary events that belong to both A and B.

(3) The complement of a set Ā. This is the set of all elementary events not in the set A.

To determine the probability of any subset of the sample space, we use Proposition 1:

Proposition 1

The probability of any subset of S is the sum of the probabilities associated with the elementary events in the subset. Thus

$$Pr(A) = \sum_{e_i \text{ in } A} Pr(e_i)$$

To illustrate these concepts, we define the following subsets using the die experiment. Let

A represent set of all elementary events that represent odd number of spots appearing on the face of die

B represent set of all elementary events that represent two or four spots appearing on face of die

C represent set of elementary events that represent number of spots appearing being less than or equal to 3

These three sets are represented in the following Venn diagram:

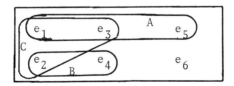

We can then use the definitions and Proposition 1 to find the following probabilities:

$$Pr(A) = Pr(e_1) + Pr(e_3) + Pr(e_5) = \frac{1}{6} + \frac{1}{6} + \frac{1}{6} = \frac{1}{2}$$

$$Pr(B) = Pr(e_2) + Pr(e_4) = \frac{1}{6} + \frac{1}{6} = \frac{1}{3}$$

147

$$Pr(C) = Pr(e_1) + Pr(e_2) + Pr(e_3) = \frac{1}{6} + \frac{1}{6} + \frac{1}{6} = \frac{1}{2}$$

$$Pr(A \cup B) = Pr(e_1) + Pr(e_2) + Pr(e_3) + Pr(e_4) + Pr(e_5) = \frac{5}{6}$$

$$Pr(A \cap B) = Pr(\Phi) = 0$$

$$Pr(A \cap C) = Pr(e_2) = \frac{1}{6}$$

$$Pr(\bar{A}) = Pr(e_2) + Pr(e_4) + Pr(e_6) = \frac{3}{6} = \frac{1}{2}$$

Likewise,

$$Pr(A \cup C) = \frac{4}{6} = \frac{2}{3}$$

$$Pr(B \cup C) = \frac{4}{6} = \frac{2}{3}$$

$$Pr(B \cap C) = Pr(e_2) = \frac{1}{6}$$

$$Pr(\bar{B}) = \frac{4}{6} = \frac{2}{3}$$

$$Pr(\bar{C}) = \frac{3}{6} = \frac{1}{2}$$

The probability of the union of two sets can also be found by using Proposition 2:

Proposition 2

Given any two sets A and B,

$$Pr(A \cup B) = Pr(A) + Pr(B) - Pr(A \cap B)$$

We can verify this proposition by using the die example, where

$$Pr(A \cup B) = Pr(A) + Pr(B) - Pr(A \cap B) = \frac{1}{2} + \frac{1}{3} - 0 = \frac{5}{6}$$

$$Pr(B \cup C) = Pr(B) + Pr(C) - Pr(B \cap C) = \frac{1}{3} + \frac{1}{2} - \frac{1}{6} = \frac{2}{3}$$

In this example with A ∩ B = Φ we have illustrated the concepts of two sets being mutually exclusive. Two sets are said to be mutually exclusive if there are no elementary events common to both sets. Thus in our illustration A and B are mutually exclusive events, but B and C are not. This property leads to Proposition 3:

Proposition 3

If two sets A and B are mutually exclusive, then

$$Pr(A \cup B) = Pr(A) + Pr(B)$$

We might also be interested in knowing the probability of the set B if it is known that the event A has already occurred. This is called <u>conditional</u> probability and is denoted by $Pr(B/A)$.

In our die illustration $Pr(C/A)$ denotes the probability of obtaining an odd number of spots on the face of a die when it is already known that the number of spots on the die is less than or equal to 3. To determine the conditional probability we use Proposition 4:

Proposition 4

The conditional probability $Pr(B/A)$ is given by the equation

$$Pr(B/A) = \frac{Pr(A \cap B)}{Pr(A)}$$

assuming that $Pr(A) \neq 0$.

In the die example, we find that

$$Pr(C/A) = \frac{Pr(A \cap C)}{Pr(A)} = \frac{1/3}{1/2} = \frac{2}{3}$$

This value confirms our intuition, which notes that of the three elementary events that could have occurred, two have the property C.

Another way of considering the conditional probability $Pr(C/A)$ is to recognize that the sample space S is reduced to the set A. We wish to find the number of elementary events in C that are contained in the set A, as illustrated by the following Venn diagram:

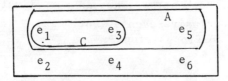

We can now introduce the concept of independence by the definition that two sets A and B are independent if $\Pr(A) = \Pr(A/B)$ or $\Pr(B) = \Pr(B/A)$. This means that the elementary events of one set are not affected by the elementary events of the other set and leads to Proposition 5:

Proposition 5

If two sets A and B are independent, then

$$\Pr(A \cap B) = \Pr(A) \cdot \Pr(B)$$

In the die example, none of the sets A, B, or C are independent of each other, but the set D consisting of elementary events e_1, e_2, e_3, e_4 is independent of A since

$$\Pr(A) = \Pr(e_1) + \Pr(e_3) + \Pr(e_5) = \frac{1}{2}$$

$$\Pr(D) = \Pr(e_1) + \Pr(e_2) + \Pr(e_3) + \Pr(e_4) = \frac{2}{3}$$

and

$$\Pr(D/A) = \frac{\Pr(A \cap D)}{\Pr(A)} = \frac{\Pr(e_1) + \Pr(e_3)}{1/2}$$

$$= \frac{2/6}{1/2} = \frac{4}{6} = \frac{2}{3}$$

$$= \Pr(D)$$

Likewise

$$Pr(A/D) = \frac{Pr(A \cap D)}{Pr(D)} = \frac{2/6}{2/3} = \frac{1}{2} = Pr(A)$$

We can thus summarize the four fundamental propositions of probability:

(1) Given any set A and B, $Pr(A \cup B) = Pr(A) + Pr(B) - Pr(A \cap B)$.

(2) If the sets A and B are mutually exclusive, then $Pr(A \cap B) = 0$, and thus $Pr(A \cup B) = Pr(A) + Pr(B)$.

(3) The conditional probability of the set B given the set A is

$$Pr(B/A) = \frac{Pr(A \cap B)}{Pr(A)} \qquad Pr(A) \neq 0$$

(4) If two sets A and B are independent, then $Pr(A/B) = Pr(A)$ and $Pr(A \cap B) = Pr(A) \cdot Pr(B)$.

We can now define a random variable more mathematically. A random variable is any numerical function defined over the set of elementary events. The probability distribution for each value of the random variable is then found by summing the appropriate probabilities of the corresponding elementary events.

Thus in the die example we could have defined a random variable by the numerical function Y = number of spots showing on the die. Therefore Y can take the values of 1, 2, 3, 4, 5, or 6. The probability distribution for the random variable Y can then be written as

$$p(Y=y) = \frac{1}{6}, \qquad y = 1, 2, 3, 4, 5, \text{ or } 6$$

However, if our interest was in being able to throw a 6, we could have defined the random variable Y using the following numerical function. Let

$$Y = 0 \qquad \text{if } e_1, e_2, e_3, e_4, \text{ or } e_5 \text{ occurs}$$

and

$$Y = 1 \qquad \text{if } e_6 \text{ occurs}$$

In this case, Y can have only two values, and the probability distribution of Y is given by

$$p(Y=y) = \begin{cases} \frac{5}{6} & \text{if } y = 0 \\ \frac{1}{6} & \text{if } y = 1 \end{cases}$$

Thus the probability distribution of a random variable can be given by either a general formula or by specifying the probability for each possible value of the random variable.

In the remaining sections we shall consider some discrete random variables and possible probability distributions (mathematical models) that can be used to determine the probability of each value of the random variable.

7.3 Discrete Mathematical Models

In presenting observed data for a discrete random variable we introduced the simple frequency distribution that associated a frequency f_j with each possible value y_j of the discrete random variable Y. In addition, we introduced the relative frequencies $f_j / \Sigma_j f_j$ since they indicated the proportion of observations that had a value equal to y_j. In fact, we can think of each relative frequency as being an estimate of the probability that an observation from the underlying population has a value of $y = y_j$.

A discrete mathematical model provides theoretical probabilities for each y_j; these are represented by

$$p(Y=y), \quad y = y_1, y_2, \cdots, y_j, \cdots, y_m$$

These probabilities $p(Y=y)$ obtained from any model must be such that

$$\sum_{j=1}^{m} p(Y=y_j) = 1$$

and

$$p(Y=y_j) \geq 0 \qquad \text{for all } y_j$$

Since the discrete distribution model provides us with a set of y_j's and their corresponding probabilities, we can use these probabilities to define the expected value of any numerical function $G(Y)$ of the random variable such as

$$E[G(Y)] = \sum_{j=1}^{m} G(y_j) \, p(Y=y_j)$$

The student should note the correspondence between the definition of the expected value of $G(Y)$ when a mathematical distribution model is given and the mean value of $G(Y)$ when the observed frequency distribution is given. Recall that

$$\text{1st moment of } G(Y) = \frac{\sum_{j=1}^{m} G(y_j) f_j}{\sum_{j=1}^{m} f_j} = \sum_{j=1}^{m} G(y_j) \left[\frac{f_j}{\sum_{j=1}^{m} f_j} \right]$$

and thus the observed relative frequency $f_j/\Sigma f_j$ has simply been replaced by the probability $p(Y=y_j)$ obtained from the theoretical model.

Once again we can specialize the function $G(Y)$ to develop the three classes of moments for the discrete model. They are given by

$$\mu_k = \sum_{j=1}^{m} y_j^k p(Y=y_j) \quad = \text{kth moment about the origin}$$

$$\nu_k = \sum_{j=1}^{m} (y_j - \mu_1)^k \, p(Y=y_j) = \text{kth moment about the mean}$$

$$\alpha_k = \sum_{j=1}^{m} \left[\frac{y_j - \mu_1}{\sqrt{\nu_2}} \right]^k p(Y=y_j) = \text{kth standard moment}$$

The various relationships that were used to find the characteristics of observed data still hold for the characteristics of a mathematical model. The numerical characteristics of interest for any model are

$$\mu = \text{mean} = E[Y] = \mu_1$$

$$\sigma^2 = \text{variance} = E[Y - \mu]^2 = \nu_2 = \mu_2 - \mu^2$$

$$\sigma = \text{standard deviation} = \sqrt{\nu_2}$$

$$\alpha_3 = \text{skewness} = \frac{\nu_3}{\nu_2^{3/2}} = \frac{(\mu_3 - 3\mu_2\mu_1 + 2\mu_1^3)}{\left(\mu_2 - \mu_1^2\right)^{3/2}}$$

$$\alpha_4 = \text{kurtosis} = \frac{\nu_4}{\nu_2^2} = \frac{(\mu_4 - 4\mu_3\mu_1 + 6\mu_2\mu_1^2 - 3\mu_1^4)}{\left(\mu_2 - \mu_1^2\right)^2}$$

Students should appreciate that there is no standard notation used to represent the numerical characteristics of either observed or mathematical models of a distribution, but generally Greek letters are used to represent the characteristics of mathematical models, and Latin letters are used for observed distributions.

Most mathematical models are given in a general form in which the model parameters have not been assigned specific values. For example, the simple point binomial (Bernoulli) model is given as

$$p(Y=y) = \theta^y \cdot (1 - \theta)^{1-y}, \quad y = 0 \text{ or } 1$$

The variable is denoted by Y, and θ is the parameter. For any particular problem a value must be given to the parameter to have the general model fit a particular observed distribution.

There are several methods for determining the value that might best be assigned to the parameter(s). For example, theoretical or physical considerations may be used to assign a value. If such considerations do not exist, then other methods of determining the value must be used.

We shall introduce one of the statistical methods that takes advantage of the information contained in the observed data. This method is called the method of moments. It is based on the concept that if the moments of the model are close to the moments of the observed distribution, a good fit of the model to the observed data is realized. The technique consists of equating as many lower-order moments as there are parameters to be estimated. For example, if there is but one parameter in the model, it should be selected so that the mean of the model is equal to the observed mean. Thus $\bar{y} = \mu$. If there are two parameters, then the values would be selected so that the mean and the standard deviation would agree, and so on. To use this technique we must know what the moments of the general model are in terms of its parameters. Other techniques for determining parameter values will be given in Chaps. 13 and 14.

Since the model will be used to make inferences or decisions, the question naturally arises as to how well the model representation fits the observed data. We shall introduce three techniques that can be used to check this goodness of fit:

(1) We can compare the higher-order moments of the model (those that have not been used to determine values of the parameters) with the corresponding moments of the observed data. For example, we may compare the skewness and kurtosis.

(2) We could superimpose the graph of the model on the relative frequency graph of the observed data and visually note the goodness of fit.

(3) We could find the expected frequency for each value of the variable (using model probabilities multiplied by the total number of observations) and compare these with the corresponding observed frequencies.

In the remainder of the chapter we shall consider different mathematical models in their general form and show how they can be used to fit observed distributions and how the resulting fit can be checked.

7.4 The Uniform Distribution

To apply these concepts to a particular model, let us first consider the uniform distribution. A general form of this model is given by

$$p(Y=y) = \frac{1}{m} \, , \quad y = 1, 2, \ldots , m$$

This model assigns to each possible value of Y the same probability of $1/m$; thus the name "uniform" is used. The unknown constant m is generally not called a parameter because its value is exactly determined by the number of possible values of Y and does not need to be estimated.

Since m is determined by the experimental design we need only the moments to test the goodness of fit of the model to a set of data. The first moment about the origin is

$$\mu_1 = E(Y) = \sum_{j=1}^{m} y_j p(Y=y_j) = \sum_{j=1}^{m} j\left(\frac{1}{m}\right)$$

$$= \frac{1}{m} \left(1 + 2 + 3 + \cdots + m\right)$$

$$= \frac{1}{m} \frac{m(m + 1)}{2}$$

$$= \frac{m + 1}{2}$$

Mathematics provides us with the formula for the sum of the first m natural numbers, which is $\sum_{i=1}^{m} i = m(m + 1)/2$. By using similar formulas for the sums of powers of the first m natural numbers, we can find that

$$\sigma^2 = E(Y - \mu)^2 = \sum_{j=1}^{m} \left(y_j - \frac{m + 1}{2}\right)^2 \frac{1}{m} = \mu_2 - \mu_1^2$$

$$= \left[\frac{(m + 1)(2m + 1)}{6} - \left(\frac{m + 1}{2}\right)^2\right] = \frac{m^2 - 1}{12}$$

where $\sum_{i=1}^{m} i^2 = m(m + 1)(2m + 1)/6$.

The skewness and kurtosis are similarly obtained as:

$$\alpha_3 = E\left[\frac{Y-\mu}{\sigma}\right]^3 = \frac{\nu_3}{\nu_2^{3/2}} = \frac{0}{\nu_2^{3/2}} = 0$$

$$\alpha_4 = E\left[\frac{Y-\mu}{\sigma}\right]^4 = \frac{\nu_4}{\nu_2^2} = 3 - \frac{6}{5}\left[\frac{m^2+1}{m^2-1}\right]$$

Following is an example of a situation in which the uniform distribution can be used. If we are interested in the distribution that is obtained when a fair die is tossed, and the variable of interest is the number of spots showing, then the uniform distribution is an appropriate mathematical model. It is given by

$$p(Y=y) = \frac{1}{6}, \quad y = 1, 2, 3, 4, 5, 6$$

In this case, we can numerically find the characteristics of this distribution model.

$$\mu = \sum_{j=1}^{6} y_j p(Y=y_j)$$

$$= 1 \cdot \frac{1}{6} + 2 \cdot \frac{1}{6} + 3 \cdot \frac{1}{6} + 4 \cdot \frac{1}{6} + 5 \cdot \frac{1}{6} + 6 \cdot \frac{1}{6}$$

$$= \frac{21}{6} = \frac{7}{2}$$

$$\sigma^2 = \sum_{j=1}^{6} (y_j - 3.5)^2 p(Y=y_j) = \nu_2$$

$$= \left(-\frac{5}{2}\right)^2 \frac{1}{6} + \left(-\frac{3}{2}\right)^2 \frac{1}{6} + \left(-\frac{1}{2}\right)^2 \frac{1}{6} + \left(\frac{1}{2}\right)^2 \frac{1}{6} + \left(\frac{3}{2}\right)^2 \frac{1}{6} + \left(\frac{5}{2}\right)^2 \frac{1}{6}$$

$$= \frac{35}{12}$$

Likewise,

$$\nu_3 = \sum_{j=1}^{6} (y_j - 3.5)^3 p(Y=y_j)$$

$$= \frac{1}{6}\left[\left(-\frac{5}{2}\right)^3 + \left(-\frac{3}{2}\right)^3 + \left(-\frac{1}{2}\right)^3 + \left(\frac{1}{2}\right)^3 + \left(\frac{3}{2}\right)^3 + \left(\frac{5}{2}\right)^3\right]$$

$$= \frac{1}{6}[0] = 0$$

$$\nu_4 = \sum_{j=1}^{6} (y_j - 3.5)^4 p(Y=y_j)$$

$$= \frac{1}{6}\left[\left(-\frac{5}{2}\right)^4 + \left(-\frac{3}{2}\right)^4 + \left(-\frac{1}{2}\right)^4 + \left(\frac{1}{2}\right)^4 + \left(\frac{3}{2}\right)^4 + \left(\frac{5}{2}\right)^4\right]$$

$$= \frac{1}{6}\left[\frac{707}{8}\right] = 14.7290$$

Thus $\alpha_3 = 0$, and

$$\alpha_4 = \frac{\nu_4}{\nu_2^2} = \frac{14.7290}{(35/12)^2} = 1.7314$$

The student can confirm that these same values are obtained if the general formulas are used with m = 6.

Let us demonstrate how the adequacy of the uniform distribution model for the toss of one die may be checked. Let Y = the number of spots showing. Toss the die 200 times. Assume that we have done this experiment and have already condensed the individual observations into the following simple frequency table:

Number of Spots	Number of Tosses
Y	F
1	30
2	35
3	33
4	36
5	29
6	37

Thus the characteristics of the observed data are

$$\bar{y} = \frac{710}{200} = 3.55$$

$$s^2 = \frac{579.5}{200} = 2.8975$$

$$a_3 = \frac{0.012725}{(2.8975)^{3/2}} = 0.0026$$

$$a_4 = \frac{14.6815}{(2.8975)^2} = 1.7487$$

For the uniform model $p(Y=y) = 1/6$ to be descriptive of the underlying observed distribution (a) the die used must be fair (each face must be equally likely to occur) and (b) each of the tosses must be independent. Thus we wish to check on the adequacy of the model fit. We shall use all three techniques.

1. Comparison of the higher-order characteristics
Since the constant m is obtained from physical considerations, we can compare all the characteristics

$$\bar{y} = 3.55 \qquad\qquad \mu = \frac{7}{2} = 3.5$$

$$s = 2.8975 \qquad\qquad \sigma^2 = \frac{35}{12} = 2.9167$$

$$a_3 = 0.0026 \qquad\qquad \alpha_3 = 0$$

$$a_4 = 1.7487 \qquad \alpha_4 = 1.7314$$

Although we have yet to develop a method of evaluation for such comparisons, it can be seen that the moments are very similar.

2. Graphical comparison

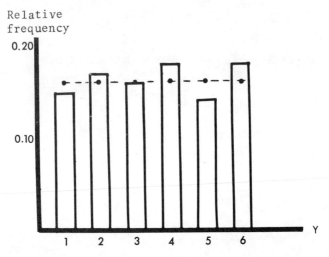

Relative
frequency

Once again, the closeness of the fit is indicated by the graphical representation. The observed relative frequencies are given by the bar graph; the theoretical frequencies are given by a frequency polygon.

3. Comparison of observed and expected frequencies

Number of Spots Y	Observed Frequency f_o	Expected Frequency f_e	Discrepancy $f_o - f_e$
1	30	33.33	-3.33
2	35	33.33	1.67
3	33	33.33	-0.33
4	36	33.33	2.67
5	29	33.33	-4.33
6	37	33.33	3.67

The common expected frequency is calculated by the formula

$$f_e = n \cdot p(Y=y) = 200\left(\frac{1}{6}\right) = 33.33$$

In looking at the above descrepancies in the frequencies, the goodness of fit of the model is not as clearly indicated as was the case using the first two techniques. This may be due to the numerical nature of the individual comparisons when comparing large frequencies.

We now need to have an objective method for deciding upon the adequacy of a fit. Such techniques will be considered in Chap. 14.

7.5 The Binomial Distribution

Let us now consider the binomial distribution model as a possible candidate for fitting an observed set of discrete data. The general form of the model is

$$p(Y=y) = C_y^n \, \theta^y \, (1-\theta)^{n-y}, \quad y = 0, 1, \ldots, n$$

This model is an appropriate model for a discrete random variable when

(1) An experiment can have only two possible outcomes.
(2) θ is the probability of a success (one of the outcomes), and $(1-\theta)$ is the probability of a failure (the other outcome).
(3) The experiment is repeated n times and the repeated trials are independent.
(4) The probability of a success on each trial remains constant.
(5) The random variable Y represents the number of successes in n trials.

We note that the formula for the binomial model has two undetermined constants: n and θ. However, we generally think of θ as being the only parameter because n is determined from underlying conditions (the number of trials).

The term C_y^n is defined to be

$$C_y^n = \frac{n!}{y! \ (n - y)!}$$

where $n! = 1, 2, 3, \ldots (n - 1) \cdot n$, as already defined in Chap. 5.

To characterize the distribution of the binomial model, we consider its moments:

$$\mu = E[Y] = n\theta$$

$$\sigma^2 = E[Y - n\theta]^2 = n\theta(1 - \theta)$$

$$\alpha_3 = E\left[\frac{(Y - n\theta)}{\sigma}\right]^3 = \frac{(1 - 2\theta)}{\sqrt{n\theta(1 - \theta)}}$$

$$\alpha_4 = E\left[\frac{(Y - n\theta)}{\sigma}\right]^4 = 3 + \frac{1 - 6\theta \ (1 - \theta)}{n\theta \ (1 - \theta)}$$

It should be noted that once n and θ are determined, all the above characteristics are known. Since n is obtained from physical considerations, we have only one free parameter, θ, that can be used to fit the general model to observed data. To find the value of θ we use the method of moments and determine θ so that $\bar{y}$ is equal to μ.

To demonstrate the application of the binomial model to an actual problem, let us consider the true-false reading examination. The observed y for each individual was the number of incorrect responses on 16 questions (trials). In selecting the binomial model we are assuming that an individual's response to one question is independent of his response to any other one, although this assumption is somewhat doubtful. We also assume that the probability of each individual's success on each question remains constant; this assumption surely is questionable. However, we still elect to continue the consideration of the binomial model for the observed data.

The number of questions is 16, so we set n = 16. To find the estimate of the parameter θ, we equate

$$\mu = n\theta = \bar{y}$$

Thus

$$\hat{\theta} = \frac{\bar{y}}{n}$$

In the true-false examination data given in Fig. 2.2, Chap. 2, the sample mean $\bar{y}$ was 2.5282, and thus the value assigned to θ is given by

$$\hat{\theta} = \frac{2.5282}{16} = 0.1580$$

The binomial model we shall use to represent the distribution of incorrect responses on the true-false reading examination is

$$p(Y=y) = C_y^{16}(0.1580)^y(0.8420)^{16-y} \quad y = 0, 1, \ldots, 16$$

This determined binomial model has the following characteristics:

$$\mu = n\theta = 16(0.1580) = 2.5280$$

$$\sigma = \sqrt{n\theta(1 - \theta)} = \sqrt{16(0.1580)(0.8420)} = 1.4590$$

$$\alpha_3 = \frac{1 - 2\theta}{\sqrt{n\theta(1 - \theta)}} = \frac{1 - 2(0.1580)}{1.4590} = 0.4688$$

$$\alpha_4 = 3 + \frac{1 - 6\theta(1 - \theta)}{n\theta(1 - \theta)}$$

$$= 3 + \frac{1 - 6(0.1580)(0.8420)}{(2.128576)} = 3.0948$$

To consider how good the fit is, we can compare the observed characteristics with those of the determined model:

Characteristic	Observed	Model
Mean	2.5282	2.5280
Standard deviation	1.4404	1.4590
Skewness	0.4379	0.4688
Kurtosis	3.3183	3.0948

Although we are still in the position that it is hard for us to evaluate the closeness of agreement for such numerical characteristics, we cannot help but note with some satisfaction that the model characteristics seem to agree closely with those of the observed distribution, and hence we feel that the fit is "fairly good." The fact that the peakedness of the data is a little greater than the model might lead us to expect that the model may slightly underestimate the highest frequencies and overestimate the adjacent ones.

This last conjecture leads us to consider the second and third technique for further evaluation of the adequacy of the fit. To accomplish such comparisons we need for our true-false example the binomial probabilities associated with $y = 0, 1, 2, \ldots, 16$. Since such evaluations require a considerable amount of computation, we shall use the computer to make the evaluations (see next section). The results of the computer evaluation of $p(Y=y)$ and the relative observed frequencies (from Chap. 2) are

Y	f/Σf	p(Y=y)
0	0.0897	0.0638
1	0.1153	0.1916
2	0.3076	0.2697
3	0.2871	0.2362
4	0.1076	0.1440
5	0.0589	0.0649
6	0.0256	0.0223
7	0.0051	0.0060
8	0.0025	0.0013
9	0.0000	0.0002
10	0.0000	0.0000
11	0.0000	0.0000
12	0.0000	0.0000
13	0.0000	0.0000
14	0.0000	0.0000
15	0.0000	0.0000
16	0.0000	0.0000

This tabulation is equivalent to the third technique used to determine the goodness of fit. In this case we are comparing relative frequencies rather than the frequencies themselves. We could convert the probabilities p(Y=y) to expected frequencies by multiplying them by n = 390 and then comparing these expected frequencies with the observed ones given in Chap. 2.

With the tabulation given above we can graph the observed relative frequencies and the probabilities. We choose to represent the observed relative frequencies by a bar graph and the theoretical binomial probabilities by a frequency polygon:

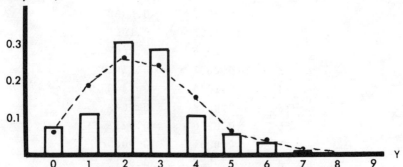

Although we may be disturbed a little with the apparent lack of fit, we note that the effect of the discrepancy between the kurtosis of the observed distribution and the kurtosis of the binomial distribution which was predicted did actually materialize.

Assuming that the binomial model gives an adequate fit for the underlying distribution of the variable, the model can be used to answer four questions relative to the underlying population such as those given below:

(1) What is the probability that an individual selected at random from the population has a value which falls within a prescribed set of values?

(2) What percentage of a randomly selected group of n_o individuals would have values falling within a prescribed set of values?

(3) From a randomly selected group of n_o individuals, how many would be expected to have values falling within a prescribed set of values?

(4) What value of A is such that $Pr[Y \leq A] = \alpha$, where α is a specified probability.

All these questions can be answered by using a mathematical model. We can sum the function over the prescribed set to obtain the probability (1), which is equivalent to the percentage (2). If this probability is multiplied by n_o we obtain the expected number (3), and (4) is obtained by finding the value of A associated with the probability α.

We could simply resort to our ability to use available FORTRAN routines or to write a routine that would evaluate p(Y=y). It may, however, not be reasonable to resort to a computer program every time such a question arises. This is especially true for certain standard distributions such as the binomial distribution, so we shall also illustrate the use of tables in answering the questions.

Let us demonstrate the mechanism used in answering such questions by utilizing the binomial model that was previously fitted to the number of incorrect responses on the true-false test, with its parameters n = 16 and θ = 0.1580.

(1) What is the probability of an individual making more than three incorrect responses on the test?

$$\Pr[Y>3] = \sum_{y=4}^{16} p(Y=y)$$

$$= \sum_{y=4}^{16} C_y^{16}(0.1580)^y(0.8420)^{16-y}$$

$$= C_4^{16}(0.1580)^4(0.8420)^{12} + C_5^{16}(0.1580)^5(0.8420)^{11}$$

$$+ \cdots + C_{16}^{16}(0.1580)^{16}(0.8420)^0$$

$$= 1 - \sum_{y=0}^{3} C_y^{16}(0.1580)^y(0.8420)^{16-y}$$

$$= 1 - C_0^{16}(0.1580)^0(0.8420)^{16} - C_1^{16}(0.1580)^1(0.8420)^{15}$$

$$- C_2^{16}(0.1580)^2(0.8420)^{14} - C_3^{16}(0.1580)^3(0.8420)^{13}$$

$$= 1 - 0.0638 - 0.1913 - 0.2695 - 0.2358$$

$$= 0.2396$$

In the above computation we could use the recursive relationship for the binomial:

$$p(Y=y+1) \ = \ p(Y=y) \left(\frac{n - y}{y + 1}\right) \cdot \left(\frac{1}{1 - \theta}\right)$$

(2) How many individuals from a randomly selected group of 150 would be expected to have scores between Y = 2 and Y = 5 inclusive?

$$\text{Expected number} = n_o \cdot \Pr[2 \leq Y \leq 5]$$

$$= n_o \cdot \sum_{y=2}^{5} C_y^{16} (0.1580)^y (0.8420)^{16-y}$$

$$= (150) \left[C_2^{16} (0.1580)^2 (0.8420)^{14} \right.$$

$$+ \ C_3^{16} (0.1580)^3 (0.8420)^{13}$$

$$+ \ C_4^{16} (0.1580)^4 (0.8420)^{12}$$

$$+ \ \left. C_5^{16} (0.1580)^5 (0.8420)^{11} \right]$$

$$= (150) \ [0.2695 + 0.2358$$

$$+ \ 0.1438 + 0.0646]$$

$$= (150) \ [0.7136]$$

$$= 107.04$$

(3) Ninety percent of the students would be expected to make fewer than how many mistakes?

$$\Pr[Y < k] \ = \ 0.90$$

Since $\Pr[Y < 9] = 0.98$ and $\Pr[Y < 8] = 0.85$, we would conclude that approximately 85 percent of the students would make fewer than eight mistakes.

To reduce the arithmetic or computing involved, the cumulative binomial tables shown in Appendix C-1 can be used. The tables give

$$Pr[0 \leq Y \leq k] = \sum_{y=0}^{k} C_y^n \theta^y (1 - \theta)^{n-y}$$

for specific values of n and θ.

To illustrate the use of the tables, let us find $Pr[Y > 3]$ when $n = 15$ and $\theta = 0.20$. We use the fact that

$$Pr[Y > 3] = 1 - Pr[Y \leq 3]$$

$$Pr[Y > 3] = 1 - \sum_{y=0}^{3} C_y^{15} (0.20)^y (0.80)^{15-y}$$

$$= 1 - 0.648$$

$$= 0.352$$

where 0.648 is obtained from the table for $n = 15$, $\theta = 0.20$, and $Y = 3$.

Likewise, we can use the tables to find

$$Pr[2 \leq Y \geq 5] = \sum_{y=0}^{5} p(Y=y) - \sum_{y=0}^{1} p(Y=y)$$

$$= 0.939 - 0.167$$

$$= 0.772$$

A mathematical model is useful in answering questions of the type just given. Although these particular questions related only to the distribution of individuals in the particular population, later when we consider statistical inference we shall use such mathematical models to describe istributions of quantities derived from the observed sample alues.

7.6 The Use of FORTRAN FUNCTIONS
in Model Fit Evaluation

In the previous chapters we used the functions SQRT(X) to evaluate the square root of a number ALOG(X) to find the natural logarithm and EXP(X) to evaluate e^x. We would like to have the capability to add other such FORTRAN functions as we need them. Although the programmer must write such a function subprogram himself, once it is available and is added to the subroutine program library he can act as if this function is like one of the standard ones previously mentioned.

To be more specific, let us say we want a function to evaluate n! = $1 \cdot 2 \cdot 3 \cdots n$ for any positive integer n. To write such a function subprogram we simply take the program written in Program 6.2, and add as the first statement of the subprogram a card with the word FUNCTION followed by the name and the argument; we also replace the STOP statement by the RETURN statement as shown in Program 7.1. The **read** and write statements are eliminated.

```
      FUNCTION FACT(N)
      FACT = 1.
      IF(N.LE.1) GO TO 10
      DO 1 I=1,N
      AI = I
   1  FACT = FACT*AI
  10  RETURN
      END
```

Program 7.1 FUNCTION FACT — a factorial function program.

The argument N is a dummy argument used only for writing the function subprogram. In **using** the function any integer constant or variable can be used as the argument.

To have the function available in a main program, the main program with the function subprogram behind it is compiled. Then in the writing of the main program FACT( ) can be used like any of the standard functions, SQRT ( ) or EXP ( ). There can be any number of functions placed behind the main program, and the order of the subprograms does not matter.

We demonstrate this concept by writing a program that will evaluate the binomial distribution. The binomial distribution has the form

$$p(Y=y) = C_y^n \theta^y (1 - \theta)^{n-y} , \quad y = 0, 1, 2, \ldots , n$$

where

$$C_y^n = \frac{n!}{y! \ (n - y)!}$$

If we want to write a program to evaluate $p(Y=y)$ for the $(n + 1)$ values of Y, we also need to evaluate three factorials. The program BINOM and subprogram FACT to evaluate the binomial distribution for any n and θ are

```
C      BINOM - EVALUATION OF BINOMIAL PROBABILITIES
       READ(5,1) N,THETA
     1 FORMAT(I3,F6.6)
       N1 = N+1
       DO 2 J=1,N1
       J1 = J-1
       I1=N-J1
       PR=FACT(N)*THETA**J1*(1.-THETA)**I1/(FACT(J1)*FACT(I1))
     2 WRITE(6,4)J1,PR
     4 FORMAT(20X,I3,10X,F10.4)
       STOP
       END
       FUNCTION FACT(N)
       FACT=1.
       IF(N.LT.1) TO TO 10
       DO 1 I=1,N
       AI=I
     1 FACT=FACT*AI
    10 RETURN
       END
```

Program 7.2 BINOM — evaluation of binomial probabilities.

The student should note that the statement numbers in the function subprogram are completely independent from those in the main program. Therefore the same numbers can be used in the main routine and the subprogram without causing errors. The same concept also applies to variable names; that is, names can be assigned in one subprogram without regard to the names used in another subprogram. The only com-

171

munication between the routines is via the calling arguments. A main program can be followed by as many subprograms as required as long as the capacity of the memory is not exceeded.

The program BINOM is easy to follow because the probability is evaluated using the defining relationship. This is not the most efficient or accurate method for evaluation, particularly if N is large. Another method will be given as an exercise at the end of this chapter.

7.7. Other Discrete Models

Although it is not our intention to provide students with a large number of distribution models, there are two additional discrete distribution models that we introduce as future reference.

1. Poisson distribution

$$p(Y=y) = \frac{e^{-\lambda} \lambda^y}{y!}, \qquad y = 0, 1, 2, 3, \ldots$$

$$\mu = \lambda \qquad\qquad \sigma = \sqrt{\lambda}$$

$$\alpha_3 = \frac{1}{\sqrt{\lambda}} \qquad\qquad \alpha_4 = 3 + \frac{1}{\lambda}$$

This model is applicable when there are an infinite number of independent trials of an experiment with the probability of success on any one trial being constant but very small and λ being the average number of occurrences. The variable Y represents the number of successes that occur.

2. Hypergeometric distribution

$$p(Y=y) = \frac{C_y^{M_1} C_{n-y}^{N-M_1}}{C_n^N} \qquad y = a, a + 1, \ldots, M_1$$

where

$$a = \begin{cases} 0 & \text{if } n \leq N - M_1 \\ n - (N - M_1) & \text{otherwise} \end{cases}$$

The characteristics are given in terms of p and q, where $p = M_1/N$ and $q = 1 - p = (N - M_1)/N$.

$$\mu = np \qquad\qquad \sigma = \sqrt{\frac{npq(N - n)}{N - 1}}$$

$$\alpha_3 = \frac{(q-p)(N - 2n)}{(N - 2)\sqrt{\frac{npq(N - n)}{N - 1}}}$$

$$\alpha_4 = \frac{(N - 1)}{(N - n)(N - 2)(N - 3)}\,[N(N +1) - 6n(N - n)$$

$$+ 3pq\{N^2(n - 2) - Nn^2 + 6n(N - n)\}]$$

This model is applicable when a finite population of N items is being considered and there are two kinds of items. The number of items of one kind is M_1; therefore the number of items of the second kind is $M_2 = N - M_1$. The model assumes that a random sample of n items is drawn (either all at one time or one at a time without replacement), and Y is the number of items (out of the n drawn) that are of the first kind.

The student should note the similarity of the characteristics of the hypergeometric distribution with the binomial (particularly in the formulas for μ and σ). In fact, if each item is replaced after it is drawn, then the appropriate distribution would be the binomial with $\theta = M_1/N$ and $1 - \theta = M_2/N$ and the same value of n. Another way to consider the interrelationship of the two models is to recognize that the failure to replace an item after it is drawn affects the independence of the trials when N is finite because θ changes from trial to trial.

7.8 Summary

Let us review the steps that were followed in the mathematical model selection process.

(1) Observations were made on the variable of interest, and the observed distribution was characterized by its numerical measures.

(2) A possible mathematical model was selected by taking advantage of any information available on the underlying natural mechanisms that are active in determining the values of the variable.

(3) The undetermined constants or parameters of the model were assigned numerical values by the method of moments. In other words, the values of the parameters were determined so that the lower-order moments of the distribution are equal to those of the observed data.

(4) The goodness of fit was then studied by (a) comparing the values of the high-order moments or (b) comparing the theoretical or model frequencies (probabilities) with the observed frequencies (relative frequencies).

We are left with the problem of deciding whether the fit using the selected model is adequate or whether we should continue searching for a better model. We must delay consideration of this question until later. We should also recognize that we need to acquire facility in selecting mathematical models to represent observed distributions of random variables, for continuous random variables in particular. In our search for an appropriate distribution model for any observable variable we also need to have at least an alternative set of available mathematical models from which to choose. Finally we need to consider how we might use a mathematical model once one has been acquired. All these problems will be considered in the next chapters.

EXERCISES

1. Find the number of three letter words that can be formed from the letters A to E when

   (a) A letter can occur any number of times in the word
   (b) A letter can only occur once in each word

2. Enumerate the sample space of all three letter words for Exercises 1(a) and (b). Let the event A represent the elementary events in which the letter A occurs at least once in the word and the event B represent the elementary events for which the letter B occurs at least once. Find the probabilities

   $Pr(A)$ $Pr(A \cup B)$ $Pr(A \cap B)$ $Pr(B)$ $Pr(\bar{A})$ $Pr(\overline{A \cup B})$
   for each of the two cases, 1(a) and 1(b).

3. In Exercise 1(a) the selection of the A letter does not influence the subsequent letter to be selected, and thus the selection of letters is independent. If A_k represents the selection of A as the kth letter, B_k represents the selection of B as the kth letter, and so forth, then we can assume that

   $$Pr(A_k) = Pr(B_k) = Pr(C_k) = Pr(D_k) = Pr(E_k) = \frac{1}{5}$$

   for k = 1, 2, 3. Under this assumption compute

   $Pr(A_1 \cap A_2 \cap A_3)$ $Pr(A_1 \cup A_2 \cup A_3)$ $Pr(A_2/B_1)$

4. Given the discrete uniform distribution

   $$p(Y=y) = \frac{1}{8} \quad y = 1, 2, 3, \ldots, 8$$

   numerically verify the formulas for its mean, standard deviation, skewness, and kurtosis.

5. Toss 10 pennies 100 times, and count the number of heads observed in each toss. Tabulate the observed number of heads into a simple frequency distribution. Compute the first four moments of your distribution, and compare these observed characteristics with the moments of the binomial with n = 10 and θ = 0.5; that is, $p(Y=y)$ = $C_y^{10} \left(\frac{1}{2}\right)^y \left(\frac{1}{2}\right)^{10-y}$. What value of θ would better fit your distribution? If you use $\theta = \bar{y}/10$ because $\mu = n\theta$, why do the observed moments other than m_1 not agree with the binomial moments?

6. Change Program 7.2 into a FUNCTION subprogram that evaluates the binomial probabilities

$$p(Y=y) = C_y^n \; \theta^y \; (1 - \theta)^{n-y}$$

for any n, θ, and y.

(a) Write a main program that calls the FUNCTION to verify the binomial probabilities for the true-false data with n = 16 and θ = 0.1580.

(b) Use the program to evaluate the binomial probabilities for θ = 0.5, n = 10 and 20; θ = 0.3333, n = 10 and θ = 0.1, n = 10. Plot the results on the same set of axes, and notice the changes in the central tendency, standard deviation, and skewness.

7. Write a main program that calls the FUNCTION to evaluat the binomial probabilities for any n and θ and then cal culates the four characteristics μ, σ, α_3, and α_4. Use the program to verify the general formulas by evaluatir the case n = 16 and θ = 0.1580.

8. Given the binomial model $p(Y=y) = C_y^n \theta^y (1 - \theta)^{n-y}$, y = 0, 1, 2, ... n, and the numerical function $G(Y)$ = $(Y + 1)(Y + 2)$. Write a computer program that determines the expected value of $G(Y)$ with respect to any

given binomial. Then verify that

$$E[G(Y)] = E[(Y + 1)(Y + 2)] = E(Y^2) + 3E(Y) + 2$$

$$= \mu_2 + 3\mu_1 + 2$$

for the case of $n = 16$ and $\theta = 0.1580$.

9. Make up your own discrete model whose probabilities follow some simple geometrical shape, i.e., right triangle, equilateral triangle, trapezoid. Include at least one undetermined constant in your model. Remember that $\sum_y p(Y=y) = 1$ for the model to be a probability distribution. Write a computer program that evaluates the numerical characteristics in terms of the undetermined constants of your model.

10. Write a program to evaluate the Poisson distribution for different values of λ.

   (a) Use the program to generate Poisson probabilities to plot the Poisson distributions for $\lambda = 2.0$, 1.0, 0.5, and 0.1.

   (b) Use the binomial program to generate values to plot the binomial distribution for $n = 10$, $\theta = 0.5$; $n = 50$, $\theta = 0.1$; $n = 100$, and $\theta = 0.05$, and compare the resulting graphs with the Poisson distribution with $\lambda = 5$.

   Because the possible values of Y for the Poisson model are infinite, a stopping rule must be developed based on the value of $Pr(Y=y)$ so that when $Pr(Y=y) < \epsilon$, the calculations are stopped.

CHAPTER 8

MATHEMATICAL MODELS FOR CONTINUOUS RANDOM VARIABLES

8.1 Introduction

In Chap. 7 we considered mathematical models that are appropriate for representing the distribution of a discrete observable random variable. In our earlier considerations of descriptive statistics we also encountered continuous observable random variables. In this chapter we shall consider models that are appropriate for this type of variable.

We shall consider the general properties of continuous mathematical models and then introduce several different continuous models and illustrate how they can be used to represent the distribution of observed continuous variables. Since the general form of a continuous model will also involve parameters, we must once again determine how to assign values to the parameters to best fit the observed data.

We shall also present computer programs that can be used to determine probabilities for any continuous mathematical model and show how these programs are useful in evaluating the goodness of fit of the particular model to the observed data.

8.2 Continuous Mathematical Models

To appreciate the distinction between discrete and continuous random variables, recall that a discrete variable can take values only from a completely specified set of values. Under this restriction, we can associate a probability $p(Y=y)$ with every possible value of Y. However, the possible values associated with a continuous random variable cannot be specified because between any two possible values there is always another possible value. Thus continuous variables can take on any value within a specified range. The range of values are generally intervals of the real line such as a < Y < b, where it is possible that a = -∞ and b = ∞.

The following illustration shows a possible mathematical representation for a distribution of a continuous random variable. Note that f(Y) is a continuous function defined over the entire range of Y, which in this case is $0 \leq Y < \infty$.

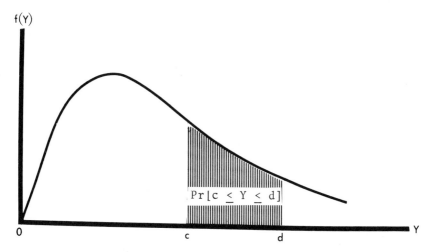

When Y is a continuous random variable we cannot simply associate a probability p(Y=y) with all values of Y. Thus the function f(Y), which represents the distribution of Y, is called a probability density function because f(Y) does not have a probability interpretation but is simply the ordinate or density of the function for that value of Y.

For a continuous random variable, probability is defined in the following manner:

(1) The probability that Y = y, Pr(Y=y), is zero for all values of y

(2) The probability that $c \leq Y \leq d$ (c and d being in the range of Y), Pr($c \leq Y \leq d$), is the shaded area under f(Y) between c and d as seen in the previous figure. We can represent this area symbolically by

179

$$\Pr[c \le Y \le d] \quad = \quad \int_c^d f(y)\ dy$$

The integral symbol $\int$ plays a comparable role for continuous models that the Σ symbol plays for the discrete models. The $\int$ symbol represents the area under the function f(Y), which is a probability of obtaining a value in the interval of interest while the Σ symbol represents the sum of the probabilities for the set of y values of interest. Since the probability of a single value is equal to zero for a continuous variable, it is immaterial whether less than (<) or less than or equal to ($\le$) is used in determining the probability to be associated with any interval.

A mathematical model for a continuous random variable f(Y) defined on the range $a \le Y \le b$ must have the following properties to be considered a probability density function:

(1) f(Y) must be a continuous function

(2) $f(Y) \ge 0$ for all $a \le Y \le b$

(3) $\int_a^b f(y)\ dy = 1$

As a simple illustration of a continuous probability density function, we can consider the continuous uniform distribution (rectangular distribution) given by

$$f(Y) = \frac{1}{2}, \quad 0 \le Y \le 2$$

which has the following graphical representation:

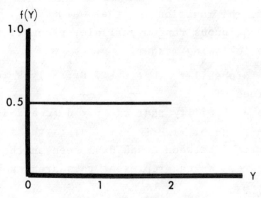

The probability $Pr[0.5 \leq Y \leq 1.25]$ can graphically be represented by

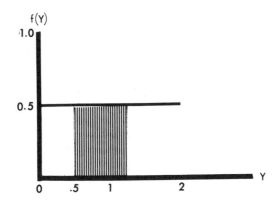

The required probability is given by the shaded rectangular area. In this simple case it can easily be found by multiplying the length of the interval by the height so that

$$Pr[0.5 \leq Y \leq 1.25] = \int_{0.5}^{1.25} \frac{1}{2} \, dy = (1.25 - 0.5) \frac{1}{2} = 0.375$$

The expected value of a numerical function $G(Y)$ of a continuous random variable Y having a probability density function $f(Y)$ is defined to be

$$E[G(Y)] = \int_{a}^{b} G(y)f(y) \, dy$$

Thus the expected value of $G(Y)$ is the area under the product function $H(Y) = G(Y)f(Y)$ between $Y = a$ and $Y = b$. The product function $H(Y)$ weights the values of $G(Y)$ by the probability density factor $f(Y)$. Finding such areas is the problem of calculus, although the computer can be used to find numerical approximations.

We can once again specialize the function $G(Y)$ to obtain the three classes of moments for a continuous distribution model. We let, in turn, $G(Y) = Y^k$, $G(Y) = (Y-\mu)^k$, and $G(Y) = [(Y-\mu)/\sigma]^k$ to obtain the three classes of moments. Thus we define

181

$$\mu_k = \int_a^b y^k f(y) \, dy = \text{kth moment about the origin}$$

$$\nu_k = \int_a^b (y - \mu_1)^k \, f(y) \, dy = \text{kth moment about the mean}$$

$$\alpha_k = \int_a^b \left[\frac{y - \mu_1}{\sqrt{\nu_2}} \right]^k f(y) \, dy = \text{kth standard moment}$$

To illustrate how the moments can be found for a given $f(Y)$, we again use the rectangular model given by $f(Y) = 1/2$, $0 \leq Y \leq 2$. If we want to find the first moment about the origin, μ_1 = expected value of Y, we need to evaluate the following integral:

$$\mu_1 = E[Y] = \int_0^2 y \, \tfrac{1}{2} \, dy$$

In this case the product function $H(Y) = G(Y)f(Y) = Y/2$ is a straight line. The graph of the product function is

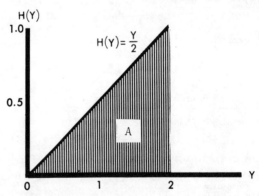

By definition μ_1 is the area under the function $H(Y) = Y/2$, and because this area is a right triangle, we find that the area is

$$\mu_1 = A = \int_0^2 \frac{Y}{2} \, dy = \tfrac{1}{2}(\text{base})(\text{height}) = \tfrac{1}{2}(2 - 0) \, 1 = 1$$

When we consider the second moment about the origin $\mu_2 = E[Y^2]$ the product function becomes

$$H(Y) = G(Y)f(Y) = Y^2 \frac{1}{2}$$

which is a parabola. The graphical representation is

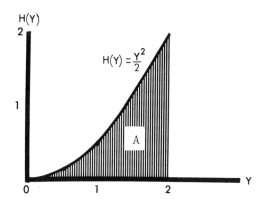

The second moment is equal to the area under the parabola

$$\mu_2 = E[Y^2] = \int_0^2 y^2 \frac{1}{2} \, dy = A$$

but now we are not able to determine the value by simple geometry. Thus we must turn to the mathematicians or the computer.

Once we obtain the moments about the origin, the characteristics of a continuous model can be found by using the same relationships that were used in the discrete case. Thus

μ = mean = μ_1

σ^2 = variance = ν_2 = $\mu_2 - \mu_1^2$

σ = standard deviation = $\sqrt{\nu_2}$

α_3 = skewness = $\dfrac{\nu_3}{\nu_2^{3/2}}$ = $\dfrac{(\mu_3 - 3\mu_2\mu_1 + 2\mu_1^3)}{(\mu_2 - \mu_1^2)^{3/2}}$

$$\alpha_4 = \text{kurtosis} = \frac{\nu_4}{\nu_2^2} = \frac{(\mu_4 - 4\mu_3\mu_1 + 6\mu_2\mu_1^2 - 3\mu_1^4)}{(\mu_2 - \mu_1^2)^2}$$

Mathematical models for continuous variables are also given in a general form containing undetermined constants or parameters. Once again, either theoretical and physical considerations can be used to assign values to these parameters for a particular distribution or statistical techniques such as the method of moments can be used. Once all the parameters have been determined we can use the same three techniques introduced in Chap. 7 to examine the goodness of fit of the model to observed data. These techniques were (1) a comparison of the higher-order characteristics of the fitted model with those of the observed distribution, (2) a graphical comparison of the model with the observed relative frequencies, and (3) a comparison of the expected frequencies (model probabilities multiplied by the total number of observations) with the observed frequencies.

Before applying these concepts to specific mathematical models we must be able to evaluate integrals to find probabilities assigned by the model. In the next section we give a computer program that will approximate any areas under a continuous function.

8.3 FORTRAN Program for Numerical Integration

We wish to write a computer program to evaluate the definite integral for any continuous function f(Y) given by

$$\int_c^d f(y) \, dy$$

Different numerical integration techniques have been evolved to evaluate the above integral. All of them divide the interval of integration [c,d] into m subintervals, usually of equal length, and for each subinterval approximate the area under the function using a simple geometric figure. The simpliest technique uses rectangles whose heights correspond to a value of the function at the midpoint of each interval and whose bases are the subinterval length. We can represent the technique by means of the following diagram:

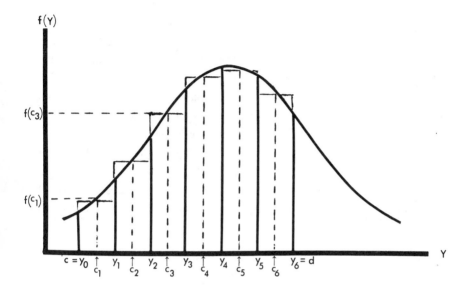

It can be seen that the area of each rectangle is

$$A_i = f(c_i) \, \Delta y \qquad A_{i+1} = f(c_{i+1}) \, \Delta y \qquad \text{with } c_{i+1} = c_i + \Delta y$$

As the diagram indicates, we shall approximate the area under the function by summing the rectangular areas A_i. So we shall compute

$$\int_c^d f(y) \, dy \doteq f\left(y_0 + \frac{\Delta y}{2}\right) \Delta y + f\left(y_1 + \frac{\Delta y}{2}\right) \Delta y + \cdots + f\left(y_{m-1} + \frac{\Delta y}{2}\right) \Delta y$$

$$\doteq \sum_{i=0}^{m-1} f\left(y_i + \frac{\Delta y}{2}\right) \Delta y = \sum_{i=1}^{m} f(c_i) \, \Delta y$$

The accuracy of the numerical approximation of the integral depends upon the length of the subintervals (Δy); thus the smaller the subintervals, the more accurate the approximation.

The interval [c,d] of integration may be such that c = $-\infty$ and/or d = $+\infty$. In such cases, for any given value of Δy the number of possible summations will become infinite. Some type of stopping rule must be introduced if the summation process is to be completed. We shall take advantage of the properties of all continuous distribution models that are (1) $f(Y) \geq 0$ for all Y in the range and (2) as Y $\rightarrow \pm\infty$, $f(Y) \rightarrow 0$ because the area under the function must equal 1. Thus we shall not attempt to have the subintervals cover an infinite range but shall start and end the intervals at a point in the range where $f(Y)$ is very small.

The computer program therefore must have the lower and upper values of the interval y_0 and y_m specified. The sub-interval length Δy must also be known. However, rather than specifying its value directly, we choose to specify the number of intervals m and then calculate the value of Δy using the relationship

$$\Delta y = \frac{y_m - y_0}{m}$$

The function $f(Y)$ that is to be integrated must also be given.

We shall write the integration program as a FUNCTION subprogram. The routine is called DEFINT because it evaluates the definite integral of any function. The necessary inputs are y_0 = Y0, y_m = YM, and m = number of intervals. Let us use for illustrative purposes the function

$$f(Y) = \frac{1}{\sqrt{2\pi}} e^{-\frac{1}{2}y^2}$$

and find its definite integral

```
FUNCTION DEFINT (YØ,YM,M)
F(Z) = 1./SQRT(2.*PI) * EXP(-Z**2/2.)
PI = 3.14159265
DEFINT = Ø
XM = M
DELTAY = (YM - YØ)/XM
Y = YØ - DELTAY/2.
DO 1 I=1,M
Y = Y + DELTAY
AREA = F(Y) * DELTAY
1 DEFINT = DEFINT + AREA
RETURN
END
```

Program 8.1 FUNCTION DEFINT — numerical integration
using the rectangular method.

Since we want to write a computer program to evaluate
different functions f(Y) with a minimum number of changes,
we introduced a new FORTRAN statement, the ARITHMETIC STATE-
MENT FUNCTION, given by F(Z) = 1./SQRT(2.*PI) * EXP(-Z**2/2.).
If we wished to integrate a different function f(Y), we
could simply replace the program card that defines the function.

The arithmetic statement function must be defined by
a single FORTRAN arithmetic expression and is available on
most FORTRAN compilers. There may be several such statement
functions, and one statement function can involve any available FORTRAN function or a previously defined arithmetic
statement function. The statement functions must be the
first executable statements in the program so that only such
nonexecutable statements as comment cards and format statements can be found in front of them.

Although the statement function F(Z) given in Program
8.1 appears to have a subscripted variable on the left side
of the equal sign, this notation designates the argument of
the function that can be integer or real. In fact, the
function name itself can be integer or real, and its mode is
independent of the mode of the arguments. The statement
function must have at least one argument and can have many
arguments (separated by commas), as in the following example:

I(A,B,C) = 2. *B - SQRT(B**2 - 4.*A*C).

All of the rules for FORTRAN expressions apply to the right-hand side of the statement function.

No execution occurs when the computer encounters the arithmetic statement function itself. However, when the statement function is called in subsequent program statements, the function is evaluated for the specific values of the arguments given in the calling statement. For this reason we could define the statement function using Z as the argument since this is only a dummy name; at the time of evaluation the appropriate value of Y is used.

Variables that are not arguments of the statement function assume their current value when the function is being executed. Thus in Program 8.1 PI is not an argument and must therefore be defined before the statement AREA = F(Y) * DELTAY is encountered. Of course, we could have evaluated the expression $1/\sqrt{2\pi}$, which is equal to 0.39894228, and used this value in the statement function itself.

There are more sophisticated methods that use more complex geometric figures to approximate a definite integral. Thus they may be more accurate and efficient than the simple DEFINT subprogram given in Program 8.1. We would naturally like to be able to take advantage of such methods and programs without having to appreciate their particular logic. Since one of the advantages of the stored program digital computer is the availability of general-purpose computer programs, we shall often utilize such programs already written by an "expert." These general programs as well as another integration program also called DEFINT can be found in Appendix B. The student can compare the accuracy of the subprogram given in this chapter with the one in the appendix. For illustrative purposes, we shall utilize Program 8.1 in the programs given in this chapter, although the FUNCTION DEFINT in the appendix can be substituted.

8.4 The Rectangular Distribution

We now want to apply the concepts developed for continuous distribution models in Sec. 8.2 to specific models. Let us first consider the rectangular or uniform model, which is given by

$$f(Y) = \frac{1}{h} , \quad 0 \le Y \le h$$

The general expression for the kth moment about the origin for this model is

$$\mu_k = \int_0^h y^k \frac{1}{h} \, dy$$

Using calculus techniques, one can obtain the general result that

$$\mu_k = \frac{h^k}{k + 1}$$

Likewise, the general expression for the moments about the mean is

$$\nu_k = \int_0^h \left(y - \frac{h}{2} \right)^k \frac{1}{h} \, dy$$

Rather than using this defining relationship to determine the moments about the mean, we choose to evaluate these moments by using the relationships between the moments about the mean and the moments about the origin:

$$\nu_2 = \mu_2 - \mu_1^2 = \frac{h^2}{3} - \left(\frac{h}{2} \right)^2 = \frac{h^2}{12}$$

$$\nu_3 = \mu_3 - 3\mu_3\mu_1^2 + 2\mu_1^3 = 0$$

$$\nu_4 = \mu_4 - 4\mu_3\mu_1 + 6\mu_2\mu_1^2 - 3\mu_1^4 = \frac{h^4}{80}$$

and the characteristics of the model can then be found since

$$\mu \;=\; \text{mean} \;=\; \frac{h}{2}$$

$$\sigma^2 \;=\; \text{variance} \;=\; \frac{h^2}{12}$$

$$\sigma \;=\; \text{standard deviation} \;=\; \frac{h}{\sqrt{12}}$$

$$\alpha_3 \;=\; \text{skewness} \;=\; \frac{\nu_3}{\nu_2^{3/2}} \;=\; 0$$

$$\alpha_4 \;=\; \frac{\nu_4}{\nu_2^2} \;=\; \frac{h^4/80}{h^4/144} \;=\; \frac{144}{80} \;=\; 1.8$$

The rectangular model can be generalized by changing the range to $a \le Y \le a + h$. Thus the moments about the origin are given by

$$\mu_k \;=\; \int_a^{a+h} y^k \,\frac{1}{h}\, dy \;=\; \frac{(a+h)^k}{k+1} \;-\; \frac{a^k}{k+1}$$

The student can verify that the only characteristic that changes is the mean, which becomes $\mu_1 = a + h/2$. The other characteristics σ^2, α_3, and α_4 retain their same values.

The rectangular model can be used to represent the distribution of a continuous variable when the probability of the variable falling within any interval is simply proportional to the length of the interval. For example, if your variable of interest is the time of arrival of individuals and the probability of an individual arriving during any period of time between $Y = 0$ and $Y = 10$ min is proportional to the length of the time period, the distribution model would be

$$f(Y) \;=\; \frac{1}{10}, \quad 0 \le Y \le 10$$

Since the geometry of the rectangular distribution is so straightforward and the constant h is generally available from physical considerations, we shall not spend time considering the application of the model to an actual example.

8.5 The Normal Distribution

The normal model has the distribution function given by

$$f(Y) = \frac{1}{\sqrt{2\pi}\,\sigma}\, e^{-\frac{1}{2}(Y - \mu)^2/\sigma^2} \qquad -\infty < Y < \infty$$

There are two parameters, μ and σ, in the model, and, as the notation suggests, they correspond to the mean and standard deviation of the model. The model has the familiar bell-shaped distribution as shown in the following graph:

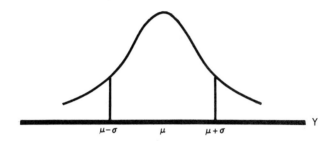

Distributions that are symmetric about a mean value and whose frequency of occurrence follows a bell-shaped curve, for example, height, reading speed, and measurement errors, can be effectively approximated by the normal distribution. Such variables are often encountered.

The general moments about the origin can be obtained by evaluating the integrals

$$\mu_k = \int_{-\infty}^{\infty} y^k \frac{1}{\sqrt{2\pi}\,\sigma}\, e^{-\frac{1}{2}(y - \mu)^2/\sigma^2}\, dy$$

Since there is not general expression available to represent the results of such integrations, we shall accept the results as obtained by mathematicians:

$$\mu_1 = \int_{-\infty}^{\infty} y \, f(y) \, dy = \mu$$

$$\mu_2 = \int_{-\infty}^{\infty} y^2 \, f(y) \, dy = \sigma^2 + \mu^2$$

$$\nu_3 = \int_{-\infty}^{\infty} (y - \mu)^3 \, f(y) \, dy = 0$$

$$\nu_4 = \int_{-\infty}^{\infty} (y - \mu)^4 \, f(y) \, dy = 3\sigma^4$$

Thus the characteristics of the normal model are

$$\mu = \text{mean} = \mu$$
$$\sigma^2 = \text{variance} = \nu_2 = \left(\sigma^2 + \mu_1^2\right) - \mu_1^2 = \sigma^2$$
$$\sigma = \text{standard deviation} = \sqrt{\nu_2}$$
$$\alpha_3 = \text{skewness} = \frac{\nu_3}{\nu_2^{3/2}} = 0$$
$$\alpha_4 = \text{kurtosis} = \frac{\nu_4}{\nu_2^2} = \frac{3\sigma^4}{\sigma^4} = 3$$

We shall demonstrate the techniques used in fitting the model to an observed distribution of a continuous variable by again using an actual example. Consider the now familiar observed frequency distribution of 390 reading times.

Reading Times of a Group of 390 New Employees

Interval	F	Class Boundaries	Y
2.0 - 2.4	21	1.95 - 2.45	2.2
2.5 - 2.9	43	2.45 - 2.95	2.7
3.0 - 3.4	68	2.95 - 3.45	3.2
3.5 - 3.9	97	3.45 - 3.95	3.7
4.0 - 4.4	72	3.95 - 4.45	4.2
4.5 - 4.9	53	4.45 - 4.95	4.7
5.0 - 5.4	21	4.95 - 5.45	5.2
5.5 - 5.9	11	5.45 - 5.95	5.7
6.0 - 6.4	4	5.95 - 6.45	6.2

We have already determined the numerical characteristics of this observed distribution to be

$$\bar{y} = 3.81 \qquad s = 0.86 \qquad a_3 = 0.25 \qquad a_4 = 2.79$$

and thus we can determine the model parameters μ and σ by setting

$$\mu = \bar{y} = 3.81$$

$$\sigma = s = 0.86$$

For the normal model the shape parameters α_3 and α_4 are not dependent upon the value used for μ and σ since α_3 is always equal to zero (the normal curve is symmetrical) and α_4 is equal to 3 (our standard for peakedness). We can thus use the comparison of these two high-order moments of the normal model with those of the observed distribution as a first indication of the goodness of fit.

Characteristic	Observed	Model
Mean	3.81	3.81
Standard deviation	0.86	0.86
Skewness	0.25	0.00
Kurtosis	2.79	3.00

Comparing the characteristics, we may anticipate that the observed data have a small positive skewness. This lack of symmetry will mean that the symmetrical model may not exactly reproduce the frequencies associated with the larger values of Y. In addition, the lack of exact correspondence between a_4 and α_4 may cause difficulties in fitting the larger frequencies. The extent to which these two discrepancies will interact is difficult to judge, and thus we find it hard to appreciate the goodness of fit using only the differences in the higher-order moments.

To obtain a graphical representation of the goodness of fit we can superimpose the graph of the model f(Y) upon the histogram of the observed distribution. We can write a com-

puter program to produce the required ordinates f(Y) needed
to plot the fitted normal model. The program, called EVALF,
is given at the end of this chapter. When it is used with
μ = YMU = 3.81 and σ = SIGMA = 0.86, the ordinates f(Y)
obtained are given in Fig. 8.1.

Y	f(Y)	Scaled f(Y)
1.00	0.0021	0.4183
1.50	0.0121	2.3609
2.00	0.0487	9.5036
2.50	0.1399	27.2839
3.00	0.2865	55.8628
3.50	0.4183	81.5717
4.00	0.4326	84.9484
4.50	0.3235	63.0915
5.00	0.1714	33.4184
5.50	0.0647	12.6241
6.00	0.0174	3.4011
6.50	0.0034	0.6535
7.00	0.0005	0.0895

Fig. 8.1 Normal ordinates and normal scaled ordinates.

To graph the histogram and the fitted normal model on
the same set of axes to observe the goodness of fit, we must
standardize the areas of the two graphs. Since the area of
the histogram is equal to n · w, and the area of the normal
model is equal to 1, we must multiply each normal ordinate
f(Y) by n · w = 390·(0.5) in order to compare observed and
expected ordinates. The corresponding graph is given
given in Fig. 8.2.

We observe from Fig. 8.2 that both of our earlier con-
jectures based on the discrepancies of the shape parameters
of the data with those of the normal model are borne out.
Although the normal model fits well, it fails to reproduce
the larger frequencies and the excessive number of large
observations.

The third technique for checking the goodness of fit is
to compare the theoretical frequencies for each of the intervals by using the normal model with the observed frequencies. We need to determine the probability of being in each
interval of the frequency distribution. For the jth

Employees

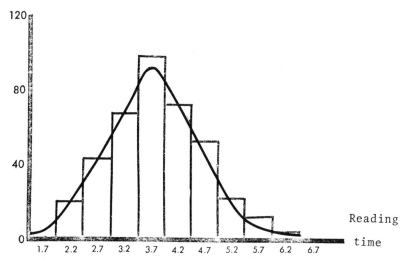

Fig. 8.2 Normal model superimposed on histogram of
reading times.

interval, this probability is given by

$$Pr\left[\ell_j < Y < u_j\right] = \int_{\ell_j}^{u_j} f(y) \ dy$$

The expected frequency for the jth interval is then given by

$$f_{ej} = n \cdot Pr\left[\ell_j < Y < u_j\right]$$

where $n = \Sigma f_j$ = the total number of observations in the
sample.

We used a computer program called NORMFREQ, which is
given at the end of this chapter, to obtain these expected
frequencies. The output of the program using the following
input data together with the observed frequencies is given
in Fig. 8.3. The inputs are m = M = 9, w = W = 0.5, b =
YLTB = 1.95, n = N = 390, μ = YMU = 3.81, σ = SIGMA = 0.86.

Class Boundaries	Obs. F	Exp. F	Difference
1.95 - 2.45	21	22.27	-1.25
2.45 - 2.95	43	39.62	3.38
2.95 - 3.45	68	69.62	-1.62
3.45 - 3.95	97	88.30	8.70
3.95 - 4.45	72	82.17	-10.17
4.45 - 4.95	53	51.60	-1.40
4.95 - 5.45	21	25.47	-4.47
5.45 - 5.95	11	8.46	2.54
5.95 - 6.45	4	2.50	1.50
Total	390	390.01	

Fig. 8.3 Observed and expected normal frequencies for reading times.

Once again we are faced with the question of the adequacy of the fit obtained, but we shall still defer this decision until later.

If we assume that the normal model adequately describes the underlying population distribution of reading time of the new employees, we can use the model to answer questions relative to the nature of the distribution. To answer such questions we generally need to be able to find probabilities associated with the normal model. We can use the FUNCTION DEFINT program or FUNCTION YORMX (Appendix D-2), to approximate the probabilities, or resort to the use of the table of probabilities given in Appendix C-2. This table gives the area under the standard normal curve, which has a mean of zero and a standard deviation of 1, and can be represented by

$$AREANORM(z_o) = \int_{-\infty}^{z_o} \frac{1}{\sqrt{2\pi}} e^{-z^2/2} dz$$

for the values of z_o from 0.00 to 3.59 in increments of 0.01. This table can be used to find probabilities associated with the general normal model with mean μ and standard deviation σ since any interval on the y axis has a corre-

196

sponding interval on the z axis. The probability of Y falling in the y interval is equal to z falling in the corresponding z interval:

$$Pr[c \leq Y \leq d] = Pr\left[\frac{c - \mu}{\sigma} < Z < \frac{d - \mu}{\sigma}\right]$$

Thus the use of the transformation $z = (y - \mu)/\sigma$ transforms
the general normal model into the standard normal model,
which enables us to use a single table of probabilities
rather than a separate table for each μ and σ being considered.

The tabulated areas given in Appendix C-2 are for the
interval $(-\infty, z_o)$. We must learn how to use these tabulated
areas to find areas for other intervals. The standard normal distribution is symmetric about zero with the total area
equal to one so that the area for the interval (z_o, ∞) is
equal to 1 minus the area for $(-\infty, z_o)$, as seen below.

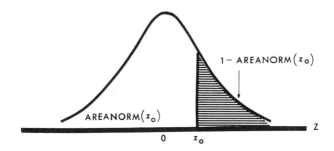

Let us now consider how the standard normal model can
be used to answer questions about the distribution of the
variable.

(1) What is the probability that a randomly selected
individual would have a reading time less than or equal to
4.5 min?

Both the DEFINT FUNCTION and the normal tables in
Appendix C-2 require that the problem be stated in terms of
the standard normal distribution ($\mu = 0$ and $\sigma = 1$). Thus we
must transform y = 4.5 into its corresponding z value. We

obtain $z = (y - \mu)/\sigma = (4.5 - 3.61)/0.86 = 0.80$; thus

$$Pr[Y \leq 4.5] = Pr\left[\frac{Y - \mu}{\sigma} \leq \frac{4.5 - \mu}{\sigma}\right] = Pr[Z \leq 0.80]$$

The normal tables give

$$Pr[Z \leq z_o] = \int_{\infty}^{z_o} \frac{1}{\sqrt{2\pi}} \; e^{-z^2/2} \; dz = \text{AREANORM}(z_o)$$

The whole number and the first decimal place of the **argument** z_o locates the row of the table, and the second decimal place locates the column. Thus for $z_o = 0.80$ we enter the ninth row and go across to the first column, obtaining 0.7881. This probability result is approximately 79/100, which means that approximately 79 out of 100 employees would be expected to have reading times less than 4.5 min. To find

$$Pr[Y > 4.5] = Pr[Z > 0.80]$$
$$= 1 - Pr[Z \leq 0.80] = 1 - \text{AREANORM}(0.80)$$
$$= 1 - 0.7881$$
$$= 0.2119$$

(2) How many out of a randomly selected group of 225 individuals would you expect to have reading times between 3.0 and 4.0 min?

To answer this question we have two values of Y: $y_1 = 3.0$ and $y_2 = 4.0$. These must be converted to corresponding z values to use the tables. We have

$$z_1 = \frac{y_1 - \mu}{\sigma} = \frac{3.0 - 3.81}{0.86} = -0.94$$

$$z_2 = \frac{y_2 - \mu}{\sigma} = \frac{4.0 - 3.81}{0.86} = 0.22$$

and thus $Pr[3.0 \leq Y \leq 4.0] = Pr[-0.94 \leq Z \leq 0.22]$.

Often a simple sketch of the normal curve, showing the integral and area of interest, will be sufficient to indicate how the tabulated areas can be used to find the desired probability.

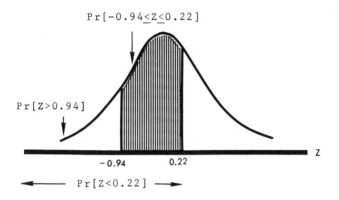

We see that to find the required probability we use the relationship

$$Pr[-0.94 \leq Z \leq 0.22] = Pr[Z \leq 0.22] - Pr[Z > 0.94]$$

$$= AREANORM(0.22) - [1 - AREANORM(0.94)]$$

$$= .5871 - (1 - .8264) = 0.4135$$

and

$$\text{Expected number} = n_o \cdot Pr[3.0 \leq Y \leq 4.0]$$

$$= (255)(0.4135) = 93$$

(3) What is the probability of obtaining an individual selected at random whose reading time falls in the interval $[\mu - \lambda\sigma \leq Y \leq \mu + \lambda\sigma]$? In particular, if $\mu = 3.81$ and $\sigma = 0.86$ with a $\lambda = 2$, we wish to find $Pr[2.09 \leq Y \leq 5.53]$. Since

$$z_1 = \frac{y_1 - \mu}{\sigma} = \frac{2.09 - 3.81}{0.86} = -2.00 = -\lambda$$

$$z_2 = \frac{y_2 - \mu}{\sigma} = \frac{5.53 - 3.81}{0.86} = 2.00 = +\lambda$$

$$
\begin{aligned}
\Pr[2.09 \le Y \le 5.53] &= \Pr[-2.00 \le Z \le 2.00] \\
&= \Pr[-2.00 \le Z \le 0] + \Pr[0 \le Z \le 2.00] \\
&= 2 \cdot \Pr[0 \le Z \le 2.00] \\
&= 2 \cdot [\text{AREANORM}(2.00) - \text{AREANORM}(0.00)] \\
&= 2[0.9772 - 0.5000] = 0.9544
\end{aligned}
$$

(4) We can also consider the inverse problem of finding the value of λ and thus the values of y_1 and y_2 for a specified probability, say $1 - \alpha$. We can now use the AREANORM tables to locate the appropriate probability and from the corresponding z value of λ. From the graphical representation

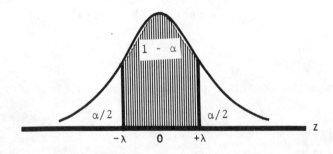

we recognize that we can find λ because $\text{AREANORM}(\lambda) = 1 - \alpha/2$. Knowing the value of λ we can then use the relationship between z and y to find the values of y_1 and y_2.

$$\Pr[-\lambda \le Z \le \lambda] = 1 - \alpha$$

$$\Pr[\mu - \lambda\sigma \leq Y \leq \mu + \lambda\sigma] = 1 - \alpha$$

For example, if we let $1 - \alpha = 0.98$, then $\alpha/2 = 0.01$ and $1 - \alpha/2 = 0.99$. The value of λ is then found by looking in the normal table for the z value that corresponds to an area of 0.9900. We find that AREANORM(2.32) = 0.9898 and AREA-NORM(2.33) = 0.9901 and thus elect to use $\lambda = 2.33$. For the reading-time distribution, we then obtain the interval

$$\Pr[-2.33 \leq Z \leq 2.33] = 0.98$$

$$\Pr[\mu - \lambda\sigma \leq Y \leq \mu + \lambda\sigma] = 0.98$$

$$\Pr[3.81 - (2.33)(0.86) \leq Y \leq 3.81 + (2.33)(0.86)] = 0.98$$

$$\Pr[1.81 \leq Y \leq 5.81] = 0.98$$

and thus can state that the probability of obtaining an observation in this interval is 0.98. These types of probability problems will become especially important when we consider the inference problems of statistics.

8.6 The Pearson Type III Distribution

In this section we shall consider a mathematical model that may be useful in representing an asymmetrical observed distribution. The model is called the Pearson type III distribution and has two forms, depending upon the sign of the parameter A.

1. $\underline{A > 0}$

$$f(Y) = \frac{A^{A^2} e^{-A^2}}{\Gamma(A^2)\sigma} e^{-A[(Y-\mu)/\sigma]} \left[\frac{A\sigma + Y - \mu}{\sigma}\right]^{A^2-1}$$

where the range of Y is given by $\mu - A\sigma < Y < \infty$.

2. $A < 0$

$$f(Y) = \frac{A^{A^2} e^{-A^2}}{\Gamma(A^2)\sigma} \, e^{A[(Y-\mu)/\sigma]} \left[\frac{A\sigma - Y + \mu}{\sigma}\right]^{A^2-1}$$

where the range of Y is given by $-\infty < Y < \mu + A\sigma$.

Note that a new symbol $\Gamma(x)$ is introduced; it represents the gamma function. Its argument x can be any value greater than zero $(x > 0)$. When x is an integer, then $\Gamma(x) = (x-1)!$ Since the gamma function is also defined for noninteger values of x, it can be considered as a generalization of the factorial function. We shall not concern ourselves with the formula needed to evaluate this gamma function because computer routines are available to evaluate it for any positive value of x.

The representation used for the parameters indicate that other than A they represent the corresponding moments of the distribution. Rather than consider the mathematical problem of evaluating the characteristics of the type III distribution, we simply give the results

$$\text{Mean} = \mu$$
$$\text{Variance} = \sigma^2$$
$$\text{Standard deviation} = \sigma$$
$$\text{Skewness} = \frac{2}{A}$$
$$\text{Kurtosis} = \frac{3}{2}\left[2 + \frac{\sqrt{2}}{|A|}\right]$$

Attention should be given to the range of the variable. The range is only infinite in one direction. It is this feature that produces the skewness in the model.

Let us demonstrate the process of fitting the type III model to an observed frequency distribution. Consider the following distribution of annual income:

Annual Income
(measured to the nearest $100)

Interval	F
2000 - 5900	8
6000 - 9900	73
10000 - 13900	157
14000 - 17900	218
18000 - 21900	172
22000 - 25900	133
26000 - 29900	92
30000 - 33900	63
34000 - 37900	37
38000 - 41900	19
42000 - 45900	6
46000 - 49900	2
Sum	980

The numerical characteristics of the observed distribution are

$$\bar{y} = 17,323 \qquad s_y = 7429 \qquad \alpha_3 = 0.8961 \qquad \alpha_4 = 3.3036$$

Note that α_3 is nearly equal to 1 and thus the normal model would not be expected to adequately represent the distribution. To fit the type III distribution to the observed data, we set

$$\mu = \bar{y} = 17,723$$
$$\sigma = s = 7,429$$
$$A = \frac{2}{\alpha_3} = \frac{2}{0.8961} = 2.2318$$

Since $A > 0$ we use the first form of the type III model, and thus the range of Y is

$$\mu - A\sigma < Y < \infty$$

$$17723 - 2.2318(7429) < Y < \infty$$

$$1143 < Y < \infty$$

Although we did not observe any salary below \$2,000, this model indicates that we might find a person with an income as low as \$1,143.

Another approach would be to use the knowledge that the annual income must be positive and thus the range of Y is from zero to infinity. This would require that the lower end point of the range given by $\mu - A\sigma$ be set to zero. Thus we have the relationship $\mu - A\sigma = 0$, and by equating $\mu = \bar{y} = 17323$ and $\sigma = s = 7429$ we obtain $A = \mu/\sigma = 17723/7429 = 2.3856$.

We now have two competitive type III models that could be used to represent the distribution of annual income. We shall compare the two models by first checking the higher-order characteristics of the model with the observed ones:

Characteristic	Observed	Model 1	Model 2
Mean	$\bar{y} = 17723$	$\mu = 17723$	$\mu = 17723$
Standard deviation	$\sigma = 7429$	$\sigma = 7429$	$\sigma = 7429$
Skewness	$a_3 = 0.8961$	$\alpha_3 = 0.8961$	$\alpha_3 = 0.8383$
Kurtosis	$a_4 = 3.3036$	$\alpha_4 = 3.9504$	$\alpha_4 = 3.8892$

Although we may be concerned as to which model is really "best," we choose to continue our considerations by using Model 1 since α_3 equals a_3 and there is not much difference in the α_4 values. However, if the range used in Model 2 is an important consideration, then it might be the more desirable model in spite of the lack of agreement of α_3 with a_3.

We may choose to superimpose the graph of the type III model upon the histogram of the observed data to further check its goodness of fit. Once again we can write a program to obtain the appropriate ordinates of f(Y), including the scaled ordinates [$\Sigma f \cdot w \cdot f(Y)$]. We shall adapt the program we used to evaluate the normal model; this revised program is also found at the end of the chapter. The results are given in Fig. 8.4:

Y	f(Y)	Scaled f(Y)
1950.00	0.0000000	0.1407
5950.00	0.0000131	51.4577
9950.00	0.0000440	172.3323
13950.00	0.0000587	230.0713
17950.00	0.0000521	204.1283
21950.00	0.0000366	143.5929
25950.00	0.0000222	86.9475
29950.00	0.0000121	47.4062
33950.00	0.0000061	23.9196
37950.00	0.0000029	11.3705
41950.00	0.0000013	5.1554
45950.00	0.0000006	2.2493
49950.00	0.0000002	0.9506
53950.00	0.0000001	0.3911

Fig. 8.4 Ordinates and scaled ordinates for the
Pearson type III distribution.

We have superimposed the graph of the fitted type III
model on the histogram and see that it gives a better fit
than the symmetrical normal model.

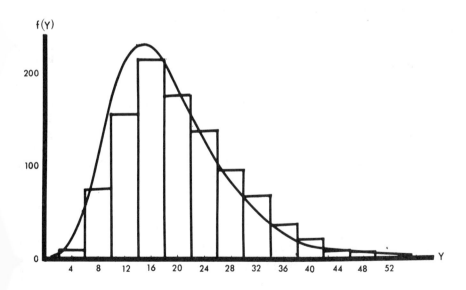

Fig. 8.5 Pearson type III model superimposed on
histogram of income (thousands of $).

We may choose to graduate the observed distribution
with the type III curve using expected frequencies for each
interval, but this is left as an exercise for the student.

The fitting of the Pearson type III model again serves to demonstrate the steps involved in selecting a model and determining its goodness of fit to observed data. Once we are willing to make the assumption that our mathematical model adequately represents the underlying distribution of the variable, we can use the model to answer questions about the nature of the underlying distribution. If the type III model is selected, tabulated areas (probabilities) under the type III model must allow for variations in parameter A even when one uses the transformation $z = (y - \mu)/\sigma$. Since the tabulated areas for different values of A would be quite voluminous, FUNCTION DEFINT could be used to find the probabilities, and this may be preferred.

8.7 Computer Programs

The following programs have been mentioned in different sections. A brief description of the program is given together with the sample input that was used to generate the output used in the chapter.

Program EVALF is a main program that can be utilized to evaluate the ordinate of any function f(Y) and the scaled ordinate [f(Y) · w · n] for any continuous mathematical distribution having two parameters. The parameters of the distribution as well as the lower and upper limits of Y to be evaluated must be read in. The number of points to be evaluated must also ge given, and the program evaluates f(Y) for the given number of points equally spaced over the given range.

$$n = AN = \text{number of observations}$$
$$w = W = \text{width of histogram interval}$$
$$\mu = YMU = \text{1st parameter}$$
$$\sigma = SIGMA = \text{2nd parameter}$$
$$YLP = \text{lower limit of Y}$$
$$YUP = \text{upper limit of Y}$$
$$NS = \text{number of points to be evaluated}$$

This program (called EVALF) is

```
C    EVALF - EVALUATION OF ORDINATES OF DIST. WITH TWO PARA
     READ(5,1) YMU,SIGMA
   1 FORMAT(2F1Ø.2)
```

```
      READ(5,2) NS,YLP,YUP,AN,W
    2 FORMAT (I5,4F1Ø.2)
      WRITE(6,6) YMU,SIGMA
    6 FORMAT(1H1,3HP1=,F1Ø.2,2X,3HP2=,F1Ø.2)
      WRITE(6,7) NS,YLP,YUP,AN,W
    7 FORMAT(1HØ,I5,4F12.3)
      WRITE(6,8)
    8 FORMAT(1HØ,12X,1HY,1ØX,4HF(Y),12X,5HSF(Y))
      SN = NS- 1
      YSTEP = (YUP - YLP)/SN
      DO 3 I=1,NS
      YI = I
      Y = YLP + (YI-1.)*YSTEP
      FY = F(Y,YMU,SIGMA)
      SFY = FY*AN*W
      WRITE(6,9) Y,FY,SFY
    9 FORMAT(1HØ,5X,F1Ø.2,5X,F1Ø.7,5X,F1Ø.4)
    3 CONTINUE
      STOP
      END
{Function F}
```

Program 8.2 EVALF — two parameter model ordinates and
scaled ordinates.

The function F(Y,YMU,SIGMA) can be changed for each
mathematical distribution. The correct function to evaluate
the normal distribution is

```
      FUNCTION F(Y,YMU,SIGMA)
      Z = (Y - YMU)/SIGMA
      F = .39894228 * EXP(-Z**2/2.)/SIGMA
      RETURN
      END
```

Program 8.3 FUNCTION F — normal ordinate evaluation.

Program 8.2, EVALF, was run with the function F given
by Program 8.3, using the following input. (The results
have already been given in Fig. 8.1.).

$$3.81 \qquad .86$$
$$12 \quad 1.00 \qquad 7.00 \qquad 390. \qquad .5$$

To evaluate the ordinates for the Pearson type III distribution, the program EVALF and FUNCTION F were modified to
accommodate the three parameters: μ, σ, and A. These modi-

fied programs are given in Programs 8.4 and 8.5.

```
C      EVALF - EVALUATION OF ORDINATES OF DISTRIBUTION WITH
C      THREE PARAMETERS, THE SAME TWO AS BEFORE AND
C      THE THIRD PARAMETER BEING A.
       READ(5,1) YMU,SIGMA,A
     1 FORMAT(3F1Ø.2)
       READ(5,2) NS,YLP,YUP,AN,W
     2 FORMAT(I5,4F1Ø.2)
       WRITE(6,6) YMU,SIGMA,A
     6 FORMAT(1H1,3HP1=,F1Ø.2,2X,3HP2=,F1Ø.2,2X,3HP3=,F1Ø.2
       WRITE(6,7) NS,YLP,YUP,AN,W
     7 FORMAT(I5,4F12.3)
       WRITE(6,8)
     8 FORMAT (1HØ,12X,1HY,1ØX,4HF(Y),12X,5HSF(Y))
       SN = NS - 1
       YSTEP = (YUP - YLP)/SN
       DO 3 I=1,NS
       YI = I
       Y = YLP + (YI-1.)*YSTEP
       FY = F(Y,YMU,SIGMA,A)
       SFY = FY*AN*W
       WRITE(6,4) Y,FY,SFY
     9 FORMAT(1HØ,5X,F1Ø.2,5X,F1Ø.7,5X,F1Ø.4)
     3 CONTINUE
       STOP
       END
{Function F}
{Function ALGAMA}
```

Program 8.4 EVALF — three parameter model ordinates
and scaled ordinates.

A FORTRAN function could easily be written to evaluate
the ordinates of the type III distribution by using the definition given in Sec. 8.6. A closer look at the expression
will, however, indicate computer difficulties in the evaluation. If for example, we have A = 8, the constant term would
be

$$
\frac{A^{A^2} e^{-A^2}}{\Gamma(A^2)} = \frac{8^{64} e^{-64}}{63!}
$$

The student can readily see that both the numerator and denominator separately will be very large even though the final result is not unreasonably large.

To avoid such numerical difficulties we shall evaluate the $\ln f(Y)$, where

$$\ln f(Y) = A^2 \cdot \ln A - A - A\left[\frac{Y - \mu}{\sigma}\right] + (A^2 - 1) \ln\left[\frac{A\sigma + Y - \mu}{\sigma}\right]$$

$$- \ln\Gamma(A^2) - \ln\sigma$$

and then take the antilog. We need only a function subprogram for $\ln\Gamma(\ )$; this is given in Appendix B and is called ALGAMA.

The routine to determine the ordinate for the type III model is given in Program 8.5. The sample input to attain the output used to plot Fig. 8.5 was

```
17723.      7429.      2.2318
    14      ]950.     53950.      980.      4000.
```

The routine is

```
      FUNCTION F(Y,YMU,SIGMA,A)
      F=Ø.
      IF (Y .LE. (YMU-A*SIGMA)) GO TO 1Ø
      AA = A*A
      Z1 = A*(Y-YMU)/SIGMA
      Z2 = (A*SIGMA + Y - YMU)/SIGMA
      ALF = AA*ALOG(A)-AA-Z1+(AA-1.)*ALOG(Z2)
    1  -ALGAMA(AA) - ALOG(SIGMA)
      F = EXP(ALF)
 1Ø  RETURN
      END
```

Program 8.5 FUNCTION F — Pearson type III ordinate evaluation.

The program NORMFREQ calculates expected frequencies of the normal probability distribution with μ = YMU and σ = SIGMA. The range of the distribution must be divided into m intervals of equal width w, and the lower true interval boundary b_1 must be given, as well as the total number of observations n.

$$m = \text{NINT} = \text{number of intervals}$$
$$n = \text{AN} = \text{number of observations}$$
$$\mu = \text{YMU} = \text{mean of distribution}$$
$$\sigma = \text{SIGMA} = \text{standard deviation of distribution}$$

b_1 = YLTB = lower true interval boundary

w = CW = class width

The program is

```
C  NORMFREQ - EXPECTED FREQUENCY FOR NORMAL DISTRIBUTION
   READ(5,1) NINT,CW,YLTB
 1 FORMAT( I2,2F1Ø.2)
   READ(5,2) AN,YMU,SIGMA
 2 FORMAT(3F1Ø.2)
   WRITE(6,3) AN,YMU,SIGMA
 3 FORMAT(1H1, 9HNO OF OBS,F1Ø.Ø,1ØHPARAMETERS,2F1Ø.2)
   WRITE(6,4)
 4 FORMAT(1HØ,1ØX, 8HINTERVAL,15X,5HEXP F)
   M = 1ØØ
   Y1 = YLTB
   Y2 = YLTB +CW
   Z2 =(Y2 -YMU)/SIGMA
   AREA = 1. - DEFINT(Z2,6.,M )
   EXPF = AN*AREA
   WRITE(6,5) Y1,Y2,EXPF
 5 FORMAT(1HØ,5X,F1Ø.2,3X,F1Ø.2,3X,F1Ø.2)
   NINT2 = NINT - 2
   DO 6  I=1,NINT2
   Y1 = Y2
   Y2 = Y2 + CW
   Z1 = (Y1 - YMU)/SIGMA
   Z2 = (Y2 - YMU)/SIGMA
   AREA = DEFINT(Z1,Z2,M)
   EXPF = AN*AREA
   WRITE(6,5) Y1,Y2,EXPF
 6 CONTINUE
   Y1 = Y2
   Y2 = Y2 + CW
   Z1 = (Y1 - YMU)/SIGMA
   AREA = DEFINT(Z1,6.,M)
   EXPF = AN*AREA
   WRITE(6,5) Y1,Y2,EXPF
   STOP
   END
{Function DEFINT}
```

Program 8.6 NORMFREQ — expected normal frequencies.

The sample input for Program 8.6 needed to obtain the output given in Fig. 8.3 is

```
    9      .5    1.95
   390.    3.81      .86
```

8.8 Summary

A mathematical model for a continuous variable Y is given as a function f(Y) defined over a prescribed range [a,b]. The probabilities associated with the occurrence of a continuous random variable can be found by the use of integrals. The notation used for these probabilities is

$$Pr[c \leq Y \leq d] = \int_c^d f(y) \, dy$$

and the value of the integral equals the area from Y = c to Y = d under the function f(Y).

The value of these definite integrals for any function f(Y) can be approximated by numerical integration techniques. A computer program FUNCTION DEFINT that accomplishes such an approximation has been written.

The characteristics of a continuous model can be found by using the same relationships used for discrete models once required moments are known. These moments are found by specializing the general definition of the expected value of a continuous model given by

$$E[G(Y)]^k = \int_a^b G(y)^k f(y) \, dy$$

so that the moments about the origin are

$$\mu_k = \int_a^b y^k f(y) \, dy$$

the moments about the mean are

$$\nu_k = \int_a^b (y - \mu_1)^k f(y) \, dy$$

and the standard moments are

$$\alpha_k = \int_a^b \left[\frac{y - \mu_1}{\sqrt{\nu_2}} \right]^k f(y) \, dy$$

Three continuous models have been considered in the chapter; (1) the rectangular distribution, (2) the normal distribution, and (3) the Pearson type III distribution. Their characteristics were given in terms of their parameters. To fit the model to an observed distribution the values of the parameters must be found; thus in this introductory text we choose the method of moments to determine their values, although several other techniques are available.

Three different techniques are available to check on the goodness of fit of the model to represent the observed distribution:

(1) Comparison of the higher-order characteristics of the model with those of the observed distribution.
(2) Graphical comparison of the model ordinates with the relative frequencies of the observed data.
(3) A comparison of the observed frequencies with the expected frequencies using the model.

If the fit of the model is satisfactory, it can be used to answer questions about the distribution of the variable. The uses of tabled areas for the normal distribution in answering such questions were shown.

The computer programs that can be used to test the goodness of fit were also given, together with the appropriate sample input and output. The programs were

(1) EVALF — Evaluation of ordinates of any two parameter function $f(y; \mu, \sigma)$
(2) EVALF — Evaluation of ordinates of any three parameter function $f(y; \mu, \sigma, A)$
(3) NORMFREQ — Evaluation of the expected frequency of the normal model.

By introducing mathematical models to represent the distribution of an observable random variable, we are now able to consider the various problems of statistics from an analytical point of view. In the next chapter we consider the problem of estimation when a random sample from the underlying population is used to determine one or more of the population parameters.

EXERCISES

1. Use the DEFINT program to compute the area of a right triangle with height 5 and base 10.

2. Use the NORMFREQ program to check the expected frequencies given in Fig. 8.3 for the reading-time distribution.

3. Consider the data and the resulting frequency distribution obtained in Exercise 2, Chap. 2. Using the tables for the area under the normal model, determine
   (a) The probability of obtaining a measurement error between 0.50 and 1.00
   (b) The expected number of errors out of 50 that would exceed 1.50
   (c) The probability of obtaining an error less than 0.20
   (d) The value of λ such that $Pr[\ |Y - \mu| > \lambda] = 0.10$
   Check the accuracy of these model values by using the observed y's to obtain the actual observed results associated with (a) to (d).

4. Use the EVALF three-parameter computer program to determine the ordinates for the Pearson type III model for different values of μ, σ, and A. Use the following values for the parameters, and plot the distributions on the same set of axes:

   (a) $\mu = 50$ $\sigma = 10$ $\alpha_3 = 1.0$
   (b) $\mu = 50$ $\sigma = 10$ $\alpha_3 = -1.0$
   (c) $\mu = 50$ $\sigma = 10$ $\alpha_3 = 0.5$
   (d) $\mu = 50$ $\sigma = 10$ $\alpha_3 = 0.1$

   Note that the equation for the type III is not defined for $\alpha_3 = 0$; however, $\lim_{\alpha_3 \to 0}$ type III$(\mu, \sigma, \alpha_3) = $ normal(μ, σ).

5. Write a computer program to determine the area under the type III curve. Use the program to generate the expected frequencies for the annual income data. (Hint: Use the method given to determine the expected frequencies utilizing the normal curve.)

6. Given that

$$\int_0^h y^k \, dy = \frac{h^{k+1}}{k+1} \qquad k = 0, 1, 2, \ldots$$

Write a computer program that will evaluate the first four moments about the origin for the rectangular distribution. Using these moments, determine the moments about the mean and the standard moments for the distribution. Note that the rectangular distribution has the form

$$f(Y) = \frac{1}{h} \qquad 0 \le Y \le h$$

and thus

$$\mu_k = \int_0^h y^k \frac{1}{h} \, dy = \frac{1}{h} \int_0^h y^k \, dy$$

Use the program to verify the moments given in the term for the special rectangular with h = 2.

7. Using the Pearson type III model (A>0) for the annual income distribution, find the expected frequencies and ordinates using the parameter values determined in the example, and compare the resulting values with the observed frequencies and histogram.

CHAPTER 9

FUNDAMENTALS OF SAMPLING THEORY

9.1 Introduction

Basic to all statistical inferences and the decisions based upon them is the uncertainty introduced by the use of a sample instead of the entire population of interest. For example, in experimentation, where the population of observations might be infinite, man's inability to observe "all nature" is obvious. In the social and behavioral sciences or other applications involving a finite population, the large size of these finite populations still dictates that samples be taken from the population. We need to have an approach which ensures that the sample is representative of the population but at the same time uses an economically feasible subset of it.

Although there are several mechanisms that can be used, we shall assume that randomized selection is used; the sampling theory in this chapter is based upon this selection mechanism. In fact, randomization is an underlying assumption in most experimental designs and sample survey approaches. This randomized selection technique ensures that any of the possible samples are equally likely to occur so that we do not play favorites by arbitrarily excluding or making it harder for some observations to be included in our sample. Since our approach in this text is to learn by doing, it is hoped that the student will come to appreciate the concept of random selection by studying examples. However, it is believed that students inherently appreciate the concept of randomization since it is encountered in the selection by lot for a game, the shuffling of the deck of cards, including the cut, and the shaking of dice before the throw.

9.2 Sampling Distribution of the Mean

Let us generate a simplified problem to illustrate certain fundamental aspects of sampling theory. Suppose that the population of interest consists of five men and we are interested in the average weight of the five men; the variable of interest Y is weight.

If we had no constraints, we could weigh each individual and compute the mean of the total population. Let us first do this by taking an actual population of observations and determining the first four numerical characteristics, which we shall also need later:

The Population — Weights of Five Men
(measured to the nearest pound)

i	Y	Numerical Characteristics of the Population		
1	132	μ_y	=	170.4000
2	177	σ_y	=	23.4231
3	201	$\alpha_{3:y}$	=	-0.4228
4	183	$\alpha_{4:y}$	=	2.0867
5	159			

Note that we use Greek letters because we are dealing with population characteristics, which are thus <u>parameters</u> of the population. The parameter of current interest is the mean, μ = 170.4 lb, although we shall assume that it is unknown to us.

We shall assume that we have only three pennies and we must use a penny scale that can weigh only one individual at a time. Our experiment or sample will therefore be limited to three observations out of five. The obvious estimate of μ, the average population weight, is the mean of the sample $\bar{y}$. There are other possibilities, such as the median (the central value), the midrange (the average of the smallest and largest observation), or the smallest observation +50 lb.

In fact, through inspiration any number of possible estimators could be imagined. One of the problems of theoretical statistics is to evolve good or even "best" estimators.
This problem becomes more difficult when the parameter of
interest changes from the mean to some other characteristic
of the population. We shall ignore such problems at present
and agree that we shall use the sample mean $\bar{y}$.

The number of possible different samples of size 3 that
could be obtained from a population of 5 is the number of
ways of selecting three objects from a set of five objects,
$C_3^5 = 5!/(3!2!) = 10$. We can enumerate the 10 possible samples and determine the mean $\bar{y}$ of each.

Sample Number	Identification	Observations	Mean $(\bar{y})$
j	(i_1, i_2, i_3)		
1	(1,2,3)	132, 177, 201	170
2	(1,2,4)	132, 177, 183	164
3	(1,2,5)	132, 177, 159	156
4	(1,3,4)	132, 201, 183	172
5	(1,3,5)	132, 201, 159	164
6	(1,4,5)	132, 183, 159	158
7	(2,3,4)	177, 201, 183	187
8	(2,3,5)	177, 201, 159	179
9	(2,4,5)	177, 183, 159	173
10	(3,4,5)	201, 183, 159	181

Fig. 9.1 All possible samples of size 3 from a population of size 5.

From this enumeration we see that $\bar{y}$ is a variable whose
value depends on the particular sample that is taken. Thus
a person who "happens" to get the first sample would pro-
claim the estimate of the mean to be $\bar{y} = 170$ lb. Persons
who obtain sample 2 will contest and say that the estimate
is too high and give $\mu = \bar{y}_2 = 164$ lb as their estimate.
Those with the third sample would laugh and say both of
these are too high because the mean is about 156 lb. Of
course, since we know that the population means is 170.4 lb,
we realize that the persons who obtain sample 1 were lucky

to be so close, but they would not know that they were any closer than the third group with their low estimate of 156 lb.

Because indivdual sample estimates usually differ from the parameter value, one has to compromise between the desire to make a definitive statement and the desire to make a correct statement. For example, a very definitive statement about μ would be "The population mean is equal to $\bar{y}$ (the sample mean observed), and I don't believe there is any estimation error." Just looking at the set of $\bar{y}$'s that we enumerated above would lead us to conclude that though the statement is very definitive (admits but a single value), it would never be true. On the other hand, one may choose to say: "The mean is around $\bar{y}$, but I may be in error by as much as 15 lb one way or the other ($\mu = \bar{y} \pm 15$)." Such an estimate is not very definite or useful since we could probably have simply guessed the answer, without wasting the 3 cents; but at least the statement would always be correct regardless of the set of three observations obtained in the sample. Something between these two extremes is required; we want to examine the underlying nature of the problem so that reasonable inferences about the population parameter can be made from one sample.

Let us examine the distribution of all possible sample means ($\bar{y}$'s) that can be obtained. The students should appreciate that if random sampling techniques are used, each of the 10 different samples are equally likely. From our work in descriptive statistics we know that one method of describing the distribution of $\bar{y}$'s is by determining its numerical characteristics. We are in a fortunate position because we have enumerated all possible samples and thus can find the numerical characteristics of the 10 sample means. The characteristics of the variable $\bar{y}$ are

$$
\begin{aligned}
\mu_{\bar{y}} &= 170.40000 \\
\sigma_{\bar{y}} &= 9.56242 \\
\alpha_{3:\bar{y}} &= +0.1151 \\
\alpha_{4:\bar{y}} &= 1.9623
\end{aligned}
$$

We note immediately that the distribution of $\bar{y}$'s has
the desirable property that its mean is equal to the parameter being estimated. This will always be the case in random sampling when using the sample mean to estimate the population mean. Therefore the <u>mean</u> of the sample <u>means</u> is the
<u>mean</u> of the population, and the sample mean is thus said to
be an unbiased estimator of the population mean. How do the
other characteristics of the sample means relate to the
characteristics of the distribution of the population? Log-
ically there must be some relationship since the $\bar{y}$'s were
generated from the y's.

Mathematically we can state the following proposition
in which the population size is denoted by N and the sample
size is denoted by n.

Proposition 1

The distribution of all possible sample means from a
finite population has the following characteristics:

$$\mu_{\bar{y}} = \mu_y$$

$$\sigma_{\bar{y}} = \sqrt{\frac{N - n}{n(N - 1)}} \; \sigma_y$$

$$\alpha_{3:\bar{y}} = \frac{N - 2n}{N - 2} \sqrt{\frac{N - 1}{n(N - n)}} \; \alpha_{3:y}$$

$$\alpha_{4:\bar{y}} = \frac{N - 1}{n(N - n)(N - 2)(N - 3)} \cdot$$

$$\{(N^2 - 6nN + N + 6n^2)\alpha_{4:y} + 3N(n - 1)(N - n - 1)\}$$

Since we have not gone through the algebra to establish the
above relationships, let us at least verify them for our example, where N = 5 and n = 3.

We already have noted that

$$\mu_{\bar{y}} = \mu_y = 170.4$$

Consider next the relationships

$$\sigma_{\bar{y}} = \sqrt{\frac{N-n}{n(N-1)}} \; \sigma_{\bar{y}}$$

$$\sigma_{\bar{y}} = \sqrt{\frac{5-3}{3(5-1)}} \; (23.42306) \quad = \sqrt{\frac{2}{12}} \; (23.42306)$$

$$= 9.562422$$

$$\alpha_{3:\bar{y}} = \frac{N-2n}{N-2} \sqrt{\frac{N-1}{n(N-n)}} \; \alpha_{3:y} \quad = \frac{5-6}{3} \sqrt{\frac{4}{3.2}} \; (0.4228)$$

$$= + \; 0.1151$$

$$\alpha_{4:\bar{y}} = \frac{N-1}{n(N-n)(N-2)(N-3)} \cdot$$

$$\{(N^2 - 6nN + N + 6n)\alpha_{4:y} + 3N(n-1)(N-n-1)\}$$

$$= 1.9623$$

Thus the truth of the relationships between the characteristics of the distribution of y and the sampling distribution of $\bar{y}$ are at least substantiated by our illustrative example.

In many situations the population size is very large as compared to the sample. In fact, as we have already noted in scientific experimentation, the population is almost always infinite. If N is infinite, the formulas can be simplified by considering the limit of the relationships as N becomes infinite.

Proposition 2

As the population size approaches infinity, the following relationships hold:

$$\text{(a)} \quad \lim_{N\to\infty} \mu_{\bar{y}} = \mu_{\bar{y}} = \mu_y$$

since N is not present, as N gets larger and larger the equality will remain true.

$$(b) \quad \lim_{N \to \infty} \sigma_{\bar{y}} = \lim_{N \to \infty} \sqrt{\frac{N - n}{n(N - 1)}} \; \sigma_y$$

$$= \lim_{N \to \infty} \sqrt{\frac{1 - n/N}{n - n/N}} \; \sigma_y$$

$$= \frac{\sigma_y}{\sqrt{n}}$$

since for fixed n,

$$\lim_{N \to \infty} \frac{n}{N} = 0$$

Similarly, it can be shown that

$$(c) \quad \lim_{N \to \infty} \alpha_{3:\bar{y}} = \frac{\alpha_{3:y}}{\sqrt{n}}$$

and

$$(d) \quad \lim_{N \to \infty} \alpha_{4:\bar{y}} = 3 + \frac{\alpha_{4:y} - 3}{n}$$

In addition to the numerical characteristics we would like to have a mathematical model to describe the distribution of the sample means $\bar{y}$. We recognize that the characteristics of the distribution of a variable help us to determine the nature of the mathematical model that can be used to describe the distribution.

Consider the characteristics of the distribution of the sample means when the observations y are drawn from a population having a normal distribution with mean μ and standard deviation σ. In this case the population is infinite; thus we shall use the relationships given by Proposition 2.

The mean and standard deviation of the sample means $\bar{y}$ are

$$\mu_{\bar{y}} = \mu_y \qquad \sigma_{\bar{y}} = \frac{\sigma_y}{\sqrt{n}}$$

Since $\alpha_{3:y} = 0$ and $\alpha_{4:y} = 3$ when the distribution of y is normal, the shape parameters for $\bar{y}$ are:

$$\alpha_{3:\bar{y}} = \frac{\alpha_{3:y}}{\sqrt{n}} = 0$$

and

$$\alpha_{4:\bar{y}} = 3 + \frac{\alpha_{4:y} - 3}{n} = 3$$

Therefore the shape parameters of the distribution of the sample means $\bar{y}$ have the same values as those associated with the normal distribution. This leads us to appreciate proposition 3:

Proposition 3

If random samples are drawn from a normally distributed population with mean μ and standard deviation σ, the distribution of the sample means $\bar{y}$ will also be normally distributed with mean μ and standard deviation $\sigma/\sqrt{n}$.

This proposition does not help us in determining a model for the distribution of $\bar{y}$ if the population is not normally distributed. To consider this question, let us now consider what happens as the sample size n is allowed to increase toward infinity. In particular, consider the shape parameters $\alpha_{3:\bar{y}}$ and $\alpha_{4:\bar{y}}$:

$$\lim_{n \to \infty} \alpha_{3:\bar{y}} = \lim_{n \to \infty} \frac{\alpha_{3:y}}{\sqrt{n}} = 0$$

and

$$\lim_{n \to \infty} \alpha_{4:\bar{y}} = \lim_{n \to \infty} \left[3 + \frac{\alpha_{4:y} - 3}{n} \right] = 3$$

As the size of sample n becomes large, $\alpha_{3:\bar{y}} \to 0$ and $\alpha_{4:\bar{y}} \to 3$. This behavior is referred to in theoretical statistics as the <u>central limit theorem</u>, which states that the distribution of $\bar{y}$ tends to normality as the sample size increases under rather general conditions on the distribution of the underlying population from which the sample is drawn. It is this property that is most useful in making meaningful statements relative to the use of $\bar{y}$ as an estimate of μ.

The central limit theorem

If random samples are drawn from a population with finite mean μ and finite standard deviation σ, then as n increases the distribution of the sample means $\bar{y}$ tends toward normal distribution with mean μ and standard deviation $\sigma/\sqrt{n}$.

Before we apply these propositions to the inference problems associated with parameter estimation, let us use the computer to help to confirm the truth of the above propositions. The student should appreciate that each of the above propositions can be rigorously established by the use of mathematics. We prefer, however, to use the computer, although it cannot really <u>prove</u> such propositions.

9.3 FORTRAN SUBROUTINES

We want to use the computer to generate the C_n^N sample means from a population of size N. We also want to check the validity of the relationships given by Proposition 1, using the generated distribution of sample means. The flow-chart that indicates the necessary steps is given in Fig. 9.2.

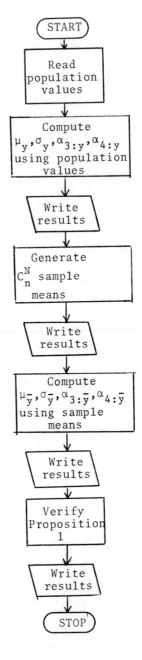

Fig. 9.2 A general flowchart for a sampling
distribution program.

We notice that the four characteristics are calculated once using the population values and once using the sample means. The FORTRAN statements for each calculation would be identical except that different variable names would be used. In addition, we have already written programs to perform such calculations and would prefer not to keypunch a new set of statements. Since we frequently encounter such situations in writing computer programs, the concept of the FORTRAN SUBROUTINE subprogram has been developed.

A SUBROUTINE subprogram, in many respects, is the same as a FUNCTION subprogram. The main difference is that a subroutine can be used to return more than one value to the calling program, while the function can return only one value, which is designated by the function name. In fact, subroutine can even return an array of values to the calling program. Since we need a subprogram that will compute the four numerical characteristics of a distribution, let us demonstrate the concept of a subroutine by writing one using the same logic as was used in Program 6.4. We shall then discuss the programming problems associated with writing and using such SUBROUTINES.

```
      SUBROUTINE CHAR(Y,N,YMU,YSIGMA,ALPHA3, ALPHA4)
      DIMENSION Y(1), SUM(4)
      AN=N
      DO 1  K=1,4
    1 SUM(K)=Ø
      DO 2  I=1, N
    2 SUM(1) = SUM(1) + Y(I)
      YMU = SUM(1)/AN
      DO 3  I=1,N
      YDEV = Y(I) - YMU
      DO 3  K=2,4
    3 SUM(K) = YDEV **K
      YSIGMA = SQRT(SUM(2)/AN)
      ALPHA3 = SUM(3)/(AN*YSIGMA**3)
      ALPHA4 = SUM(4)/(AN*YSIGMA**4)
      RETURN
      END
```

Program 9.1 CHAR SUBROUTINE — calculation of
four characteristics.

You will note that a subroutine subprogram must begin with the program statement SUBROUTINE. This is followed by the subroutine name, which must have between one and six characters. The inputs and outputs are designated within the parentheses. The words used to represent these inputs and outputs can be considered as dummy names, as in the case of a function program, because when the subroutine is "called," the symbols used in the calling statement are used in place of the dummy variable names. If an array is to be passed to a subroutine, only the array name needs to be designated in the subroutine statement. In this case the array must be dimensioned in the calling program, and the dimension used in the subroutine must be less than or equal to the dimension given in the calling program and can be a token (1). If an array is used only by the subroutine, it must be given the full dimension in the subroutine and does not need to be dimensioned in the calling program.

In SUBROUTINE CHAR, the input arguments are the real array $Y(I)$ and the integer variable N, and the output arguments are YMU, YSIGMA, ALPHA3, ALPHA4, which will be calculated by the subroutine.

A RETURN statement followed by an END statement is required at the end of each subroutine. FORTRAN subroutine decks must follow the main program deck if the subroutine is to be used in the main program.

To show the use of a subroutine we could write the following program that reads in an array, uses the subroutine to compute the four numerical characteristics, and then prints out these characteristics.

```
C     MAIN PROGRAM
      DIMENSION A(1ØØ)
      READ(5,1) NY
  1   FORMAT (I3)
      READ(5,2) (A(I), I=1,NY)
  2   FORMAT(F1Ø.2)
      CALL  CHAR(A,NY,YMU,YSIG,A3,A4)
      WRITE(6,3) YMU,YSIG,A3,A4
  3   FORMAT (1HØ,1ØX,4HYMU=F1Ø.2/1ØX,5HYSIG=,F1Ø.2,/1ØX,
     18HALPHA3=,F1Ø.4/1ØX,7HALPHA4=,F1Ø.4)
      STOP
      END

{Subroutine Deck}
{data}
```

226

A subroutine is called simply by using the word CALL followed by the name of the subroutine. The variables names that are to be passed to the subroutine or returned to the main program are specified in the call statement. For example, the real array A and the integer variable NY are passed to the subroutine, while the real variables YMU, YSIG A3 and A4 will contain the appropriate values after the subroutine is executed. Of course the order is important, so one must know what the names used in the subprograms represent. The variable names used in the CALL statement must have the same mode (integer or real) as the dummy names.

9.4 A Sampling Distribution Program

We have already written the subroutine to calculate the four characteristics of any array of values. This subroutine is used twice according to the flowchart given in Fig. 9.1, once to compute μ_y, σ_y, $\alpha_{3:y}$, and $\alpha_{4:y}$ for the population values and once to calculate $\mu_{\bar{y}}$, $\sigma_{\bar{y}}$, $\alpha_{3:\bar{y}}$, and $\alpha_{4:\bar{y}}$ for the C_n^N sample means.

We also need to generate the C_n^N sample means; we choose to use another subroutine for this purpose. It is easier to write a separate subroutine for each sample size n than to write a general routine for all values of n. We shall write a SUBROUTINE for n = 3 (sample of size 3) and call it SAMPT3. We use nested DO loop to generate the samples.

```
SUBROUTINE SAMPT3(Y,NPOP,YMEAN)
DIMENSION Y(1), YMEAN(1)
L=∅
N2=NPOP-2
N1=NPOP-1
DO 1    I=1,N2
JJ=I+1
DO 1   J=JJ,N1
KK=J+1
DO 1   K=KK,NPOP
L=L+1
1 YMEAN(L)=(Y(I)+Y(J)+Y(K))/3.
RETURN
END
```

Program 9.2 SAMPT3 SUBROUTINE — generation of all samples of size 3.

We note that the above subroutine requires that the array of population values and the array of sample means must be dimensioned in the main program. In addition, we should note that the end values of the DO loops must be carefully defined so that a DO loop is not entered when the initial value is already greater than the ending value, as for example, DO 1 I=7,6. In such cases the loop will be executed once, and one cannot be cerain as to just what values will be processed within the loop.

The main program can now be written:

```
C       SAMPLEDIST - A SAMPLING DISTRIBUTION PROGRAM
        DIMENSION Y(2Ø), YMEAN(114Ø)
        READ(5,1) NPOP
      1 FORMAT(I2)
        ANPOP = NPOP
        READ(5,3) (Y(I),I=1,NPOP)
      3 FORMAT( F1Ø.2)
        CALL SAMPT3 (Y,NPOP,YMEAN)
        CALL CHAR(Y,NPOP,YMU,YSIG,ALPHA3,ALPHA4)
        WRITE(6,5) (Y(I),I=1,NPOP)
      5 FORMAT(1H1,1ØX,17HPOPULATION VALUES/(1X,6(F1Ø.2,2X)))
        WRITE(6,6) YMU,YSIG,ALPHA3,ALPHA4
      6 FORMAT(1HØ,4HYMU=,F1Ø.2/1X,5HYSIG=,F1Ø.2/
      1   1X,7HALPHA3=,F1Ø.4/ 1X,7HALPHA4=,F1Ø.4)
C       SAMPLE SIZE IS DENOTED BY SN AND MUST BE CHANGED IF
C          SUBROUTINE SAMPT3 IS CHANGED
        SN = 3.
        N= FACT(NPOP)/(FACT(NPOP-3)*FACT(3)) +.5
        CALL CHAR(YMEAN,N,YMBAR,YMSTD,AM3,AM4)
        WRITE(6,7) (YMEAN(J),J=1,N)
      7 FORMAT(1HØ,///,1ØX,12HSAMPLE MEANS/(1X,6(F1Ø.2,2X)))
        WRITE(6,8) YMBAR,YMSTD,AM3,AM4
      8 FORMAT(1HØ,6HYMBAR=,F1Ø.2/1X,6HYMSTD=,F1Ø.2/
      1   1X,4HAM3=,F1Ø.4/ 1X,4HAM4=,F1Ø.4)
        YMCAL = YMU
        ANN = ANPOP - SN
        AN1 = ANPOP - 1.
        AN2 = ANPOP - 2.
        AN3 = ANPOP - 3.
        YSTDC = SQRT( ANN/( SN*AN1))*YSIG
        AM3CAL = (ANPOP-2.*SN)/AN2*SQRT(AN1/( SN*ANN))*ALPHA3
        AM4CAL = AN1*((ANPOP**2 -6.*SN*ANPOP+ANPOP+6.*SN**2)
      1 *ALPHA4 + (3.*ANPOP*(SN -1.)*(ANN-1.)))
      2  /( SN*ANN*AN2*AN3)
        WRITE(6,9)
      9 FORMAT(1HØ,26HCALCULATED CHARACTERISTICS)
        WRITE(6,1Ø) YMCAL,YSTDC,AM3CAL,AM4CAL
```

```
1Ø FORMAT(1HØ, 2F1Ø.2,2F1Ø.4)
   STOP
   END
```

{Subroutine SAMPT3}
{Subroutine CHAR}
{Function FACT}
{Data}

Program 9.3 SAMPEDIST — a sampling distribution program.

This main program is followed by subroutine given in Programs 9.1 and 9.2 and FUNCTION FACT given in Program 7.1. The computer verification of Proposition 1 stating the relationship of the distribution of the sample means to the distribution of the population can now be made. We shall use the population of weights (N = 5) given in Sec. 9.2. The corresponding input for the program is

$$5$$
$$132.$$
$$177.$$
$$201.$$
$$183.$$
$$159.$$

The output is:

POPULATION VALUES

132.00	177.00	201.00	183.00	159.00

YMU	=	170.40
YSIG	=	23.42
ALPHA3	=	-0.4228
ALPHA4	=	2.0565

SAMPLE MEANS

170.00	164.00	156.00	177.00	164.00	158.00
187.00	179.00	173.00	181.00		

YMBAR	=	170.40
YMSTD	=	9.56
AM3	=	0.1151
AM4	=	1.9623

229

170.40   9.56   0.1151   1.9623

Since we obtained these same values in Sec. 9.2 by
using the sampling formulas, the above computer program has
enabled us to verify Proposition 1. The student can now use
the program to verify Proposition 1 for other population
values. The same program can also be used to generate samples of different sizes (n) from any population by simply
changing the input data and the SAMPT3 routine.

9.5 The Limiting Sampling Distributions of the Mean

We now turn our attention to the verification of the
four limiting relationships for $\mu_{\bar{y}}$, $\sigma_{\bar{y}}$, $\alpha_{3:\bar{y}}$, and $\alpha_{4:\bar{y}}$ as
$N \to \infty$ given in Proposition 2. We can readily write a pro-
gram to determine the exact values given by the formulas in
Proposition 1 for the characteristics and the limiting values given by Proposition 2. Although the writing of the
program is given as one of the exercises, we can consider
two cases.

In the first case, samples of size 25 (n = 25) are
taken from a population of size N = 50, and in the second
case samples of the same size (n = 25) are taken from a population of size N = 500. We can use Propositions 1 and 2 to
evaluate the standard deviation of the sample mean given
by σ.

Exact	Limiting Approximation
(Proposition 1)	(Proposition 2)

$$\sigma_{\bar{y}} = \sqrt{\frac{N - n}{n(N - 1)}}\ \sigma_y \qquad\qquad \sigma_{\bar{y}} = \frac{\sigma_y}{\sqrt{n}}$$

$$N = 50 \quad \sigma_{\bar{y}} = \sqrt{\frac{50 - 25}{25(49)}}\ \sigma_y = \frac{\sigma_y}{\sqrt{49}} \qquad\qquad \sigma_{\bar{y}} = \frac{\sigma_y}{\sqrt{25}}$$

$$N = 500 \quad \sigma_{\bar{y}} = \sqrt{\frac{500 - 25}{25(499)}}\ \sigma_y = \frac{\sigma_y}{\sqrt{26.20}} \qquad\qquad \sigma_{\bar{y}} = \frac{\sigma_y}{\sqrt{25}}$$

We can see that when N = 50 the exact formula is needed because the approximation is not good. When N = 500, however, there is little difference in the two results.

We choose to verify the central limit theorem before considering Proposition 3. Recall that this theorem tells us that as n (the sample size) becomes larger, the distribution of the sample means ($\bar{y}$) from any population approaches normality as long as the underlying population distribution from which the sample was drawn has a finite mean and standard deviation.

We shall use the computer to verify the theorem using the uniform 0-1 distribution as the underlying nonnormal population distribution. Random observations from the 0-1 distribution are generated by computer subprogram SUBROUTINE URANDN, given in Appendix B. Program 9.4, given at the end of the chapter, calculates sample means using the subroutine to generate the random observations from the uniform distribution. We can then examine the effect of the sample size on the distribution of sample means by taking n = 2, 10, and 50.

To check on the "fit" of the normal distribution model for the sample means we can use any of the three approaches introduced in the previous chapters: (1) the determination of the numerical characteristics, (2) the superimposition of the graph of the model on the observed histogram using Program 8.2 (EVALF), and (3) the determination of the expected frequencies for each interval of the observed frequency distribution using Program 8.6 (NORMFREQ). Since the total number of sample means that are generated for a particular value of n should be greater than 50 (to obtain an adequate observed frequency distribution), we shall find it advantageous to have a computer subprogram that tabulates the sample means into a frequency distribution as they are generated; the subprogram is given in Program 9.5 at the end of the chapter.

Program 9.4 and the appropriate subprograms were run using m = 200 sample means. Each set of 200 sample means used a different sample size: n = 1, 2, 10, 50.

231

Numerical Characteristic of 200 Sample Means

	n = 1	n = 2	n = 10	n = 50
Mean	0.46	0.50	0.51	0.50
Standard deviation	0.30	0.20	0.09	0.04
Skewness	0.1813	-0.0352	0.1253	-0.0585
Kurtosis	1.8249	2.4020	3.2234	2.9308

Frequency Distribution

	n = 1	n = 2	n = 10	n = 50
0 - 0.1	26	3	0	0
0.1 - 0.2	28	14	0	0
0.2 - 0.3	25	23	0	0
0.3 - 0.4	11	21	25	4
0.4 - 0.5	23	37	57	93
0.5 - 0.6	19	35	88	102
0.6 - 0.7	20	29	26	1
0.7 - 0.8	12	25	4	0
0.8 - 0.9	17	9	0	0
0.9 - 1.0	19	4	0	0

We note that when n = 50, the effect of the central limit theorem is apparent since a_3 is close to zero and a_4 is close to 3.

To graphically demonstrate the effectiveness of the central limit theorem as the sample size n increases, we show the histograms of the generated observations from the uniform distribution (n = 1) and the three distributions of $\bar{y}$ for n = 2, 10, and 50. To better illustrate the distribution for n = 50, we shall use smaller intervals than those given in the above frequency table.

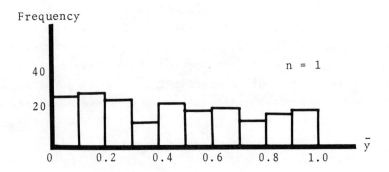

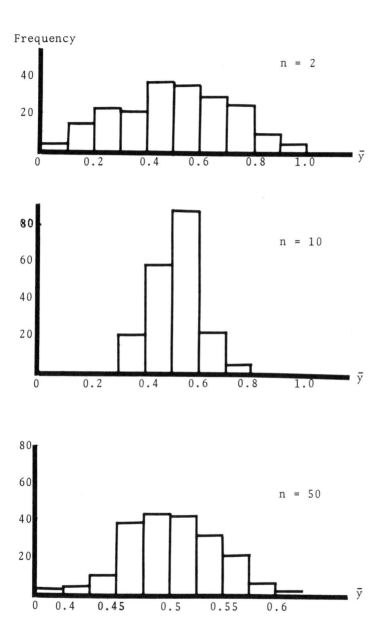

Fig. 9.3 Sampling distribution of the mean when the underlying population distribution is uniform.

We now turn our attention to Proposition 3, which states that the distribution of sample means from a normal distribution will also be normal. To verify the proposition we need a subprogram that will select a random observation from any specified normal distribution. The normal generator subprogram RANNOR is given at the end of the chapter (Program 9.6). Note that the technique to generate random normal numbers is based on the central limit theorem. To check on normality of the sampling distribution we shall graph the histograms for means generated by the computer for sample sizes of n = 1, 2, 5, and 10 and calculate the four characteristics. The student should notice the normality exhibited for all values of n.

Numerical Characteristics of 200 Sample Means

	n = 1	n = 2	n = 5	n = 10
Mean	-0.15	0.94	0.02	0.01
Standard deviation	1.09	0.73	0.49	0.29
Skewness	-0.0390	-0.1893	-0.0262	-0.0453
Kurtosis	3.1979	3.1181	3.3941	2.5883

Frequency Distribution

Interval	n = 1	n = 2	n = 5	n = 10
-3.0 - -2.6	3	0	0	0
-2.6 - -2.2	3	1	0	0
-2.2 - -1.8	8	1	0	0
-1.8 - -1.4	10	5	1	0
-1.4 - -1.0	19	11	3	0
-1.0 - -0.6	25	15	17	4
-0.6 - -0.2	22	35	45	46
-0.2 - 0.2	35	54	64	95
0.2 - 0.6	21	38	54	52
0.6 - 1.0	26	18	12	3
1.0 - 1.4	15	16	3	0
1.4 - 1.8	6	5	1	0
1.8 - 2.2	3	1	0	0
2.2 - 2.6	2	0	0	0
2.6 - 3.0	2	0	0	0

As can be seen from the four histograms in Fig. 9.4, the distribution of sample means for small and large values of n reflects the normal distribution from which the samples are selected.

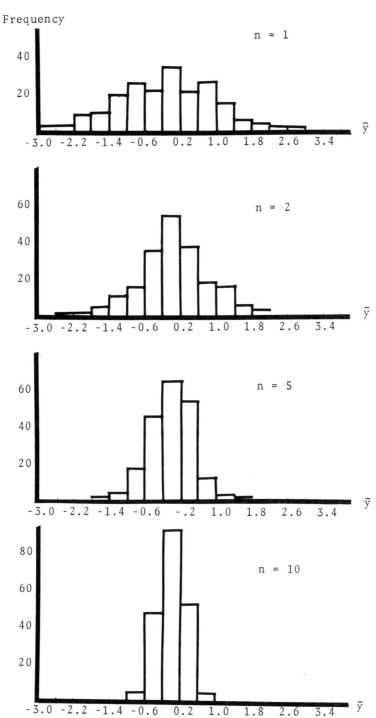

Fig. 9.4 Sampling distribution of the mean when the
underlying population distribution is normal.

9.6 The Use of Sampling Distributions

We can now appreciate that if the population is close to being normal, or if the sample size is large (say n > 30), the statistic $\bar{y}$ obtained by random sampling is approximately normally distributed, with mean equal to μ_y and standard deviation equal to

$$\sigma_{\bar{y}} = \frac{\sigma_y}{\sqrt{n}}$$

This means that our observed $\bar{y}$ will randomly occur according to the probabilities given by the normal curve.

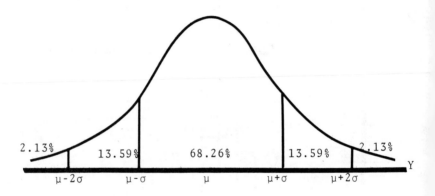

Fig. 9.5 The normal curve.

That is, about 68 percent of the time, $\bar{y}$ will fall within one $\sigma_{\bar{y}}$ of μ, 95 percent of the time $\bar{y}$ will fall within $2\sigma_{\bar{y}}$ of μ, and $\bar{y}$ is almost certain (99.7 percent of the time) to fall within $3\sigma_{\bar{y}}$. How do we make use of these probabilities? Let us consider three types of sampling problems by using the reading-time example encountered previously.

1. Tests of hypothesis

A new recruiting program has been instituted, and the first group of 50 new employees has arrived for training. The production manager is concerned as to whether or not the new program will attract new employees as good as those attracted by the old program (as measured by reading time).

236

He has had enough experience under the old program to "know" that the distribution of reading times was essentially normal with μ_y = 2.52 min and σ_y = 0.67 min.

To solve this problem, the production manager (more probably the research analyst on his staff) sets up two hypotheses: a null and an alternative hypothesis. The null hypothesis is that the new program generates the same type of employee as the old one did. The alternative hypothesis is that the new system generates individuals whose average reading time is greater than that realized by the old method.

If the null hypothesis is true, then the sample mean $\bar{y}$ of new employee reading times should be close to μ_y = 2.52, and under the alternative hypothesis the sample mean should be larger. To test this hypothesis a sample of 50 new employees is selected at random; its mean is $\bar{y}$ = 2.64 min. Does this large $\bar{y}$ lead one to conclude that the new method recruits people who are inferior to those obtained under the older practice? Is the observed sample mean $\bar{y}$ = 2.64 too high to make the null hypothesis tenable? To answer this we determine the probability of obtaining an observed sample mean as large or larger than that observed ($\bar{y}$ = 2.64) under the null hypothesis that μ = 2.52. If this probability is very small, we doubt that our null hypothesis is really true and discard it for the more feasible (alternative) hypothesis.

To determine this probability the distribution of $\bar{y}$ must be found. We assume that distribution of y is normal with mean μ_y and standard deviation σ_y; $\bar{y} \sim N(\mu_y, \sigma_{\bar{y}})$, where

$$\sigma_{\bar{y}} = \frac{\sigma_y}{\sqrt{n}} = \frac{0.63}{\sqrt{50}} = 0.096$$

Under the null hypothesis H : μ_y = 2.52 and the distribution of the $\bar{y}$ looks like

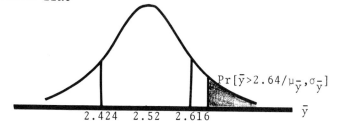

$$Pr[\bar{y} > 2.64 / \mu_{\bar{y}}, \sigma_{\bar{y}}]$$

2.424   2.52   2.616 $\bar{y}$

The alternative hypothesis states that $H_a:\mu_y > 2.52$. The production manager is considering only the "one-sided" alternative hypotheses, that the new program is inferior to the old one. Thus $H_a:\mu_y > 2.52$. The shaded area of the right tail corresponds to the probability of obtaining $\bar{y} > 2.64$ under the null hypothesis, that is given that $\mu_y = 2.52$ and $\sigma_{\bar{y}} = 0.096$. To find the probability we must transform the $\bar{y}$ value to the standard z value by

$$Pr[\bar{y} > 2.64/\mu_y, \sigma_{\bar{y}}] = Pr\left[\frac{\bar{y} - \mu_y}{\sigma_{\bar{y}}} > \frac{2.64 - 2.52}{0.096}\right]$$

$$= Pr[Z > 1.24] = 1 - AREANORM(1.24) = 0.1075$$

The manager is faced with the fact that there are about 11 chances in 100 that the new program is operating as well as the old one (relative to distribution of the reading time of new employees). Should he elect to discard the null hypothesis or not? Classical statistical methods generally require the probability to be less than 0.05 or even 0.01 before the null hypothesis is discarded. Thus if the manager follows the 0.05 convention, he would not worry about the new recruiting program.

The problem just illustrated relates to the testing of a hypothesis. We note that the manager had a hypothesis about the quality of the new employees obtained under the new system, namely, that they were as good as the old ones, and used the sample (experiment) to test the hypothesis. To emphasize the methodology let us formalize the procedure followed in testing hypotheses.

A hypothesis test is usually formulated in terms of a parameter of a theoretical distribution. The null hypothesis is that the parameter is equal to a specific value. For example, when the population mean is being tested the null hypothesis is

$$H_o: \quad \mu = \mu_o$$

The alternative hypothesis, also stated in terms of the parameter, could be any one of the following:

$$H_{a1}: \quad \mu > \mu_o$$
$$H_{a2}: \quad \mu < \mu_o$$
$$H_{a3}: \quad \mu \neq \mu_o$$

The first two alternative hypotheses are one-sided; the last one is said to be two-sided.

A null hypotheses is tested using a sample statistic that in our case is the sample mean $\bar{y}$ whose distribution is known when the null hypothesis is true. Therefore we shall assume that $\bar{y}$ has a normal distribution with mean μ and standard deviation $\sigma_{\bar{y}}$. Although $\bar{y}$ can be used as the test statistic, as illustrated by the previous example, it is simplier to use the standardized z values given by

$$z = \frac{\bar{y} - \mu}{\sigma_{\bar{y}}}$$

in making the test. The distribution of z is also known since μ and $\sigma_{\bar{y}}$ are given, and thus under the null hypothesis $Z \sim N(0,1)$.

The probability of rejecting the null hypothesis when it is true must also be given. This probability is usually denoted by α and is called the level of significance. The standard values of α are $\alpha = 0.05$ and $\alpha = 0.01$.

The probability α and the alternative hypothesis determine the set of z's for which the null hypothesis is rejected. This set of values is called the critical region and is found by determining the value of z that corresponds to the appropriate area under the normal curve α. This value of z denoted by z_α is called the critical value for the level of significance α. The illustration below gives the critical region and value for each of the alternative hypotheses.

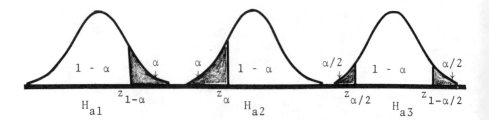

Using the 5 percent level of significance as an illustration ($\alpha = 0.05$), the critical value of z for each alternative is determined by

$$H_{a1}: Pr(Z > z_{1-\alpha}) = \alpha = 0.05 \qquad\qquad z_{1-\alpha} = 1.645$$

$$H_{a2}: Pr(Z < z_{\alpha}) \quad = \alpha = 0.05 \qquad\qquad z_{\alpha} = -1.645$$

$$H_{a3}: Pr(Z < z_{\alpha/2}) = Pr(Z > z_{1-\alpha/2}) = \frac{\alpha}{2} \qquad z_{\alpha/2} = -1.96$$

$$= 0.025 \qquad z_{1-\alpha/2} = 1.96$$

Recall that because of the symmetry of the normal distribution, $z_{\alpha} = -z_{1-\alpha}$.

Thus we can summarize the steps to be followed in testing a hypothesis about the value of a population mean. In this test we assume that $\bar{y}$ is normally distributed and σ_y is known.

(1) Identify the null hypothesis $H_o: \mu = \mu_o$.
(2) Determine the appropriate alternative hypothesis.
(3) Determine the level of significance to be used, α.
(4) Find the critical value of z, either $z_{1-\alpha}$, z_{α} or $z_{\alpha/2}$.
(5) Compute the test statistic

$$z = \frac{\bar{y} - \mu_o}{\sqrt{\frac{N-n}{n(N-1)}}\,\sigma_y} \qquad \text{if N is finite}$$

$$z = \frac{\bar{y} - \mu_o}{\sigma_y/\sqrt{n}} \qquad \text{if N is infinite}$$

240

(6) If the alternative is given by

$$H_{a1}: \mu > \mu_o \qquad \text{then reject } H_o \text{ if } z > z_{1-\alpha}$$

$$H_{a2}: \mu < \mu_o \qquad \text{then reject } H_o \text{ if } z < z_\alpha$$

$$H_{a3}: \mu \neq \mu_o \qquad \text{then reject } H_o \text{ if } z < z_{\alpha/2}$$

$$\text{or } z > z_{1-\alpha/2}$$

Using this approach to the testing of a hypothesis one would discard the null hypothesis when it is true 100α percent of the time.

We repeat the test on reading time by using the six steps:

(1) $H_o: \mu = 2.52$

(2) $H_a: \mu > 2.52$ (one-sided)

(3) $\alpha = 0.05$

(4) $z_{1-\alpha} = 1.65$

(5) $z = \dfrac{\bar{y} - \mu_o}{\sigma_y/\sqrt{n}} = \dfrac{2.64 - 2.52}{0.63/\sqrt{50}} = 1.24$

(6) Since $z < z_{1-\alpha}$, we do not discard the null hypothesis.

This formulation leads to the same conclusion as before since

$$\Pr[\bar{y} > 2.52/\mu_y, \sigma_{\bar{y}}] = \Pr(Z > 1.24) = 0.1072$$

and since $0.1072 > 0.05$, the hypothesis was not rejected.

Let us apply this concept to another problem. A group of 250 new employees has just completed a special training program. Their achievements have been evaluated by means of a true-false examination consisting of 50 questions. The average number of correct responses of the 250 recruits is

$\mu = 32.4$. One instructor thinks that the employees in the afternoon did better than the group as a whole and indicates this to the manager. The manager (a man of action) does not agree but performs a quick check by selecting (at random) the tests of 15 trainees from the afternoon class and computing their average grade, which is 34.7. Should the manager retain his hypothesis that the afternoon group did no better than the entire group or discard it and investigate why a difference exists?

This problem differs from the reading-time problem in two respects: (1) the population of interest is finite because it consists of only 250 trainees and (2) the variable of interest is discrete. Since our standard hypothesis procedure assumes that the distribution of $\bar{y}$ is normal, we need to determine if the normal distribution is a good enough representation of the distribution of $\bar{y}$, the average number of correct responses on the true-false examination. We therefore investigate the underlying distribution of the 250 y's by calculating the four population characteristics of the 250 observations, obtaining

$$\mu_y = 32.4$$
$$\sigma_y = 3.8$$
$$\alpha_{3:y} = -0.14$$
$$\alpha_{4:y} = 2.85$$

Since α_3 is close to zero and α_4 is close to 3, we are willing to assume that the distribution of y is normally distributed even though the variable is discrete and the population finite. However, it is the sampling distribution of $\bar{y}$ that is critical, and we should appreciate that our normal approximation will be even better for this distribution because of Proposition 1.

We can now apply the six steps:

(1) $H_o : \mu = \mu_o = 32.4$

(2) $H_a : \mu > 32.4$

(3) $\alpha = 0.10$

(4) $z_{1-\alpha} = 1.28$

(5) $z = \dfrac{\bar{y} - \mu_o}{\sqrt{(N-n)/n(N-1)}\ \sigma} = \dfrac{34.7 - 32.4}{\sqrt{(250-15)/15(250-1)}\ (3.8)}$

$$= 2.42$$

(6) Since $z > z_{1-\alpha}$, we discard the null hypothesis.

Thus the manager suspects that the afternoon classes have better results.

The instructor might simply have stated that the two groups were not the same instead of specifying that the morning class was better. The alternative hypothesis in this case would be a two-sided alternative, where the null hypothesis is rejected if the sample mean is either too low or too high. Step 4 now uses the critical value

$$z_{1-\alpha/2} = 1.65$$

and we would discard H_o if $z > 1.65$ or $z < -1.65$.

2. Confidence intervals

Let us now consider another type of problem. Suppose that the manager must learn to live with the type of recruit now being obtained. He is therefore interested in the distribution of reading times under the new program. For simplicity let us assume that σ_y has not changed but remains equal to 0.63 and the prime characteristic of interest is μ_y, the mean reading time.

A group of 100 (a random sample) new employees is tested to estimate the average reading time under the new program. If the manager wants a "point" estimate for μ, the sample mean $\bar{y} = 2.85$ is obviously a good estimator to use. However, μ can also be estimated by an interval $\bar{y} - \varepsilon < \mu < \bar{y} + \varepsilon$. That is, we could say

243

$$\mu = 2.85 \pm 2.00$$

Such a large value of ε is not useful because the manager probably already knows that μ must be somewhere between 0.85 and 4.85. On the other hand, if ε is very small (which thus makes the interval narrow), we are almost certain to be wrong if we assume that μ falls within the prescribed interval.

To determine a useful value of ε, we again assume that $\bar{y}$ is distributed normally with mean μ and standard deviation $\sigma_{\bar{y}}$. We can then use the following probability statement:

$$\Pr\left[-z_{1-\alpha/2} < \frac{\bar{y} - \mu}{\sigma_{\bar{y}}} < z_{1-\alpha/2}\right] = 1 - \alpha$$

to construct the confidence interval for μ

$$\Pr\left[\bar{y} - z_{1-\alpha/2}\ \sigma_{\bar{y}} < \mu < \bar{y} + z_{1-\alpha/2}\ \sigma_{\bar{y}}\right] = 1 - \alpha$$

This confidence interval can be written as

$$\left[\bar{y} - z_{1-\alpha/2}\ \sigma_{\bar{y}} < \mu < \bar{y} + z_{1-\alpha/2}\ \sigma_{\bar{y}}\right]$$

or as

$$\mu = \bar{y} \pm z_{1-\alpha/2}\ \sigma_{\bar{y}}$$

For an infinite population $\sigma_{\bar{y}} = \sigma_y/\sqrt{n}$, and the confidence interval becomes

$$\mu = \bar{y} \pm z_{1-\alpha/2}\ \sigma_y/\sqrt{n}$$

For a finite population $\sigma_{\bar{y}} = \sqrt{(N-n)/n(N-1)}\ \sigma_y$, and the confidence interval becomes

$$\mu = \bar{y} \pm z_{1-\alpha/2}\ \sqrt{\frac{N-n}{n(N-1)}}\ \sigma_y$$

This confidence interval is an interval estimate of μ. Furthermore, we can assert that the population mean lies within the computed interval; this assertion will be correct $100(1 - \alpha)$ percent of the time.

To find the value of $z_{1-\alpha/2}$ we use the normal tables as illustrated in Sec. 8.5.

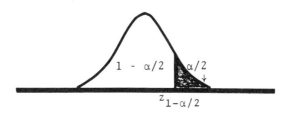

Thus we find the probability $1-\alpha/2$ in the table and read the corresponding value of z.

To illustrate the approach, let us return to the example. If we choose the 95 percent level of confidence, then the appropriate value of z is $z_{1-\alpha/2}$, where AREANORM $(z_{1-\alpha/2})$ = $1-\alpha/2$ = 0.975. Thus $z_{1-\alpha/2} = z_{.975}$ = 1.96. The 95 percent confidence interval is given by

$$\left[\bar{y} - z_{1-\alpha/2} \frac{\sigma_y}{\sqrt{n}} < \mu_y < \bar{y} + z_{1-\alpha/2} \frac{\sigma_y}{\sqrt{n}} \right]$$

$$\left[2.85 - 1.96 \frac{0.63}{\sqrt{100}} < \mu_y < 2.85 - 1.96 \frac{0.63}{\sqrt{100}} \right]$$

$$[2.73 < \mu_y < 2.97]$$

We assert with 95 percent assurance that μ lies within the computed interval, although we do not know whether this particular interval contains μ.

The above confidence interval assumes that the population is large or infinite. Consider another example when the population is finite. The manager has to plan a work schedule for 120 employees and thus needs to know the average time it would take an employee to assemble a piece of equipment. He knows from experience that σ_y = 3.81 min. A sample of 20 of the 120 employees is tested, and the average assembly time is found to be $\bar{y}$ = 16.73 min. What is the 99 percent confidence interval for the average assembly time?

In this case α = 0.01, $1-\alpha/2$ = 0.995 and $z_{.995}$ = 2.58. Thus

$$\mu_y = \bar{y} \pm z_{1-\alpha/2} \sqrt{\frac{N - n}{n(N - 1)}} \; \sigma_y$$

$$= 16.73 \pm 2.58 \sqrt{\frac{120 - 20}{20(119)}} (3.81)$$

$$= 16.73 \pm 2.015$$

3. Sample size determination

Let us present another basic application of the sampling propositions. The manager wants an estimate of μ_y such that he can be 95 percent confident that $\bar{y}$ does not differ from μ_y by more than 1.5 min. In other words, he wants the sampling error ε of the confidence interval to be 1.5 or

$$Pr[\bar{y} - 1.5 < \mu < \bar{y} + 1.5] = 0.95$$

The question is how large a sample size n should be used. From the previous considerations, we know that ε is equal to

$$\varepsilon = z_{1-\alpha/2} \sqrt{\frac{N - n}{n(N - 1)}} \; \sigma_y$$

Thus we need to solve this equation for n. The degree of confidence is reflected by the value of $z_{1-\alpha/2}$. Algebra yields

$$n = \frac{N z_{1-\alpha/2}^2 \sigma_y^2}{z_{1-\alpha/2}^2 \sigma_y^2 + (N - 1)\varepsilon^2}$$

Or if N is large,

$$n = \frac{z_{1-\alpha/2}^2 \sigma_y^2}{\varepsilon^2}$$

In our problem we have

$$n = \frac{(120)(1.96)^2(3.81)^2}{(1.96)^2(3.81)^2 + (119)(1.5)^2}$$

$$= 20.7 = 21$$

and the sampling percentage is $n/N = 21/120 = 0.175 = $ 18 percent. For example, if we double the population size to $N = 240$, we would find that

$$n = \frac{(240)(1.96)^2(3.81)^2}{(1.96)^2(3.81)^2 + (239)(1.5)^2}$$

$$= 22.55 = 23$$

and the sampling percentage is $23/240 = 0.096 = 9.6$ percent.

Thus it does not make sense to take a fixed percentage sample of the population. When $N = 210$, an 18 percent sample should be used, and when $N = 240$, a 9.6 percent sample should be used. In fact, if $N = 2500$, the sample size is

$$n = \frac{(2500)(1.96)^2(3.81)^2}{(1.96)^2(3.81)^2 + (1.5)^2(2499)}$$

$$= 24.55 = 25$$

so that a 1 percent sample would suffice.

A similar study of the effect of changing α or ε can be made. For example, for a large N the sample size n is directly proportional to $z_{1-\alpha/2}^2$ and inversely proportional to ε^2. The student can consider different values of α and ε and use the sample size formula to determine the changes in the sample size.

9.7 The Generation of Sampling Distributions and Associated Computer Programs

To study the sampling distribution of a statistic (estimator) of a population parameter obtained from random samples from an underlying distribution model, the mathematical statistician attempts to evolve the mathematical distribution of the estimator. For example, he would evolve the mathematical form of the distribution model. However, in our orientation to statistical analysis we attempt to utilize the computational power and speed of the digital computer in place of the power of mathematical analysis by generating approximations of the mathematical distribution.

One technique for generating such approximations is to simulate the sampling process by using a random number generator. A basic random number generator obtains uniform random numbers between zero and 1. Once such a generator is available we can easily generate random observations from normal and other related distributions.

We include a uniform random number generator in Appendix B (URANDN), although there are also other techniques for generating uniform random numbers on a digital computer that can be used. None of these techniques, however, will generate true random numbers because such a requirement exceeds man's mathematical capabilities, and therefore computer-generated random numbers are called psuedorandom numbers.

To call the uniform random number generator, we use

CALL URANDN (IODD,IY,Y).

IODD is a large odd number that must be defined in the program calling URANDN, IY is a random integer that is calculated by the subroutine and is used by the calling program to replace the IODD value, and Y is the uniform random number that is generated.

If the same starting value of IODD is used, the same sequence of random numbers is generated. To avoid this difficulty, different values of IODD should be used to generate different samples.

To illustrate the use of the subroutine we shall write a program that generates samples from a uniform distribution, calculates each sample mean, and tabulates the sample means into a frequency table.

The inputs that must be given are

m = M = total number of sample means to be generated

n = NSAM = number of observations to be used for each sample mean

NINT = number of intervals in frequency distribution

IODD = random starting digit (any odd integer)

```
C     UNIFORMSM - SAMPLE MEANS FROM UNIFORM DISTRIBUTION
      DIMENSION F(16), YM(3ØØ)
      READ(5,1Ø) M,NSAM,NINT,IODD
   1Ø FORMAT( 4I5)
      SAM = NSAM
      DO  1  I=1,M
      YM(I) = Ø.
      DO 2  J=1,NSAM
      CALL URANDN(IODD,IY,Y)
      IODD = IY
    2 YM(I) = YM(I) + Y
    1 YM(I) = YM(I)/SAM
      CALL FREQ( YM,M,F,Ø.,.1,NINT)
      WRITE(6,11)
   11 FORMAT(1H1,3HINT,4X,9HFREQUENCY)
      WRITE(6,12) (J,F(J),J=1,NINT)
   12 FORMAT(1HØ,I5,3X,F1Ø.Ø)
      CALL CHAR(YM,M,YMU,YSIG,A3,A4)
      WRITE(6,13)  YMU,YSIG,A3,A4
   13 FORMAT(1HØ,4HMEAN,F12.2/1H ,3HSTD,F12.2/
     11H ,2HA3,F12.4/1H ,2HA4,F12.4)
      STOP
      END
{Subroutine URANDN}
{Subroutine CHAR}
{Subroutine FREQ}
```

Program 9.4 UNIFORMSM — generation of m sample means from a uniform distribution.

The program decks for the subroutines URANDN, FREQ, and CHAR should be placed behind the main program and before the data cards. The FORTRAN statements for URANDN and CHAR have already been given; thus only FREQ, which tabulates an array into a frequency distribution, needs to be given here.

The inputs for FREQ are

y_i = Y(I) = array to be tabulated into frequency distribution

m = M = size of Y array

f_j = F(J) = array of frequencies obtained

YL = lower interval boundary of frequency distribution

w = W = interval width for frequency distribution

NINT = total number of intervals

```
SUBROUTINE FREQ(Y,M,F,YL,W,NINT)
DIMENSION Y(1), F(1)
DO 1 J=1,NINT
1 F(J) = Ø.
INT1 = NINT -1
DO 2 I=1,M
DO 3 J=1, INT1
XJ= J
EP = YL + W*XJ
IF( Y(I).GT.EP) GO TO 3
F(J) = F(J) +1.
GO TO 2
3 CONTINUE
F(NINT) = F(NINT) +1.
2 CONTINUE
RETURN
END
```

Program 9.5 FREQ SUBROUTINE — frequency tabulation.

The input data cards that were used to generate the output given in Sec. 9.5 were

200	1	10	113413
200	2	10	67949
200	10	10	14977
200	50	10	10345

We also would like to generate random numbers from a normal distribution with a specified mean μ and standard deviation σ. We can use the central limit theorem to obtain a normal random number by averaging n uniform random numbers. Let u_i be a uniform random number. Then

$$x = \sum_{i=1}^{n} u_i$$

is approximately normal distributed with mean $u = n/2$ and standard deviation $\sigma = \sqrt{n/12}$. The use of $n = 12$ as a number is based upon the power of the central limit theorem, which is already quite effective when $n = 10$ (see Sec. 9.5). In addition, when $n = 12$, $\mu_x = 6$ and $\sigma_x^2 = 1$. Thus a random normal number from a distribution with mean $\mu = 0$ and $\sigma = 1$ is found by using the transformation

$$z = \frac{(x - 6)}{1}$$

The normal random number with mean μ and standard deviation σ can be obtained from the relationship

$$y = z \cdot \sigma + \mu$$

The appropriate subroutine RANNOR is

```
      SUBROUTINE RANNOR( IODD,YN,YMU,SIGMA)
      SUM = Ø.
      DO 1 I=1,12
      CALL URANDN(IODD,IY,Y)
      IODD = IY
    1 SUM = SUM +Y
      YN = (SUM -6.) *SIGMA +YMU
      RETURN
      END
```

Program 9.6 RANNOR SUBROUTINE — a random normal number generator.

To generate a frequency table of sample means from a normal population, we simply substitute in Program 9.4

```
      CALL RANNOR( IODD,Y,YMU,SIGMA)
```

in place of

```
      CALL URANDN( IODD,IY,Y)
      IODD = IY
```

change the calling arguments of FREQ to

CALL FREQ (YM,M,F,-3., .4, NINT)

and define the values of YMU and SIGMA. The following
input data cards were used to generate the outputs given
in Sec. 9.5:

200	1	15	13613
200	2	15	14765
200	5	15	51463
200	10	15	67819

Random uniform and normal observations are generated using
specialized techniques that cannot apply to any distribution.
Since we are interested in investigating different distributions, let us investigate a more general technique to generate random observations.

If we have a discrete distribution model, say $p(Y = y)$,
$y = y_1, \ldots, y_j, \ldots, y_m$ we can determine the cumulative distribution

$$Pr(Y \leq y_j) = \sum_{y=y_1}^{y_j} p(Y=y)$$

We can then generate a uniform random number u and compare
it with each y_j until the following condition is met:

$$Pr[Y \leq y_{j-1}] \leq u < Pr[Y \leq y_j]$$

and then we assign the value y_j to our random observation.

To illustrate the procedure with a simple case, let us
use the model

$$p(Y=0) = \frac{1}{4} \quad p(Y=1) = \frac{1}{2} \quad p(Y=2) = \frac{1}{4}$$

We then generate the cumulative probabilities

$$Pr[Y \leq 0] = 0.25 \quad Pr[Y \leq 1] = 0.75 \quad Pr[Y \leq 2] = 1.00$$

252

```
If 0    ≤ u < 0.25      set y = 0
If 0.25 ≤ u < 0.75      set y = 1
Or 0.75 ≤ u < 1.00      set y = 2
```

Note the application of the equality sign since u falls between 0.000 · · · and 0.999 · · ·.

In the case of a continuous variable, the procedure
must be geared to the continuity of the underlying distribution models. Thus we have the distribution model f(Y),
which graphically is represented as

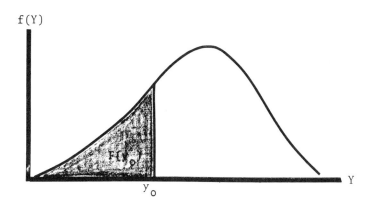

Fig. 9.6 A distribution model f(Y) and its
cumulative distribution $F(y_o)$.

As in the discrete case, we need to find the cumulative distribution $F(y_o)$, which is the area under the curve f(Y) up
to y_o, as seen in Fig. 9.7.

$$F(y_o) = \int_{-\infty}^{y_o} f(y) \, dy$$

where $0 \le F(y_o) \le 1$.
Graphically, the cumulative distribution can be represented
by

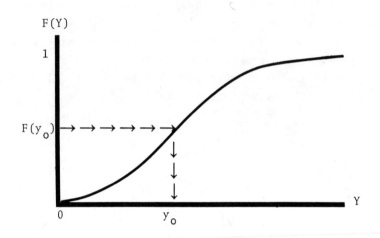

Fig. 9.7 The cumulative distribution F(Y).

To obtain a random observation of our variable of interest distributed according to the model $f(Y)$, we first have the computer generate a 0-1 random number called $u = F(y_o)$ and then obtain the corresponding value of y_o by solving the integral equation

$$u = F(y_o) = \int_{-\infty}^{y_o} f(y) \, dy$$

Let us illustrate the procedure for a continuous model by using the standard normal distribution with $u = 0$ and $\sigma = 1$:

$$f(Y) = \frac{1}{\sqrt{2\pi}} e^{-Y^2/2}$$

and

$$F(y_o) = \int_{-\infty}^{y_o} \frac{1}{\sqrt{2\pi}} e^{-y^2/2} dy$$

254

A random uniform number u is generated, say u = 0.4394. Then the normal tables can be used to find the value of y_o so that the probability $Pr(Y \leq y_o) = u$, which in this case is $y_o = -0.16$. A computer program can be written to generate the uniform random number and then the inverse normal function subprogram in the Appendix D can be called to determine the corresponding value of y_o.

This technique makes it possible to generate a random observation from any distribution model if a subroutine is available to solve the integral equation. The subroutines for certain probability distributions are given in Appendix D; thus the student can duplicate the studies we have made in this chapter for any of these distributions. If no subroutine is available to solve the integral equation for a given distribution, the technique used in the discrete case can always be used. The range of Y can be broken down into a set of small intervals, and the area for each interval under the distribution could then be assigned to the midpoint of the interval as its probability of occurrence.

With enough exposure to the concepts, even the nonmathematical student should become a believer in the four prop - sitions. If you have not been convinced as yet, try to disprove any one of them by finding a counter example because according to the principles of logic one should either accept the truth of a statement or show that it is false by means of an example.

9.8 Summary

We illustrated the fundamentals of sampling theory by investigating the sampling distribution of the mean and were able to identify three propositions and a theorem relative to the distribution of $\bar{y}$:

Proposition 1

The distribution of the sample mean has the following characteristics:

$$\mu_{\bar{y}} = \mu_y$$

$$\sigma_{\bar{y}} = \sqrt{\frac{N-n}{n(N-1)}} \; \sigma_y$$

$$\alpha_{3:\bar{y}} = \frac{N - 2n}{N - 2} \sqrt{\frac{N - 1}{n(N - n)}} \; \alpha_{3:y}$$

$$\alpha_{4:\bar{y}} = \frac{N - 1}{n(N - n)(N - 2)(N - 3)} \left\{ (N^2 - 6nN + N + 6n^2)\alpha_{4:y} \right.$$

$$\left. + 3N(n - 1)(N - n - 1) \right\}$$

Proposition 2

As the population approaches infinity the above relationships can be approximated by

$$\mu_{\bar{y}} = \mu_y$$

$$\sigma_{\bar{y}} = \frac{\sigma_y}{\sqrt{n}}$$

$$\alpha_{3:\bar{y}} = \frac{\alpha_{3:y}}{\sqrt{n}}$$

$$\alpha_{4:\bar{y}} = 3 + \frac{(\alpha_{4:y} - 3)}{n}$$

Proposition 3

If random samples are drawn from a normal population, the distribution of the sample means is also normal.

Central limit theorem

If random samples are drawn from a population having a finite mean and variance, then the distribution of the sample mean approaches the normal distribution as the sample size n increases.

We verified these propositions and the theorem using random number generators and demonstrated how the results could be used to answer questions of the following types:

1. Testing of a hypothesis

To test the null hypothesis that a sample mean $\bar{y}$ comes from a specified population, we use the statistic

$$z = \frac{\bar{y} - \mu_y}{\sigma_y \sqrt{(N-n)/n(N-1)}} \qquad \text{N finite}$$

$$= \frac{\bar{y} - \mu_y}{\sigma_y/\sqrt{n}} \qquad \text{N infinite}$$

and compute the probability of obtaining more extreme values of z. If the probability is small (less than 1 or 5 percent, we reject the null hypothesis and accept the alternative hypothesis. We can, of course, simply compare the computed z with z_α, $z_{\alpha/2}$, or $z_{1-\alpha/2}$.

2. Confidence interval estimation

We can find an interval estimate for the unknown population mean μ having a specified level of confidence, α. The interval is given by

$$\mu = \bar{y} \pm z_{1-\alpha/2} \sqrt{\frac{N - n}{n(N - 1)}} \; \sigma_y \qquad \text{N finite}$$

$$\mu = \bar{y} \pm z_{1-\alpha/2} \frac{\sigma_y}{\sqrt{n}} \qquad \text{N infinite}$$

and $z_{1-\alpha/2}$ is found by the relationship $\text{AREANORM}(z_{1-\alpha/2}) = 1 - \alpha/2.$

3. Size of sample

We determine the sample size needed to have less than a prescribed sampling error with a specified level of confidence.

$$n = \frac{N z_{1-\alpha/2}^2 \sigma_y^2}{z_{1-\alpha/2}^2 \sigma_y^2 + (N - 1)\epsilon^2} \qquad \text{N finite}$$

$$= \frac{z_{1-\alpha/2}^2 \sigma_y^2}{\epsilon^2} \qquad \text{N infinite}$$

In each of the applications, we have assumed that the variance of the population σ_y^2 is known from past experience and that either the population is normally distributed or n is sufficiently large.

In a subsequent chapter we shall consider alternative approaches when one or the other of these conditions do not prevail. In addition, we need to consider what can be done if the parameter is not the mean.

1. Consider a population of N = 6 individuals, and use the computer to determine all possible samples of size 2, n = 2, and their corresponding means. Verify the formulas for the characteristics of the distribution of sample means by computing the values for these C_2^6 = 6·5/2 = 15 sample means. Compare the results with formula values.

2. Generate 100 sample means of size n = 9 from the rectangular model f(Y) = 1, $0 \leq Y \leq 1$. Verify the relationship between the characteristics of the distribution of the population and characteristics of the distribution of the sample means

$$\mu_{\bar{y}} = \mu_y \quad \sigma_{\bar{y}} = \frac{\sigma_y}{\sqrt{n}} \quad \alpha_{3:\bar{y}} = \frac{\alpha_{3:y}}{\sqrt{n}} \quad \alpha_{4:\bar{y}} = 3 + \frac{(\alpha_{4:y} - 3)}{n}$$

Note that the verification is only approximate because these relationships are true only if you have all possible samples of a certain size. In this case, since there are an infinite number of possible samples, we can never expect to obtain a complete list.

3. Modify the computer program given in Program 9.3 to generate samples of size n = 4 from a population of size N. Have the computer determine the four characteristics of the distribution of sample means generated from a given population.

4. Use Program 9.4 (modified to generate random normal numbers using RANNOR) to study the validity of Proposition 3 under varying conditions for the following values:

(a) μ = 30 σ = 10   n = 10   m = 5,10,25,50
(b) μ = 100 σ = 5   n = 2,10,25   m = 50
(c) μ = 50 σ = 2,5,10   n = 10   m = 50
(d) μ = 10,25,50 σ = 10   n = 10   m = 50

5. Consider the population

Y	p(Y=y)
0	0.3
1	0.7

Write a computer program to generate 500 samples of size
$n = 2, 5, 10, 20, 50$ from this population. (Hint: Use
the routine URANDN to generate a uniform random number u.
If $u < 0.3$, then $Y = 0$; otherwise $Y = 1$).

6. Write a computer program to generate samples from the
Poisson model

$$p(Y=y) = \frac{e^{-\lambda}\lambda^y}{y!} \qquad y = 0, 1, 2, \ldots$$

Use the program to select 100 samples, and compute the
numerical characteristics of the distribution of such
sample means for

(a) $n = 2$ $\lambda = 0.50$
(b) $n = 10$ $\lambda = 0.50$
(c) $n = 30$ $\lambda = 0.50$
(d) $n = 2$ $\lambda = 1.50$

(Note: One must first determine the possible values of
Pr(Y=y) and then compare a uniform random number with
each of the probabilities to see what value of y should
be assigned to the random observation.

7. Consider the sample size formula for a finite population

$$n = \frac{N \cdot z_{1-\alpha/2}^2 \cdot \sigma_y^2}{z_{1-\alpha/2}^2 \cdot \sigma_y^2 + (N - 1)\varepsilon^2}$$

Write a FUNCTION subprogram to determine the sample size
for given values of σ_y, z, N, and ε. Find n when $\varepsilon = 1$,
$z = 2.58$, $\sigma_y^2 = 10$, $N = 10, 25, 50, 100, 1000, 10000$.

8. The manager of a branch considers that the number of sick-leave days are an indication of the morale of the branch. He knows that the average for all the branches in the company is 4.4 days per employee per year, with a standard deviation of 0.85. The manager would like to test the hypothesis that the average number of sick leave days for his branch is equal to the company average. Thus $H_o:\mu = 4.4$ and $H_a:\mu > 4.4$.

There are 250 employees in his branch, so the manager decides to take a random sample of 40 employees. The sample data was obtained, and the following sample characteristics were calculated:

$$n = 40 \quad \bar{y} = 5.3 \quad s_y = 0.95 \quad a_3 = 1.3 \quad a_4 = 3.8$$

(a) Would the manager accept or reject the null hypothesis? Why?

(b) If the null hypothesis were true, what is the probability of rejecting the null hypothesis?

(c) If the actual average number of sick-leave days for the branch is 5.6 days, what is the probability that the manager would reject the null hypothesis?

(d) What assumptions are made to test the null hypothesis? Are you willing to make some of those assumptions in view of the a_3 and a_4 values?

9. Determine the 98 percent confidence interval for the average number of sick-leave days in the branch based on sample results in Exercise 8. Determine the 98 percent confidence interval for $n = 75$; for $n = 100$.

10. Using the same problem presented in Exercise 8, determine the sample size needed to have 95 percent assurance that the sample mean will be within 0.5 days of the true mean. Determine the sample size needed if there are 500 employees; 5,000 employees. Compute the ratio of the required sample size to the population size for each of the above three population sizes and compare them.

CHAPTER 10

APPLICATIONS OF SAMPLING THEORY

10.1 Introduction

In the preceding chapter we introduced the concept of sampling distributions. We presented some basic propositions and showed how they could be applied to solve three types of sampling problems: (1) testing of hypothesis, (2) estimation of population parameters, and (3) determination of the sample size to have an interval estimate of a fixed width.

The applications of the propositions were limited by the following constraints: (1) the parameter was always taken to be μ, the mean of the population, (2) the sample was either considered to be large or the population was assumed to have a normal distribution, and (3) the variance of the parent population was known. The only degree of generality was that the population could be finite or infinite. In this chapter we shall extend the sampling theory results by relaxing the three constraints so that our sampling theory results will be more generally applicable.

10.2 The Known Population Variance Assumption

In practice the population variance is not known, and thus the techniques presented in Chap. 9, even when the parameter of interest is μ, are not applicable. The obvious approach is to estimate the population variance from the sampled observations. We could use

$$s^2 = \frac{\sum_{i=1}^{n} (y_i - \mu)^2}{n}$$

as the estimate of σ^2. When, however, the population mean μ is not known we use

$$s^2 = \frac{\sum\limits_{i=1}^{n} (y_i - \bar{y})^2}{n}$$

This estimator is a biased one because the mean of its sampling distribution is less than the parameter value σ^2. To appreciate this fact, let us show that $\sum\limits_{i}(y_i - \bar{y})^2 \leq \sum\limits_{i}(y_i - \mu)^2$, if $\bar{y} \neq \mu$.

$$\sum_{i=1}^{n} (y_i - \mu)^2 = \Sigma(y_i - \bar{y} + \bar{y} - \mu)^2$$

$$= \Sigma\left[(y_i - \bar{y}) + (\bar{y} - \mu)\right]^2$$

$$= \Sigma\left[(y_i - \bar{y})^2 - 2(y_i - \bar{y})(y - \mu) + (\bar{y} - \mu)^2\right]$$

$$= \Sigma(y_i - \bar{y})^2 + \Sigma 2(\bar{y} - \mu) \cdot (y_i - \bar{y}) + \Sigma(\bar{y} - \mu)^2$$

$$= \Sigma(y_i - \bar{y})^2 + n(\bar{y} - \mu)^2$$

because $\Sigma(y_i - \bar{y}) = 0$ and $\Sigma(\bar{y} - \mu)^2 = n(\bar{y} - \mu)^2$. The quantity $n(\bar{y} - \mu)^2 \geq 0$ and zero only when $\bar{y} = \mu$ so that

$$\Sigma_i(y_i - \bar{y})^2 \leq \Sigma_i(y_i - \mu)^2$$

We could check this bias emperically by generating random samples ($j = 1, 2, \ldots, m$) from a population with known σ^2 and computing s_j^2 for each sample. Individual s_j^2's may be larger than σ^2, but generally the average of the generated s_j^2's,

$$\bar{s}^2 = \sum_{j=1}^{m} \frac{s_j^2}{m}$$

will be smaller than σ^2. However because of sampling variation, $\bar{s}^2$ may sometimes exceed the parameter σ^2.

Program 10.1 was written to generate m samples of a
given size n from a normal distribution $N(\mu,\sigma)$ and to compute s_j^2 for each sample and to then find $\bar{s}^2$. The use of the
normal distribution is for convenience since the bias exists
for any distribution. The result of the emperical study
using m = 200 and different population variances and sample
sizes is summarized in Fig. 10.1:

n	σ^2	$\bar{s}^2$
4	1	0.74
30	1	0.99
50	1	1.00
4	25	17.62
30	25	24.73
50	25	25.03
4	100	73.28
30	100	76.88
50	100	100.86

Fig. 10.1 The bias of the variance estimator s^2.

We note that the average sample variance $\bar{s}^2$ is consistently less than σ^2, although the discrepancy diminishes as
the sample size n becomes larger. Because the population
mean has no effect on the sampling distribution of s^2, we
let $\mu = 0$ in all the cases studied. The student is asked to
verify the independence of μ and the bias as an exercise.

Since s^2 is a biased estimator of σ^2 (consistently too
small), the question arises as to what sample estimator
should be used. The answer is to adjust s^2 by dividing the
sum of squares of the deviations by n - 1 rather than n.
We define the new estimator

$$s'^2 = \frac{\sum\limits_{i}^{n}(y_i - \bar{y})^2}{n - 1}$$

and use this as our estimator of σ^2 since this new estimator is unbiased. In the exercises the student is asked to modify the sample generation computer program and compute s'^2 to verify the fact that the expected value of s'^2 is σ^2.

10.3 The Sampling Distribution of $\bar{y}$
for an Unknown Variance

When the population variance is unknown and the random variable y has a normal distribution, we could use the unbiased estimator s'^2 in lieu of σ^2 in the procedures given in Chap. 9. This substitution would yield the following test statistic, which we shall denote by t:

$$t = \frac{\bar{y} - \mu}{\sqrt{\frac{N - n}{n(N - 1)}}\, s'} \qquad N \text{ finite}$$

$$t = \frac{\bar{y} - \mu}{s'/\sqrt{n}} \qquad N \text{ infinite}$$

If the distribution of t is the same as the standard normal, we could use the same critical values used previously in Chap. 9. However, the sampling variation of t, which has the additional random variable s' in the denominator, can be significantly different from the distribution of z, which has only one random variable, $\bar{y}$.

To study the difference between t and z difference emperically, Program 10.2 was written to generate m samples, each of size n, from a normal population with mean μ and standard deviation σ. The program computes $\bar{y}$ and s', calculates t and z for each sample, and then tabulates the m sample results into two frequency distributions. The inputs of m = 200, μ = 0, σ = 1, and n = 2, 10, 30, 100 were used to generate the results given in Fig. 10.2.

It can be seen that for any fixed value of n the sampling variation of the t distribution is larger than that of the z distribution. This means that $\Pr[z \geq a] < \Pr[t \geq a]$ for any a, and thus the use of the normal distribution as

	n = 2		n = 10		n = 30		n = 100	
	t	z	t	z	t	z	t	z
Mean	0.42	0.04	0.00	-0.01	-0.04	-0.05	0.00	0.01
Standard deviation	13.42	1.04	1.14	1.02	1.08	1.04	0.99	0.99
Skewness	5.88	0.25	0.08	0.01	0.08	0.07	0.03	0.07
Kurtosis	87.94	2.76	3.25	2.59	2.66	2.47	2.80	2.78
Intervals:								
below -3.0	25	0	2	0	0	0	0	0
-3.0 - -2.5	2	0	2	1	0	0	0	0
-2.5 - -2.0	6	2	3	3	7	5	5	4
-2.0 - -1.5	7	11	14	14	16	16	6	8
-1.5 - -1.0	14	23	16	16	13	20	22	23
-1.0 - -0.5	20	27	25	25	32	24	32	30
-0.5 - 0.0	26	37	39	42	34	37	35	35
0.0 - 0.5	21	31	38	40	41	41	37	37
0.5 - 1.0	28	35	25	27	22	22	31	32
1.0 - 1.5	10	16	17	17	17	19	19	20
1.5 - 2.0	10	10	9	10	11	11	10	7
2.0 - 2.5	5	4	4	5	5	5	2	3
2.5 - 3.0	2	4	6	0	2	0	0	0
3.0 and above	24	0	0	0	0	0	1	1

Fig. 10.2 Sampling distribution of t and z for
varying value of the sample size n.

the distribution model for the t ratio would result in incorrect statements of the probability levels for tests of
hypothesis, confidence intervals, or sample size determination. However, it can also be seen that the t distribution
approaches the z distribution as the sample size n increases.
This latter observation allows one to use the normal distribution and probabilities for t when the sample size is
greater than 30.

The lack of correspondence between the z and the t distribution when sampling from a normal population for small
sample sizes is evident from the frequency tabulations in
Fig. 10.2. Thus it is apparent that we need a mathematical
model which is different from the normal model to represent
the t distribution. The distribution should also have the
property that as the sample size increases, the distribution
approaches the normal distribution of z. Such a model was

developed in 1908 by Gossett and is given by

$$f(t,d) = \frac{\Gamma[(d+1)/2]}{\sqrt{d\pi}\,\Gamma(d/2)}\left(1 + \frac{t^2}{d}\right)^{-(d+1)/2} \qquad -\infty < t < \infty,\ d > 0$$

The symbol $\Gamma(\ )$ is the same gamma function that was introduced with the type III distribution in Chap. 8. The parameter $d = n - 1$ in the distribution is called the degrees of freedom. The model is not called Gossett's t distribution because he published under the pseudoynm Student; hence we call the distribution the Student's t distribution. In the exercises the student is asked to plot the ordinates of t for different values of d and to compare the distributions to the standard normal.

Let us use our three techniques to check how well the Student's t distribution fits the observed distributions given in Fig. 10.2.

(1) We first compare the model characteristics with those of the observed distribution. We need the formulas for the characteristics of the t distribution. For $d > 4$ they are

$$\mu = 0 \qquad \sigma = \sqrt{\frac{d}{d-2}}$$

$$\alpha_3 = 0 \qquad \alpha_4 = \frac{3(d-2)}{d-4}$$

In Fig. 10.2 we tabulated the characteristics for each observed distribution, so we can compare the characteristics by simply giving d the appropriate value. Thus for $n = 10$ we have $d = n - 1 = 9$ and obtain

Characteristic	Observed	t distribution
Mean	0.02	0
Standard deviation	1.14	1.1338
Skewness	0.08	0
Kurtosis	3.25	4.2

We have come to appreciate that the higher-order characteristics may show rather large discrepancies just by chance, so we should not be too disturbed over the lack of correspondence we have in this case. We have not used the n = 2 case because when d is equal to 1, even σ fails to exist, and our characteristic comparison approach fails. The student can check that as d → ∞, the characteristics of the t distribution approach those of the normal distribution by substituting large values of d in the formulas.

(2) The second technique is to graphically compare the observed histogram with the ordinates of the fitted model. To obtain the required model ordinates for selected values of the variable we can adapt Program 8.2, EVALF, using the appropriate FUNCTION F subprogram, which is given by Program 10.3, Sec. 10.9. The resulting superimposed mathematical model is presented graphically in Fig. 10.3; the values are given in Fig. 10.4. The results show that other than for the case n = 2, the agreement between the two graphs is satisfactory. In the n = 2 case the pathological nature of the distribution makes any empirical verification difficult.

(3) We can also compare the expected t distribution with the observed frequencies. For this purpose we adapt Program 8.6, NORMFREQ, to evaluate the probabilities for the t distribution; the resulting program, Program 10.4 is presented in Sec. 10.9. The expected frequencies for the normal model are calculated to compare the goodness of fit of both models. The results are given in Fig. 10.5:

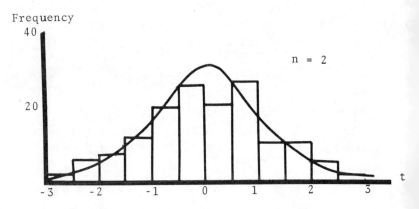

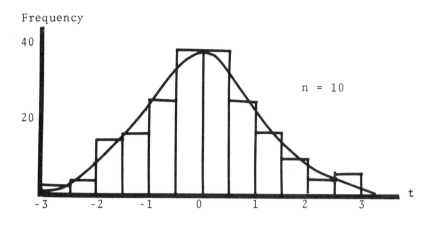

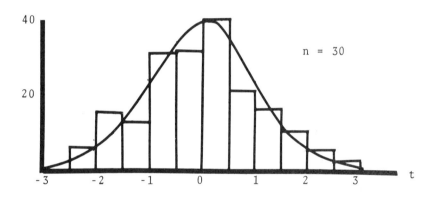

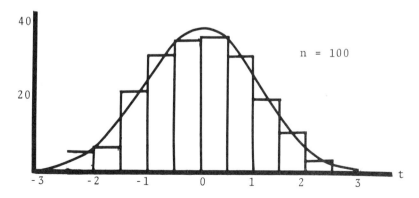

Fig. 10.3 Histograms for sampling distribution of t
with superimposed ordinates of t.

Y	n = 2 f(t)	sf(t)	n = 10 f(t)	sf(t)	n = 30 f(t)	sf(t)	n = 100 f(t)	sf(t)	Normal f(t)	sf(t)
-4.00	0.0187	1.8724	0.0023	0.2346	0.0005	0.0543	0.0002	0.0222	0.0001	0.0129
-3.50	0.0240	2.4023	0.0053	0.5288	0.0020	0.2003	0.0012	0.1166	0.0008	0.0840
-3.00	0.0318	3.1831	0.0121	1.2126	0.0068	0.6861	0.0051	0.5133	0.0043	0.4265
-2.50	0.0439	4.3905	0.0278	2.7780	0.0212	2.1172	0.0186	1.8641	0.0169	1.8667
-2.00	0.0637	6.3662	0.0617	6.1712	0.0569	5.6942	0.0549	5.4918	0.0519	5.1955
-1.50	0.0979	9.7941	0.1272	12.7152	0.1289	12.8941	0.1294	12.9364	0.1246	12.4634
-1.00	0.1592	15.9154	0.2291	22.9132	0.2379	23.7862	0.2408	24.0754	0.2328	23.2847
-0.50	0.2546	25.4646	0.3384	33.8358	0.3477	34.7740	0.3508	35.0795	0.3388	33.879⁰
0.00	0.3183	31.8308	0.3880	38.8036	0.3955	39.5521	0.3979	39.7924	0.3839	38.3900
.50	0.2546	25.4646	0.3384	33.8358	0.3477	34.7740	0.3508	35.0795	0.3388	33.8790
1.00	0.1592	15.9154	0.2291	22.9132	0.2379	23.7862	0.2408	24.0754	0.2328	23.2847
1.50	0.0979	9.7941	0.1272	12.7152	0.1289	12.8941	0.1294	12.9364	0.1246	12.4634
2.00	0.0637	6.3662	0.0617	6.1712	0.0569	5.6942	0.0549	5.4918	0.0519	5.1955
2.50	0.0439	4.3905	0.0278	2.7780	0.0212	2.1172	0.0186	1.8641	0.0169	1.8667
3.00	0.0318	3.1831	0.0121	1.2126	0.0068	0.6861	0.0051	0.5133	0.0043	0.4265
3.50	0.0240	2.4023	0.0053	0.5288	0.0020	0.2003	0.0012	0.1166	0.0008	0.0840
4.00	0.0187	1.8724	0.0023	0.2346	0.0005	0.0543	0.0002	0.0222	0.0001	0.0129

Fig. 10.4 Ordinates and scaled ordinates for t and
z distribution.

Intervals	Exp. F for t model n = 2	n = 10	n = 30	n = 100	Exp. F Normal Model
Below -3.0	20.48	1.50	0.55	0.34	0.25
-3.0 - -2.5	3.74	1.89	1.28	1.06	0.97
-2.5 - -2.0	5.29	4.27	3.66	3.41	3.30
-2.0 - -1.5	7.91	9.12	8.94	8.85	8.80
-1.5 - -1.0	12.56	17.55	18.11	18.29	18.37
-1.0 - -0.5	20.48	28.57	29.53	29.85	29.98
-0.5 - 0.0	29.53	37.11	37.93	38.20	38.31
0.0 - 0.5	29.53	37.11	37.93	38.20	38.31
0.5 - 1.0	20.48	28.57	29.53	29.85	29.98
1.0 - 1.5	12.56	17.55	18.11	18.29	18.37
1.5 - 2.0	7.91	9.12	8.94	8.85	8.80
2.0 - 2.5	5.29	4.27	3.66	3.41	3.30
2.5 - 3.0	3.74	1.89	1.28	1.06	0.97
Above 3.0	20.48	1.50	0.55	0.34	0.25

Fig. 10.5 Comparison of expected frequencies of t
and z model.

The student should compare these frequencies with the observed frequencies in Fig. 10.2 to note the goodness of fit.

Although our empirical studies of the goodness of fit of the Student's t distribution to observed t ratio do not provide as convincing an argument in favor of the t distribution as one might like, let us accept the mathematical analysis that proves its appropriateness. We shall now consider how the t ratio

$$t = \frac{\bar{y} - \mu}{s'/\sqrt{n}}$$

and the associated Student's t distribution can be used in making inferences about the mean μ of a normally distributed variable Y. These applications follow the same pattern as those used when we considered the z ratio and we shall therefore parallel these applications in the following sections.

10.4 The Use of the t Distribution

Before we proceed we need to be able to determine the area under the Student's t distribution. The DEFINT subprogram can be modified to evaluate the areas under that distribution, or the subprogram FUNCTION TTX(DX,DF) given in Appendix D can be used, where $TTX(DX,DF) = \int_{-\infty}^{DX} f(t,d)\ dt$.
We note that for the t distribution the degrees of freedom DF = d = n - 1 must be given in addition to the value of DX.

Tables of the t distribution are given in Appendix C-3. Since two arguments, t and d, are involved, the tables give the values of t associated with specific probabilities for different degrees of freedom. These values are also called the critical values of t and are denoted by $t_{1-\alpha}$, where

$$\int_{-\infty}^{t_{1-\alpha}} f(t,d)\ dt = 1 - \alpha$$

The 1 - α values that are tabulated range from 0.60 to).999 because for smaller values we can use the symmetry of he t distribution since t_α value is simply the negative of he recorded value $t_{1-\alpha}$; that is

271

$$t_\alpha = -t_{1-\alpha}$$

For example, if d = 9 and α = 0.05, then the appropriate value of $t_{9,1-\alpha}$ is found by looking up

$$t_{9,1-\alpha} = t_{9,.95} = 1.833$$

and then

$$t_{9,\alpha} = -t_{9,1-\alpha} = -1.833$$

Using these tabled values for the t distribution, we can work the three types of sampling problems. To facilitate the presentation we shall use actual examples.

1. Testing hypotheses

Let us assume that standards to regulate the pollution of rivers have been established. The standards specify that the materials discharged into a river shall not impede the growth rate of the native fish by more than 0.5 cm/month. A factory owner has asked for permission to discharge a given type of material into the river at a specified rate, claiming it will not have an adverse effect upon the growth of fish. To make a preliminary evaluation of the owner's claim we study the growth response of 10 native fish when they are in the contaminated water. We obtain the following differences between the expected and actual monthly growth rates for each fish: 0.37, 0.16, 0.67, 0.45, 0.71, 0.63, 0.44, 0.81, 0.66, and 0.48. At the 5 percent level of significance would we discard the hypotheses that the average change in the growth rate μ_y is equal or less than 0.5 (H_0: $\mu_y \leq 0.5$) in favor of the alternative H_a: $\mu_y > 0.5$? If we discard the hypothesis, we would not allow the discharge in its present form.

The steps followed in testing the hypothesis are

(a) The sum and sum of squares of the observations are determined; they are found to be

$$n = 10 \qquad \Sigma y_i = 5.38 \qquad \Sigma y_i^2 = 3.2290$$

(b) From these data the sample mean and the standard deviation are established; they are

$$\bar{y} = \frac{\Sigma y_i}{n} = \frac{5.38}{10} = 0.538$$

$$s' = \sqrt{\frac{\Sigma y_i^2 - (\Sigma y_i)^2/n}{n - 1}} = \sqrt{\frac{3.2290 - (5.38)^2/10}{9}}$$

$$= \sqrt{0.3718} = 0.6097$$

(c) The t statistic is found:

$$t = \frac{\bar{y} - \mu_o}{s'/\sqrt{n}} = \frac{0.538 - 0.50}{0.6097/\sqrt{10}} = \frac{0.038}{0.1930} = 0.197$$

Although it should be appreciated that such a small value is hardly significant, we complete the steps.

(d) The degrees of freedom d are determined: $n - 1 = 9$. The critical area is shown in the following sketch and indicates that to find $t_{9,1-\alpha}$ we must enter the table for $t_{9,.95}$. We find that $t_{9,.95} = 1.8333$. The corresponding critical area is the shaded region of the diagram.

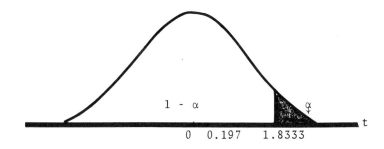

(e) Apply the rule: If $t > t_{1-\alpha}$, reject H_o. Since 0.197 < 1.833, we do not reject the hypothesis and thus allow the discharge to take place at least until a more extensive study is made.

2. Confidence interval

We could obtain an interval estimate for the average reduction in growth caused by the discharge. For example, a 99 percent confidence interval for the average reduction in the expected growth caused by the discharge (using the same 10 data points used for the hypothesis) is obtained by the following steps:

(a) Find summations

$$n = 10 \qquad \Sigma y_i = 5.38 \qquad \Sigma y_i^2 = 3.2290$$

(b) Compute the parameter estimators from the sample

$$\bar{y} = \frac{\Sigma y_i}{n} = \frac{0.538}{10} = 5.38$$

$$s' = \sqrt{\frac{\Sigma y_i^2 - (\Sigma y)^2/n}{n-1}} = \sqrt{\frac{3.2290 - (0.538)^2/10}{9}}$$

$$= \sqrt{0.3718} = 0.6097$$

(c) Since our confidence level is 0.99, $\alpha = 0.01$ and we enter the tables for the tabulated area with $1 - \alpha/2 = 0.995$, obtaining $t_{9,.995} = 3.250$.

(d) Determine the confidence interval determination

$$\mu = \bar{y} \pm t_{d,1-\alpha/2} \frac{s'}{\sqrt{10}}$$

$$= 0.538 \pm (3.250) \frac{(0.6097)}{\sqrt{10}}$$

$$= 0.538 \pm 0.627$$

3. Size of sample

We now wish to design a testing program to be used in evaluating the growth rates of fish. The requirements are to determine how many fish should be studied to estimate μ with a 95 percent confidence level of having the sample mean estimator of μ be in error by less than $\varepsilon = 0.2$ cm.

We have the relationship

$$n = \frac{t^2_{d,1-\alpha/2} \; s'^2}{\varepsilon^2}$$

which is obtained from the t ratio

$$t = \frac{\bar{y} - \mu}{s'/\sqrt{n}}$$

by solving for n. To find n we must know ε, t, and s'. The value of t depends upon the n we use, and because n is unknown, t cannot be found even when α is known. In addition, we have no value for s', although we could use either the range estimate R/6 or previous experience to determine a value.

To determine the value of t for a given confidence level, we follow an iterative procedure. The steps are (1) assume a value of n and (2) find the corresponding $t_{d,1-\alpha/2}$ and then use this t to compute the new n value. Repeat the process until the procedures converges (the two n values become equal). It usually takes only a few iterations because the convergence is rapid. We demonstrate the approach:

Given that $\alpha = 0.05$, $1 - \alpha/2 = 0.0 = 0.975$, $\varepsilon = 0.2$, and $s'^2 = 0.3718$,

$$n_o = 15 \qquad\qquad t_{14} = 2.145$$

$$n_1 = \frac{t^2_{14} s'^2}{0.04} = 42.77 \qquad\qquad t_{42} = 2.018$$

$$n_2 = \frac{t^2_{42} s'^2}{0.04} = 37.85 \qquad\qquad t_{37} = 2.030$$

$$n_3 = \frac{t_{36}^2 s'^2}{0.04} = 38.30 \qquad t_{38} = 2.025$$

$$n_4 = \frac{t_{38}^2 s'^2}{0.04} = 38.12$$

$$n = 39$$

Since the appropriateness of the Student's t distribution depends upon the assumption that the underlying distribution of the variable is normal, we shall consider this assumption in the next section.

10.5 The Normal Distribution Assumption

How do we make inferences about a population mean when the population is not normal, the variance is unknown, and the sample is small? There are five approaches that may be used when this type of difficulty is encountered:

(1) Ask a mathematical statistician to emulate the work done by Gossett and mathematically develop the appropriate statistic and its distribution model. Experience has demonstrated that Gossett's approach requires time and money because such theoretical distribution problems are not easily solved.

(2) Hope that the "robustness" of the t ratio is such that the lack of normality in your case does not preclude your using the Student's t distribution. That is, we know the Student's distribution gives correct probabilities only if the population distribution is normal. However, for certain underlying distributions the correct but unknown probabilities for the t ratio may not differ significantly from the t distribution probabilities, so they can be used in place of the exact ones. In this case robustness studies are needed to determine how different the true sampling distribution of the t ratio is as compared with the Student's t distribution.

(3) Attempt to transform the variable y into a variable that has a normal distribution. Thus, we would work with a

new variable such as $y^* = \log y$, $y^* = \sqrt{y}$, $y^* = y^2$. The use of a transformation may mean that we believe the lack of normality is caused by using the wrong variable and that if we transform the variable (change the measure) we could essentially have a normal population. For example, if our observational units were small disks made by some production process, we might measure their diameters, thicknesses, top surface areas, volumes, weights, and so forth. Surely if one of these measures had a normal distribution because of random variations that occur in the manufacturing processes, the others would not. For illustration, Fig. 10.6 plots the distribution of a normally distributed variable when it is squared.

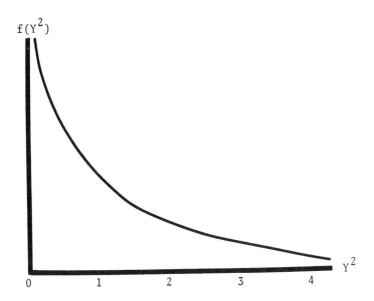

Fig. 10.6 The distribution of Y^2 when Y is a normal (0,1) variable.

The lack of symmetry in the distribution of y^2 is proof enough of the lack of normality of the y^2 variable.

However, with approach 3 it is difficult to find the appropriate transformation and to properly interpret the re-

277

sults in light of the transformation. This requires sufficient data for testing the effects of different transformations upon the observed distribution — a trial-and-error approach is often necessary. If we are to consider different transformations, we need to be able to select the best one to use. A possible selection criterion is

$$SSE = \sum_{j=1}^{m} (0_j - E_j)^2$$

where 0_j is the observed frequency for the jth class of the frequency distribution for the observed data and E_j is the theoretical frequency associated with the class using the normal curve. We might decide that the transformation yielding the smaller SSE will be chosen as the better transformation. As an exercise, the student is asked to write a program that will transform data according to any given transformation, tabulate the data using the frequency subroutine, graduate the distribution using the normal curve, and compute SSE that can then be used for such a comparison.

(4) Use the simulation power of the computer to generate the sampling distribution of t and determine the appropriate probabilities. We have already shown how this concept can be used: Several of our computer programs generate many samples of a given size from a mathematical model and tabulate the resulting sampling distribution of these sample observations. Thus if one had the mathematical form for the population distribution, we could modify these programs to generate random samples from the particular distribution model. Although approach 4 can be time consuming and expensive, it may be easier to use than the mathematical approach suggested in approach 1. We want to estimate $\Pr[t \geq t_o]$ for the t ratio from a particular underlying distribution. We can find the estimate by using the sample generation program and counting the number of t's of the total number generated that were greater than or equal to t_o:

$$\Pr[t \geq t_o] = \frac{\text{number of } t \geq t_o}{\text{total number}}$$

If the total number is large, this ratio should give a good approximation to the probability.

(5) Use a nonparametric technique to make the test. A nonparametric technique does not require the model for the population distribution to be specific since such model assumptions introduce parameters into our considerations. Thus a nonparametric approach is applicable regardless of the form of the population distribution. It can be argued that such techniques should therefore always be used because man rarely knows the exact form of the population. However, such techniques may not be too effective because they must necessarily be very general in their nature. Fortunately there are many situations when the assumption of the normal distribution is justified, and thus a parametric analysis can be used.

The student may be disturbed by the lack of a well-defined approach to solving the small-sample nonnormal-type problem, even when the interest is restricted to the population mean. Such difficulties are common in data analysis, and we prefer to leave the student with the feeling that he will have to work and think to solve his data analysis problem. There is no cookbook approach to data analysis. All we can do is illustrate techniques and approaches that have proven useful.

10.6 Another Parameter — The Population Percentage

Recall the restrictions that were imposed on the sampling theory developments in the previous chapter to make inferences about population parameters:

(1) The population distribution was normal or the sample size was large.
(2) The population variance was known.
(3) The mean was the parameter of interest.

We discussed alternative approaches when the first two restrictions were not met. We shall now consider the third restriction: The parameter of interest is not the mean.

Suppose the parameter of interest is the fraction (p)

of individuals in the population that have some prescribed attribute. For example, a purchasing agent buys a manufactured item in large lots. Let us assume that a new lot of 1,000 items (N = 1,000) has arrived. He would like to know the fraction of items p that are defective to decide whether to accept or reject the lot. His particular approach to this problem might be based on one of the three basic sampling considerations:

(1) He might want to test the hypothesis that $p = p_o$ (the old experience) against the alternative $p \neq p_o$ (two-sided), $p > p_o$ (one-sided), or $p < p_o$ (one-sided).

(2) He might want to determine a confidence interval for p.

(3) He might want to know how large a sample of items he should test to obtain an estimate of p that meets his needs.

To solve this problem, we create a new variable Y by assigning a zero value to Y if an item in the population is not defective and a value of 1 to an item that is defective. Thus the population model can be represented by the simple distribution

Y	$p(Y = y)$
0	$1 - p$
1	p

Let us examine the numerical characteristics of this distribution. We see that $p(Y = y) \geq 0$ for all y and that

$$\sum_{y=0}^{1} p(Y=y) = (1 - p) + p = 1$$

Thus the distribution is a valid discrete model. We can determine the characteristics by direct computation:

$$\mu = E(Y) = \sum_{y=0}^{1} yp(Y=y) = 0 \cdot (1 - p) + 1 \cdot p = p$$

and

$$\sigma^2 = E(Y - \mu)^2 = \sum_{y=0}^{1} (y - p)^2 \cdot p(Y=y) = (0 - p)^2 \cdot (1 - p)$$

$$+ (1 - p)^2 \cdot p = p(1 - p)$$

We see that the mean of this synthetic distribution is the parameter of interest; thus we have transferred the percentage problem into a problem about the population mean. Any sample from the population will have y's that are either zero or 1. Therefore we can calculate the sample mean $\bar{y} = f$ and use it as the estimate of p.

The only real complication in the percentage problem is that the variance σ^2 is equal to $p(1 - p)$ and thus depends upon the value of the parameter of interest p. However, for large samples we can simply estimate the population variance from the sample so that $s' = \sqrt{f(1 - f)}$ and proceed as if it were the true variance. Since the central limit theorem is effective even for this discrete distribution, when the sample size is large we can use the statistic

$$z = \frac{f - p}{\sqrt{f(1 - f)\dfrac{N - n}{n(N - 1)}}} \qquad \text{N finite}$$

$$z = \frac{f - p}{\sqrt{f(1 - f)/n}} \qquad \text{N = infinite}$$

We shall demonstrate the approach by using the above-mentioned problem, where the parameter of interest is the percentage of defective items in a population of N = 1,000 items.

1. Testing hypothesis

To test the hypothesis that p = 0.08 (the upper value specified by a purchase contract against the alternative p > 0.08), a sample of n = 100 items is tested, with a resulting $\bar{y} = f = 10/100 = 0.10$. So

$$z = \frac{f - p}{\sqrt{f(1 - f)\dfrac{N - n}{n(N - 1)}}} = \frac{0.10 - 0.08}{\sqrt{0.10(.90)\left(\dfrac{1000 - 100}{100(999)}\right)}}$$

$$= \frac{0.02}{0.0285} = 0.70$$

Since the critical value of z using $\alpha = 0.05$ is $z_{.95} = 1.645$ and $0.70 < 1.645$, there is no evidence to reject the $H_o:p = 0.08$ in favor of the alternative: $H_a:p > 0.08$.

2. Confidence intervals

The purchasing agent's research assistant is given a sample, $n = 125$, of items and finds that there are 20 defective items among the 125; that is, $f = 20/125 = 0.16$. The agent would like to know what the 95 percent confidence interval for p is. The assistant computes

$$p = f \pm z_{0.975} \sqrt{(N - n)/n(N - 1)} \ \sqrt{f(1 - f)}$$

$$= 0.16 \pm (1.96) \sqrt{(1,000 - 125)/125(999)} \ \sqrt{(0.16)(0.84)}$$

$$= 0.16 \pm 0.0601$$

The assistant then reports with 95 percent assurance that the percentage of defective items in the shipment is between 10 and 22.

3. Sample size

If the requirement is to obtain an estimate of p that will have but one chance in 100 of being in error by as much as 0.005, we have sample size formula by replacing σ_y with $p(1 - p)$:

$$n = \frac{p(1 - p) \ z_{1-\alpha/2}^2 N}{p(1 - p) \ z_{1-\alpha/2}^2 + (N - 1)\varepsilon^2}$$

which is obtained from the general sample size formula with $\sigma_y^2 = p(1 - p)$. The question is then what value of p to use

in determining n. Lacking any information, one often takes
p = 0.50 since this will maximize the required value of n.
Of course, if some better approximation of p is available,
it should be used. Say we believe that p = 0.10. We would
then have

$$n = \frac{(0.10)(0.90)(2.58)^2(1000)}{(0.10)(0.90)(2.58)^2 + (999)(0.0005)^2}$$

$$= \frac{599}{0.599 + 0.024975} = \frac{599}{0.623975}$$

$$= 960$$

Note that in this extreme case sampling as many as 960 out
of 1,000 would probably not save enough money to pay for the
cost of the sampling.

We shall not consider the small sampling problem for
percentages of a population at this time, but the student
should appreciate that in such cases the normal model can
no longer be used. Since the population is not normally
distributed, the t distribution also cannot be used; the
appropriate technique in this case is beyond the scope of
this book and thus will not be given.

10.7 A Large Sample Approach for Other Parameters

The concept of standard error theory enables us to
solve the three types of sampling problems using large sample sizes for many estimators of different parameters. The
concept assumes that the test statistic z, which is obtained
by dividing the difference between an estimator of the
parameter and its true value by the standard deviation of
the estimator (the standard error), has a standard normal
distribution. Thus to consider any population parameter we
need an estimator of the parameter that is unbiased, the
formula for the standard error of the estimator, and the
assurance that the central limit theorem is applicable.

We shall illustrate the approach by presenting the parallel approaches for the mean parameter, variance parameter,

and any general parameter θ. We shall assume that the population is infinite and that a sample of size n is available. The notation for the three cases being considered is

	Case 1	Case 2	Case 3
Parameter	mean = μ	variance = σ^2	general θ
Statistic	$\bar{y}$	s'^2	c
Standard error	$\dfrac{\sigma}{\sqrt{n}}$	$\dfrac{\sqrt{2}\sigma^2}{\sqrt{n-1}}$	$se(\sigma,n)$

The functional notation for the general case $se(\sigma,n)$ indicates that the standard error of a statistic generally depends upon the population standard deviation and sample size.

1. Tests of hypothesis

Null hypothesis	$\mu = \mu_o$	$\sigma^2 = \sigma_o^2$	$\theta = \theta_o$

Alternative hypothesis:

one-sided	$\mu<\mu_o$ or $\mu>\mu_o$	$\sigma^2<\sigma_o^2$ or $\sigma^2>\sigma_o^2$	$\theta<\theta_o$ or $\theta>\theta_o$
two-sided	$\mu \neq \mu_o$	$\sigma^2 \neq \sigma_o^2$	$\theta \neq \theta_o$
Estimate of standard error	$\dfrac{s'}{\sqrt{n}}$	$\dfrac{\sqrt{2}\,s'^2}{\sqrt{n-1}}$	$se(s',n)$
Test statistic	$z = \dfrac{\bar{y} - \mu}{s'/\sqrt{n}}$	$z = \dfrac{s'^2 - \sigma_o^2}{\sqrt{2}s'^2/\sqrt{n-1}}$	$z = \dfrac{c - \theta_o}{\sqrt{se(s',n)}}$

Critical value:

one-sided	z_α or $z_{1-\alpha}$	z_α or $z_{1-\alpha}$	z_α or $z_{1-\alpha}$
two-sided	$z_{\alpha/2}$ and $z_{1-\alpha/2}$	$z_{\alpha/2}$ and $z_{1-\alpha/2}$	$z_{\alpha/2}$ and $z_{1-\alpha/2}$

It should be noted that this procedure yields approximate tests and in many cases better tests can be devised.

2. Confidence intervals

The $100(1-\alpha)$ percent confidence interval for the parameter using standard error theory is obtained by the formula

$$\theta = c \pm z_{\alpha/2} \cdot se(s',n)$$

The student should appreciate that this is the same procedure we used to obtain the confidence interval for the mean of an infinite population when the sample size was large.

3. Sample size

To determine the appropriate sample size, the following equation must be solved for n, given $\varepsilon = c - \theta$ and α:

$$z_{1-\alpha/2} = \frac{\varepsilon}{se(\sigma,n)}$$

We once again have to contend with the fact that σ is unknown, but we can adopt one of the approaches used in estimating σ when we encountered the same problem in considering the mean as the parameter of interest.

Although the use of standard error theory provides a general approach, given a large sample for solving inference problems about an unknown parameter, it is restricted to situations when the variance of the parameter estimate is known and the central limit theorem is applicable. When only a small sample is available and the underlying population is normal, the t distribution can be used. It should be noted that the standard error theory yields approximate tests and in many cases better tests can be found.

10.8 Computer Programs

The first computer program in this chapter is needed to determine whether the variance estimator s^2

$$s^2 = \frac{\Sigma (y_i - \bar{y})^2}{n}$$

285

is biased. Since a bias exists if the mean of the sampling distribution of s^2 is not equal to σ^2, a program can be written to generate random samples from a population. These samples form an approximate sampling distribution of s^2. The true sampling distribution would require all possible sample of size n from the population, and if we use the normal population, there are an infinite number of samples. The following program requires the values to be specified:

$$m = M = \text{number of samples to be generated, } m \leq 300$$
$$n = \text{NSAM} = \text{sample size}$$
$$\text{IODD} = \text{positive odd number to be used as starting value for random number generator}$$
$$\mu = \text{YMU} = \text{population mean}$$
$$\sigma^2 = \text{YVAR} = \text{population variance}$$

The computer program is

```
C     S2 DISTRIBUTION - GENERATION OF SAMPLING DIST. OF S2
      DIMENSION  S2(3ØØ)
 1ØØ  READ(5,1Ø) M,NSAM,IODD
  1Ø  FORMAT(3I5)
      READ(5,9) YMU,YVAR
   9  FORMAT(2F1Ø.2)
      WRITE(6,11) M,NSAM,IODD
  11  FORMAT( 1H1,13HNO OF SAMPLES, I5,11HSAMPLE SIZE,I5,
     1 16HSTARTING INTEGER, I5)
      WRITE(6,12)YMU,YVAR
  12  FORMAT(1HØ,3HYMU,F12.2/1H,4HYVAR,F12.2)
      SIGMA = SQRT(YVAR)
      SAM = NSAM
      DO 1 I=1,M
      SUMY = Ø
      SUMY2=Ø
      DO 2 J=1,NSAM
      CALL RANNOR(IODD,Y,YMU,SIGMA)
      SUMY = SUMY + Y
   2  SUMY2 = SUMY2 + Y**2
   1  S2(I) =(SUMY2-SUMY*SUMY/SAM)/SAM
      CALL CHAR( S2,M,YM,YSIG,A3,A4)
      WRITE(6,13)YM,YSIG,A3,A4
  13  FORMAT(1HØ,4HMEAN, F12.2/1H ,3HSTD,F12.2/
     1 1H ,2HA3,F12.4/1H ,2HA4,F12.4)
      STOP
      END
{Subroutine RANNOR}
{Subroutine URANDN}
{Subroutine CHAR}
```

Program 10.1 S2 DISTRIBUTION — sampling distribution of s^2.

The input cards used to obtain the results given in Fig. 10.1 are

(1)	200	4	13
	0.		1.
(2)	200	30	945
	0.		1.
(3)	200	50	147
	0.		1.
(4)	200	4	157
	0.		25.
(5)	200	30	149
	0.		25.
(6)	200	50	2145
	0.		25.
(7)	200	4	19
	0.		100.
(8)	200	30	17
	0.		100.
(9)	200	50	133
	0.		100.

We also want to generate an approximation for the distribution of t and compare this with the corresponding distribution of z. Again we can generate m random samples from a normal distribution with mean μ and standard deviation σ. For each sample of size n we set up a frequency table and finally determine the characteristics of the tabulated values. Inputs to the program are

      m = M = number of samples, $m \le 300$

      n = NSAM = sample size

    IODD = positive odd number to be used as starting value for random number generator

    NINT = number of intervals for frequency table, ≤ 30

      μ = YMU = population mean

      σ = YSIGMA = population standard deviation

```
C      T DISTRIBUTION -SAMPLING DISTRIBUTION OF T
       DIMENSION T(3ØØ),F(3Ø),Z(3ØØ)
  1ØØ  READ(5,1Ø) M,NSAM,IODD,NINT
   1Ø  FORMAT(4I5)
       READ(5,11) YMU,YSIGMA
   11  FORMAT(2F1Ø.2)
       WRITE(6,2Ø) M,NSAM,IODD
   2Ø  FORMAT(1H1, 2HM=,I6,3X,2HN=,I6,3X,5HIODD=,I6)
       WRITE(6,21) YMU,YSIGMA
   21  FORMAT(1H ,8HPOP MEAN,F1Ø.2,3X,7HPOP STD,F1Ø.2)
       SAM = NSAM
       DO 1 I=1,M
       SUMY = Ø.
       SUMY2 = Ø.
       DO 2 J=1,NSAM
       CALL RANNOR(IODD,Y,YMU,YSIGMA)
       SUMY = SUMY + Y
    2  SUMY2 = SUMY2 + Y**2
       YMEAN = SUMY/SAM
       YSTD = SQRT((SUMY2-SUMY*SUMY/(SAM- 1.))
       T(I) = (YMEAN-YMU)*SQRT(SAM)/YSTD
    1  Z(I) = (YMEAN-YMU)*SQRT(SAM)/YSIGMA
       WRITE(6,22)
   22  FORMAT(1HØ,14HZ DISTRIBUTION/1H ,3HINT,4X,9HFREQUENCY)
       CALL FREQ(Z,M,F,-6.,.5,NINT)
       WRITE(6,23)(J,F(J), J=1,NINT)
   23  FORMAT(1H ,I5,3X,F1Ø.Ø)
       CALL CHAR(Z,M,YM,YSIG,A3,A4)
       WRITE(6,24) YM,YSIG,A3,A4
   24  FORMAT(1HØ,4HMEAN,F12.2/1H ,3HSTD,F12.2/
    1   1H ,2HA3,F12.4/1H ,2HA4,F12.4///)
       WRITE(6,25)
   25  FORMAT(1HØ,14HT DISTRIBUTION/1H ,3HINT,4X,9HFREQUENCY)
       CALL FREQ(T,M,F,-6.,.5,NINT)
       WRITE (6,23) (J,F(J),J=1,NINT)
       CALL CHAR(T,M,YM,YSIG,A3,A4)
       WRITE(6,24) YM,YSIG,A3,A4
       STOP
       END
```

{Subroutine RANNOR}
{Subroutine URANDN}
{Subroutine FREQ}
{Subroutine CHAR}

   Program 10.2 T DISTRIBUTION — sampling distribution of
                    t and z.

       The input values used to generate the outputs of
Fig. 10.3 are

(1)   200   2   125   24
      0.   1.
(2)   200   10   64579   24

	0.		1.	
(3)	200	30	949	24
	0.		1.	
(4)	200	100	1347	24
	0.		1.	

To evaluate the ordinates of the t distribution, Program 8.2 EVALF, can be used. In this case, however, there is only one parameter, and thus the READ and WRITE statement for the parameters must be changed to

```
READ(5,1) DF
```

and

```
    WRITE(6,6) DF
6   FORMAT(4X,3P1=,F10.2)
```

The CALL statement must also be changed to

```
FY = F(Y,DF)
```

A new FUNCTION F must be written for the t distribution. Since we must evaluate $\Gamma(X)$ we shall use the same technique used to evaluate type III model, which is to convert to natural logarithms. Thus we have

$$\ell nf(t_d) = \ell n\Gamma\left(\frac{d+1}{2}\right) - \ell n\sqrt{d\pi} - \ell n\Gamma\left(\frac{d}{2}\right) - \left(\frac{d+1}{2}\right) \ell n \ (1 + t^2/d)$$

Once this function is evaluated we can take the antilog (exp) to determine the ordinate of the t distribution for a specific value of t. The subprogram is

```
      FUNCTION F(T,DF)
      D1 =(DF+1.)/2.
      SQ = SQRT(DF*3.141593)
      F = ALGAMA(D1) - ALOG(SQ) - ALGAMA(DF/2.)-D1*
    1 ALOG(1.+T**2/DF)
      F = EXP(F)
      RETURN
      END
```

Program 10.3 FUNCTION F — evaluation of t ordinates.

The input data used in plotting the ordinates given in Fig. 10.4 are

(1) 1.
     16   -4.00   4.00   200.   .5
(2) 9.
     16   -4.00   4.00   200.   .5
(3) 29.
     16   -4.00   4.00   200.   .5
(4) 99.
     16   -4.00   4.00   200.   .5

Program 8.6 can be used to determine the expected frequencies of the t distribution. However, certain changes must be made; thus the program and the integration routine DEFINT are given:

$$NINT = \text{number of intervals}$$
$$w = CW = \text{interval width}$$
$$b_1 = YLTB = \text{true lower boundary of first interval}$$
$$d = DF = \text{degrees of freedom for t distribution}$$
$$n = AN = \text{number of observations}$$

The program is

```
C       TFREQ - DETERMINATION OF EXPECTED T FREQUENCIES
        READ(5,1) NINT,CW,YLTB
    1   FORMAT(I2,2F1Ø.2)
        READ(5,2) AN,DF
    2   FORMAT(2F1Ø.2)
        WRITE(6,3)
    3   FORMAT(1H1, 9HNO OF OBS,F1Ø.Ø,3X,18HDEGREES OF
       1  FREEDOM,F1Ø.Ø)
        WRITE(6,4)
    4   FORMAT(1HØ,1ØX,8HINTERVAL,15X,5HEXP F)
        M = 1ØØ
        Y2 = YLTB + CW
        AREA = 1.-DEFINT(Y2,24.,M,DF)
        EXPF = AN*AREA
        WRITE(6,5) YLTB,Y2,EXPF
    5   FORMAT(1HØ,5X,F1Ø.2,3X,F1Ø.2,3X,F1Ø.2)
        NINT2 = NINT - 2
        DO 6 I=1,NINT2
        Y1 = Y2
        Y2 = Y2 + CW
        AREA = DEFINT(Y1,Y2,M,DF)
        EXPF = AN*AREA
```

```
      WRITE(6,5) Y1,Y2,EXPF
   6  CONTINUE
      Y1 = Y2
      Y2 = Y2 + CW
      AREA = DEFINT(Y1,24.,M,DF)
      EXPF = AN*AREA
      WRITE(6,5) Y1,Y2,AREA
      STOP
      END
```

{Function DEFINT}
{Function F}
{Function ALGAMA}

Program 10.4 TFREQ — expected t frequencies.

The subprogram DEFINT is

```
      FUNCTION DEFINT (Y1,YM,M,DF)
      DEFINT = 0
      XM = M
      DELTAY = (YM - Y1)/XM
      Y = Y1 - DELTAY/2.
      DO 1 I=1,M
      Y = Y + DELTAY
      AREA = F(Y,DF) * DELTAY
   1  DEFINT = DEFINT + AREA
      RETURN
      END
```


Program 10.5 FUNCTION DEFINT — numerical integration
routine for the t distribution.

The input data used to produce the output given in
Fig. 10.6 are

(1)   16   .5   -4.00
      200.   1.
(2)   16   .5   -4.00
      200.   9.
(3)   16   .5   -4.00
      200.   29.
(4)   16   .5   -4.00
      200.   99.

The student should note that the t distribution has a large variance when the degrees of freedom are small. Thus, to evaluate the $\int_{y_1}^{\infty} f(t)\,dt$ the value assigned to the upper limit must be large enough so that the $f(t) > 0.00001$ for the t value used as the upper limit. This limit is specified in the statement AREA = 1.-DEFINT(Y2,24.,M,DF) and AREA=DEFINT(Y1,24.,M,DF) by the 24.. This upper limit is sufficient in most cases. However, when the degrees of freedom are equal to 1(n=2), the use of 24 creates an underestimate of the area because a significant amount of the area is beyond 24. In this case the program could be changed so that the upper limit is 100.

10.9 Summary

The following concepts were presented in this chapter:

(1) The unbiased sample estimator of the population variance σ^2 is

$$ s'^2 = \frac{\Sigma(y_i - \bar{y})^2}{n - 1} $$

(2) When the population variance is unknown and the random variable y has a normal distribution, then

$$ t = \frac{\bar{y} - \mu}{s'/\sqrt{n}} $$

has a Student's t distribution, with d = n - 1 degrees of freedom. This distribution can then be used to solving the standard statistical inference problems. To determine the sample size an iterative technique is required because the distribution of t implicitly involves n.

(3) When the population is nonnormal and the sample size is small so that the central limit theorem cannot be used, one can (a) evolve the "distribution," (b) use the t ratio and hope it is a good approximation, (c) transform the original variable so that it is approximately normally distributed, (d) approximate the sampling distribution through sampling simulation using the computer, or (e) use a non-

parametric technique.

(4) To make inferences about a population percentage, a 0-1 artificial population can be introduced. For large samples the standard z ratio for the mean can then be used.

(5) To make inferences about parameters other than the mean, standard error theory can be used. The critical ratio

$$z = \frac{c - \theta}{se(\sigma,n)}$$

where c is the statistic used to estimate θ and se (σ,n) is the standard deviation of the statistic, can be assumed to follow a normal distribution if n is sufficiently large. The bias of the statistic relative to the estimation of the parameter must be given consideration.

EXERCISES

1. Modify the sample generation program, Program 10.1, to compute s'^2. Use the modified program to demonstrate that s'^2 is an unbiased estimator of σ^2.

2. Use the modified program written in Exercise 1 to illustrate that the sampling distribution of s'^2 is independent of the population mean. (Hint: Run the program for different values of μ, and show that regardless of the value of μ used, the generated s'^2 distribute about σ^2 value in the same fashion.

3. The statistic $t = (\bar{y} - \mu)/s'$ has a Student's t distribution. Program 10.2 generates samples from a normal population with mean μ and standard deviation σ and can be used to observe the differences in the sampling distribution of $z = (\bar{y} - \mu)/\sigma$ and t.

   (a) Run the program using a small sample size.
   (b) Run the program using a large sample size to show that as the sample sizes increases, the t distribution approaches the z distribution.
   (c) Repeat (a) and (b) for $\mu \neq 0$ and $\sigma \neq 1$ to show that the sampling distributions of t or z are not affected by the mean and standard deviation.

4. To further demonstrate that the t distribution approaches the normal distribution as the sample size increases, adapt Program 8.2, EVALF, to obtain ordinates of the t distribution by writing and using a FUNCTION subprogram for the t distribution. Plot the ordinates of the t distribution for degrees of freedom = 2, 5, 10, and 30 on the same graph. Also plot the z distribution on the same graph, and compare the five distributions.

5. Write a FUNCTION subprogram to determine the $(1 - \alpha)100$ percent confidence interval for μ from a large sample,

given $z_{1-\alpha/2}$, $\bar{y}$, s', and n. Write a program that uses the FUNCTION subprogram to study how the size of the confidence interval is affected by changing the value of n.

6. Write a program to estimate the sample size required to meet a certain precision. The value of σ^2, the precision requirement ε, and the critical value for the specified confidence level $z_{1-\alpha/2}$ must be read in by the program. Use the program to study how the sample size is affected by changing the requirements.

7. The DDT standards require that soil has no more than 2.0 parts per million (ppm) of DDT. If the observations given in Exercise 1, Chap. 2 represent random samples of soil after an application of DDT, would you be willing to accept the hypothesis that $\mu = 2$ against the alternative hypothesis that $\mu > 2$, using the 5 percent level of significance?

8. The calibration program for radar location of a stationary communication satellite requires that observations of the satellite location are such that the horizontal component error has a mean of less than one nautical mile. Using the 1 percent level of significance and the observations given in Exercise 2, Chap. 2, would you conclude that the radar should be recalibrated?

9. Using the observations given in Exercise 1, Chap. 2 as a random sample, find the 97 percent confidence interval for μ, the mean level of DDT (ppm) in the soil.

10. If the observations given in Exercise 2, Chap. 2 represent your past experience, and it has been shown that the variance does not change with time, how large is a sample needed to estimate the mean of the horizontal error component such that there would be only 95 chances in 100 that the difference between the sample mean and true mean would be greater than 0.025?

11. Write a program that will generate samples of size n from the rectangular distribution, using the FUNCTION URANDN. Determine the value of t for each sample, and group these t values into a frequency distribution. Print out the resulting frequency distribution. Since the central limit theorem enables us to use the normal model for the distribution of the t ratio for n > 30, even when sampling from the rectangular distribution, we can use the program to see the same holds true for n = 10.

12. Write a subroutine that will read in an array of observations and perform a specified transformation on them. Use a code to indicate the desired transformation. The choices of transformations should include

(a) log Y
(b) $\sqrt{Y}$
(c) arctan Y

(Note: to perform the log and $\sqrt{Y}$ transformations, the observations must all be greater than or equal to zero. Print out error messages, if the desired transformations are not possible. Also compute the moments of the original and transformed observations, and print out the two sets of moments. Since the transformations are generally used to produce a normal variable, make up some data that have a positive skewness, and use the program to see which transformations are most effective.)

CHAPTER 11

INTRODUCTION TO STATISTICAL DECISION MAKING

11.1 Introduction

In real life decisions dominate our activities. Every day we must decide what time to awake, what to wear, whether to shower or not, what to eat for breakfast, and how to get to work. We could become like Pavlov's dog and simply following a ritualistic response or decision pattern, never even concerning ourselves as to why we follow the particular pattern. Such a mechanistic way of life may suffice for the more mundane aspects of life, and, in fact, it is perhaps best not to worry about the unessential details. There are, however, major problems that are encountered which warrant something more than a routine response. Many such problems are present in the analysis of data.

In the previous chapters decisions had to be made in order to perform the required analyses. When collecting new data, for example, we needed to determine how many observations to use. When the data were grouped, the intervals to be used for the tabulation had to be determined. In selecting a mathematical model to represent the distribution of an observable random variable, decisions had to be made about the form of the mathematical model and about how to estimate the parameters. In statistical inferences about model parameters, decisions as to what confidence level to use had to be made. Statistical decision theory will provide us with the means of selecting the "best" number of observations, intervals, estimation technique, or confidence level. Thus it seems appropriate to consider this aspect of statistics before proceeding with other aspects of statistical methodology.

297

11.2 An Illustration

To introduce statistical decision theory, let us change our illustration from technical problems to one that is closer to everyday life. Consider the decision problem of an individual who has a diabetic condition that is being controlled through a diet program. His decision problem is to determine whether to (1) continue his present diet, (2) modify the diet himself, or (3) go see the doctor for a more thorough checkup. These decisions, which can also be called actions, will be referred to as a_1, a_2, and a_3, respectively.

Assume that the diabetic's present condition can be described by one of the following three categories, which are called states of nature: (1) the diabetic condition is being controlled by the present diet, (2) the diabetic condition is slightly out of control, or (3) the condition has drastically changed. If the diabetic knew his state of nature, he could easily determine the proper action; however, because he does not, he must perform some experiment that will provide him with information as to which state of natur prevails.

Let us assume that the experiment he decides to use consists of obtaining three independent urine tests by using chemically treated strips of tape. He quantifies the experimental results by simply counting the number of positive readings obtained. A reading is considered positive when the tape changes color because this indicates the presence of sugar in the urine. If we let Y be the number of positive readings observed, then Y can have only one of four possible outcomes or values: Y = 0, 1, 2, or 3.

If the diabetic wants to base the decision upon the value of Y he observes (after his three tests), he must set up a correspondence between the possible outcomes Y = 0, 1, 2, or 3 and the three available actions a_1, a_2, or a_3. Such a correspondence is called a strategy. The patient might evolve the strategy $[a_1, a_2, a_2, a_3]$, which indicates that if Y = 0, he will take action a_1; if Y = 1 or 2 he will take action a_2; and if Y = 3 he will take action a_3. The set of all possible strategies can be obtained by permuting the actions associated with each value of Y. In this case there are 3^4 = 81 different possible strategies.

One of the tasks of statistical decision theory is to develop a system for evaluating these competitive strategies. To illustrate the procedure, let us numerically follow the steps required in making a strategy evaluation, although we shall restrict ourselves to comparing the strategy given by $[a_1, a_2, a_2, a_3]$ with the strategy given by $[a_1, a_2, a_3, a_3]$.

Step 1

Identification of the set of possible actions or decisions that can be made:

a_1 = continue present diet
a_2 = modify diet
a_3 = go see doctor

Step 2

Designation of the finite set of possible outcomes for the observable random variable Y:

y_1 = 0
y_2 = 1
y_3 = 2
y_4 = 3

Step 3

Representation of the finite set of possible states of nature. In statistical decision theory, the states of nature must be described in terms of parameter values of the distribution model for the random variable. In our case we can use the binomial model given by

$$p(Y=y) = C_y^3 \, p^y (1 - p)^{3-y} \quad y = 0, 1, 2, 3$$

where p is the probability of obtaining a positive reading on a single test strip and can have any value between zero and 1.

We have assumed that there are only three possible
states of nature. Each state must be represented by a particular value of p obtained from past experience; thus

$\theta_1:p_1$ = 0.08   state 1 = controlled
$\theta_2:p_2$ = 0.50   state 2 = slightly out of control
$\theta_3:p_3$ = 0.85   state 3 = out of control

The value of p_1 is not zero because there is always a
chance of obtaining a positive reading due to poor quality
tape or an error in interpretation.

Step 4

Determination of the probabilities of occurrences.
By using the binomial model we can find the probability of
occurrence for each of the four possible outcomes of the
random variable, given a particular state of nature. This
conditional probability is represented by

$$Pr[Y=y/\theta_k]$$

which is read as "the probability of Y being equal to y
given the state of nature θ_k."
We thus have a different binomial distribution for each
state of nature:

$$Pr[Y=y/\theta_1] = Pr(Y=y/p_1 = 0.08) = C_y^3(0.08)^y(0.92)^{3-y}$$

$$Pr[Y=y/\theta_2] = Pr(Y=y/p_2 = 0.50) = C_y^3(0.50)^y(0.50)^{3-y}$$

$$Pr[Y=y/\theta_3] = Pr(Y=y/p_3 = 0.85) = C_y^3(0.85)^y(0.15)^{3-y}$$

$$Y = 0, 1, 2, 3$$

If we evaluate these probabilities for the four possible
outcomes, we obtain the required probabilities given in Fig.
11.1:

| | Possible Outcomes for Y | | | |
States of Nature	$Y = 0$	$Y = 1$	$Y = 2$	$Y = 3$
$\theta_1 : p_1 = 0.08$	0.7787	0.2031	0.0177	0.0005
$\theta_2 : p_2 = 0.50$	0.1250	0.3750	0.3750	0.1250
$\theta_3 : p_3 = 0.85$	0.0034	0.0574	0.3251	0.6141

Fig. 11.1 Probabilities of occurrence.

These are really theoretical probabilities; for example, the entry 0.0177 means that the probability of obtaining two positive tests when the system is being controlled ($\theta_1 = p_1 = 0.08$) is equal to 0.0177 (less than 2 chances out of 100).

Step 5

Determination of the losses. A numerical value that represents the loss of taking each action when a particular state of nature prevails must be attained.

Consider the loss when action a_3 (see doctor) is taken and when the state of nature is θ_1 (system under control). In this case the patient must needlessly arrange for an appointment, take the time to keep the appointment, undergo the required tests, pay the bill, and even run the risk that the doctor may not diagnose his problem. The student should consider the losses that can occur for each state of nature (Fig. 11.2):

| | Actions | | |
| | a_1 | a_2 | a_3 |
States of Nature	(continue diet)	(change diet)	(see doctor)
θ_1 control	1	5	15
θ_2 slightly out of control	8	3	12
θ_3 out of control	25	20	10

Fig. 11.2 Losses.

The losses do not need to represent actual costs because it is their relative values that are important in evaluating strategies. Therefore we often assign a 1 to the cell having the smallest loss.

The diagonal values (1,3,10) represent the preferred actions for the indicated state of nature. These losses are nonzero since even when the correct action is taken there is a direct cost. The off-diagonal values represent the loss associated with the improper actions; they are larger because they include not only the direct costs but also the penalty that one would eventually have to pay for such a wrong decision.

Step 6

Designation of the competitive strategies. For the diabetic illustration we restrict the evaluation problem to that of choosing between the two strategies

$$s_1 = [a_1, a_2, a_2, a_3] \text{ and } s_2 = [a_1, a_2, a_3, a_3]$$

Step 7

Evaluation of the action probabilities. The probability of taking each action, given each state of nature, is required for each strategy. Strategy $s_1 = [a_1, a_2, a_2, a_3]$ indicates that action a_1 should be taken if $Y = 0$, a_2 if $Y = 1$ or $Y = 2$, and a_3 if $Y = 3$. To evaluate the probability of taking each action, the probabilities of occurrence must be used. We obtain

$$Pr[a_1/\theta_1] = Pr[Y = 0/\theta_1] = 0.7787$$

$$Pr[a_2/\theta_1] = Pr[Y = 1/\theta_1] + Pr[Y = 2/\theta_1]$$

$$= 0.2031 + 0.0177 = 0.2208$$

$$Pr[a_3/\theta_1] = Pr[Y = 3/\theta_1] = 0.0005$$

Similar probability summations yield the $Pr[a_i/\theta_2]$ and $Pr[a_i/\theta_3]$, which are given in Fig. 11.3:

States of Nature	Actions		
	a_1	a_2	a_3
θ_1	0.7787	0.2208	0.0005
θ_2	0.1250	0.7500	0.1250
θ_3	0.0034	0.3825	0.6141

Fig. 11.3 Action probabilities for s_1.

The action probabilities for $s_2 = [a_1, a_2, a_3, a_3]$ for the state of nature θ_1 are

$$Pr[a_1/\theta_1] = Pr[Y = 0/\theta_1] = 0.7787$$

$$Pr[a_2/\theta_1] = Pr[Y = 1/\theta_1] = 0.2031$$

$$Pr[a_3/\theta_1] = Pr[Y = 2/\theta_1] + Pr[Y = 3/\theta_1]$$

$$= 0.0177 + 0.0005 = 0.0182$$

The same procedure is used to obtain the action probabilities when the state of nature is θ_2 and θ_3. These probabilities are given in Fig. 11.4:

States of Nature	Actions		
	a_1	a_2	a_2
θ_1	0.7787	0.2031	0.0182
θ_2	0.1250	0.3750	0.5000
θ_3	0.0034	0.0571	0.9395

Fig. 11.4 Action probabilities for s_2.

Step 8

Determination of the average loss. For each competing strategy we need to determine the average loss for each state of nature. These average losses are found by weighting each possible loss by its corresponding action probability. Thus for strategy s_1 we have

$$\bar{\ell}_{\theta_1}(s_1) = (0.7787)1 + (0.2208)5 + (0.0005)15 = 1.8902$$

$$\bar{\ell}_{\theta_2}(s_1) = (0.1250)8 + (0.7500)3 + (0.1250)12 = 4.7500$$

$$\bar{\ell}_{\theta_3}(s_1) = (0.0034)25 + (0.3825)20 + (0.6141)10 = 13.8760$$

and for strategy s_2 we have

$$\bar{\ell}_{\theta_1}(s_2) = (0.7787)1 + (0.2031)5 + (0.0182)15 = 2.0672$$

$$\bar{\ell}_{\theta_2}(s_2) = (0.1250)8 + (0.3750)3 + (0.5000)12 = 8.1250$$

$$\bar{\ell}_{\theta_3}(s_2) = (0.0034)25 + (0.0571)20 + (0.9395)10 = 10.6220$$

To better consider the next step we display these average losses in a single tabulation in Fig. 11.5:

	Strategies	
States of Nature	s_1	s_2
θ_1	1.8902	2.0672
θ_2	4.7500	8.1250
θ_3	13.8760	10.6220

Fig. 11.5 Average losses.

If we knew which state of nature prevailed, we could easily determine the strategy to follow. If the first state of nature prevails, s_1 is superior to s_2 because its average loss is smaller, but if the third state of nature prevails, one should prefer s_2. We do not know which state of nature prevails, and it is this uncertainty that led to the testing procedure. The problem is how to select a good strategy when the actual state of nature is unknown; we shall consider three approaches (criteria) to making such a selection

1. The minimax criterion

To use the minimax criterion we determine the maximum average loss that could be realized for each particular strategy and then select the strategy whose maximum average loss is the smallest (minimum). In our case, s_1 has a maximum average loss of 13.8760, and s_2 has a maximum average loss of only 10.6220. The minimax criterion thus selects strategy s_2, and under this criterion the diabetic would follow the actions designated by strategy s_2.

2. The minimum expected loss (Bayes criterion)

If the relative frequency of occurrence for each state of nature is available prior to obtaining any observations (prior probabilities), it can be used to weight the average loss for each state to obtain an expected loss for each strategy. These prior probabilities can be thought of as "no-data" or prior probabilities that we are willing to associate with the state of nature.

Let q_1, q_2, and q_3 be the prior probabilities for θ_1, θ_2, and θ_3, respectively. The expected loss for strategy s_1 is thus given by

$$EL(s_1) = q_1 \bar{\ell}_{\theta_1}(s_1) + q_2 \bar{\ell}_{\theta_2}(s_1) + q_3 \bar{\ell}_{\theta_3}(s_1)$$

For example, if $q_1 = 0.50$, $q_2 = 0.35$, and $q_3 = 0.15$, then

$$EL(s_1) = (0.50)(1.8902) + (0.35)(4.7500) + (0.15)(13.8760)$$

$$= 4.6890$$

and similarly the expected loss for s_2 uses the average losses for s_2; we obtain

$$EL(s_2) = (0.50)(2.0672) + (0.35)(8.1250) + (0.15)(10.6220)$$

$$= 5.4707$$

The minimum expected loss (Bayes) criterion selects the strategy whose expected loss is a minimum. We would thus select s_1 as the Bayes strategy for the given set of prior probabilities. The student should note that the Bayes strategy is sensitive to the set of q's. If we had used the set of q's given by (0.35,0.10,0.55) instead of the set (0.50, 0.35,0.15), s_2 would be the Bayes strategy.

3. The minimum expected regret or minimum risk criteria

A criterion comparable to the Bayes criterion considers regrets rather than losses in comparing the strategies. Regrets are obtained from the losses by simply subtracting the minimum loss in each row from the other losses that occur in the row. For example, in our problem the losses are

	Actions		
States of Nature	a_1	a_2	a_3
θ_1	1	5	15
θ_2	8	3	12
θ_3	25	20	10

If we take action a_1, and θ_3 prevails, we should "regret" it with a weight of 25 - 10 = 15 rather than the full 25 units because even if we had taken the correct action a_3, this action would result in a loss of 10. The regrets obtained are shown in Fig. 11.6:

	Actions		
States of Nature	a_1	a_2	a_3
θ_1	0	4	14
θ_2	5	0	9
θ_3	15	10	0

Fig. 11.6 Regrets.

We must generate the average regrets and then weigh them according to the prior probabilities as we did in the Bayes

procedure. The average regrets and the expected regrets obtained are

States of Nature	Prior Probabilities	Average Regret	
		s_1	s_2
θ_1	0.50	0.8902	1.0672
θ_2	0.35	1.7500	5.1250
θ_3	0.15	3.8760	0.6220
Expected regret		1.6390	2.42065

In this case s_1 would still be the preferred strategy because it has the minimum regret.

The selection of a single strategy from a set of
competing strategies depends upon not only the inputs used
in the evaluation but also upon whether losses or regrets
are used and the particular criterion that is selected.
Often the decision problem will dictate the approach to be
used. For example, when one state of nature has losses that
are significantly larger than those for other states, regrets may be used to dampen the influence of the large
losses. If, however, certain losses are large enough to
cause bankruptcy or ruin, then the minimax criterion may be
used to avoid the ruin. This situation is often encountered
in military strategy problems or in long-range decision making for small businesses.

11.3 Two-dimensional Arrays

Although the above procedures and computations are not
too involved, it is easy to appreciate how such strategy
evaluations can become very burdensome. In the simplified
diabetic decision problem we could easily increase the number of strategies to 12 or even to the total number of
possible strategies, which is $3^4 = 81$. In addition, the
number of states of nature could be increased because the
continuous nature of the parameters of the mathematical
model allows the introduction of additional states of nature

to represent intermediate values of the parameter.

In this section we shall present the necessary algebraic notiation to formulate the general decision theory problem and the corresponding FORTRAN concepts that will then be used to write a general program to make the evaluations.

We already introduced the algebraic representation y_1, y_2, $\ldots$,y_j, $\ldots$,y_n, which lists the possible outcomes of the random variable Y. Such a list is called a vector and can be written as $[y_1, y_2, \ldots , y_j, \ldots , y_n]$ or in condensed form as $\{y_j : j = 1, n\}$, where y_j represents the typical element and there is a total of n elements. To store and manipulate such vectors in a FORTRAN program, one-dimensional arrays are used (introduced in Chap. 6). In fact, we have already used the FORTRAN array (Y(J),J=1,N) to store vectors.

In the diabetic decision theory problem the losses were given by

	a_1	a_2	a_3
θ_1	1	5	15
θ_2	8	3	12
θ_3	25	20	10

Such a two-way representation is called a matrix. To represent the elements of this loss matrix in general, we use the letter such as the letter ℓ. Each element can then be uniquely identified by two indices (subscripts), one to designate the state of the nature and the other to designate the action:

$$
\begin{array}{c}
\begin{array}{cccccc}
 & a_1 & \cdots & a_i & \cdots & a_n
\end{array} \\
\begin{array}{c}
\theta_1 \\ \vdots \\ \theta_k \\ \vdots \\ \theta_m
\end{array}
\left[
\begin{array}{cccccc}
\ell_{11} & \cdots & \ell_{1i} & \cdots & \ell_{1n} \\
\vdots & & \vdots & & \vdots \\
\ell_{k1} & \cdots & \ell_{ki} & \cdots & \ell_{kn} \\
\vdots & & \vdots & & \vdots \\
\ell_{m1} & \cdots & \ell_{mi} & \cdots & \ell_{mn}
\end{array}
\right]
\end{array}
$$

Fig. 11.7 Loss matrix.

In our particular example both m and n are equal to 3.

We can designate the matrix by the following condensed notation: $\{\ell_{ki}; i = 1,n; k = 1,m\}$, where ℓ_{ki} represents the typical element. The dimension of the matrix is determined by the number of rows(m) and the number of columns(n) and is represented by [m,n]. The total number of elements in the matrix is m · n.

Such matrices can be stored as two-dimensional FORTRAN arrays. For example, the losses can be stored in an array called ((ACTLOS(K, I), I=1,N) K=1,M). The two subscripts K and I must be integer variables or constants.

As with one-dimensional arrays, a DIMENSION statement is needed to reserve the maximum number of storage locations needed for the elements of the array. The values used in the DIMENSION statement must be greater than or equal to the number of rows and columns of the matrix. The following FORTRAN statements could be used to define and store all the values of the loss matrix for the diabetic example:

```
DIMENSION ACTLOS(3,3)
ACTLOS(1,1) = 1.
ACTLOS(1,2) = 5.
ACTLOS(1,3) = 15.
ACTLOS(2,1) = 8.
ACTLOS(2,2) = 3.
ACTLOS(2,3) = 12.
ACTLOS(3,1) = 25.
ACTLOS(3,2) = 20.
ACTLOS(3,3) = 10.
```

We can manipulate this array in the same manner as the one-dimensional one, except that both subscripts must be specified. Thus we can use DO loops to find the sum of all the elements in the matrix, as seen below.

```
  SUMLOS=0                            SUMLOS=0
  DO 1 K=1,3                          DO 1 K=1,3
  DO 2 I=1,3                          DO 1 I=1,3
2 SUMLOS=SUMLOS+ACTLOS(K,I)         1 SUMLOS=SUMLOS+ACTLOS(K,I)
1 CONTINUE
```

Both procedures go through identical steps. The first
time, K is set to 1, then I is set to 1, and SUMLOS=0 +
ATLOS(1,1). Then K remains equal to 1, I is incremented by
1, and ACTLOS(1,2) is added to SUMLOS. The index I is
again incremented by 1 so that I=3, and thus ACTLOS(1,3) is
added to SUMLOS. The inner loop has now been completed for
K=1. K is then incremented to 2, and all the values of I
are again stepped through. The procedure is repeated for
K=3 until the last element ACTLOS(3,3) is added to SUMLOS
and the two loops are completed.

It should be appreciated that the nested DO loops used
in determining SUMLOS are equivalent to double summation
notation. Algebraically we would write the following:

$$\text{SUMLOS} = \sum_{k=1}^{3} \sum_{i=1}^{3} \ell_{ki} = \ell_{11} + \ell_{12} + \ell_{13} + \ell_{21} + \ell_{22} + \ell_{23}$$
$$+ \ell_{31} + \ell_{32} + \ell_{33}$$

We also need to know how to read and write multidimensional arrays or lists. We could read the array ACTLOS(K,I)
by any of the following three methods:

(1) Two DO loops

```
      DO 2   K=1,3
      DO 2   I=1,3
    2 READ(5,1)  ACTLOS(K,I)
    1 FORMAT(1X,F10.2)
      DO 3   K=1,3
      DO 3   I=1,3
    3 WRITE(6,1) ACTLOS(K,I)
```

(2) One DO loop and one implied loop

```
      DO 2   K=1,3
    2 READ(5,1)(ACTLOS(K,I),I=1,3)
    1 FORMAT(1X,F10.2)
      DO 3   K=1,3
    3 WRITE(6,1)(ACTLOS(K,I),I=1,3)
```

(3) Two implied loops

```
        READ(5,1) ((ACTLOS(K,I),I=1,3),K=1,3)
      1 FORMAT(1X,F10.2)
        WRITE(6,1) ((ACTLOS(K,I),I=1,3),K=1,3)
```

All these methods will read in the same set of punch cards
(one value per card) and print the values, one per line. If
there is more than one value per data card, then method (2)
or (3) should be used with the appropriate format. If method
(2) is used, the values for a new row (when the value of K
changes) must always start on a new card. If method (3) is
used, all the values for the entire matrix could be read
from the same data card (i.e., if there is sufficient space).
We leave it to the student to experiment with the effects of
using these different methods.

Multidimensional arrays can be passed to subroutines
and functions in the same manner as single-dimensioned
arrays. Thus we can have a subroutine called SUML that calculates the sum of all the losses while the main program
reads the input data and writes the result. In this case
the main program could simply consist of the following
FORTRAN statements:

```
        DIMENSION ACTLOS(10,10)
        READ(5,1) ((ACTLOS(K,I),I=1,3),K=1,3)
        CALL SUML(ACTLOS,SUMLOS)
        WRITE(6,1) SUMLOS
      1 FORMAT(1X,F10.3)
        STOP
        END
```

The array ACTLOS that is passed as a subroutine argument
must be dimensioned in the subroutine, although the dimension can be less than or equal to the dimension of the same
array specified in the main program.

The following dimension statement can be used in the subroutine SUML:

```
SUBROUTINE SUML(ACTLOS,RESULT)
DIMENSION ACTLOS(10,10)
    .
    .

(Calculation of the sum of the losses)
    .
    .

RETURN
END
```

The same DIMENSION requirement exists when passing values of a matrix to a FUNCTION subprogram.

The use of dummy dimension of 1 as introduced in passing one-dimensional arrays (vectors) to subprograms can also be used in multidimensional arrays. The use of this dummy dimension is, however, restricted to the last index. Thus we could have used the statement DIMENSION ACTLOS(10,1) in the above subroutine.

11.4 The Generalized Decision Theory Formulation

We shall now take advantage of the vector and matrix notation to represent the general decision theory problem of selecting the best strategy from a set of competing strategies. We shall proceed through the general formulation, using the same sequence used in the diabetic example. This example will be used throughout this section to relate each step of the general formulation to a particular example.

Since we also want to write a computer program to perform the required evaluations, we shall give FORTRAN names to the variables and arrays along with the necessary FORTRAN statements to perform the calculations. The complete computer program will then be given in the next section.

Step 1

Identification of the actions. In general, the actions can be represented algebraically by the vector $[a_1, a_2, \cdots, a_i, \cdots, a_{n_a}]$ or $\{a_i : i = 1, n_a\}$, where n_a is the total number of actions. In the diabetic example the vector is $[a_1, a_2, a_3]$ and $n_a = 3$.

However, an action can also be identified by simply using the index i. Thus in the FORTRAN program we need only specify the total number of actions by NA and use the capital letter I to denote the index that identifies the action.

Step 2

The possible outcomes of the random variable. The possible outcomes of the random variable Y can be given by the vector $[y_1, y_2, \cdots, y_j, \cdots, y_{n_y}]$ or $\{y_j : j = 1, n_y\}$, where n_y is the total number of possible outcomes. In the illustrative example, the vector is $[0, 1, 2, 3]$ and $n_y = 4$. Since the possible outcomes can be any set of real numbers (not always integer values in a certain sequence), we shall use a one-dimensional FORTRAN array, (Y(J),J=1,NY), to store the NY values.

Step 3

The states of nature. The kth state of nature is represented by θ_k. Each of these states of nature must have an associated mathematical model (with specified parameter values) that gives the distribution of the random variable Y. Thus we shall let θ_k represent the parameter value of the particular model that is to be associated with that state of nature. We therefore have the vector $[\theta_1, \theta_2, \cdots, \theta_k, \cdots, \theta_{n_\theta}]$ or $\{\theta_k : k = 1, n_\theta\}$, where n_θ is the total number of states of nature. In the illustrative example, $n_\theta = 3$, and the vector represents the values of the binomial parameter p. Thus $[\theta_1, \theta_2, \theta_3] = [p_1, p_2, p_3] = [0.08, 0.50, 0.85]$.

We shall restrict our program to states of nature that are specified by one model parameter; thus we can use a one-dimensional FORTRAN array, (THETA(K),K=1,NT), to store these parameter values.

Step 4

Determination of the probabilities of occurrence. The probabilities of occurrence depend upon the state of nature and the value of the random variable, so we must use a matrix representation. In the illustrative example, these probabilities were given by the following matrix:

$$
\begin{array}{ccccc}
 & Y = 0 & Y = 1 & Y = 2 & Y = 3 \\
\theta_1 & \begin{bmatrix} 0.7787 & 0.2031 & 0.0177 & 0.0005 \\ \theta_2 \\ \theta_3 \end{bmatrix} \\
\end{array}
$$

$$
\begin{array}{c c c c c}
 & Y = 0 & Y = 1 & Y = 2 & Y = 3 \\
\theta_1 & 0.7787 & 0.2031 & 0.0177 & 0.0005 \\
\theta_2 & 0.1250 & 0.3750 & 0.3750 & 0.1250 \\
\theta_3 & 0.0034 & 0.0574 & 0.3251 & 0.6141
\end{array}
$$

In general, each element of the probability of occurrence matrix will be given by f_{kj} (k for the state of nature and j for the outcome). The matrix will have the dimension $[n_\theta, n_y]$; it is given by

Outcomes of Y

$$
\begin{array}{c}
\theta_1 \\
\vdots \\
\theta_k \\
\vdots \\
\theta_{n_\theta}
\end{array}
\begin{bmatrix}
f_{11} & \cdots & f_{1j} & \cdots & f_{1n_y} \\
\vdots & & \vdots & & \vdots \\
f_{k1} & \cdots & f_{kj} & \cdots & f_{kn_y} \\
\vdots & & \vdots & & \vdots \\
f_{n_\theta 1} & \cdots & f_{n_\theta j} & \cdots & f_{n_\theta n_y}
\end{bmatrix}
$$

Fig. 11.8 Probability of occurrence matrix.

or $\{f_{kj} : j = 1, n_y ; \ k = 1, n_\theta\}$.

Each element of the matrix represents the conditional probability

$$
f_{kj} = \Pr[Y = y_j / \theta_k]
$$

and this probability depends upon the particular mathematical model that is used to represent the distribution of the random variable Y.   In the diabetic example, the binomial model is used, with the number of trials being three.   Thus

$$f_{kj} = C^3_{y_j} \ (p_k)^{y_j} \ (1 - p_k)^{3-y_j}$$

We shall use a two-dimensional FORTRAN array to store the probabilities of occurrence and use the name $((YPR(K,J), J=1,NY),K=1,NT)$.   The FORTRAN arrays and variables needed to calculate the probabilities of occurrence are

(1) $(Y(J),J=1,NY)$
(2) $(THETA(K),K=1,NT)$
(3) NY
(4) NT

and the output is the array $((YPR(K,J),J=1,NY),K=1,NY)$.

Since the calculation of these probabilities depends upon the mathematical model for the distribution of the random variable, we decide to calculate them in a subroutine. Thus when the model is changed, the subroutine can simply be replaced by another one, without any changes having to be made in the main program.   The subroutine called YPROB, which is needed to calculate the probabilities of occurrence for the binomial model, is given by Program 11.1:

```
       SUBROUTINE YPROB(Y,THETA,NY,NT,YPR)
       DIMENSION Y(1),THETA (1),YPR(10,1)
       N = NY-1
       DO 1 K=1,NT
       P = THETA(K)
       DO 1 J=1,NY
       IY = Y(J)
   1   YPR(K,J) = BINOM(N,IY,P)
       RETURN
       END
```

Program 11.1 SUBROUTINE YPROB — calculation
                   of probabilities of occurrence
                   for a binomial model.

All the arrays used in SUBROUTINE YPROB are also needed in the main program; thus dummy dimension values can be used in the subroutine. The subroutine calls a function BINOM, which calculates the binomial probabilities for a value of N, IY, and P. This routine is

```
FUNCTION BINOM (N,IY,P)
Q = 1. -P
BINOM = FACT(N)/(FACT(N-IY)*FACT(IY)) *P**IY
1  *Q**(N-IY)
RETURN
END
```

Program 11.2 FUNCTION BINOM — calculation
of binomial probabilities.

The FUNCTION BINOM in turn calls the FUNCTION FACT, which has already been given in Program 7.1, Chap. 7. This procedure for calculating the binomial probabilities is not the optimal method, but because it is easy to follow it is used in this program. There are many faster and more accurate methods to obtain the probabilities without calling two additional functions, but such improvements are left to the student.

Step 5

Determination of the losses. For each state of nature and for each action the losses must be specified; thus we use the matrix $\{\ell_{ki} : i = 1, n_a; k = 1, n_\theta\}$. The illustration of this matrix was given when we introduced the concept of matrices in the previous section.

We use the two-dimensional FORTRAN array ((ACTLOS(K,I), I=1,NA),K=1,NT) to store the values of the losses.

Step 6

Designation of the competitive strategies. In the two-strategy evaluation in Sec. 11.2 we had the following matrix:

$$
\begin{array}{cccc}
Y = 0 & Y = 1 & Y = 2 & Y = 3 \\
\end{array}
$$

$$
\begin{array}{c}
s_1 \\
s_2
\end{array}
\begin{bmatrix}
a_1 & a_2 & a_2 & a_3 \\
a_1 & a_2 & a_3 & a_3
\end{bmatrix}
$$

316

In general we shall use the index m to represent the mth stragegy and n_s to represent the total number of strategies being considered. If we let d_{mj} represent the action specified by strategy s_m when y_j is observed, we can represent all the strategies by the following matrix or the condensed form $\{d_{mj} : j = 1, n_y; \; m = 1, n_s\}$:

Outcomes of Y

$$
\begin{array}{c}
s_1 \\
\vdots \\
s_m \\
\vdots \\
s_{n_s}
\end{array}
\left[
\begin{array}{ccccc}
d_{11} & \cdots & d_{1j} & \cdots & d_{1n_y} \\
\vdots & & \vdots & & \vdots \\
d_{m1} & \cdots & d_{mj} & \cdots & d_{mn_y} \\
\vdots & & \vdots & & \vdots \\
d_{n_s 1} & \cdots & d_{n_s j} & \cdots & d_{n_s n_y}
\end{array}
\right]
$$

Fig. 11.9 Matrix of strategies.

The dimension of this matrix is $[n_s, n_y]$, and in the diabetic example $n_s = 2$ and $n_y = 4$. The set of values that can be used for d_{mj} is, of course, restricted to the set of possible actions.

We recognize once again that each d_{mj} can be represented simply by the index i because its value is sufficient to indicate the appropriate action to be taken. Thus the different strategies can be designated in FORTRAN by the two-dimensional integer array ((ID(M,J),J=1,NY),M=1,NS). In the particular example, we have $s_1 = [a_1, a_2, a_2, a_3]$ and $s_2 = [a_1, a_2, a_3, a_3]$ so that

ID(1,1) = 1   ID(1,2) = 2   ID(1,3) = 2   ID(1,4) = 3
ID(2,1) = 1   ID(2,2) = 2   ID(2,3) = 3   ID(2,4) = 3

The maximum number of possible strategies is restricted, with $n_s \leq (n_a)^{n_y}$. In the diabetic problem, the maximum value of n_s is $3^4 = 81$.

Step 7

Evaluation of the action probabilities. A matrix of action probabilities must be calculated for each strategy. In the example, we obtained the following matrix for strategy s_1:

$$\begin{array}{c} \\ \theta_1 \\ \theta_2 \\ \theta_3 \end{array} \begin{array}{c} \begin{array}{ccc} a_1 & a_2 & a_3 \end{array} \\ \begin{bmatrix} 0.7787 & 0.2031 & 0.0182 \\ 0.1250 & 0.3750 & 0.5000 \\ 0.0034 & 0.0571 & 0.9395 \end{bmatrix} \end{array}$$

and for strategy s_2 we obtained

$$\begin{array}{c} \\ \theta_1 \\ \theta_2 \\ \theta_3 \end{array} \begin{array}{c} \begin{array}{ccc} a_1 & a_2 & a_3 \end{array} \\ \begin{bmatrix} 0.7787 & 0.2208 & 0.0005 \\ 0.1250 & 0.7500 & 0.1250 \\ 0.0034 & 0.3825 & 0.6141 \end{bmatrix} \end{array}$$

Thus to represent the action probabilities for the mth strategy we use a matrix with the typical element $g_{ki}(m)$, as given in Fig. 11.10:

$$\begin{array}{c} \\ \theta_1 \\ \vdots \\ \theta_k \\ \vdots \\ \theta_{n_\theta} \end{array} \begin{array}{c} \begin{array}{ccccc} a_1 & & a_i & & a_{n_a} \end{array} \\ \begin{bmatrix} g_{11}(m) & \cdots & g_{1i}(m) & \cdots & g_{1n_a}(m) \\ \vdots & & \vdots & & \vdots \\ g_{k1}(m) & \cdots & g_{ki}(m) & \cdots & g_{kn_a}(m) \\ \vdots & & \vdots & & \vdots \\ g_{n_\theta 1}(m) & \cdots & g_{n_\theta i}(m) & \cdots & g_{n_\theta n_a}(m) \end{bmatrix} \end{array}$$

Fig. 11.10 Action probability matrix for strategy s_m.

The condensed notation for the matrix is $\{g_{ki}(m): i = 1, n_a; k = 1, n_\theta\}$. The parenthetical notation (m) indicates that we have a different matrix for each strategy. Thus there are

n_s different matrices.

Recall that the action probabilities for strategy $s_1 = [a_1, a_2, a_2, a_3]$ when the state of nature θ_1 prevails were found by summing the appropriate probabilities of occurrence. That is, if the particular action was indicated when a certain outcome was observed, its corresponding probability was included in the sum. This technique yielded the following sums in the diabetic example:

$$g_{11}(1) = Pr[Y = 0/\theta_1] = 0.7787$$

$$g_{12}(1) = Pr[Y = 1/\theta_1] + Pr[Y = 2/\theta_1]$$

$$= 0.2031 + 0.0177 = 0.2208$$

$$g_{13}(1) = Pr[Y = 3/\theta_1] = 0.005$$

In general, the action probability for the kth state of nature and the ith action is found by evaluating the sum

$$f_{ki}(m) = \sum_{\{j : d_{mj} = a_i\}} f_{kj}$$

where the symbol $\{j : d_{mj} = a_i\}$ means that we sum over all values of j for which $d_{mj} = a_i$. This general selection process will become clearer when we write the subroutine to accumulate the action probabilities.

In representing the action probabilities algebraically we use the parenthetical notation $g_{ki}(m)$ instead of three indices g_{mki} because it is easier to keep track of what the indices represent. In FORTRAN programming, however, we do not have this option, so we must use a three-dimensional array (((APR(M,K,I),I=1,NA),K=1,NT),M=1,NS) to represent the action probabilities. The rules for reading, writing, and manipulating three or higher dimensional arrays (the maximum number of subscripts depends upon the particular computer) are the same as for the one- and two-dimensional arrays.

We choose to calculate the action probabilities in a subroutine called ACTPR. The required inputs are

(1) (Y(J),J=1,NY)
(2) ((ID(M,J),J=1,NY),M=1,NS)
(3) ((YPR(K,J),J=1,NY),K=1,NT)
(4) NA
(5) NT
(6) NY
(7) NS

The output of the subroutine is the set of n_s matrices where each matrix gives the action probabilities for a given strategy. The FORTRAN array that is used to store the action probabilities is called (((APR(M,K,I),I=1,NA),K=1,NT),M=1,NS). The subroutine is given by Program 11.3:

```
      SUBROUTINE ACTPR(Y,ID,YPR,NA,NT,NY,NS,APR)
      DIMENSION Y(1),ID(10,1),YPR(10,1),APR(10,10,1)
      DO 1 M=1,NS
      DO 1 K=1,NT
      DO 1 I=1,NA
    1 APR(M,K,I) = 0.
      DO 2 M=1,NS
      DO 2 K=1,NT
      DO 3 J=1,NY
      DO 3 I=1,NA
      IF(ID(M,J) .NE. I) GO TO 3
      APR(M,K,I) = APR(M,K,I) + YPR(K,J)
    3 CONTINUE
    2 CONTINUE
      RETURN
      END
```

Program 11.3 SUBROUTINE ACTPR — calculation of action probabilities.

In the subroutine the action probabilities are first initialized to zero. Then the decision as to whether a probability of occurrence is to be accumulated to the appropriate APR(M,K,I) is made by the statement

IF(ID(M,J) .NE. I)   GO TO 3

320

If the logical statement is true, the routine continues searching through the set of I's because statement 3 is a CONTINUE statement. If the logical statement is not true, that is, ID(M,J) = I, then the corresponding probability of occurrence f_{kj} will be added to the action probability for the stragegy, the kth state of nature, and the ith action before the loop is continued. This process is repeated for all strategies, all states of nature, and all actions.

Step 8

Determination of average losses. The average losses for the diabetic example were

$$\begin{array}{cc} & \begin{array}{cc} s_1 & \quad s_2 \end{array} \\ \begin{array}{c} \theta_1 \\ \theta_2 \\ \theta_3 \end{array} & \left[\begin{array}{cc} 1.8902 & 2.0677 \\ 4.7500 & 8.1250 \\ 13.8760 & 10.6220 \end{array}\right] \end{array}$$

In general, we can represent the matrix of average losses by

$$\begin{array}{c} \begin{array}{cccc} s_1 & \cdots & s_m & \cdots & s_{n_s} \end{array} \\ \begin{array}{c} \theta_1 \\ \vdots \\ \theta_k \\ \vdots \\ \theta_{n_\theta} \end{array} \left[\begin{array}{cccc} \bar{\ell}_1(1) & \cdots \bar{\ell}_1(m) & \cdots \bar{\ell}_1(n_s) \\ \vdots & \vdots & \vdots \\ \bar{\ell}_k(1) & \cdots \bar{\ell}_k(m) & \cdots \bar{\ell}_k(n_s) \\ \vdots & \vdots & \vdots \\ \bar{\ell}_{n_\theta}(1) & \cdots \bar{\ell}_{n_\theta}(m) \cdots \bar{\ell}_{n_\theta}(n_s) \end{array}\right] \end{array}$$

Fig. 11.11 Average loss matrix.

where the typical element is $\bar{\ell}_k(m)$. The dimension of the matrix is $[n_\theta, n_s]$.

Recall that in the diabetic example the average loss for θ_1 and s_2 was given by

$$\bar{\ell}_1(2) = (0.7787)\,1 + (0.2031)\,5 + (0.0182)\,15 = 2.0672$$

Thus each loss is weighted by the corresponding action probability for a particular strategy. In general, the average losses are found by

$$\bar{\ell}_k(m) = \sum_{i=1}^{n_a} g_{ki}(m) \cdot \ell_{ki}$$

$$= g_{k1}(m) \cdot \ell_{k1} + g_{k2}(m) \cdot \ell_{k2} + \cdots + g_{kn_a}(m) \cdot \ell_{kn_a}$$

In the program we shall designate the average loss matrix by (AVGLS(K,M),M=1,NS),K=1,NT). Since the average losses will be calculated in the main program, we shall merely give the program section that is used to find them:

```
            .
            .
            .
      DO 11 K=1,NT
      DO 11 M=1,NS
      AVGLS(K,M) = Ø
      DO 11 I=1,NA
   11 AVGLS(K,M) = AVGLS(K,M) + APR(M,K,I)*ACTLOS(K,I)
            .
            .
            .
```

Step 9

Selection of strategy.

(a) For the selection of the minimax strategy we determine the strategy s_{min} such that

$$s_{min} = \min_m [\max_k \bar{\ell}_k(m)]$$

(b) For the selection of the minimum expected loss (Bayes) strategy s_B we first determine the expected loss (EL) for each strategy:

$$EL(s_m) = \sum_{k=1}^{n_\theta} q_k \cdot \bar{\ell}_k(m)$$

where $[q_1, q_2, \ldots, q_k, \ldots, q_{n_\theta}]$ is a vector of prior probabilities associated with each state of nature. We select
the strategy s_B such that

$$EL(s_B) = \min_m [EL(s_m)]$$

(c) For the selection of the minimum regret strategy we
replace ℓ_{ki} by

$$r_{ki} = \ell_{ki} - \min_i \ell_{ki}$$

and proceed in the same manner as when we find the Bayes
strategy.

We shall not give the FORTRAN program steps needed to
apply any of the above criterion in selecting a particular
strategy but will discuss the necessary steps when we consider the main program in the next section.

11.5 A Statistical Decision Theory Program

We now want to write a program that determines the best
strategy from a set of competing strategies according to a
specified criterion, such as the minimax, Bayes, or minimum
risk criterion. A program could be written so that the
decision criterion to be used can be designated as part of
the input. We feel that this flexibility would unnecessar-
ily increase the complexity of the program for this first
presentation, so we shall simply write a program to select
the minimax strategy.

Although the steps that must be performed by the program were given in the previous section, let us present them
in the form of a flow diagram to indicate the inputs, calcu-
ations, and outputs of the program.

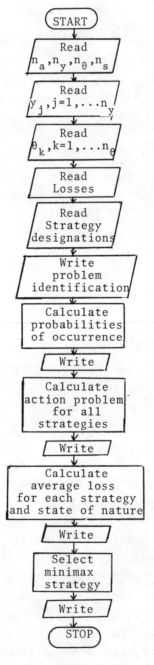

Fig. 11.12 Flow diagram for minimax decision program.

We could include the evaluations of the probabilities of occurrence and the action probabilities in the main program. We chose, however, to use subroutines for these calculations to set them apart and thus enable the student to better appreciate the steps involved.

The main program is given by Program 11.4. The student should follow the steps presented in the previous section and compare them with the corresponding program statements. He should also notice that the maximum allowable dimension for each array is given in the main program. However, if problems require larger arrays, the dimension statement in both the main program and the subroutines must be changed accordingly. The fact that APR(10,10,10) already reserves 1000 storage locations should, however, be kept in mind so that the dimensions are not increased beyond the computer capabilities. Presently the following limitations for each input are

(1) NA must be less than or equal to 10.
(2) NT must be less than or equal to 10.
(3) NS must be less than or equal to 10.
(4) NY must be less than or equal to 10.

```
C       PROGRAM MINIMAX-  A STATISTICAL DECISION THEORY PROGRAM
C       DETERMINATION OF THE MINIMAX STRATEGY
        DIMENSION Y(1Ø),THETA(1Ø),YPR(1Ø,1Ø,APR(1Ø,1Ø,1Ø),
       1 ACTLOS(1Ø,1Ø),ID(1Ø,1Ø),AVGLS(1Ø,1Ø),XAVGL(1Ø)
        WRITE(6,5Ø)
   5Ø   FORMAT(1H1,1ØX,33HDETERMINATION OF MINIMAX STRATEGY)
        READ(5,1) NA,NY,NT,NS
    1   FORMAT(4I2)
        WRITE(6,51) NA
   51   FORMAT(1HØ, 22HTOTAL NO OF ACTIONS IS, I5)
        READ(5,2) (Y(J),J=1,NY)
    2   FORMAT(1ØF8.2)
        WRITE(6,52) (Y(J),J=1,NY)
   52   FORMAT(1HØ, 22HOBSERVATION VALUES ARE/(1H
       1 ,1Ø(F1Ø.2,2X)))
        READ(5,2)(THETA(K),K=1,NT)
        WRITE(6,53)(THETA(K),K=1,NT)
   53   FORMAT( 1HØ,2ØHSTATES OF NATURE ARE/(1H
       1 ,1Ø(F1Ø.2,2X)))
        DO 3 K=1,NT
    3   READ(5,2) (ACTLOS(K,I),I=1,NA)
        WRITE(6,54)
   54   FORMAT(1HØ,1ØHLOSS TABLE/1HØ,6HS OF N,1ØX, 7HACTIONS)
        DO 4 K=1,NT
```

```
   4   WRITE(6,55) THETA(K),(ACTLOS(K,I),I=1,NA)
  55   FORMAT(1H ,F1Ø.2,2X,1Ø(F1Ø.2,2X))
       DO 5 M=1,NS
   5   READ(5,6) (ID(M,J),J=1,NY)
   6   FORMAT(1ØI2)
       WRITE(6,56)
  56   FORMAT( 1HØ, 8HSTRATEGY,1ØX, 6HACTION)
       DO 7 M=1,NS
   7   WRITE(6,57) M,(ID(M,J),J=1,NY)
  57   FORMAT(1H ,I3,1ØX,1Ø(I3,2X))
C      CALCULATION OF PROBABILITIES OF OCCURRENCE
       CALL YPROB(Y,THETA,NY,NT,YPR)
       WRITE(6,58)
  58   FORMAT(1HØ,41HPROB FOR EACH STATE OF NATURE AND OBSERV.
      1 /1H ,6HS OF N,1ØX,12HOBSERVATIONS)
       DO 8 K = 1,NT
   8   WRITE(6,55) THETA(K),(YPR(K,J),J=1,NY)
C      CALCULATION OF PROBABILITIES OF OCCURRENCE COMPLETED.
C      NOW CALCULATE ACTION PROB FOR EACH STRATEGY M
       CALL ACTPR(Y,ID,YPR,NA,NT,NY,NS,APR)
       DO 9 K=1,NS
       WRITE(6,59) M
  59   FORMAT(1HØ,38HACTION PROBABILITY MATRIX FOR STRATEGY,
      1 I5/1H , 6HS OF N, 1ØX, 7HACTIONS)
       DO 1Ø K=1,NT
  1Ø   WRITE(6,55) THETA(K),(APR(M,K,I),I=1,NA)
   9   CONTINUE
C      CALCULATION OF ALL ACTION PROB TABLES COMPLETED, NOW
C      COMPUTE AVERAGE LOSSES FOR EACH STRATEGY AND ST OF N.
       DO 11 K =1,NT
       DO 11 M =1,NS
       AVGLS(K,M) = Ø.
       DO 11 I=1,NA
  11   AVGLS(K,M) = AVGLS(K,M) + APR(M,K,I)*ACTLOS(K,I)
       WRITE(6,6Ø)
  6Ø   FORMAT(1HØ,19HAVERAGE LOSS MATRIX/1H ,6HS OF N,1ØX,
      1 1ØHSTRATEGIES)
       DO 12 K=1,NT
  12   WRITE(6,55) THETA(K),(AVGLS(K,M),M=1,NS)
C      FIND THE MAXIMUM AVERAGE LOSS OF EACH STRATEGY
       DO 13 M=1,NS
       XAVGL(M) = AVGLS(1,M)
       DO 13 K=2,NT
       IF(AVGLS(K,M) .LE. XAVGL(M) ) GO TO 13
       XAVGL (M) = AVGLS(K,M)
  13   CONTINUE
C      THE ARRAY XAVGL(M) NOW HAS THE MAXIMUM AVERAGE LOSS FOR
C      EACH STRATEGY M ( MAXIMIZED OVER THE STATES OF NATURE).
C      NOW FIND THE MINIMUM AVERAGE LOSS AMONG THIS ARRAY
C      AMONG THE DIFFERENT STRATEGIES
       MINMAX = 1
       SLOSS = XAVGL(1)
       DO 14 M=2,NS
       IF( XAVGL(M) .GE. SLOSS ) GO TO 14
       MINMAX = M
       SLOSS = XAVGL(M)
  14   CONTINUE
```

```
      WRITE(6, 62) MINMAX,SLOSS
  62  FORMAT(1HØ,19HMINIMAX STRATEGY IS, I5/
   1   1H ,18HASSOCIATED LOSS IS, F1Ø.2)
      STOP
      END
```

Program 11.4 MINIMAX — determination of minimax strategy
for decision problems.

The inputs used to determine the outputs given for the diabetic example are

```
03040402
   0.          1.      2.      3.
   .08         .50     .85
   1.          5.      15.
   8.          3.      12.
   25.         20.     10.
   01020203
   01020303
```

The student should notice the technique used to determine the minimum or maximum values of an array. Set the
minimum (maximum) variable name equal to the first element
in the array. Then compare each element in the array to the
minimum (maximum) value. If the array element is less than
(greater than) the value, it is then taken as the minimum
(maximum) value. At the end of the loop, when all comparisons are completed, the minimum (maximum) value has been
found. This technique can also be extended to order an entire array in either ascending or descending order; such a
sort subroutine is presented in Appendix B.

11.6 Determination of the Bayes Strategy

In our earlier considerations we introduced the Bayes
criterion as a method for picking a particular strategy from
a set of competing strategies. If we let the set of competing strategies be the set of all possible strategies, we
could find the strategy that has the minimum expected loss —
he Bayes strategy. However, this complete enumeration ap-

proach is not efficient, so in this section we shall present a more direct technique that can be used to find the Bayes strategy.

To find the Bayes strategy we must first determine the best action (minimum expected loss also called the Bayes action) when only the prior probabilities for each state of nature $\{\Pr[\theta_k] = q_k\}$ and the losses are available. In finding the Bayes action we consider the prior probability $\Pr[\theta_k]$ to be actually the conditional probabilities: $\Pr[\theta_k/$ all information prior to the acquisition of the "experimental" data].

To illustrate the technique, we shall continue to use the medical problem for illustrative purposes. Thus we have

Actions	Prior Probabilities
a_1:continue diet	$q_1 = 0.50$
a_2:change diet	$q_2 = 0.35$
a_3:see doctor	$q_3 = 0.15$

The losses are

	Actions		
State of Nature	a_1	a_2	a_3
θ_1	1	5	15
θ_2	8	3	12
θ_3	25	20	10

In this case we compute the expected loss for each action, which is

$$EL(a_i) = \Pr[\theta_1] \cdot \ell(a_i/\theta_1) + \Pr[\theta_2] \cdot \ell(a_i/\theta_2)$$

$$+ \Pr[\theta_3] \cdot \ell(a_i/\theta_3)$$

$$= \sum_{k=1}^{3} q_k \cdot \ell(a_i/\theta_k)$$

so that

$$EL(a_1) = (0.50)\ 1 + (0.35)\ 8 + (0.15)\ 25 = 7.05$$

$$El(a_2) = (0.50)\ 5 + (0.35)\ 3 + (0.15)\ 20 = 6.55$$

$$EL(a_3) = (0.50)\ 15 + (0.35)\ 12 + (0.15)\ 10 = 13.20$$

The minimum expected loss of Bayes action is thus a_2. This action can also be called the no-data solution because it is based upon only the prior probabilities and the losses. We should recognize that a_2 is also the minimum expected loss strategy or Bayes strategy because even if the diabetic wants to take another action on occasion, there are no data available to tell him when such a change is indicated. The best he could do would be to randomly pick another action; since the other actions have larger losses, the expected loss of this randomized strategy (using different actions) would exceed $EL(a_2)$.

To determine the Bayes strategy when experimental data are available, we use the no-data approach but first modify the prior probabilities so that they reflect the information given by the experiment. For each possible outcome of Y we obtain the posterior probabilities, which are modified prior probabilities and are represented by $Pr[\theta_k/Y]$. We then use these posterior probabilities in place of the prior probabilities in finding the expected loss for a given action, where

$$EL(a_i) = Pr[\theta_1/Y] \cdot \ell(a_i/\theta_1) + Pr[\theta_2/Y] \cdot \ell(a_i/\theta_2)$$

$$+ Pr[\theta_3/Y] \cdot \ell(a_i/\theta_3)$$

he action having the minimum expected loss is the Bayes ction for that value of Y. This procedure is repeated for ach possible outcome of Y; the set of Bayes actions is the ayes strategy.

We thus need to be able to find these posterior probabilities, $\Pr[\theta_k/Y]$. For this we use Bayes theorem, which states that

$$\Pr[\theta_k/Y] = \frac{\Pr[\theta_k] \ \Pr[Y/\theta_k]}{\displaystyle\sum_k \Pr[\theta_k] \ \Pr[Y/\theta_k]}$$

The probabilities on the right-hand side of the Bayes equation are known since they are the prior probabilities and the probabilities of occurrence.

To illustrate this procedure, let us take the case when the patient observes two positive tests, $Y = 2$. We must first obtain the posterior probabilities $\Pr[\theta_1/Y = 2]$, $\Pr[\theta_2/Y = 2]$, and $\Pr[\theta_3/Y = 2]$, using Bayes theorem. In this case we have

$$\Pr[\theta_k/Y=2] = $$

$$\frac{\Pr[\theta_k] \ \Pr[Y=2/\theta_k]}{\Pr[\theta_1] \ \Pr[Y=2/\theta_1] + \Pr[\theta_2] \ \Pr[Y=2/\theta_2] + \Pr[\theta_3] \ \Pr[Y=2/\theta_3]}$$

where $\Pr[\theta_k]$ is the prior probability of the state of nature θ_k and $\Pr[Y=2/\theta_k]$ is the probability of occurrence for θ_k when $Y = 2$. These posterior probabilities can be interpreted as the probabilities that a diabetic is in condition θ_k when he obtained two positive tests; they are equal to

$$\Pr[\theta_1/Y=2] = \frac{(0.50)(0.0177)}{(0.50)(0.0177) + (0.35)(0.3750) + (0.15)(0.3251)}$$

$$= \frac{0.008850}{0.188865} = 0.0469$$

$$\Pr[\theta_2/Y=2] = \frac{(0.35)(0.3750)}{(0.50)(0.0177) + (0.35)(0.3750) + (0.15)(0.3251}$$

$$= \frac{0.131250}{0.188865} = 0.6949$$

$$Pr[\theta_3/Y=2] = \frac{(0.15)(0.3251)}{(0.50)(0.0177) + (0.35)(0.3750) + (0.15)(0.3251)}$$

$$= \frac{0.048765}{0.188865} = 0.2582$$

When Y = 2 is observed, $Pr[\theta_1] = 0.50$ is modified to $Pr[\theta_1/Y=2] = 0.0469$; thus it is very unlikely for the diabetic to be in condition θ_1, given that he has obtained Y = 2.

Using these posterior probabilities, we can find the Bayes action (the action that has a minimum expected loss) when Y = 2 by evaluating

$$El(a_i) = Pr[\theta_1/Y=2] \cdot \ell(a_i/\theta_1) + Pr[\theta_2/Y=2] \cdot \ell(a_i/\theta_2)$$

$$+ Pr[\theta_3/Y=2] \cdot \ell(a_i/\theta_3)$$

and picking the action with the smallest EL.

The expected losses for each action are

$$EL(a_1) = (0.0469)\ 1 + (0.6949)\ 8 + (0.2582)\ 25 = 12.0611$$

$$EL(a_2) = (0.0469)\ 5 + (0.6949)\ 3 + (0.2582)\ 20 = 7.4832$$

$$EL(a_3) = (0.0469)15 + (0.6949)\ 12 + (0.2582)\ 10 = 11.6243$$

Thus, if Y = 2, the action which has a minimum expected loss is a_2. Similarly, when the patient observes Y = 0 the posterior probabilities are

$$Pr[\theta_1/Y=0] = \frac{0.389356}{0.433610} = 0.8979$$

$$Pr[\theta_2/Y=0] = \frac{0.043750}{0.433610} = 0.1009$$

$$Pr[\theta_3/Y=0] = \frac{0.000510}{0.433610} = 0.0012$$

nd

$$EL(a_1) = (0.8979) \ 1 + (0.1009) \ 8 + (0.0012) \ 25 = 1.7351$$

$$EL(a_2) = (0.8979) \ 5 + (0.1009) \ 3 + (0.0012) \ 20 = 4.8162$$

$$EL(a_3) = (0.8979) \ 15 + (0.1009) \ 12 + (0.0012) \ 10 = 14.6913$$

The minimizing action when $Y = 0$ is observed is therefore a_1.

By finding the best action or Bayes action to take for each outcome of the random variable Y, we obtain the Bayes strategy. A convenient approach for determining all the Bayes actions is shown in Fig. 11.13:

1. Losses

State of Nature	Actions		
	a_1	a_2	a_3
θ_1	1	5	15
θ_2	8	3	12
θ_3	25	20	10

$\ell(a_i/\theta_k)$

2. Probabilities of occurrence

	Observations				Priors
	y_1	y_2	y_3	y_4	q_k
θ_1	0.7787	0.2031	0.0177	0.0005	0.50
θ_2	0.1250	0.3750	0.3750	0.1250	0.35
θ_3	0.0034	0.0574	0.3251	0.6141	0.15

$f_{kj} = Pr[Y=y_j/\theta_k]$

3. Weighted probabilities

θ_1	0.389350	0.101550	0.008850	0.000250
θ_2	0.043750	0.131250	0.131250	0.043750
θ_3	0.000510	0.008610	0.048765	0.092115

$q_k \cdot f_{kj} = Pr[\theta_k]Pr[Y=y_j/\theta_k]$

4. Sum of weighted probabilities

D_j	0.433610	0.241410	0.188865	0.136115

$\sum_k q_k \cdot f_{kj} = \sum_k Pr[\theta_k]Pr[Y=y_j/\theta_k]$

5. Posterior probabilities

θ_1	0.8979	0.4207	0.0469	0.0018
θ_2	0.1009	0.5437	0.6949	0.3215
θ_3	0.0012	0.0356	0.2582	0.6767

$Pr[\theta_k/Y=y_j] = \dfrac{q_k \ f_{kj}}{D_j}$

6. Weighted losses

Action				
a_1	1.7351	5.6603	12.0611	19.4913
a_2	4.8162	4.4466	7.4832	14.5075
a_3	14.6913	13.1909	11.6243	10.6520

$\sum_k \ell(a_i/\theta_k)Pr[\theta_k/Y=y_j]$

7. Bayes action

$$a_B = (a_1 \qquad a_2 \qquad a_2 \qquad a_3)$$

8. Expected loss for each action

$$EL(a_B) = (1.7351 \quad 4.4466 \quad 7.4832 \quad 10.6520)$$

Fig. 11.13 Tabular scheme for Bayes strategy computation.

The Bayes strategy that the patient should follow is

Observation	Y = 0	Y = 1	Y = 2	Y = 3
Bayes action	a_1	a_2	a_2	a_3
Expected loss	1.7351	4.4466	7.4832	10.6520

so that the Bayes strategy is $s_B = (a_1, a_2, a_2, a_3)$. The expected loss for the entire strategy is found by the following equation:

$$EL(s_B) = Pr[Y=0] \cdot EL(a_B/Y=0) + Pr[Y=1] \cdot EL(a_B/Y=1)$$

$$+ Pr[Y=2] \cdot EL(a_B/Y=2) + Pr[Y=3] \cdot EL(a_B/Y=3)$$

The probabilities of obtaining a particular observation
$Pr[Y=y_j]$ are given by the D_j in the tabulation since

$$Pr[Y=y_j] = Pr[\theta_1] \, Pr[Y=y_j] + Pr[\theta_2] \, Pr[Y=y_j/\theta_2] + \cdots$$

$$+ Pr[\theta_{n_\theta}] \, Pr[Y=y_j/\theta_{n_\theta}]$$

Thus the expected loss for s_B is

$$EL(s_B) = (0.4336)(1.7351) +$$

$$+ (0.1889)(7.4832) + (0.1361)(10.6520)$$

$$= 4.6891$$

and this strategy has the minimum expected loss with respect
to all other possible strategies.

The formula to find the typical element is given in the
tabular scheme to facilitate the utilization of the technique in general. As the dimensions of the problems, n_s, n_a,
or n_y, change, the scheme can easily be extended or con-
tracted as needed, though the steps in the computation remain the same.

We can verify that the procedure just given does produce the Bayes strategy by using the computer program to evaluate the set of all possible strategies. However, Program 11.4 has to be modified so that it picks the strategy having the minimum expected loss rather than the minimax strategy.

One advantage of the computational technique given above is that it can be used to determine the appropriate Bayes action for a given value of the random variable Y without developing the entire Bayes strategy. This means that one can wait until after the particular observation on the random variable is obtained and then perform only those computations that are associated with the particular observation to find the appropriate Bayes action.

Let us illustrate the procedure using the diabetic decision problem. We shall assume that the diabetic has obtained the observation $Y = 1$ (only one of the three strips showed a positive response); now he wonders what action he should take, using the minimum expected loss (Bayes) criterion.

The diabetic need determine only the Bayes action for $Y = 1$. Thus we can develop a simplified tabular array by rotating the comprehensive tabulation given in Fig. 11.13.

1. Losses:

Actions	State of Nature		
	θ_1	θ_2	θ_3
a_1	1	8	25
a_2	5	3	20
a_3	15	12	10

2. $\Pr[Y=1/\theta_k]$ 0.2031   0.3750   0.0534
3. $\Pr[\theta_k]$ 0.50   0.35   0.15
4. $\Pr[\theta_k]\ \Pr[Y=1/\theta_k]$ 0.101550   0.131350   0.008610 $D = 0.241410$
5. $\Pr[\theta_k/Y=1]$ 0.4207   0.5437   0.0356
6. Expected losses:

    a_1: 5.6603 = (0.4207) 1 + (0.5437) 8 + (0.0356) 25
    a_2: 4.4466 = (0.4207) 5 + (0.5437) 3 + (0.0356) 20
    a_3: 13.1909 = (0.4207) 15 + (0.5437) 12 + (0.0356) 10

7. Bayes action a_B is a_2, and the minimum loss is 4.4466.

Fig. 11.14 Tabular scheme for Bayes action computation.

The student should notice that the matrix of losses has been transposed so that the rows represent the actions and the columns represent the states of nature. The representation is useful in this simplified computational scheme because each row of the loss matrix is simply multiplied by the posterior probabilities to obtain the minimum loss for each action.

11.7 The Multivariate Extension for the Bayes Strategy Determination

Although there are many extensions and applications of the technique introduced in the last section, we shall give only the extension that enables us to consider the multivariate case when there is more than one random variable. For example, in the medical problem the random variable was the number of positive responses to the urine tests for the presence of sugar. To illustrate the multivariate case, let us now assume that the diabetic is to determine not only the number of positive test responses on the three urine tests but is also asked to perform tests which detect the presence of acid bodies in the urine. The diabetic is required to perform three sugar tests, so that Y (the number of positive responses on the urine test) can equal 0, 1, 2, or 3, but he is also required to perform two acid tests, so that X (the number of positive test responses on the acid test) can be equal to 0,1, and 2.

Recall that the states of nature are

θ_1: condition controlled
θ_2: condition slightly out of control
θ_3: condition out of control

When we considered only the variable Y, the value of the parameter p (the probability of obtaining a positive response on each test) associated with each state of nature was assumed to be

$$\theta_1: p_y = 0.08$$
$$\theta_2: p_y = 0.50$$
$$\theta_3: p_y = 0.85$$

The test for the presence of acid bodies is assumed to be independent of the level of the sugar in the urine, although both tests indicate how well the food is being metabolized by the system. If acid bodies are present in the system, the situation is more serious because they act as poisons and indicate that the diabetic's situation is out of control. However, as with the sugar test, it is possible to obtain a positive response even when there are no acid bodies in the system. Let us assume that the probability of detecting acid bodies associated with each state of nature is

$$\theta_1: p_x = 0.02$$
$$\theta_2: p_x = 0.25$$
$$\theta_3: p_x = 0.90$$

We can thus find the probabilities of occurrence for the random variable X by utilizing the binomial model $Pr[X=x] = C_x^2(p_x)^x(q_x)^{2-x}$. The outcomes of these evaluations are

State of Nature	Outcomes		
	X=0	X=1	X=2
θ_1	0.9801	0.0198	0.0001
θ_2	0.5625	0.3750	0.0625
θ_3	0.0100	0.1800	0.8100

Fig. 11.15 Probabilities of occurrence for X.

Since the random variables are assumed to be independent, we can obtain the joint probability of each pair (X=x, Y=y) for each state of nature by multiplying the appropriate individual probabilities

$$Pr[X=x,Y=y] = Pr[X=x] \cdot Pr[Y=y]$$

336

since X and Y are independent. These joint probabilities of occurrence for each state of nature can be displayed in matrix form, as shown in Figs. 11.16 to 11.18:

Acid Test Responses	Sugar Test Responses				$Pr[X=x/\theta_1]$
	$Y=0$	$Y=1$	$Y=2$	$Y=3$	
$X=0$	0.7632	0.1991	0.0173	0.0005	0.9801
$X=1$	0.0154	0.0040	0.0004	0.0000	0.0198
$X=2$	0.0001	0.0000	0.0000	0.0000	0.0001
$Pr[Y=y/\theta_1]$	0.7787	0.2031	0.0177	0.0005	1.0000

Fig. 11.16 Joint probabilities of occurrence $\theta = \theta_1$.

Acid Test Responses	Sugar Test Responses				$Pr[X=x/\theta_2]$
	$Y=0$	$Y=1$	$Y=2$	$Y=3$	
$X=0$	0.0703	0.2110	0.2109	0.0703	0.5625
$X=1$	0.0469	0.1406	0.1406	0.0469	0.3750
$X=2$	0.0078	0.0234	0.0235	0.0078	0.0625
$Pr[Y=y/\theta_2]$	0.1250	0.3750	0.3750	0.1250	1.0000

Fig. 11.17 Joint probabilities of occurrence $\theta = \theta_2$.

Acid Test Responses	Sugar Test Responses				$Pr[X=x/\theta_3]$
	$Y=0$	$Y=1$	$Y=2$	$Y=3$	
$X=0$	0.0000	0.0006	0.0033	0.0061	0.0100
$X=1$	0.0006	0.0103	0.2585	0.1106	0.1800
$X=2$	0.0028	0.0465	0.2633	0.4974	0.8100
$Pr[Y=y/\theta_3]$	0.0034	0.0574	0.3251	0.6141	1.0000

Fig. 11.18 Joint probabilities of occurrence $\theta = \theta_3$.

If the variables are not independent, the joint probabilities cannot be obtained by multiplication; a bivariate distribution model would be required to generate the joint probabilities. It should be noted that the sum of the joint probabilities over one of the random variables always results in the univariate probabilities associated with the other variable. Thus

$$\sum_{y=0}^{3} \Pr[X=x,Y=y/\theta_k] = \Pr[X=x/\theta_k]$$

and

$$\sum_{x=0}^{2} \Pr[X=x,Y=y/\theta_k] = \Pr[Y=y/\theta_k]$$

These univariate probabilities are called marginal probabilities.

To determine the Bayes action we need to determine the posterior probabilities of each state of nature for each possible outcome from the joint probability distribution and the prior probabilities. These posterior probabilities are denoted by $\Pr[\theta_1/X,Y]$, $\Pr[\theta_2/X,Y]$ and $\Pr[\theta_3/X,Y]$ and can be found by using the Bayes theorem.

$$\Pr[\theta_k/X,Y] = \frac{\Pr[\theta_k] \ \Pr[X,Y/\theta_k]}{\sum_k \Pr[\theta_k] \ \Pr[X,Y/\theta_k]}$$

For example, when X=1 and Y=2, we obtain

$$\Pr[\theta_1/X=1,Y=2] = \frac{\Pr[\theta_1] \ \Pr[X=1,Y=2/\theta_1]}{\sum_k \Pr[\theta_k] \ \Pr[X=1,Y=2/\theta_k]}$$

$$= \frac{(0.50)(0.0004)}{(0.50)(0.0004) + (0.35)(0.1406) + (0.15)(0.0585)}$$

$$= \frac{0.000200}{0.058185} = 0.0034$$

Likewise, we obtain

$$\Pr[\theta_2/X=1,Y=2] = \frac{(0.35)(0.1406)}{0.058185} = 0.8458$$

$$Pr[\theta_3/X=1,Y=2] = \frac{(0.15)(0.0585)}{0.058185} = 0.1508$$

To then determine the Bayes action for the bivariate observation X=1 and Y=2, we evaluate as before the expected loss
for each action, using the posterior probabilities as
weights. That is, we evaluate

$$EL(a_1) = Pr[\theta_1/X=1,Y=2]\cdot\ell(a_1/\theta_1) + Pr[\theta_2/X=1,Y=2]\cdot\ell(a_1/\theta_2)$$

$$+ Pr[\theta_3/X=1,Y=2]\cdot\ell(a_1/\theta_3)$$

$$= (0.0034)\ 1 + (0.8458)\ 8 + (0.1508)\ 25 = 10.5398$$

and similarly

$$EL(a_2) = (0.0034)\ 5 + (0.8458)\ 3 + (0.1508)\ 20 = 5.5704$$

$$EL(a_3) = (0.0034)\ 15 + (0.8458)\ 12 + (0.1508)\ 10 = 11.7086$$

Thus the Bayes action when X=1 and Y=2 is a_2, which states
that the diet should be modified. We could find the Bayes
action for each of the other 11 bivariate observations in a
similar fashion.

When the variables are independent we can use a sequential approach in finding the Bayes strategy. First use the
variable Y to modify the prior probabilities into posterior
probabilities; then use the probabilities of occurrence for
the variable X to obtain a new set of posterior probabilities.

To illustrate this sequential approach we can use the
posterior probabilities obtained for variable Y in the medical example given in Sec. 11.8. These posterior probabilities for each state of nature, given Y, $Pr[\theta_k/Y]$, are re-
produced in Fig. 11.19:

States of Nature	Outcomes for Y			
	Y=0	Y=1	Y=2	Y=3
θ_1	0.8979	0.4207	0.0469	0.0018
θ_2	0.1009	0.5437	0.6949	0.3215
θ_3	0.0012	0.0356	0.2582	0.6767

Fig. 11.19 Posterior probabilities for each state of
nature given Y.

We now use these posterior probabilities as prior probabilities and follow the procedure to find the Bayes strategy, using the probabilities of occurrence for X. We can
thus set up a separate tabular array for each possible outcome of the first variable Y because each outcome has different posterior probabilities associated with the states of
nature.

For example, when Y=2, the tabular array to find the
Bayes actions when X=0, 1, or 2 is shown in Fig. 11.20. We
note that the posterior probabilities for Y=2 given in Fig.
11.19 are used as the prior probabilities in step 2. We see
that the Bayes actions are (a_2, a_2, a_3), which means that if
Y=2 and X=0, action a_2 should be taken; if Y=2 and X=1, again,
action a_2 should be taken; and if Y=2 and X=2, action a_3
should be taken. It can be seen that except for round-off
errors, the losses and actions are the same, using the sequential approach as using the bivariate approach. The advantage of the sequential approach is that additional variables can be added as they are needed, and the appropriate
action can be found for the desired combination of variables.

For the medical example, when Y=2 the tabular scheme is
given by

1. Losses:

State of Nature	Actions		
	a_1	a_2	a_3
θ_1	1	5	15
θ_2	8	3	12
θ_3	25	20	10

2. Probabilities of occurrence: $\Pr[X=x/\theta_k]$

	Outcomes for X			Posterior Probabilities
	x_1	x_2	x_3	$\Pr[\theta_k/Y=2]\}$
θ_1	0.9801	0.0198	0.0001	0.0469
θ_2	0.5625	0.3750	0.0625	0.6949
θ_3	0.0100	0.1800	0.8100	0.2582

3. Weighted probabilities: $\Pr[\theta_k/Y=2]\cdot\Pr[X=x/\theta_k]$

θ_1	0.0459669	0.00092862	0.00000469
θ_2	0.3908813	0.26058750	0.04343125
θ_3	0.0025820	0.04647600	0.20914200

4. Sum of weighted probabilities: $\Pr[X=x,Y=2]$

D_j	0.4394302	0.3079921	0.25257794

5. Posterior probabilities: $\Pr[\theta_k/X=x,Y=2]$

θ_1	0.1046	0.0030	0.0000
θ_2	0.8895	0.8461	0.1720
θ_3	0.0059	0.1509	0.8280

6. Weighted losses: $\Sigma\ell(a_i/\theta_k)\cdot\Pr[\theta_k/X=x,Y=2]$

a_1	7.3681	10.5443	22.0760
a_2	(3.3095)	(5.5713)	17.0760
a_3	12.3020	11.7072	(10.3440)

7. Bayes action a_B

a_2	a_2	a_3

8. Expected loss for each action: $EL(a_B/X=x,Y=2)$

3.3095	5.5713	10.3440

Fig. 11.20 Tabular scheme for Bayes actions for the multi-variate problem when Y=2 and X=0, 1, or 2.

The Bayes actions, when Y = 2 is observed and the acid test to find the value of X is also performed, are $s_{Y=2}$ = (a_2, a_2, a_3), and the expected loss for the action is

$$EL(s_{Y=2}) = (0.4394)(3.3095) + (0.3080)(5.5713)$$

$$(0.2526)(10.3440)$$

$$= .7850$$

In the particular bivariate medical example we recognize that the entire Bayes strategy requires actions to be associated with each of the 12 possible outcomes. The expected loss of this Bayes strategy can be found by multiplying the expected loss for each action by its probability of occurrence. However, the determination of the Bayes action can simply be deferred until the observations are obtained so that only the single required Bayes action needs to be determined.

11.8 The Worth of Experimental Data

The Bayes strategy determination provides a means of evaluating the worth of an experiment relative to decision making. It is apparent that if a particular variable being considered provides little or no information to aid in the decision-making process, the cost of collecting, checking, storing, and retrieving the variable can hardly be justified.

Let $W(y_\ell/\underline{y})$ represent the worth of including variable y_ℓ in the set of observed variables $\underline{y}$. This worth is defined as the difference between the minimum expected loss when the variable is included and the minimum expected loss when the variable is not included. Thus

$$W(y_\ell/\underline{y}) = EL(\underline{y}) - EL(\underline{y} \text{ without } y_\ell)$$

If this worth is less than the acquisition cost of y_ℓ, the wisdom of acquiring and using $y_{\ell y}$ can be questioned.

To illustrate this evaluation technique we shall use the bivariate diabetic problem. We wish to determine the worth of including the variable X when the urine sugar test is already being used.

From the computations in Sec. 11.7 we have the minimum expected loss when Y=2, which is 7.4832. From Sec. 11.7 we know that the minimum expected loss when Y=2 for any value of X is 5.7830. Thus

$$W(X/Y=2) = EL(Y=2) - EL(X/Y=2)$$

$$= 7.4832 - 5.7830 = 1.7002$$

Thus if the determination of the cost of obtaining X exceeds 1.7, it is questionable whether it is worthwhile to make the further determination of X. If the other two conditional losses are obtained, they could be weighted by the probabilities of obtaining each value of Y to obtain an average worth that could be used to determine if X should continue to be routinely collected.

In a still more general situation, several different combinations of variables would have to be considered to determine the appropriate subgroup of descriptors to use in the decision. For example, we might have three variables X, Y, and Z but want to use only two. We then compare the worth of each pair to make an intelligent selection.

11.9 Summary

We presented the steps required to formulate a decision-making problem in terms of statistical decision theory:

Step 1: Identification of the set of possible actions a_i

Step 2: Designation of the possible outcomes of the random variable Y, y_j

Step 3: Representation of the possible states of nature θ_k

Step 4: Evaluation of the probabilities of occurrence for each possible outcome of the variable and each state of nature

Step 5: Determination of the losses for each action and state of nature

Step 6: Designation of the set of competing strategies
Step 7: Evaluation of the action probabilities
Step 8: Determination of the average losses for each strategy
Step 9: Application of the criterion to determine the "best" strategy

Three criteria are available for selecting the best strategy from a set of competing strategies:

1. The minimax criterion

According to the minimax criterion, the strategy whose maximum average loss is a minimum among the set of competing strategies is selected.

2. The Bayes criterion

According to the Bayes criterion, the strategy that has the minimum expected loss among the competing strategies for a given set of prior probabilities is selected.

3. The minimum risk criterion

The strategy that has the minimum expected regret among the competing strategies for a given set of prior probabilities is selected. The only difference between the Bayes criterion and the minimum risk criterion is that in the latter regrets instead of losses are used.

The use of double indices or matrix notation was introduced to evolve a general representation of the statistical decision problem and the steps to be followed in the evaluation procedure.

A computer program that enables one to find the minimum strategy among a set of competitive strategies, given the appropriate inputs, was presented.

The direct method of evolving the Bayes strategy for a single random variable and discrete inputs was presented. This method uses the observations to generate posterior probabilities from a set of prior probabilities and then uses the no-data technique to find the appropriate action for each possible observation value.

The technique for finding the Bayes strategy in the uni-variate case was generalized to accommodate multivariate observations. An illustration of a technique that generates the Bayes action for a particular observation was also given.

Techniques that can be used to evaluate the worth of any variables in the decision process were presented. Using these techniques, it is possible to determine the best subset of variables to use in decision making.

1. Evaluate the two competing strategies given in Sec. 11.2, $s_1 = (a_1, a_2, a_2, a_3)$ and $s_2 = (a_1, a_2, a_3, a_3)$, using the same states of nature and actions. Substitute the following losses:

	Actions		
State of Nature	a_1	a_2	a_3
θ_1	1	5	10
θ_2	4	3	6
θ_3	8	6	5

2. Consider the following decision problem. A large shipment of electric switches has arrived at your factory, which uses the switches in manufacturing electric shavers. Since you have been having trouble with the quality of the switches in some of the lots received, you must decide which of the following actions to take relative to the shipment:

a_1: use the switches without testing them
a_2: test each switch before using it
a_3: return the shipment as being unsatisfactory

To determine what action to take, you select a random sample of four switches, test them, and determine Y, the number of defective switches. Since you have decided to base the decision upon the value of Y, you wish to evaluate the following strategies:

$s_1 = (a_1, a_1, a_2, a_2, a_3)$
$s_2 = (a_1, a_1, a_2, a_3, a_3)$
$s_3 = (a_1, a_2, a_2, a_3, a_3)$

The shipment is large, so Y can be considered a binomial random variable, and thus we can identify the states of nature by the value of p, the probability of obtaining a defective switch out of the lot. Consider the following four states of nature:

θ_1: $p = 0.03$ superior lot
θ_2: $p = 0.05$ acceptable lot
θ_3: $p = 0.08$ poor lot
θ_4: $p = 0.12$ unacceptable lot

The losses are

States of Nature	Actions		
	a_1	a_2	a_3
θ_1	1	5	8
θ_2	2	5	7
θ_3	10	5	6
θ_4	12	6	3

From past experience, the prior probabilities assigned to the states of nature are

$q_1 = 0.20$ $q_2 = 0.45$ $q_3 = 0.25$ $q_4 = 0.10$

Determine

   (a) The minimax strategy of the three competitive strategies
   (b) The Bayes strategy among the three strategies
   (c) The minimum risk strategy among the three strategies

3. Use the computer program given by Program 11.4 to verify your computations for determining the minimax strategy in Exercise 2.

4. Modify Program 11.4 so that it will calculate the Bayes strategy and the minimax strategy. Check and debug your program by applying it to the problem in Exercise 2.

5. Modify program 11.4 so that it will generate all the possible 81 strategies for the medical example given in Sec. 11.2 instead of simply reading in the set of strategies that are to be evaluated. Verify the program by applying it to the medical example and comparing the Bayes strategy and the associated expected loss with the results given in Sec. 11.7, where the Bayes strategy was directly determined. To change the program you can substitute the following statements in place of the reading for the strategy indicators II(M,J):

```
        M = 0
        DO 100 I1 = 1,3
        DO 100 I2 = 1,3
        DO 100 I3 = 1,3
        DO 100 I4 = 1,3
        M = M+1
        II(M,1) = I1
        II(M,2) = I2
        II(M,3) = I3
100    II(M,4) = I4
```

This set of four DO loops will generate all the possible strategies, and the program will then give the minimax strategy and the Bayes strategy. (Hint: be sure to change the DIMENSION statement to the appropriate values.)

6. Write a computer program to determine the Bayes strategy, using the scheme given in Sec. 11.7. Apply the program to the diabetic decision problem worked out in Sec. 11.2 to check it. Once the program is debugged, find the Bayes strategy for the problem given in Exercise 2.

7. In Sec. 11.8 the extension of the direct method to determine the Bayes strategy was extended to accommodate multivariate observations. The Bayes strategy for X=x and Y=2 was found. Applying the same approach, determine the Bayes strategy for X=x and Y=0.

8. Use the computer program written in Exercise 6 to check the answer to Exercise 7 and to find the Bayes strategy when X=x, and Y=1 and X=x and Y=3.

9. Generate your own decision problem, and find the Bayes strategy.

CHAPTER 12

STATISTICAL APPLICATIONS OF DECISION THEORY

12.1 Introduction

Applying decision theory to the statistical inference problems presented in Chaps. 9 and 10 requires generalization of the techniques developed for the discrete decision theory approach given in Chap. 11 because the outcomes of the random variable, actions, and states of nature may be infinite in number. In the introduction we shall simply outline how such generalizations can be made, and then in the following sections we shall apply these generalizations to actual statistical inference problems.

The approach we shall follow is to replace the infinite number of values for the random variable, actions, or states of nature by a finite number. Although the use of a finite number of values introduces problems in finding the optimum strategy, repeated application of the decision theory procedure will enable us to approach the optimum strategy as closely as desired. The use of this approach also enables us to utilize the computer to evaluate competing strategies. In fact, only slight modifications need to be made in Program 11.4 to use it for the problems presented in this chapter.

Let us consider in turn each of the following situations: (1) the random variable Y is continuous and has an infinite range given by $-\infty < Y < \infty$, (2) the set of possible actions is infinite, and (3) the set of parameter values is infinite. When the random variable Y is continuous and has an infinite range, the strategy designation procedure, the determination of the probabilities of occurrence, and the computation of the action probabilities must all be generalized. We assume that the finite number of actions, given by $a_1, a_2, \ldots, a_{n_a}$, can be ordered such that as the value of Y increases the more desirable it becomes to take an action with a higher index. In this case an entire class of possible strategies can be obtained by dividing the entire range of Y into n_a nonoverlapping intervals and associating a particular action with each interval. Thus a strategy can simply be designed by the interval end points. For example, the mth strategy can be represented by $s_m = [b_1(m), b_2(m), \ldots, b_{n_a-1}(m)]$, where $[b_{i-1}(m), b_i(m)]$ is the interval to be associated with action a_i. We cover the infinite range by taking $b_0(m) = -\infty$ and $b_{n_a}(m) = \infty$.

Consider the case when Y has a normal distribution and there are five possible actions: a_1, a_2, a_3, a_4, and a_5. The class of strategies can be designated by $s_m = [b_1(m), b_2(m), b_3(m), b_4(m)]$, which means that

$$
\begin{array}{lll}
\text{If} & Y \le b_1(m), & \text{take action } a_1 \\
\text{If } b_1(m) < & Y \le b_2(m), & \text{take action } a_2 \\
\text{If } b_2(m) < & Y \le b_3(m), & \text{take action } a_3 \\
\text{If } b_3(m) < & Y \le b_4(m), & \text{take action } a_4 \\
\text{If} & Y > b_4(m), & \text{take action } a_5
\end{array}
$$

Once the strategies are specified the action probabilities can be determined because they are simply the appropriate area under the normal model assumed for the particular state of nature θ_k. The areas that correspond to the action probabilities for strategy s_m and θ_k are indicated graphically in Fig. 12.1. Thus when the random variable is continuous, there is no need to determine the probabilities of occurrence since they were merely used as a vehicle to find the action probabilities.

Although this procedure is most applicable when the actions are directly related to the value of the observed random variable Y, it can easily be modified to accommodate other cases. However, the set of strategies generated by this simple interval designation technique includes most of the meaningful strategies for decision problems encountered

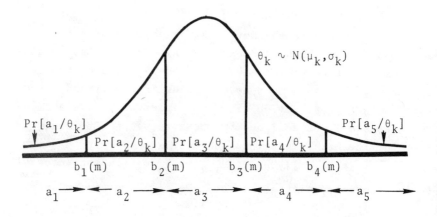

Fig. 12.1 Action probabilities for strategy s_m and θ_k.

in the real world.

　　Having considered the required generalization when the random variable is continuous, let us consider the case when the set of possible actions is infinite. This might occur when we are using decision theory to estimate a parameter because in this case an action consists of assigning a specific value a to the parameter. When there are an infinite number of possible values, we replace them by a finite number. This reduction of the allowable estimates being considered does not seriously restrict us because we can repeat the determination of the estimate by using a more refined set of values centered around the previously determined estimate. Since such a sequential approach requires the availability of losses for a continuously changing set of actions, some generalization of the loss matrix is required.

　　We assume that there are still a finite number of states of nature. In this case we can replace the loss matrix by continuous loss functions denoted by $L(a/\theta_k)$, the loss associated with action a, given the state of nature θ_k. An illustration of such continuous loss functions when there are only two states of nature is shown by Fig. 12.2; an example of the use of such loss functions will be given in the next sections.

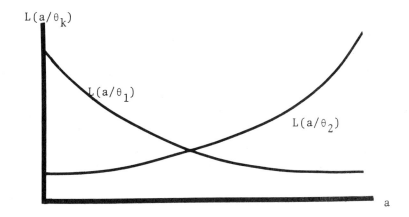

Fig. 12.2 A representation of loss functions when
the set of actions are infinite.

Another possible generalization of the discrete decision problem is to have the number of states of nature be infinite. For example, in the diabetic illustration in Chap.
11, the three states of nature were designated by a specified value of the binomial parameter p. Actually, the possible values of p in this situation could range continuously
from zero to 1. We again choose to replace the infinite set
of values by a finite number by introducing additional
states of nature with corresponding values of the parameter.

When the possible states of nature are infinite in number, the concept of continuous loss functions defined over
the parameter space θ can once again be used. To illustrate
the approach, consider the loss function for a particular
action a_i represented by $L(a_i/\theta)$ in Fig. 12.3. Once a
selected value of the parameter θ is chosen for each of the
finite number of states of nature, the loss value associated
with each θ_k and a_i can be obtained by using either the value of the loss function at θ_k or the area under the loss
function between b_{k-1} and b_k, as shown in Fig. 12.3.

353

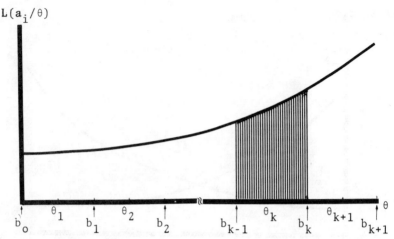

Fig. 12.3 A representation of a loss function for a
continuous parameter θ.

The generalizations discussed in this section allow us
to accommodate a large class of decision problems. Some
statistical applications using these generalizations will be
introduced in the subsequent sections and chapters.

12.2 Testing Hypotheses

In this section we consider how decision theory might
be applied to problems related to the testing of hypotheses.
We shall consider the problem introduced in Chap. 9, in which
the personnel manager is interested in the mean reading time
of new employees who have been attracted to the firm by a
new recruiting program. Using a sample of reading times of
new employees, he wants to test the hypothesis $H_o : \mu = 2.5$
against the alternative that $H_1 : \mu > 2.5$. The strategy selection problem faced by the personnel manager is whether to
use the 5 percent level or the 1 percent level of significance when making the test. He has decided to use the minimum expected loss criterion in selecting the appropriate
level of significance so that these different levels of significance are the different strategies.

Let us proceed through the nine steps required in evaluating the competing strategies in general and for a specific
example.

Step 1

Identification of the set of possible actions:

a_1: accept the null hypothesis
a_2: reject the null hypothesis

Step 2

Designation of the possible outcomes of the observable random variable. In testing hypothesis a sample statistic is used as the observable random variable. Since we are testing a hypothesis about the mean, we shall use the sample mean $\bar{y}$. This variable is continuous, and its theoretical range is from minus infinity to plus infinity.

Step 3

Representation of the possible states of nature. If we use the normal model for the distribution of $\bar{y}$, and we assume that the standard deviation σ is known (or if n is large, $\sigma = s'$), so that μ is the only undetermined parameter.

Although there are an infinite number of possible values of μ, we shall represent the range of values by a finite number of points beginning with μ_o (since the one-sided interval $\mu \geq \mu_o$ is the alternative hypothesis) and successively adding a constant increment to each μ to obtain the next μ.

Recognizing that the standard deviation of $\bar{y}$ is $\sigma/\sqrt{n}$, we could use this value as the increment and thus represent the states of nature by

$$\theta_o : \mu_o$$

$$\theta_k : \mu_k = \mu_o + \frac{k\sigma}{\sqrt{n}}$$

In the particular example we shall assume that σ is known and equal to 0.5 and the sample size is n = 25. To minimize the computations but still keep the problem meaningful, we shall use six states of nature:

355

$$\theta_0 : \mu_0 = 2.5$$
$$\theta_1 : \mu_1 = 2.6$$
$$\theta_2 : \mu_2 = 2.7$$
$$\theta_3 : \mu_3 = 2.8$$
$$\theta_4 : \mu_4 = 2.9$$
$$\theta_5 : \mu_5 = 3.0$$

This seems to provide a reasonable coverage of the values of μ that are of interest because under the null hypothesis $\bar{y}$ would be expected to fall within the range $\mu_0 + 3\sigma_{\bar{y}}$, or $2.5 \leq \bar{y} \leq 2.8$. This indicates that the states of nature μ_0 and μ_5 are surely distinct relative to the distribution of sample means that can be obtained under each state of nature. This is, if μ_5 is true, we would hardly expect the value of $\bar{y}$ to mislead us into believing μ_0 is true or conversely.

Step 4

Evaluation of the probabilities of occurrence. Since we assume that $\bar{y}$ has a normal distribution with mean μ and standard deviation $\sigma/\sqrt{n}$, the density function for $\bar{y}$ for the kth state of nature is given by

$$f(\bar{y}/\mu_k) = \frac{1}{\sqrt{2\pi}\ \sigma/\sqrt{n}}\ e^{-\frac{1}{2}\left[\frac{(\bar{y}-\mu_k)}{\sigma/\sqrt{n}}\right]^2} \quad -\infty \leq Y \leq \infty$$

We shall not use this step in our evaluations; we shall compute the action probabilities directly.

Step 5

Determination of the losses. We present the losses for action a_1 by the loss function $L(a_1/\mu)$, which gives the loss of accepting the null hypothesis ($\mu = \mu_0$) for different values of μ. Similarly, for action a_2 we use the function $L(a_2/\mu)$, which represents the loss incurred in rejecting the null hypothesis for different values of μ. We simply give a graphical representation of these two loss functions in Fig. 12.4. An analytical expression for the functions could be developed, but in this application we shall simply read the required losses from the graph for the specified values of μ_k. Thus $\ell_{ki} = L(a_i/\mu_k)$.

For each of the six states of nature specified we can read the corresponding losses from the loss functions given in Fig. 12.4 to obtain the losses given in Fig. 12.5.

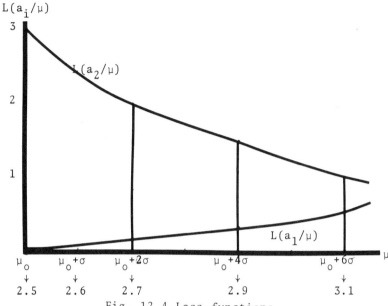

Fig. 12.4 Loss functions.

| | Actions | |
States of Nature	a_1:Accept H_o	a_2:Reject H_o
$\theta_o:\mu_o = 2.5$	0	3.0
$\theta_1:\mu_1 = 2.6$	0.1	2.5
$\theta_2:\mu_2 = 2.7$	0.2	2.0
$\theta_3:\mu_3 = 2.8$	0.3	1.5
$\theta_4:\mu_4 = 2.9$	0.4	1.3
$\theta_5:\mu_5 = 3.0$	0.5	1.0

Fig. 12.5 Losses.

Step 6

Designation of the set of competing strategies. The strategy s_1 specifies that the 5 percent level of significance should be used in testing the null hypothesis. To make this test we use the test statistic

357

$$z = \frac{\bar{y} - \mu_o}{\sigma/\sqrt{n}}$$

and reject the null hypothesis (take action a_2) if $z > 1.645$. Thus each strategy is designated by the critical value of z, which divides the range of $\bar{y}$ into two intervals.

We designate the 5 percent level of significance strategy by the following rule:

$$s_1: \left\{ \begin{array}{l} \text{if } z = \dfrac{\bar{y} - \mu_o}{\sigma/\sqrt{n}} > 1.645, \text{ take action } a_2 \\[2em] \text{if } z < 1.645, \text{ take action } a_1 \end{array} \right\}$$

This is equivalent to the rule

$$\left\{ \begin{array}{l} \text{if } \bar{y} > 1.645 \dfrac{\sigma}{\sqrt{n}} + \mu_o, \text{ take action } a_2 \\[2em] \text{if } \bar{y} < 1.645 \dfrac{\sigma}{\sqrt{n}} + \mu_o, \text{ take action } a_1 \end{array} \right\}$$

In our illustration, we thus have

$$\left\{ \begin{array}{l} \text{if } \bar{y} > 2.645, \text{ take action } a_2 \\[1em] \text{if } \bar{y} < 2.645, \text{ take action } a_1 \end{array} \right\}$$

Similarly, s_2, the 1 percent level of significance strategy, could be designated by

$$s_2: \left\{ \begin{array}{l} \text{if } z = \dfrac{\bar{y} - \mu_o}{\sigma/\sqrt{n}} > 2.33, \text{ take action } a_2 \\[2em] \text{if } z < 2.33, \text{ take action } a_1 \end{array} \right\}$$

Again, this is equivalent to

$$\left\{ \begin{array}{l} \text{if } \bar{y} > 2.33 \, \dfrac{\sigma}{\sqrt{n}} + \mu_0, \text{ take action } a_2 \\[2em] \text{if } \bar{y} < 2.33 \, \dfrac{\sigma}{\sqrt{n}} + \mu_0, \text{ take action } a_1 \end{array} \right\}$$

We can thus specify each strategy by the critical value of $\bar{y}$ or the critical value of z. We can denote these critical z values for the mth strategy by $z_o(m)$. Thus $z_o(1) = 1.645$ and $z_o(2) = 2.33$.

Step 7

Evaluation of the action probabilities. For each competing strategy we are interested in the probabilities of taking action a_1 or action a_2, given any state of nature $\theta_k : \mu_k$. For each strategy s_m and state of nature θ_k we need

$$g_{k1}(m) = Pr[a_1/\mu_k] = Pr[z \leq z_o(m)/\mu_k]$$

and

$$g_{k2}(m) = Pr[a_2/\mu_k] = Pr[z > z_o(m)/\mu_k]$$

For example, if we depict the probabilities for strategy s_1 (5 percent significance level) graphically, we have

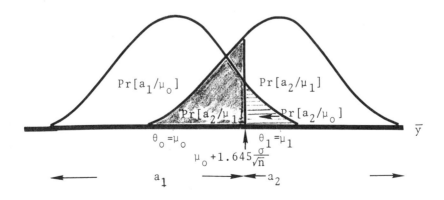

Fig. 12.6 Action probabilities for strategy s_1.

For strategy s_1 the action probabilities can be found by the following technique:

$$g_{k2}(1) = \Pr[a_2/\mu_k] = \Pr[z > 1.645/\mu_k]$$

$$= \Pr\left[\frac{\bar{y}-\mu_o}{\sigma/\sqrt{n}} > 1.645/\mu_k\right] = \Pr\left[\bar{y} > 1.645 \frac{\sigma}{\sqrt{n}} + \mu_o/\mu_k\right]$$

But since for the kth state of nature $\bar{y} \sim N(\mu, \sigma/\sqrt{n})$, we can standardize this probability so that

$$g_{k2}(1) = \Pr\left[\frac{\bar{y} - \mu_k}{\sigma/\sqrt{n}} > \frac{1.645 \sigma/\sqrt{n} + \mu_o - \mu_k}{\sigma/\sqrt{n}}\right]$$

$$= \Pr\left[z > 1.645 + \frac{\mu_o - \mu_k}{\sigma/\sqrt{n}}\right]$$

Thus

$$g_{k2}(1) = \int_{z_k(1)}^{\infty} \frac{1}{\sqrt{2\pi}} e^{-z^2/2} \, dz$$

where

$$z_k(1) = 1.645 + \frac{\mu_o - \mu_k}{\sigma/\sqrt{n}}$$

Similarly, for strategy s_2 (1 percent significance level) we have

$$g_{k2}(2) = \Pr[a_2/\mu_k] = \Pr[z > 2.33/\mu_k]$$

$$= \Pr\left[z_k > 2.33 + \frac{\mu_o - \mu_k}{\sigma/\sqrt{n}}\right]$$

$$= \int_{z_k(2)}^{\infty} \frac{1}{\sqrt{2\pi}} e^{-z^2/2} \, dz$$

where

$$z_k(2) = 2.33 + \frac{\mu_o - \mu_k}{\sigma/\sqrt{n}}$$

Thus, once we have μ_k and σ the evaluation of these action probabilities becomes a table look-up problem.

Since there are only two possible action, the action probabilities for a_1 are found by evaluating

$$g_{k1}(m) = 1 - g_{k2}(m)$$

In the particular example, the action probabilities for s_1 are

$$z_o(1) = 1.645 - \frac{0}{(0.5)/\sqrt{25}}$$

$$= 1.645$$

$$g_{o2}(1) = \int_{1.645}^{\infty} \frac{1}{\sqrt{2\pi}} e^{-z^2/2} \, dz$$

$$= 0.05$$

$$z_1(1) = 1.645 - \frac{0.1}{(0.5)/\sqrt{25}}$$

$$= 0.645$$

$$g_{12}(1) = \int_{0.645}^{\infty} \frac{1}{\sqrt{2\pi}} e^{-z^2/2} \, dz$$

$$= 0.2594$$

$$z_2(1) = 1.645 - \frac{0.2}{(0.5)/\sqrt{25}}$$

$$= -0.355$$

$$g_{22}(1) = \int_{-0.355}^{\infty} \frac{1}{\sqrt{2\pi}} e^{-z^2/2} \, dz$$

$$= 0.6787$$

$$z_3(1) = -1.355$$

$$g_{32}(1) = 1 - \text{AREANORM}(-1.355)$$

$$= 0.9123$$

$$z_4(1) = -2.355 \qquad\qquad g_{42}(1) = 1 - \text{AREANORM}(-2.355)$$

$$= 0.9907$$

$$z_5(1) = -3.355 \qquad\qquad g_{52}(1) = 1 - \text{AREANORM}(-3.355)$$

$$= 0.9995$$

For s_2 we have

$$z_k = 2.33 + \frac{\mu_1 - \mu_k}{\sigma/\sqrt{n}}$$

$$z_o = 2.33 \qquad g_{02}(2) = 1 - \text{AREANORM}(2.33) = 0.0100$$

$$z_1 = 1.33 \qquad g_{12}(2) = 1 - \text{AREANORM}(1.33) = 0.0918$$

$$z_2 = 0.33 \qquad g_{22}(2) = 1 - \text{AREANORM}(0.33) = 0.3707$$

$$z_3 = -0.67 \qquad g_{32}(2) = 1 - \text{AREANORM}(-0.67) = 0.7486$$

$$z_4 = -1.67 \qquad g_{42}(2) = 1 - \text{AREANORM}(-1.67) = 0.9525$$

$$z_5 = -2.67 \qquad g_{52}(2) = 1 - \text{AREANORM}(-2.67) = 0.9960$$

Since in this illustration there are only two actions, so that $\Pr[a_1] = 1 - \Pr[a_2]$, we summarize only the action probabilities for a_2:

States of Nature	Strategy s_1	s_2
θ_o	0.05	0.01
θ_1	0.2594	0.0918
θ_2	0.6787	0.3707
θ_3	0.9123	0.7486
θ_4	0.9907	0.9525
θ_5	0.9995	0.9960

Fig. 12.7 Action probabilities for a_2.

Step 8

Determination of average losses. Recall that the average loss for each strategy and state of nature is given by

$$\bar{\ell}_k(m) = \sum_{i=1}^{n_a} \Pr[a_i/\mu_k] \cdot L(a_i/\mu_k)$$

$$= \sum_{i=1}^{n_a} g_{ki}(m) \ell_{ki}$$

In our case $n_a = 2$ and $n_s = 2$. We thus have the simplification for the average loss for the kth state of nature and mth strategy, which is given by

$$\bar{\ell}_k(m) = g_{k1}(m)\ell_{k1} + g_{k2}(m)\ell_{k2} = [1 - g_{k2}(m)]\ell_{k1} + g_{k2}(m)\ell_{k2}$$

where $g_{ki}(m)$ now represents the action probabilities for strategy s_m, which is the probability of taking action a_i when the state of nature is $\theta_k = \mu_k$. The ℓ_{ki} are obtained from the loss curve $L(a_i/\mu)$ and are shown in Fig. 12.4.

In the illustration, the average losses for the different states of nature when strategy s_1 is used are determined by

$$\bar{\ell}_0(1) = (0.95)(0) + (0.05)(3.0) = 0.15$$

$$\bar{\ell}_1(1) = (0.7406)(0.1) + (0.2594)(2.5) = 0.72$$

For strategy s_2 we evaluate

$$\bar{\ell}_0(2) = (0.99)(0) + (0.01)(3.0) = 0.03$$

$$\bar{\ell}_1(2) = (0.9082)(0.1) + (0.0918)(2.5) = 0.32$$

and similarly for the other states of nature. We can thus obtain the average losses, which are given in Fig. 12.8:

Prior Probabilities	States of Nature	Average Losses s_1	s_2
0.30	$\theta_0:\ \mu_0$	0.15	0.03
0.25	$\theta_1:\ \mu_1$	0.72	0.32
0.20	$\theta_2:\ \mu_2$	1.35	0.87
0.10	$\theta_3:\ \mu_3$	1.39	1.20
0.08	$\theta_4:\ \mu_4$	1.29	1.26
0.07	$\theta_5:\ \mu_5$	1.00	1.00

Fig. 12.8 Average Losses.

Along the left margin of Fig. 12.8 we have recorded the values of the prior probabilities for the six states of nature being used.

Step 9

Selection of the strategy. If we use the Bayes criterion, we need the prior probabilities and the average losses to determine the expected loss for each strategy. This expected loss is found by evaluating

$$EL(s_m) = \sum_{k=1}^{n_\theta} q_k\ \bar{\ell}_k(m)$$

In the illustration, we assume that the prior probabilities are as given in Fig. 12.8. Thus we obtain

$EL(s_1) = (0.30)(0.15) + (0.25)(0.72) + (0.20(1.35)$

$\qquad + (0.10)(1.30) + (0.08)(1.24) + (0.07)(1.00)$

$\qquad = 0.81$

$EL(s_2) = (0.30)(0.03) + (0.25)(0.32) + (0.20)(0.87)$

$\qquad + (0.10)(1.20) + (0.18)(1.26) + (0.07)(1.00)$

$\qquad = 0.55$

By comparing the expected loss for s_1 and s_2 it can be seen that the 1 percent level of significance is superior to the 5 percent level of significance because the expected loss for s_2 is only 0.55, as compared with an expected loss of 0.81 for s_1. In fact, a new concept is exhibited by this example. A comparison of the average losses of s_2 with s_1 for each state of nature will show that the average loss for s_2 is less than the corresponding average loss for s_1 for all θ_k. Thus, the superiority of s_2 over s_1 would be preserved for all sets of prior probabilities. In such cases we say that strategy s_2 dominates strategy s_1. Any strategy that is dominated by any other strategy in the set can be dropped from consideration because any rational criterion of strategy selection would select the dominating strategy rather than the dominated one.

Although in this example the 1 percent level of significance strategy dominates the 5 percent level of significance, we should not hastily conclude that the 1 percent level of significance is always superior in testing hypotheses. The average losses obtained for a strategy are very sensitive to the losses for each state of nature and action as well as to the underlying probabilities of occurrences. We could easily change the above evaluations so that the 5 percent level of significance dominates the 1 percent level or neither level dominates the other.

Several interesting considerations are illustrated by the above evaluation. For example, one might ask what is the best number of states of nature that should be used since the sensitivity of the results to different input values is of interest. In addition, we could consider parameter variation studies to determine the parameter or input space that corresponds to the situation when the 5 percent level of significance would be superior. Of course, in this two-strategy situation the complement of this region would be that region where the 1 percent is superior, and the boundary between these two regions would be the region of indifference. Since the number of input values μ_0, σ, n, the loss functions, and the prior probabilities are very numerous even in this simple situation, such parameter variation considerations can involve either unsolvable mathematical difficulties or a very large number of evaluations. At this state of our consideration we simple note these interesting problems before discussing the next type of decision problem.

12.3 Estimation of Parameters

Another problem considered in Chaps. 9 and 10 was parameter estimation. Two estimation concepts were developed: point estimation and interval estimation. In both cases a point estimate of the parameter had to be obtained, and it was argued that a good characteristic of such a point estimator is that it is unbiased. Instead of using such characteristics in selecting the estimator we shall consider how decision theory can be used in selecting a point estimator.

We present the approach in general but also use a typical problem found in public health as an example. Consider the problem of measuring the health hazards when using a new DDT spraying technique. A measure of the hazard must be determined. It has been decided that the average amount of DDT assimilation by individuals using the spraying technique gives a good indication of the health hazard. The assimilation is measured by the parts per million of DDT found in the blood. We thus wish to estimate the average amount of DDT, μ, that one would find in individuals using the spraying technique.

To estimate μ we use a sample of individuals who are using the spraying technique. The observable random variable is therefore the ppm of DDT found in the blood. Although we could use all the sample observations individually, we choose to summarize the information by using the mean $\bar{y}$.

We shall use decision theory to determine what action should be taken if a particular value of $\bar{y}$ is observed. We shall one again proceed through the steps used in evaluating competing strategies.

Step 1

Possible actions:

a_1: conclude that there is no health hazard
a_2: conclude that the hazard is within safe limits
a_3: conclude that a possible hazard exists
a_4: conclude that a definite hazard exists
a_5: conclude that an extreme hazard exists

Step 2

Possible outcomes of the observable random variable. Since $\bar{y}$ is a continuous variable, there are an infinite number of possible outcomes.

Step 3

States on nature. If we assume that Y has a normally distribution, then there are two parameters μ and σ that must be specified for each state of nature. We shall also consider σ to be an undetermined parameter because in this problem it does not seem reasonable to assume that the variation remains the same over all the states of nature.

To have only a discrete number of states of nature, let us assume the medical scientists have identified the possible states as being θ_1: no hazard; θ_2: hazard within safe limits; θ_3: possible health hazard; θ_4: definite health hazard; and θ_5: extreme health hazard. In making these identifications, the medical scientists have also associated values of μ and σ with each state of nature. These are

$$\theta_1: \mu_1 = 2.0 \qquad \sigma_1 = 0.5$$
$$\theta_2: \mu_2 = 4.0 \qquad \sigma_2 = 1.0$$
$$\theta_3: \mu_3 = 7.0 \qquad \sigma_3 = 2.5$$
$$\theta_4: \mu_4 = 10.0 \qquad \sigma_4 = 7.0$$
$$\theta_5: \mu_5 = 15.0 \qquad \sigma_5 = 8.0$$

Step 4

Probabilities of occurrence. The probabilities of occurrence are not needed when a continuous random variable is used because the action probabilities can be generated directly. We shall, however, given the distribution of $\bar{y}$ for any state of nature, θ_k, which is

$$f(\bar{y}/\mu_k, \sigma_k) = \frac{1}{\sqrt{2\pi}\ \sigma_k/\sqrt{n}}\ e^{-\frac{1}{2}\left[\frac{\bar{y}-\mu_k}{\sigma_k/\sqrt{n}}\right]^2} \quad -\infty \leq \bar{y} \leq \infty$$

Step 5

Losses. Once again we encounter the problem of determining appropriate loss values. Such losses can be determined by fairly systematic techniques. Consider, for example, the loss for cell (θ_1, a_5). If the action taken when an a_5 decision is made is to introduce more costly spraying techniques, then these additional costs could be accumulated. The loss for cell (θ_5, a_1) is different because this loss is due to the possible incapacitation of the employees and could be evaluated on some actuarial basis. Now, however, we simply assign reasonable loss values to the matrix.

We shall use the following set of losses:

States of Nature	Actions				
	a_1	a_2	a_3	a_4	a_5
θ_1	1	3	12	30	50
θ_2	3	2	8	20	40
θ_3	35	5	4	15	35
θ_4	40	20	10	7	40
θ_5	75	60	40	30	25

Fig. 12.9 Losses.

Step 6

Designation of the competing strategies. Following the procedure given in Sec. 12.2, we shall consider the class of strategies that is obtained by dividing the range of $\bar{y}$ into five intervals. This class of strategies can be represented by the rule

$$\text{If } \bar{y} \leq b_1, \qquad \text{take action } a_1$$
$$\text{If } b_1 < \bar{y} \leq b_2, \qquad \text{take action } a_2$$
$$\text{If } b_2 < \bar{y} \leq b_3, \qquad \text{take action } a_3$$
$$\text{If } b_3 < \bar{y} \leq b_4, \qquad \text{take action } a_4$$
$$\text{If } \bar{y} > b_4, \qquad \text{take action } a_5$$

Since the b_i depend on the strategy s_m, we shall designate each set of interval end points by $b_i(m)$. Thus any strategy in the class can be represented by: $s_m = [b_1(m), b_2(m), b_3(m), b_4(m)]$.

We shall restrict our consideration to the following three strategies:

$$s_1 = [3.0, 5.5, 8.5, 12.5]$$
$$s_2 = [3.0, 5.0, 7.5, 11.0]$$
$$s_3 = [3.5, 6.0, 9.0, 13.5]$$

Note that the strategy s_1 simply uses the midpoints of the μ values used to designate the states of nature; s_2 is a strategy which is more likely to lead to the conclusion that a hazardous condition exists (since a smaller $\bar{y}$ would lead us to same action designated by s_1); and s_3 is less likely to lead to the conclusion that a hazardous condition exists (since the mean $\bar{y}$ must be larger to lead to the same action designated by s_1).

Evaluation of action probabilities. The distribution of $\bar{y}$ was presented in step 4. To evaluate the action probabilities, the sample size n must be specified. In our illustration we shall use n = 9. The action probabilities are areas under the normal curve determined by the state of nature and strategy as graphically illustrated for strategy s_1 in Fig. 12.10:

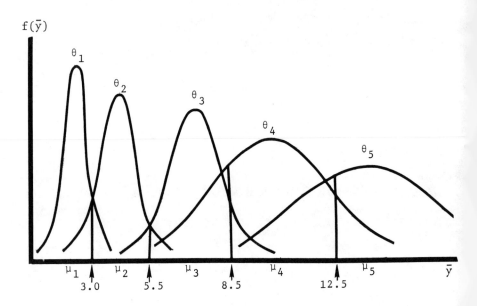

Fig. 12.10 Representation of the action probabilities.
for $s_1 = [3.0, 5.5, 8.5, 12.5]$

The action probabilities for the mth strategy are

$$g_{ki}(m) = \Pr[a_i/\theta_k] = \int_{b_{i-1}(m)}^{b_i(m)} \frac{1}{\sqrt{2\pi}\, \sigma_k/\sqrt{n}}\, e^{-\frac{1}{2}\left[\frac{\bar{y}-\mu_k}{\sigma_k/\sqrt{n}}\right]^2} d\bar{y}$$

To evaluate these integrals we use the standardizing transformation to obtain

$$g_{ki}(m) = \int_{z_{i-1}(m)}^{z_i(m)} \frac{1}{\sqrt{2\pi}} e^{-z^2/2} \, dz$$

where

$$z_{ki}(m) = \frac{b_i(m) - \mu_k}{\sigma_k/\sqrt{9}}$$

The values of $b_1(m)$, $b_2(m)$, $b_3(m)$, and $b_4(m)$ are given by the strategy representation, while $b_{k0}(m) = z_{k0}(m) = -\infty$ and $b_{k5}(m) = z_{k5}(m) = \infty$. To obtain the action probabilities for strategy $s_1 = [3.0, 5.5, 8.5, 12.5]$ and state of nature θ_1: $\mu_1 = 2.0$, $\sigma_1 = 0.5$, we evaluate $z_{1i}(1)$, which are given below.

$$z_{1_0}(1) = -\infty$$

$$z_{11}(1) = \frac{3.0 - 2.0}{(0.5)/3} = 6$$

$$z_{12}(1) = \frac{5.5 - 2.0}{(0.5)/3} = 21$$

$$z_{13}(1) = \frac{8.5 - 2.0}{(0.5)/3} = 39$$

$$z_{14}(1) = \frac{12.5 - 2.0}{(0.5)/3} = 63$$

$$z_{15}(1) = \infty$$

So,

$$g_{11}(1) = \Pr[a_1/\theta_1] = \int_{-\infty}^{6} \frac{1}{\sqrt{2\pi}} e^{-z^2/2} \, dz = 1.000$$

and the remaining four action probabilities $g_{1i}(1) = 0$. For strategy s_1 and state of nature θ_2: $\mu_2 = 4.0$ and $\sigma_2 = 1.0$, we have

$$z_{21}(1) = \frac{3.0 - 4.0}{(1.0)/3} = -3.00$$

$$z_{22}(1) = \frac{5.5 - 4.0}{(1.0)/3} = 4.5$$

$$z_{23}(1) = \frac{8.5 - 4.0}{(1.0)/3} = 13.5$$

$$z_{24}(1) = \frac{12.5 - 4.0}{(1.0)/3} = 25.5$$

$$z_{25}(1) = \infty$$

Thus the probabilities are

$$g_{21}(1) = \Pr[a_1/\theta_2] = \int_{-\infty}^{-3} \frac{1}{\sqrt{2\pi}} e^{-z^2/2} \, dz = 0.0014$$

$$g_{22}(1) = \Pr[a_2/\theta_2] = \int_{-3}^{4.5} \frac{1}{\sqrt{2\pi}} e^{-z^2/2} \, dz = 0.9986$$

$$g_{2i}(1) = 0.0000 \text{ for } i \geq 3$$

since

$$g_{2i}(1) = \int_{z_{2,i-1}(1)}^{z_{2i}(1)} \frac{1}{2} e^{-z^2/2} = 0.0000$$

Similar computations for the other strategies and states of nature yield the following values:

State of Nature	Actions a_1	a_2	a_3	a_4	a_5
θ_1	1.0000	0	0	0	0
θ_2	0.0014	0.9986	0	0	0
θ_3	0	0.0359	0.9282	0.0359	0
θ_4	0.0014	0.0254	0.2343	0.4034	0.1423
θ_5	0	0.0002	0.0071	0.1663	0.8266

Fig. 12.11 Action probabilities for s_1.

State of Nature	Actions a_1	a_2	a_3	a_4	a_5
θ_1	0.5675	0.1240	0.1297	0.1120	0.0668
θ_2	0.0014	0.9972	0.0014	0	0
θ_3	0.1151	0.7698	0.1151	0	0
θ_4	0.0014	0.0148	0.1261	0.5241	0.3336
θ_5	0	0.0001	0.0024	0.0643	0.9332

Fig. 12.12 Action probabilities for s_2.

State of Nature	Actions a_1	a_2	a_3	a_4	a_5
θ_1	1.0000	0	0	0	0
θ_2	0.0668	0.9332	0	0	0
θ_3	0	0.1151	0.0767	0.0082	0
θ_4	0.0026	0.0410	0.2900	0.5996	0.0668
θ_5	0	0.0003	0.0119	0.2755	0.7123

Fig. 12.13 Action probabilities for s_3.

Step 8

Computation of average loss. Using the losses given in Fig. 12.9, and weighting them by corresponding action probabilities, we obtain

| | Strategies | | |
State of Nature	s_1	s_2	s_3
θ_1	1.0000	9.1929	1.0000
θ_2	2.0014	2.0098	2.0668
θ_3	4.4308	8.3379	4.2053
θ_4	11.4228	18.6257	10.6932
θ_5	25.9500	25.3610	26.5665

Fig. 12.14 Average losses.

Step 9

Selection of the strategy. No strategy is dominated. Any of the three criteria for selection can now be used; for example, the minimax strategy would be s_2. Given a set of prior probabilities, the Bayes or minimum risk strategy could also be found, but this is left to the student.

12.4 The Use of Decision Theory in Sample Size Determinations

In Chaps. 9 and 10 we introduced the sample size problem and presented methods of determining the sample size so as to have a prescribed level of confidence [(1-α) percent] that the parameter estimate $\bar{y}$ has an error which is less than a prescribed amount ε. These methods, however, did not consider the cost of obtaining an observation or the cost in making an error in the estimate, and both these costs are important practical considerations. One way of introducing such considerations is to use the decision theory approach for the sample size determination.

To demonstrate the approach we shall use a numerical illustration but shall not attempt to identify the illustration with any particular problem because the computational steps are the same as in the previous two sections.

We assume that the population being sampled has a normal distribution. If the population variance σ^2 is known, then the appropriate sample size is given by

$$n = \frac{z^2_{1-\alpha/2} \cdot \sigma^2}{\epsilon^2}$$

Thus for a specified α and ϵ the appropriate sample size can be found if σ^2 is known. Although the value of σ^2 is not known, the decision theory approach uses the states of nature to specify the unknown parameter values. Let us assume that in our illustration we shall restrict the values of σ^2 to five and use $\sigma^2 = 80, 90, 100, 110,$ and 120 as the possible parameter values. If we specify $\alpha = 0.05$ and $\epsilon = 1.5$, the appropriate sample size for each value of σ^2 can be found. For example, if $\sigma^2 = 80$, the appropriate sample size would be

$$n = n_1 = \frac{z^2_{.975} \cdot \sigma^2}{\epsilon^2} = \frac{(1.96)^2 (80)}{(1.5)^2} = 136.59 = 137$$

Since there are only five states of nature, there will only be five "possible" actions. Each action specifies the appropriate sample size needed to have 95 percent assurance that the mean of the sample will estimate μ within 1.5 units.

We need a random variable upon which to base the decision as to what action to take. In this case we use the s'^2 obtained from a preliminary sample.

Let us now proceed through the decision steps.

<u>Step 1</u>

Actions (for the five states of nature already given):

a_1: use $n = n_1 = \dfrac{z^2_{1-\alpha/2} \cdot \sigma^2_1}{\epsilon^2} = 136.59 = 137$

a_2: use $n = n_2 = \dfrac{z^2_{1-\alpha/2} \cdot \sigma^2_2}{\epsilon^2} = 153.66 = 154$

a_3: use $n = n_3 = \dfrac{z^2_{1-\alpha/2} \cdot \sigma^2_3}{\epsilon^2} = 170.74 = 171$

a_4: use $n = n_4 = \dfrac{z^2_{1-\alpha/2} \cdot \sigma^2_4}{\epsilon^2} = 187.81 = 188$

$$a_5: \text{use } n = n_5 = \frac{z^2_{1-\alpha/2} \cdot \sigma^2_5}{\epsilon^2} = 204.89 = 205$$

Step 2

Possible outcomes of the random variable. The random variable is s'^2, and its range is $0 \leq s'^2 \leq \infty$.

Step 3

States of Nature. Each state of nature is identified by a value of the parameter σ^2. We shall use the five values already given above. Thus

$$\theta_1: \sigma^2_1 = 80$$

$$\theta : \sigma^2_2 = 90$$

$$\theta_3: \sigma^2_3 = 100$$

$$\theta_4: \sigma^2_4 = 110$$

$$\theta_5: \sigma^2_5 = 120$$

Step 4

Probabilities of occurrence. Since s'^2 is a continuous random variable, we can compute the action probabilities directly but need the probability distribution of s'^2. The exact distribution will be discussed in Chap. 14; thus, if the preliminary sample is small (less than 30), the completion of the illustration would have to be delayed. However, if the preliminary sample is large, we can use the central limit theorem, which states that

$$z = \frac{s'^2 - \sigma^2}{\sqrt{2/(n-1)} \ \sigma^2}$$

and has an approximate normal distribution with mean zero and variance 1.

Step 5

Losses.

States of Nature	Actions				
	$n = n_1$	$n = n_2$	$n = n_3$	$n = n_4$	$n = n_5$
$\theta_1 : \sigma_1^2 = 80$	1	3	5	7	9
$\theta_2 : \sigma_2^2 = 90$	50	2	4	8	10
$\theta_3 : \sigma_3^2 = 100$	80	40	3	5	7
$\theta_4 : \sigma_4^2 = 110$	90	60	30	4	6
$\theta_5 : \sigma_5^2 = 120$	100	70	40	20	5

Fig. 12.15 Losses.

We note from the loss table that the minimum loss is associated with the preferred action, as is to be expected. In addition, the losses increase with n, which indicates the costs associated with selecting more observations than needed. The losses associated with using too small a sample are larger and reflect a higher penalty.

Step 6

Designation of the competing strategies. Since s'^2 is a continuous random variable, we can again designate an entire class of strategies by interval designators. Let the mth strategy be represented by $s_m = [b_1(m), b_2(m), b_3(m), b_4(m)]$. For example, the strategy designated by [85,95,105,115] would be

If $s'^2 \leq 85$, take action a_1 (use $n = n_1$)
If $85 < s'^2 \leq 95$, take action a_2 (use $n = n_2$)
If $95 < s'^2 \leq 105$, take action a_3 (use $n = n_3$)
If $105 < s'^2 \leq 115$, take action a_4 (use $n = n_4$)
If $s'^2 > 115$, take action a_5 (use $n = n_5$)

Step 7

Determination of action probabilities Once an appropriate set of strategies is chosen, the action probabilities can
be found by evaluating the appropriate integral:

$$g_{ki}(m) = \Pr[a_i/\theta_k]$$

$$= \int_{b_{i-1}(m)}^{b_i(m)} \frac{1}{\sqrt{2\pi}} \, e^{-\frac{1}{2}\left[\frac{s'^2-\sigma_k^2}{\sqrt{2\sigma_k^4/(n-1)}}\right]^2} \, ds'^2$$

$$= \int_{z_{ki-1}(m)}^{z_{ki}(m)} \frac{1}{\sqrt{2\pi}} \, e^{-z^2/2} \, dz$$

where

$$z = \frac{b_i(m) - \sigma_k^2}{\sqrt{2/(n-1)} \, \sigma_k^2}$$

and

$$z_{ko}(m) = -\infty \text{ and } z_{k5}(m) = \infty$$

The completion of this approach to determine the sample size
follows the same steps as the applications covered in Secs.
12.2 and 12.3 and thus is not repeated.

12.5 A Detailed Application: The Salary
Offer Decision Problem

To show the power and scope of statistical decision
theory we shall consider a practical decision problem and
use decision theoretical concepts to find its solution.

A publishing firm needs to increase its typing output.
It can either increase its own typing force or send out the
material to a service organization that will charge a flat

rate of $1.25 per page. The company has determined that it grosses $1.50 for each typed page; thus, if the service organization is used, the management will net $0.25 per page.

By hiring their own typing force the publishing firm hopes to be able to increase the net income per typed page. However, the salary that can be paid to an applicant is flexible, and thus the management must determine what salary to offer the applicant. Since the typing accuracy is of prime importance to the firm, all applicants are screened initially on the basis of an accuracy test; only those applicants whose accuracy is above a specified value are considered for the job. After the accuracy requirement is met, the applicant's typing speed is tested by using three different manuscripts. The salary offer is to be based upon the results of these tests. During the period of economic expansion the number of qualified applicants available to the firm is not sufficient to meet the typing needs, so management knows that if an applicant turns down the offer, the typing she would have done will have to be farmed out to the service company.

In this decision problem the set of actions are the possible beginning salaries that can be offered to an applicant. On the basis of such an offer the typist may accept or reject the employment opportunity. The uncertainty lies in the unknown value of each applicant's typing speed, which is represented by θ. If the offer is too high relative to the value of θ, the typist is more likely to accept, but the low value of her θ will cause her production to be lower than is expected, and thus it is possible for the company to lose money on an individual who accepts the salary offer. On the other hand, a low offer may cause the applicant to reject the offer, and the publishing company must therefore pay the service company rates for the pages that would have been typed by the applicant.

From past experience, management knows the relationship between the number of pages per hour $m(\theta)$ that can be typed by an individual whose average typing speed is θ. This relationship is given graphically in Fig. 12.20. Management can obtain an estimate of the average typing speed θ for each applicant by averaging the applicant's typing speed on three

different manuscripts. The salary offer can then be specified by a strategy that relates a particular action (the
salary to be offer) with each estimate of θ.

To select the best strategy from a set of competing
strategies, we once again proceed through the required decision theory steps, but now we are in a position to emphasize
how to obtain a meaningful measure of regret.

Step 1

Actions. The action is to make a salary offer, which is
represented by a = the amount of the offer. The set of possible offers is infinite because a is a continuous variable
whose range is a $\geq$ 0.

Step 2

Possible outcomes of the random variable. Each applicant has a distribution of typing speeds that is caused principally by the variation of the difficulty of the material
being typed. We assume that this distribution is normal with
mean θ and standard deviation σ = 15. The typing test yields
three independent observations, y_1, y_2, and y_3, from the typing speed distribution of an applicant. We shall use the
mean $\bar{y}$ of these observed values as the random variable since
we are interested in estimating the mean θ.

Step 3

The states of nature. The state of nature is designated
by the unknown average typing speed of the applicant θ.
Although the range of θ is infinite, we shall use a finite
set of representative values in our evaluation. We use nine
states of nature ranging from θ = 30 words/minute to θ = 110
words/minute in steps of 10 words/minute.

Step 4

Probabilities of occurrence. Since we assume that Y
has a normal distribution with mean θ and σ = 15, the sample
mean $\bar{y}$ also has a normal distribution with mean θ and stand-

ard deviation $\sigma_{\bar{y}} = \sigma_y/\sqrt{n} = 15/\sqrt{3} = 8.66$. Since $\bar{y}$ is a continuous random variable, we can again calculate the action probabilities directly and thus do not need to consider the prob- of occurrence.

Step 5

Since our evaluation of regret in this case involves the strategy being followed, we have to reorder the steps to designate the competing strategies before giving the losses or regrets:

s_1: Assume that $\theta = \hat{\theta} = \bar{y}$ and offer salary that will maximize expected profit per page if applicant actually has this typing speed. Determination of this maximizing salary is shown in section on regrets.

s_2: Offer each applicant \$100/month regardless of $\bar{y}$.

Step 6

Regrets. We have a different table of regrets of each strategy. Since the action to be taken in strategy s_1 depends upon the value of $\hat{\theta}$, we give the regrets for each state of nature θ for selected values of $\hat{\theta}$:

Estimates, $\hat{\theta}$

State of Nature	30	40	50	60	70	80	90	100	110
30	0.20	2.85	24.43	34.42	39.42	54.42	60.00	67.42	69.42
40	0.50	0	4.87	19.53	24.25	39.25	49.25	54.25	54.25
50	0.28	0.28	0	2.90	9.90	27.78	37.78	42.78	42.78
60	1.13	1.13	1.13	0	0.70	14.88	24.88	29.88	29.88
70	0.61	0.61	0.61	0.24	0	10.18	21.24	24.24	24.24
80	6.51	6.51	6.51	6.51	6.51	0	10.18	15.26	15.21
90	6.68	6.68	6.68	6.68	6.68	1.48	0	4.23	4.23
100	11.70	11.70	11.70	11.70	11.70	11.70	4.20	0	0
110	12.00	12.00	12.00	12.00	12.00	12.00	6.00	0	0

Fig. 12.16 Regrets $r(\theta,\hat{\theta})$ for strategy s_1.

There is only one action for strategy s_2, so in this case there is only a single regret value for each state of nature:

State of Nature θ	Regret for all $\hat{\theta}$
30	34.42
40	19.53
50	2.90
60	0.00
70	0.24
80	6.51
90	6.68
100	11.70
110	12.00

Fig. 12.17 Regrets $r(\theta,\hat{\theta})$ for strategy s_2.

Step 7

Determination of the action probabilities. Since strategy s_1 specifies the action to be taken in terms of the estimate $\hat{\theta} = \bar{y}$ that is obtained, we shall find the probabilities of obtaining a particular estimate because this will be more convenient computationally.

Although the estimator $\hat{\theta}$ is a continuous random variable, we shall only use a set of representative values:

$$\hat{\theta} = 30, 40, 50, \ldots , 110$$

To find the probabilities we need to designate an interval that corresponds with each $\hat{\theta}$ value; we use

If $\hat{\theta}$ = 30, use interval $-\infty < \bar{y} \leq 35$
If $\hat{\theta}$ = 40, use interval $35 < \bar{y} \leq 45$
If $\hat{\theta}$ = 50, use interval $45 < \bar{y} \leq 55$
$\quad\quad\quad\quad\quad$.
$\quad\quad\quad\quad\quad$.
$\quad\quad\quad\quad\quad$.
If $\hat{\theta}$ = 100, use interval $95 < \bar{y} \leq 105$
If $\hat{\theta}$ = 110, use interval $105 < \bar{y} \leq \infty$

Using the fact that the distribution of $\bar{y}$ for each state of nature is $N(\theta, 8.66)$, we can determine the probability of obtaining a particular value of the estimate. For example, when the state of nature is $\theta = 60$ and the estimate $\hat{\theta} = 50$, the required probability is

$$\int_{45}^{55} \frac{1}{\sqrt{2\pi}(8.66)} e^{-\frac{1}{2}\left[\frac{\bar{y}-60}{8.66}\right]^2} d\bar{y}$$

To evaluate this integral we use the standard transformation, which yields

$$\int_{-1.73}^{-0.58} \frac{1}{\sqrt{2\pi}} e^{-\frac{1}{2}z^2} = 0.2392$$

where

$$z_1 = \frac{45 - 60}{8.66} = -1.73$$

$$z_2 = \frac{55 - 60}{8.66} = -0.58$$

These probabilities are given in Fig. 12.18:

State of Nature θ	$\hat{\theta} = \bar{y}$								
	30	40	50	60	70	80	90	100	110
30	0.7190	0.2392	0.0667	0.0019	0	0	0	0	0
40	0.2810	0.4380	0.2392	0.0399	0.0019	0	0	0	0
50	0.0418	0.2392	0.4380	0.2392	0.0399	0.0019	0	0	0
60	0.0019	0.0399	0.2392	0.4380	0.2392	0.0399	0.0019	0	0
70	0	0.0019	0.0399	0.2392	0.4380	0.2392	0.0399	0.0019	0.0019
80	0	0	0.0019	0.0399	0.2392	0.4380	0.2392	0.0399	0.0038
90	0	0	0	0.0019	0.0399	0.2392	0.4380	0.2392	0.0410
100	0	0	0	0	0.0019	0.0399	0.2392	0.4380	0.2810
110	0	0	0	0	0	0.0019	0.0399	0.2392	0.7190

Fig. 12.18 Action probabilities for strategy s_1.

Since the action a = 100 is always taken by strategy s_2, the action probabilities are 1.00 for action a = 100 for states of nature and 0 for all other actions.

Step 8

Average regrets. To find the average regret for strategy s_1, we must find the weighted average of the regrets in Fig. 12.18, with the corresponding action probabilities. For example, when θ = 30, the average loss for s_1 is given by

$$r_{30}(1) = 0.7190(0.20) + (0.2392)(2.85) + 0.0399(24.43)$$

$$+ (0.0019)(34.43)$$

$$= 1.87$$

The average regret for strategy s_2 is simply the regret for each state of nature given in Fig. 12.17 because the action probabilities are equal to 1 if a = 100 and 0 otherwise.

The average regrets for each strategy are

	Strategies	
State of Nature θ	s_1	s_2
30	1.87	34.42
40	1.77	19.53
50	1.22	2.90
60	0.87	0.00
70	0.72	0.24
80	5.11	6.51
90	1.82	6.68
100	1.49	11.70
110	0.26	12.00

Fig. 12.19 Average regrets.

It can readily be seen that s_2 is only superior to s_1 when the applicant's average speed is about 60 or 70 words/minute, indicating that if the prior probability of $\theta = 60$ is very large, it may not pay to give the typing test but to simply use the fixed-offer strategy. However, if a large variation is expected in the θ value of applicants, the use of the three tests and their average value to determine the offer is surely superior. In fact, s_1 is the Bayes strategy relative to the information contained in the triplet (y_1, y_2, y_3).

We forego the last steps of computing the expected regret for a given set of prior probabilities since the outcome is obvious when comparing only these two strategies.

In step 6 we simply presented the regrets for each of the nine states of nature and for the representative set of estimates. We shall now show how these regrets can actually be generated.

Determination of Regrets

The relationship between the typing speed of an individual and her productivity must be determined. In our example, we assume that the productivity is measured by the number of pages typed per week. Let $m(\theta)$ denote the number of pages typed by an individual whose average typing speed is θ. The functional relationship between θ and $m(\theta)$ is given in Fig. 12.20 and is well known to anyone experienced in operating a typing pool.

If an offer of a is made to an applicant whose typing speed is θ, she would produce $m(\theta)$ typed pages per week. If she accepts the offer, the company would gross $1.50 \cdot m(\theta)$ per week and have a net profit of $1.50 \cdot m(\theta) - a$. If the applicant does not accept the offer, the company must use the typing service for the $m(\theta)$ pages; then the net profit is $0.25 per page or $0.25 \cdot m(\theta)$ per week.

Since the applicant may or may not accept the offer, the expected net profit must be found by weighting the two possible net profits by their respective probabilities. The probability of acceptance depends upon the offer a and the

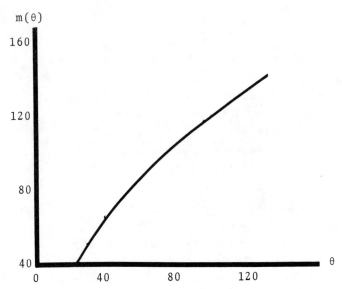

Fig. 12.20 Relationship between pages typed m(θ)
and average typing speed θ.

value of θ; this is shown in Fig. 12.21 as Pr[θ,a]. The
probability of not accepting the offer is simply (1-Pr[θ,a]).
It is assumed that these probability curves have been generated by the company from past experience.

The expected net profit V(θ,a) can be found by weighting the two net profits by their respective probabilities.
We thus have

$$V(\theta,a) = \text{Pr}[\theta,a](1.50m(\theta) - a) + (1 - \text{Pr}[\theta,a])(0.25m(\theta))$$

We generate these expected net profits numerically for
each state of nature [θ = 30,40,50,60,70,80,90,100,110] and
for different values of the actions [a = 25,30,35,40,45,
... , 145] by reading the appropriate value of m(θ) from
Fig. 12.20 and Pr[θ,a] from Fig. 12.21. Let us illustrate
the technique. Consider θ = 60 and a = 100. From the
graphs, we obtain m(60) = 85 and Pr(60,100) = 0.18. Thus

$$V(60,100) = 0.18[1.50(85) - 100] + 0.82[0.25(85)]$$

$$= 22.38$$

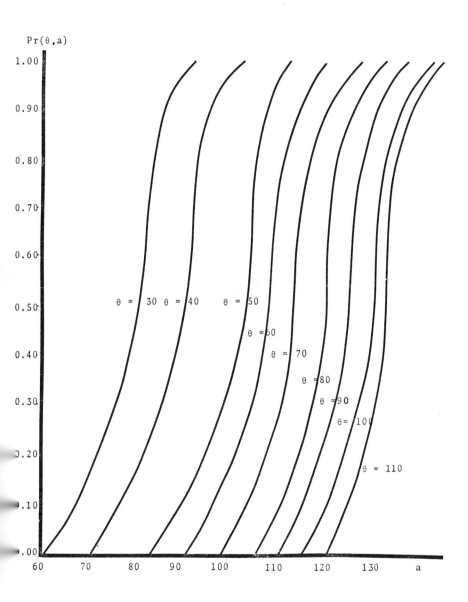

Fig. 12.21 Probability of acceptance of job if offer is a.

We have computed these values for specific values of θ and a; they are given in Fig. 12.22. The function V(θ,a) could also be graphed for values of θ and a; this would be useful if interpolation for values other than the tabled ones is necessary.

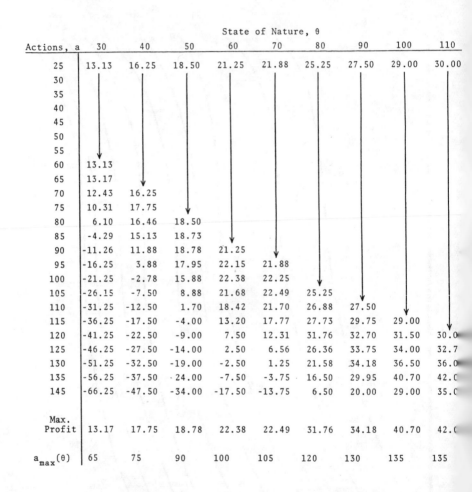

State of Nature, θ

Actions, a	30	40	50	60	70	80	90	100	110
25	13.13	16.25	18.50	21.25	21.88	25.25	27.50	29.00	30.00
30									
35									
40									
45									
50									
55									
60	13.13								
65	13.17								
70	12.43	16.25							
75	10.31	17.75							
80	6.10	16.46	18.50						
85	-4.29	15.13	18.73						
90	-11.26	11.88	18.78	21.25					
95	-16.25	3.88	17.95	22.15	21.88				
100	-21.25	-2.78	15.88	22.38	22.25				
105	-26.15	-7.50	8.88	21.68	22.49	25.25			
110	-31.25	-12.50	1.70	18.42	21.70	26.88	27.50		
115	-36.25	-17.50	-4.00	13.20	17.77	27.73	29.75	29.00	
120	-41.25	-22.50	-9.00	7.50	12.31	31.76	32.70	31.50	30.0
125	-46.25	-27.50	-14.00	2.50	6.56	26.36	33.75	34.00	32.7
130	-51.25	-32.50	-19.00	-2.50	1.25	21.58	34.18	36.50	36.0
135	-56.25	-37.50	-24.00	-7.50	-3.75	16.50	29.95	40.70	42.0
145	-66.25	-47.50	-34.00	-17.50	-13.75	6.50	20.00	29.00	35.0
Max. Profit	13.17	17.75	18.78	22.38	22.49	31.76	34.18	40.70	42.0
$a_{max}(\theta)$	65	75	90	100	105	120	130	135	135

Fig. 12.22 The expected net profit V(θ,a).

Strategy s_1 specifies that an offer of $a_{max}(\theta)$ is made; this maximizes the expected net profit, assuming that $\theta = \hat{\theta}$. Thus we need to have the value of $a_{max}(\theta)$ for different values of θ. These values are given in Fig. 12.22; they have been plotted in Fig. 12.23 and connected by a smooth curve

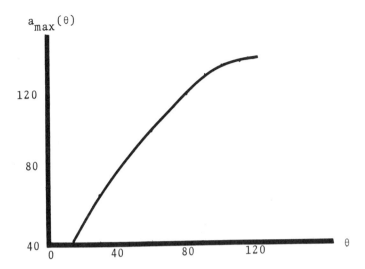

Fig. 12.23 Value of a that maximizes profit for
a given θ.

in order to be able to read $a_{max}(\theta)$ for any value of θ.

We are now in position to determine the regret values for strategy s_1 for representative values of the states of nature θ and for different estimates $\hat{\theta}$. If the true state of nature θ is known, then the salary offer of $a_{max}(\theta)$ would be made, and we would then maximize the expected net profit. Since we use $\hat{\theta}$ instead of θ, the expected net profit is not maximized, and difference between these two values is the regret. The regret is thus given by

$$r(\theta,\hat{\theta}) = V(\theta,a_{max}(\theta)) - V(\theta,a_{max}(\hat{\theta}))$$

To illustrate the procedure, consider the case when the true state of nature is θ = 60 and the estimated value is $\hat{\theta} = \bar{y} = 70$. From Fig. 12.22 we find $a_{max}(\theta) = a_{max}(60) = 100$ and the expected net profit is V(60,100) = 22.38, and $a_{max}(\hat{\theta}) = a_{max}(70) = 105$ and the expected net profit is only V(60,105) = 21.68. Thus

$$r(\theta,\hat{\theta}) = r(60,70)$$

$$= V(60,100) - V(60,105)$$

$$= 22.38 - 21.68 = 0.70$$

which can be checked against the value given in step 6. We perform one more computation to find $r(90,30)$. In this case

$$a_{max}(\theta) = a_{max}(90) = 130 \qquad \text{offer that should have been made (using } \theta = 90)$$

$$a_{max}(\hat{\theta}) = a_{max}(30) = 65 \qquad \text{offer that is actually made (using } \hat{\theta} = 30)$$

and then our two expected net profits are

$$V(\theta,a_{max}(\theta)) = V(90,130) = 34.18$$

$$V(\theta,a_{max}(\hat{\theta})) = V(90,65) = 27.50$$

and

$$r(\theta,\hat{\theta}) = r(90,30) = 34.18 - 27.50 = 6.68$$

Using this approach, the regret values given in step 6 were calculated. The regrets can also be graphed for different values of θ and $\hat{\theta}$.

In considering the regrets associated with strategy s_2, the problem is simplified because we always made the offer of $a = 100$. Thus the regret for an individual whose state of nature in represented by θ is simply

$$r(\theta) = V(\theta,a_{max}(\theta)) - V(\theta,100)$$

Thus for $\theta = 80$ we have $a_{max}(80) = 120$, and the expected net profits are

$$V(80,120) = 31.76 \quad \text{and} \quad V(80,100) = 25.25$$

Using these values, we obtain

$$r(80) = 31.76 - 25.25 = 6.51$$

as given in step 6.

Although this determination of regret is somewhat involved, the approach introduced has general applicability, especially when the action to be taken is a bid or dollar offer. The real problem in determining the regrets is having enough experience to generate the appropriate inputs used in the evaluation of the regrets.

12.6 Summary

In Chap. 11 we developed the techniques of evaluating competing strategies that might be formulated for a decision theory problem when the random variable is discrete and there are a finite number of states of nature and actions. In Chap. 12 we illustrated how these concepts can be used for the generalized decision theory problem when the random variable is continuous or the number of states of nature or actions is infinite.

The technique suggested is to convert the generalized decision theory problem into the discrete case since we have already developed the latter techniques. Specifically,

(1) When the random variable Y is continuous and the finite set of actions can be ordered relative to the value of Y, then an entire class of strategies can be formulated by dividing the range of Y into a finite number of intervals corresponding to the ordered actions. Evaluations can then be made regarding such strategies.

(2) When the number of possible states of nature and/or actions is infinite, we select a finite subset that is representative of the infinite set.

This approach to the general decision theory problem is particularly useful when the computer is used to evaluate competing strategies. In this case the use of the finite number of intervals, states of nature, or actions can be compensated by repeating the decision problem several times using more refined values for the intervals, states of nature, and actions that are closer to the previously selected optimum values. In fact, with some modification, the computer program in Chap. 11 can be used to evaluate the strategies.

The three inference problems (1) testing hypothesis, (2) parameter estimation, and (3) sample size determination were used to illustrate how these problems can be approached from a decision theory viewpoint.

In addition, a detailed example of how decision theory can be used in practice was given, including a discussion of how realistic and meaningful losses and regrets can be obtained.

1. Consider the diabetic decision problem as introduced in Sec. 11.2, Chap. 11, but assume that the diabetic makes an actual blood sugar analysis. In this case the response variable, Y = milligrams of sugar in the blood, is a continuous variable. We shall use the normal distribution model to represent each state of nature. The corresponding parameter values are

$$\theta_1: \quad \mu_1 = 80 \qquad \sigma_1 = 4$$
$$\theta_2: \quad \mu_2 = 95 \qquad \sigma_2 = 5$$
$$\theta_3: \quad \mu_3 = 110 \qquad \sigma_3 = 6$$

Each strategy will be specified in the following manner:

If $\quad Y \le b_1$, take action a_1

If $b_1 < Y \le b_2$, take action a_1

If $\quad Y > b_3$, take action a_3

(a) Using the strategies $s_1 = [90,100]$, $s_2 = [85,95]$, and $s_3 = [95,105]$, find the action probabilities for each strategy and each state of nature.

(b) Using the same prior probabilities, $q_1 = 0.50$, $q_2 = 0.35$, and $q_3 = 0.15$, determine which of the three strategies is the Bayes strategy.

2. Evaluate the Bayes loss for the test of hypothesis problem presented in Sec. 12.2, using the following strategy:

$$\text{If } \Pr\left[z > \frac{\bar{y} - \mu_0}{\sigma/\sqrt{n}} \le 0.005 \right] \text{take action } a_2$$

. Evaluate the following strategy for the problem given in Sec. 12.3: $s_2[b_1,b_2,b_3,b_4]$, where $b_1 = 2.5$, $b_2 = 6.0$, $b_3 = 9.0$, and $b_4 = 12.0$. Compare these results with these given for s_1. Consider how you would generate a new strategy superior to both s_1 and s_2.

4. Complete the sample size problem as presented in Sec.
   12.4. Assume that the states of nature are equally
   likely, and evolve the appropriate sample size decision.
   How would you refine the sample size determination using
   a decision theory approach?

5. To estimate the percentage of property owners who would
   vote in favor of a school bond issue, a random sample of
   100 property owners are canvased, and the number who
   claim they would vote for the bond issue is noted.

   (a) Use decision theory to determine what conclusion
       should be reached relative to the bond issue if there
       are two possible actions:

   a_1: the issue will pass

   a_2: the issue will be defeated

   and the states of nature are

   θ_1: p_1 = 0.25   most owners are against the
                                 issue

   θ_2: p_2 = 0.45   a bare majority are against
                                 the issue

   θ_3: p_3 = 0.55   a bare majority are in favo
                                 of the issue

   θ_4: p_4 = 0.75   most are in favor of the
                                 issue

   and the losses are

	a_1	a_2
θ_1	10	1
θ_2	3	3
θ_3	2	10
θ_4	1	20

and the prior probabilities are [0.35,0.30,0.20,0.15].

(b) Evolve competing strategies to be used, including the strategies for the classical test of hypothesis that $p = 0.50$ against the alternative $p \neq 0.50$, and use (1) the 10 percent level of significance and (2) the 5 percent level of significance, and also include some strategies of your own construction. Before making the evaluation, "guess" what the best strategy will be by looking at the loss table and prior probabilities. If the prior probabilities are [0.15,0.20, 0.30,0.35], what would be your guess? Use only a few representative values of Y in the evaluation.

(c) To estimate p, let us assume the following alternative estimates:

$$a_1: p < 0.35$$
$$a_2: .35 \leq p < 0.50$$
$$a_3: .50 \leq p < 0.65$$
$$a_4: p \geq 0.65$$

Work out the decision approach that involves strategies based on Y relative to the four conclusions (actions) possible.

(d) In either (b) or (c) test the sensitivity of the conclusions to the values of Y being used by inserting additional values of Y into the set of possible values and redoing the evaluation. Use the computer as a tool to assist you in your evaluations.

CHAPTER 13

REGRESSION AND CORRELATION

13.1 Introduction

In the previous chapter we considered how descriptive statistics can be used to find a mathematical model for the distribution of an observable random variable, how to test hypotheses relative to the parameters of such models, and how to estimate these parameters. In addition to such inferences, it is often important to be able to predict the value of a random variable when observations on the variable are difficult or expensive to determine or may not be available until some time in the future. Thus it is useful to be able to predict an unknown value of the variable using currently available information.

It is not difficult to identify prediction problems. For example, we may want to predict the quality of the work performed by a college student, using information from his high school record or from a battery of test scores; the performance of a manufactured product based on the characteristics of the components used in its manufacture; the effectiveness of a medical treatment, knowing the symptoms exhibited by a patient; or the performance of a possible new employee, using his past experience and training. In this chapter we shall introduce techniques that can be used in making such predictions.

13.2 Statistical Regression

Statistical regression theory assumes that a functional relationship exists between the variable to be predicted and the variable(s) which can be observed and that this relationship can be determined from observed data. To present the

396

technique, let us consider an actual problem. Suppose we wish to predict a college freshman's grade-point average from his combined scholastic aptitute test (SAT) scores. The variable to be predicted is the random variable, which will be denoted by Y; the SAT combined score will be denoted by X. It is assumed that the X values used in the sample have been preassigned, usually to cover the possible range of X for which predictions are to be made. The random variable Y is often called the dependent variable because it is assumed that its value depends upon X, and X is called the independent variable to distinguish it from Y.

To obtain a prediction equation (also called a regression equation), we need the grade-point average of individuals who have completed their freshman year and their SAT score obtained prior to college admission. Once the regression equation is obtained, however, it can be used to predict the grade-point average of an individual prior to college admission knowing only his SAT score.

Suppose that we have the following bivariate observations (x_i, y_i) on each of the 15 students who have completed their freshman year:

Student Identification	SAT Score	Freshman Grade point Avg.
i	X	Y
1	743	1.44
2	802	2.34
3	851	1.95
4	906	2.18
5	942	2.56
6	1004	1.95
7	1050	2.84
8	1097	2.11
9	1147	2.67
10	1208	2.48
11	1248	3.61
12	1301	2.73
13	1347	2.85
14	1406	3.44
15	1451	3.11

Fig. 13.1 SAT score and grade-point average
of college freshmen.

Since the variable X is used to predict Y, and we wish to have representative values of X in the sample, we selected the students to cover the range of the SAT score. Although the desired values of X were (750,800,850,900,950, 1000, ... , 1450), practical considerations caused us to select students whose SAT score only approximated the desired value.

The next step is to look at the data to develop some idea of the functional relationship between Y and X. We can plot the bivariate observations (x_i, y_i) in a scatter diagram and attempt to determine the functional relationship needed to fit the points.

The scatter plot for the data on the 15 students obtained by plotting the 15 observations on a graph is given in Fig. 13.2. To support our visual appraisal that the relationship between Y and X is linear, we have drawn by judgment a straight line on the graph that might best fit the plotted points:

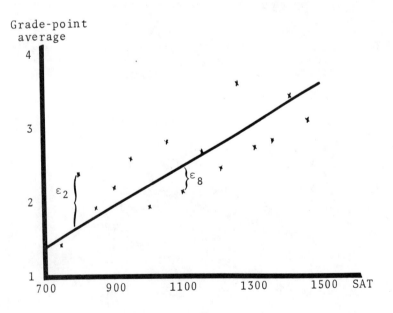

Fig. 13.2 Scatter diagram of observations.

There are many types of relationships that might be exhibited by the points; three are shown in Fig. 13.3. They represent the linear relationship given by $Y = \alpha + \beta X$, the quadratic relationship $Y = \alpha + \beta X + \gamma X^2$, and the exponential relationship $Y = \alpha e^{\beta X}$.

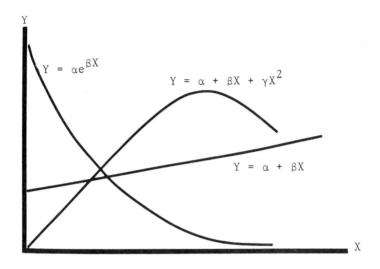

Fig. 13.3 Various   relationships between X and Y.

Returning to the illustrative example, we shall assume that the functional relationship between Y and X can be represented by the linear model

$$Y = \alpha + \beta X + \varepsilon \tag{13.1}$$

where α is the y intercept of the line and β is its slope. The ε represents the error made in fitting the linear model to the scatter of the observed data. Using the assumption that α and β are constants and $E(\varepsilon) = 0$, the expected value of y for a particular x is

$$E(y) = \alpha + \beta x_i \tag{13.2}$$

Thus the model error can be expressed by the difference be-

399

tween the observed and expected y values, or

$$\epsilon_i = y_i - E(y) = y_i - \alpha - \beta x_i \qquad (13.3)$$

Generally we do not know α and β, and so must estimate them. Although many estimation techniques such as the method of moments are available, we choose to use the method of least squares because it will determine those estimates of α and β that minimize the sums of squares due to the model error given by

$$SSE = \sum_{i=1}^{n} \epsilon_i^2 = \sum_{i=1}^{n} (y_i - \alpha - \beta x_i)^2 \qquad (13.4)$$

The development of the least squares estimates depends upon the analytical nature of the model and involves calculus. Thus we shall simply give the resulting simultaneous least squares equations (normal equations) for the estimates of the linear model given by $\hat{\alpha} = a$ and $\hat{\beta} = b$, which are

$$\sum_{i=1}^{n} y_i = na + b \sum_{i=1}^{n} x_i \qquad (13.5)$$

$$\sum_{i=1}^{n} x_i y_i = a \sum_{i=1}^{n} x_i + b \sum_{i=1}^{n} x_i^2 \qquad (13.6)$$

These two equations can be solved simultaneously for a and b to obtain

$$a = \frac{\Sigma y_i}{n} - b \frac{\Sigma x_i}{n} \qquad (13.7)$$

$$b = \frac{\Sigma x_i y_i - (\Sigma x_i \Sigma y_i)/n}{\Sigma x_i^2 - (\Sigma x_i)^2/n} \qquad (13.8)$$

To find a and b we need the appropriate sums and sums of squares for the observed data. The results for our illustrative example are

$$n = 15$$

$$\sum_{i=1}^{n} x_i = 16503.00 \qquad\qquad \sum_{i=1}^{n} x_i^2 = 18863203.00$$

$$\sum_{i=1}^{n} y_i = 38.26 \qquad\qquad \sum_{i=1}^{n} y_i^2 = 102.3704$$

$$\sum_{i=1}^{n} x_i \cdot y_i = 43550.66$$

Thus

$$b = \frac{43550.66 - (16503)(38.26)/15}{18863203 - (16503)^2/15}$$

$$= \frac{1457.01}{706602.00}$$

$$= 0.0020619 = 0.002062$$

and

$$a = \frac{38.26}{15} = (0.002062)\,\frac{16503}{15}$$

$$= 2.5507 - 2.2686$$

$$= 0.2821$$

The resulting regression equation, which can be used for prediction, is given by

$$\hat{y}_i = 0.2821 + 0.002062\, x_i \qquad\qquad (13.9)$$

To consider how well the linear model fits the observed data, we can use the regression equation to predict the grade-point average for the 15 students used in the study. The results are summarized in Fig. 13.4:

i	X	Y	$\hat{Y}$	$\varepsilon = Y - \hat{Y}$
1	743	1.44	1.81	-0.37
2	802	2.34	1.94	+0.40
3	851	1.95	2.04	-0.09
4	906	2.18	2.15	+0.03
5	942	2.56	2.22	+0.34
6	1004	1.95	2.35	-0.40
7	1050	2.84	2.45	+0.39
8	1097	2.11	2.54	-0.43
9	1147	2.67	2.65	+0.02
10	1208	2.48	2.77	-0.29
11	1248	3.61	2.86	+0.75
12	1301	2.73	2.96	-0.23
13	1347	2.85	3.06	-0.24
14	1406	3.44	3.18	+0.26
15	1451	3.11	3.27	-0.16

Fig. 13.4 Observed and predicted Y's and
the errors.

A measure of the goodness of fit is desired; it should involve the observed errors. Although we might wish to use the average or mean value of the ε's, such a measure is not useful because the sums of the errors will always equal zero independent of the goodness of fit of the model, as shown below.

$$\text{Mean error} = \bar{\varepsilon} = \frac{\sum_{i=1}^{n} \varepsilon_i}{n} = \frac{\sum_{i=1}^{n} (y_i - \hat{y}_i)}{n}$$

$$= \frac{\sum_{i=1}^{n} (y_i - a - bx_i)}{n} = \frac{\Sigma y_i - na - b \, \Sigma x_i}{n}$$

$$= 0$$

since the first least squares equation (13.5) is

$$\Sigma y_i = na + b \ \Sigma x_i$$

The standard deviation of the ε_i's can be used as a measure of the goodness of fit because the ε_i's vary about their mean of zero. Thus a large standard deviation means that the errors are numerically large. The variance and standard deviation of the errors are given by the following equations:

$$s_\varepsilon'^2 = \frac{\sum\limits_{i=1}^{n} (\varepsilon_i - \bar{\varepsilon})^2}{n - 2}$$

but since $\bar{\varepsilon} = 0$, we have

$$s_\varepsilon'^2 = \frac{SSE}{n-2} = \frac{\sum\limits_{i=1}^{n} \varepsilon_i^2}{n - 2} = \frac{\sum\limits_{i=1}^{n} (y_i - a - bx_i)^2}{n - 2}$$

$$= \frac{\sum\limits_{i=1}^{n} \varepsilon_i^2}{n - 2} \tag{13.10}$$

and

$$s_\varepsilon' = \sqrt{s_\varepsilon'^2}$$

Notice that in the regression problem the divisor $n - 2$ is used in calculating the variance and standard deviation. This divisor is necessary for s^2 to be an unbiased estimator since two parameters α and β had to be estimated, using the data.

We obtain the above results for the illustrative example. We first determine

$$\sum\limits_{i=1}^{15} \varepsilon_i = 0.01$$

and find that though it is not exactly equal to zero, the

403

discrepancy is due to round-off error. Then

$$SSE = \sum_{i=1}^{n} \varepsilon_i^2 = (-0.37)^2 + (+0.40)^2 + \cdots + (-0.16)^2$$

$$= 1.7557$$

Thus

$$s_\varepsilon'^2 = \frac{SSE}{n-2} = \frac{1.7557}{13} = 0.1351$$

and

$$s_\varepsilon' = \sqrt{0.1375} = 0.3675 = 0.37$$

It is also useful to be able to obtain $s_\varepsilon'^2$ without calculating each ε_i. The following equation is equivalent to
Eq. (13.10):

$$s_\varepsilon' = \frac{[\Sigma y_i^2 - (\Sigma y_i)^2/n] - b[\Sigma x_i y_i - (\Sigma x_i \cdot \Sigma y_i)/n]}{n-2} \qquad (13.11)$$

where the numerator is the sums of squares of error SSE.
Using this equation to calculate $s_\varepsilon'^2$ for the example, we
obtain

$$s_\varepsilon'^2 = \frac{4.7819 - 0.002062(1457.01}{13}$$

$$= \frac{1.7775}{13} = 0.1367 = 0.14$$

and

$$s_\varepsilon' = 0.3698 = 0.37$$

Once again we note the effect of round-off error in the
failure of the two computations to yield exactly the same

values.

The a and b regression values were determined to minimize SSE. The student can verify this by writing a computer program that uses other a and b values and determines SSE for those values.

Although s'_ε is a relative measure of goodness of fit, its size depends upon the variation of the random variable Y, and thus it fails to provide an absolute indication as to how well the model fits the data. However, the coefficient of determination, denoted by R^2, and defined by

$$R^2 = \frac{SSR}{SST} = \frac{b[\Sigma x_i y_i - \Sigma x_i \Sigma y_i/n]}{[\Sigma y_i^2 - (\Sigma y_i)^2/n]} = \frac{b \Sigma (x_i - \bar{x}) y_i}{\Sigma (y_i - \bar{y})^2} \qquad (13.12)$$

does provide such an index since it is the ratio of the total variability explained by the regression line to the total variability of the observations about their mean. In other words, the denominator of R^2 is the total variation exhibited by the dependent variable (the total corrected sums of squares for Y denoted by SST). This total variation can be partitioned into two parts, that variation which is explained by the regression model (SSR) and that which is unexplained or model error (SSE). Thus

$$SST = SSR + SSE$$

If there is a perfect linear relationship between Y and X, then all errors would be zero. Thus SSE = 0, and SST = SSR. If there is no relationship between Y and X (independence of Y and X), then b = 0. Thus SSR = 0, and SST = SSE. Using this information, we can see that the index R^2 indicates the percentage of total variation explained by the regression model; thus $0 \leq R^2 \leq 1$.

In the example, the coefficient of determination is

$$R^2 = \frac{3.0044}{4.7810} = 0.6283 = 0.63$$

which indicates that the regression of Y or X is fairly

405

strong since the value is closer to 1.00 than to 0.00. In fact, we can interpret this value to mean that the independent variable X accounts for 63 percent of the total variation of Y. If we are interested in improving our ability to predict the grade-point average, we could consider another regression model or include additional independent variables. Such considerations will be presented later in this chapter.

13.3 The Sampling Variation of Regression Estimates

Until now we have learned how to fit data with a linear regression model that can then be used to predict the value of the dependent variable. If we repeated the experiment and obtained a new set of observations, the estimates of the model parameters would differ from those found in our particular sample. Thus the predicted y's would be different for a given value of x. In addition, the standard deviation of the errors s_ε' and the coefficient of determination R^2 would not be the same.

To study this sampling variation we must assume some underlying population from which the samples are drawn. It seems natural to once again assume normality, but now we have bivariate observations (X,Y), and thus it would seem that we must have a bivariate model. This is not necessary, however, because we select the values of X in advance, and so they do not have an underlying probability distribution. Therefore only the y's are subject to sampling variation and have an underlying distribution. To consider the distribution of Y, let us assume that the errors ε_i have an expected value $E(\varepsilon) = 0$ and a variance σ_ε^2 and that the following linear relationship exists between Y and X: $Y = \alpha + \beta X + \varepsilon$. In this case the mean of the Y's for a given x is

$$\mu_{Y/x} = E(Y) = E(\alpha + \beta x + \varepsilon) = \alpha + \beta x \qquad (13.13)$$

since $E(\varepsilon) = 0$ and α, β, and x are fixed values. This means that the expected value of Y for a given x varies linearly with respect to x.

The variance of Y is assumed to be independent of the value of X and is given by

$$\sigma^2_{Y/x} = E(Y - \mu_{Y/x})^2 = E[\alpha + \beta x + \epsilon - (\alpha + \beta x)]^2 \qquad (13.14)$$

$$= E(\epsilon)^2 = \sigma^2_\epsilon$$

If, in addition, the distribution of the errors is normal, then the distribution of Y for a given x is

$$f(Y/x) = \frac{1}{\sqrt{2\pi}\ \sigma_\epsilon} e^{-\frac{1}{2}(Y-\alpha-\beta x)^2/\sigma^2_\epsilon} \qquad -\infty < Y < \infty \qquad (13.15)$$

This means that the distribution of Y for each x is normal with mean $\mu_{Y/x} = \alpha + \beta x$ and variance σ^2_ϵ.

One way to study the sampling distribution of the estimates a, b, and $s_\epsilon'^2$ is by sample generation. Given the parameter values α, β, and σ^2_ϵ, the procedure is

(1) Determine a fixed set of x's to be used. (The x values do not all need to be different.)

(2) Determine $\mu_{Y/x}$ for each x, using the fact that $\mu_{Y/x} = \alpha + \beta x$.

(3) Generate a random normal number Y_x for each x, where

$$Y_x \sim N(\alpha + \beta x, \sigma_\epsilon)$$

(4) Determine the regression estimates a, b, and s'^2 for the generated sample.

(5) Repeat the above sample generation procedure (1 - 4) m times, and find the characteristics (mean, standard deviation, a_3, and a_4) for the generated sampling distribution of a, b, and $s_\epsilon'^2$.

Following these steps, a program was written to generate m samples from the regression model

$$Y = 0.2821 + 0.002062 \cdot x + \epsilon$$

where the errors ε have a mean of zero and standard deviation of 0.37. The set of x's used were the same as those in the illustrative example given in the last section: (743, 802,851,906,942,1004,1050,1097,1147,1208,1248,1301,1347, 1406,1451).

We can compare the sample generation results with the theoretical results since it can be shown mathematically that the distribution of both a and b are normal and that the mean and standard deviations are given by

$$E(a) = \alpha \qquad \sigma_a = \sigma_\varepsilon \sqrt{\frac{\Sigma x_i^2}{n \Sigma (x_i - \bar{x})^2}} \qquad (13.16)$$

$$= \sigma_\varepsilon \sqrt{\frac{1}{n} + \frac{\bar{x}^2}{\Sigma (x_i - \bar{x})^2}}$$

$$E(b) = \beta \qquad \sigma_b = \frac{\sigma_\varepsilon}{\sqrt{\Sigma (x_i - \bar{x})^2}} \qquad (13.17)$$

The expected value of $s_\varepsilon'^2$, $E(s_\varepsilon'^2) = \sigma_\varepsilon^2$, but the distribution of $s_\varepsilon'^2$ is not normal, and thus\ the consideration of its sampling distribution will be postponed until Chap. 14.

For the illustration, the standard deviations of a and b are given by

$$\sigma_a = (0.37) \sqrt{\frac{1}{15} + \frac{(1100.2)^2}{706602}} = 0.4936$$

$$\sigma_b = \frac{(0.37)}{\sqrt{706602}} = 0.0004$$

with $\alpha_3 = 0$ and $\alpha_4 = 3$ for both distributions.

In Figure 13.5 we summarize the results of the sampling generation program, along with the theoretical characteristics of a and b:

	Characteristics of a		Characteristics of b	
	Theoretical	Estimated	Theoretical	Estimated
Mean	0.2821	0.3148	0.0021	0.0020
Standard dev.	0.4936	0.4887	0.0004	0.0004
Skewness	0	-0.0450	0	0.0530
Kurtosis	3	3.2423	3	3.4550

Fig. 13.5 Comparison of characteristics of sampling
distribution of a and b.

Although an infinite number of samples is required to generate the true sampling distribution of a and b, and only 200
were generated, the results obtained from the computer simulation verify the theoretical results since they are fairly
close.

We are also interested in the distribution of the predicted Y's. Again the distribution depends upon the values
of X; thus we shall denote each predicted Y by $\hat{Y}$. The predicted value of Y is found by evaluating

$$\hat{Y} = a + bx$$

It can be shown theoretically that Y has a normal distribution ($\alpha_3 = 0$, $\alpha_4 = 3$), with its mean being

$$E(\hat{Y}) = \mu_{Y/x} = \alpha + \beta x \qquad (13.18)$$

which is equal to the mean of the population from which the
observations were drawn, and its variance being

$$E(\hat{Y} - \mu_{\hat{Y}/x})^2 = \sigma_{\hat{Y}}^2 = \sigma_\epsilon^2 \left[\frac{1}{n} + \frac{(x - \bar{x})^2}{\Sigma (x_i - \bar{x})^2} \right] \qquad (13.19)$$

We note that the variance of the predicted $\hat{Y}$'s depends
upon the value of X, while the variance of the observed Y's
given by Eq. (13.14) is independent of x since it is always
equal to σ_ϵ^2. To verify these theoretical results, we also

let the computer program that generated the samples calculate each predicted $\hat{Y}$ for a given x in the sample. We then
determine the characteristics of the set of 200 $\hat{Y}_x$'s that
are obtained for each value of x. The results are given in
Fig. 13.6 for only three values of x but the results for the
other values of x can be obtained by using the computer
program.

	x = 743		x = 1097		x = 1451	
	Theoretical	Estimated	Theoretical	Estimated	Theoretical	Estimated
Mean	1.8142	1.8237	2.5441	2.5426	3.2741	3.2615
Std. dev.	0.1840	0.1796	0.0955	0.1048	0.1816	0.1987
Skewness	0	0.0152	0	0.2846	0	0.1007
Kurtosis	3	2.8334	3	3.2619	3	3.4363

Fig. 13.6 Comparison of characteristics of sampling distribution of the predicted values of Y for a given x.

As expected, we see that the variance of $\hat{Y}$ is different
for different values of x. In fact, by looking at Eq.
(13.19) we would expect the smallest variance to be found
when x = $\bar{x}$ = 1100.2 and that the variance increases as x
moves away from $\bar{x}$ in either direction.

13.4 Tests of Hypotheses, Confidence Intervals, and Sample Size Considerations

Whenever we estimate parameter values from sample data,
we may wish to make inferences about the true parameter values for the particular population being studied. Regression
is no exception, and in fact there are two types of inferences that can be made. We may wish to make inferences
about the population regression line $Y = \alpha + \beta x + \varepsilon$, using
the estimated regression line $Y = a + bx$, or we may wish to
make inferences about the population values of Y for any
given x, using the predicted $\hat{Y}$: $\hat{Y} = a + bx$.

Let us first consider the inferences about the popula-
tion regression equation $Y = \alpha + \beta x + \varepsilon$. Since there are
three parameters in the model, α, β, and ε, we could consider any one of these separately and make tests of hypothe-

410

ses about the parameter values, find confidence intervals, or make sample size determinations. Since the sampling distribution of $s_{\varepsilon}'^2$ has yet to be introduced, we shall delay considering the parameter σ_{ε}^2 until Chap. 14. If we are only interested in making inferences about one of the two paremeters α and β, we can use the approaches given in subsections 1 and 2. The problem of making a simultaneous inference about both α and β will not be considered in this introductory text because the estimate of α depends upon the estimate of β, $a = \bar{y} - b\bar{x}$, and thus the estimates a and b are not independent.

(1) Let us consider the slope parameter β and its sample estimate b. Using the fact that when $\varepsilon \sim N(0, \sigma_{\varepsilon})$, and thus

$$b \sim N\left(\beta, \ \frac{\sigma_{\varepsilon}}{\sqrt{\Sigma(x_i - \bar{x})^2}}\right)$$

we can make the following inferences:

(a) Test of hypothesis

We wish to test the null hypothesis $H_o: \beta = \beta_o$ versus any alternative hypothesis. We use the t statistic since the standard deviation of b, σ_b, is not known and must be estimated by s_b. Thus we have

$$t = \frac{b - \beta_o}{s_b} = \frac{b - \beta_o}{s_{\varepsilon}'/\sqrt{\Sigma(x_i - \bar{x})^2}}$$

which has the Student's t distribution with $d = n - 2$ degrees of freedom (recall the divisor used to find an unbiased estimate of σ^2 is $n - 2$).

Using the illustrative example, let us test the hypothesis that β is zero. Thus $H_o: \beta = 0$ versus $H_a: \beta > 0$. We compute

$$t = \frac{0.002062 - 0}{0.37/\sqrt{706602}} = \frac{0.002062}{0.00044} = 4.69$$

411

which we compare with

$$t_{d,1-\alpha} = t_{13,.95} = 1.771$$

if the 5 percent (one-sided) level of significance is used. Since 4.69 > 1.771, we conclude that β is greater than zero and thus Y does depend on X.

(b) <u>Confidence interval</u>

The $(1 - \alpha)$ percent confidence interval for β is given by

$$b - t_{d,1-\alpha/2} \frac{s_{\varepsilon}'}{\sqrt{\Sigma(x_i - \bar{x})^2}} < \beta < b + t_{d,1-\alpha/2} \frac{s_{\varepsilon}'}{\sqrt{\Sigma(x_i - \bar{x})^2}}$$

The 90 percent confidence interval for β in the example is

$$0.002062 - (1.771) \frac{0.37}{\sqrt{706602}} < \beta < 0.002062 + (1.771) \frac{0.37}{\sqrt{706602}}$$

$$0.002062 - 0.000779 < \beta < 0.002062 + 0.000779$$

$$0.001283 < \beta < 0.002841$$

(c) <u>Sample size</u>

If we wish to estimate the regression coefficient β within a prescribed sampling error $\varepsilon = b - \beta$, there is some difficulty in using the t statistic

$$t = \frac{b - \beta}{s_{\varepsilon}' \Big/ \sqrt{\Sigma'(x_i - \bar{x})^2}}$$

to determine the required sample size n because n does not appear explicitly in the equation. We modify the equation by writing

$$t = \frac{\varepsilon}{s_\varepsilon' / \sqrt{n} \sigma_x}$$

because

$$\sigma_x = \sqrt{\frac{\Sigma (x_i - \bar{x})^2}{n}}$$

reflects the variation introduced in the set of x's used. It is assumed that regardless of the size of sample, the set of x's will have about the same population standard deviation. We can now solve this modified test statistic for n. We obtain

$$n = \frac{t_{d,1-\alpha/2}^2 \, s_\varepsilon'^2}{\sigma_x^2 \varepsilon^2}$$

Thus, if we have the requirement of estimating β to within 0.0001 units with 95 percent confidence, we need to know $s_\varepsilon'^2$, σ_x^2, and t. To determine the value of t, we need to know the degrees of freedom. We may use n = 20 as a first guess; thus d = n - 2 = 18 and $t_{18,.975}$ = 2.010. Because s_ε' = 0.37 and

$$\sigma_x^2 = \frac{\Sigma (x_i - \bar{x})^2}{n} = \frac{706602}{15} = 47107$$

we obtain as our first approximation for n

$$n = \frac{(2.101)^2 (0.37)^2}{(0.0005)^2 (47107)} = \frac{0.704303}{0.011777} = 51.31 = 52$$

Thus we replace $t_{18,.975}$ by $t_{50,.975}$= 2.011 and recompute

$$n = \frac{(2.011)^2 (0.37)^2}{(0.0005)^2 (47107)} = \frac{0.553640}{0.011777} = 47.01 = 47$$

We now use $t_{45,.975} = 2.650$ and find

$$n = \frac{(2.005)^2 (0.37)^2}{(0.005)^2 (47107)} = \frac{0.550341}{0.011777} = 46.73 = 47$$

We conclude that n = 47 because further iterations would yield the same value.

(2) Let us now consider the Y-intercept parameter α and its sample estimate a. Using the fact that when $\varepsilon \sim N(0, \sigma_\varepsilon)$, then

$$a \sim N\left(\alpha, \sigma_\varepsilon \sqrt{\frac{1}{n} + \frac{\bar{x}^2}{\Sigma (x_i - \bar{x})^2}}\right)$$

Thus we can make the following inferences:

(a) <u>Test of hypothesis</u>

To test the hypothesis H_o: $\alpha = \alpha_o$ versus any alternative, we use the t statistic

$$t = \frac{a - \alpha_o}{s_a} = \frac{a - \alpha_o}{s_\varepsilon' \sqrt{\frac{1}{n} + \frac{\bar{x}^2}{\Sigma (x_i - \bar{x})^2}}}$$

This t statistic has a Student's t distribution, with d = n - 2 degrees of freedom.

For example, let us test the hypothesis H_o: $\alpha = 0$ against the alternative H_a: $\alpha \neq 0$, using the illustration. We compute

$$t = \frac{0.2821 - 0}{(0.37) \sqrt{\frac{1}{15} + (1100.2)^2/706602}} = \frac{0.2821}{(0.37)(1.7797)}$$

$$= \frac{0.2821}{0.4936} = 0.57$$

To test the hypothesis at the 1 percent level of significance, we compare the calculated t value with

$$t_{d,1-\alpha/2} = t_{13,.995} = 3.012$$

and since 0.57 < 3.012, we do not reject the null hypothesis.

(b) Confidence interval

The (1 - α) percent confidence interval for α is given
by

$$a - t_{d,1-\alpha/2}s'_\epsilon \sqrt{\frac{1}{n} + \frac{\bar{x}^2}{\Sigma(x_i - \bar{x})^2}} < \alpha <$$

$$a + t_{d,1-\alpha/2}s'_\epsilon \sqrt{\frac{1}{n} + \frac{\bar{x}^2}{\Sigma(x_i - \bar{x})^2}}$$

To find the 99 percent confidence interval for α using the
example data, we have

$$0.2821 - (3.012)(0.37)\ \sqrt{1.7797} < \alpha <$$

$$0.2821 + (2.012)(0.37)\ \sqrt{1.7797}$$

$$0.2821 - 1.4867 < \alpha < 0.2821 + 1.4867$$

$$-1.20 < \alpha < 1.77$$

(c) Sample size

Since n appears explicity in the t statistic, we can
replace the unknown sample values by sample design values,
and solve for n. Thus

$$n = \frac{\mu_2\ t^2_{d,1-\alpha/2}\ s'^2_\epsilon}{\epsilon^2\sigma^2_x}$$

where μ_2 is the second moment about the origin. It has re-placed $\Sigma x^2/n$ because we assume that $\Sigma x^2/n$ equals the μ_2 val-ue assumed in the design and $\Sigma(x_i - \bar{x})^2/n$ equals the σ_x^2 value.

The use of this formula requires an iterative approach as illustrated because once again the t value to be used depends upon the sample size n. Since the technique is the same as when the sample size for the estimate of β is deter-mined, we shall not illustrate the approach again.

(3) In regression our interest is often focused on the pre-diction of Y for a given value of x. Thus we shall want to consider inferences relative to such predictions. In fact, there are two prediction problems: (1) to predict the mean Y for a given x and (2) to predict the Y for a particular observation.

The regression prediction equation $\hat{Y} = a + bx$ can serve as an estimate of either $\mu_{Y/x}$ (the population mean of all observations having the specified x) or as the estimate of the Y value for some particular individual whose x is known. Let us denote the estimate of $\mu_{Y/x}$ by

$$\hat{\mu}_{Y/x} = \hat{Y} = a + bx$$

and the estimate for the particular Y for a given x by

$$\hat{Y}_x = \hat{Y} = a + bx$$

The expected value of both estimates is the same because

$$E(\hat{\mu}_{Y/x}) = E(a + bx) = \alpha + \beta x$$

and

$$E(\hat{Y}_x) = E(a + bx) = \alpha + \beta x$$

but the standard deviations differ because

$$\sigma_{\hat{\mu}_{Y/x}} = \sigma_{\epsilon} \sqrt{\frac{1}{n} + \frac{(x - \bar{x})^2}{\Sigma(x_i - \bar{x})^2}} \qquad (13.20)$$

$$\sigma_{\hat{Y}_x} = \sigma_{\epsilon} \sqrt{1 + \frac{1}{n} + \frac{(x - \bar{x})^2}{\Sigma(x_i - \bar{x})^2}} \qquad (13.21)$$

We have already encountered Eq. (13.20) in the sample generation study of the regression equation, and we note that the terms within the parentheses come from the sampling variation of the intercept estimate a and the slope estimate b. We note that the standard deviation of $\hat{Y}_x$ has the same two terms as Eq. (13.20), which are due to the sampling variation of the estimates a and b, but it also has an additional term σ_{ϵ}. This additional term must be included since we are now predicting the Y value of an individual. Thus this expected variation from the $\mu_{Y/x}$ must be included; it is given by σ_{ϵ}^2, (the variance of the observed Y's).

We shall consider only the inferences about $\mu_{Y/x}$ because the approach for $\hat{Y}_x$ is the same, using the appropriate standard deviation.

(a) Test of hypothesis

To test the hypothesis $H_0: \mu_{Y/x} = \mu_0$ against the alternative $H_a: \mu_{Y/x} \neq \mu_0$, we use the t ratio

$$t = \frac{\hat{\mu}_{Y/x} - \mu_0}{s_{\epsilon}' \sqrt{\frac{1}{n} + \frac{(x - \bar{x})^2}{\Sigma(x_i - \bar{x})^2}}}$$

Using the illustrative example, let us test the hypothesis that a student with a combined SAT score of 650 will make at least a grade of C (Y = 2.00). Thus

$$H_0: \mu_{Y/650} = 2.00 \text{ vs. } H_a: \mu_{Y/650} < 2.00$$

Since

$$\mu_{Y/650} = 0.2821 + 0.002062(650) = 1.6224$$

$$t = \frac{1.6224 - 2.0000}{(0.37)\sqrt{\dfrac{1}{15} + \dfrac{(650 - 1100.2)^2}{706602}}} = \frac{-0.3776}{(0.37)(0.5946)}$$

$$= \frac{-0.3776}{0.2200} = -1.72$$

Using the 5 percent level of significance (one sided), the critical value of t is

$$t_{13,.05} = -1.78$$

and since

$$-1.78 < -1.72$$

we do not reject the null hypothesis that the student with a combined SAT score of 650 can make at least a grade of C.

(b) <u>Confidence interval for $\mu_{Y/x}$</u>

The confidence interval for the true mean $\mu_{Y/x}$ is given by

$$\hat{\mu}_{Y/x} - t_{d,1-\alpha/2} s'_\varepsilon \sqrt{\frac{1}{n} + \frac{(x - \bar{x})^2}{\Sigma(x_i - \bar{x})^2}} \le \mu_{Y/x} < \hat{\mu}_{Y/x}$$

$$+ t_{d,1-\alpha/2} s'_\varepsilon \sqrt{\frac{1}{n} + \frac{(x - \bar{x})^2}{\Sigma(x_i - \bar{x})^2}}$$

Note that the term $(x - \bar{x})^2$ increases as x moves away from the mean $\bar{x}$, and thus the confidence interval increases as x moves away from the mean $\bar{x}$ in either direction.

To plot the 90 percent confidence limits for the illustrative example we first draw the estimated regression line

$$\hat{Y} = 0.2821 + 0.002062x$$

We then need the following values:

$$s'_\epsilon = 0.37 \qquad \Sigma(x_i - \bar{x})^2 = 706602 \qquad \bar{x} = 1100.2$$

$$n = 15 \qquad\qquad d = 13 \qquad t_{13,.95} = 1.78$$

The 90 percent confidence interval for $\mu_{Y/x}$ in general is

$$(0.2821 + 0.002062x) \pm (1.78)(0.37)\sqrt{\frac{1}{15} + \frac{(x - 1100.2)^2}{706602}}$$

In particular, when $x = 700$ the interval is

$$1.7256 - 0.3567 \le \mu_{Y/x} \le 1.7256 + 0.3567$$

$$1.3689 < \mu_{Y/x} \le 2.0823$$

and when $x = 1400$ the interval is

$$3.1689 - 0.2900 \le \mu_{Y/x} \le 3.1689 + 0.2900$$

$$2.8789 \le \mu_{Y/x} \le 3.4589$$

Evaluating the confidence limits for other values of x enables us to graph the regression line and the upper and lower 90 percent confidence limits, which are shown in Fig. 13.7.

We verify from the graph that the confidence interval is the smallest at $\bar{x}$ and becomes wider as x moves away from $\bar{x}$ in either direction.

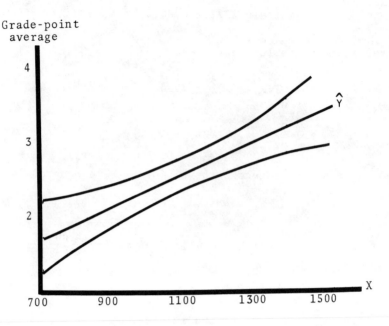

Fig. 13.7 Ninety percent confidence interval for $\mu_{Y/x}$.

(c) <u>Sample size</u>

To determine sample size n needed to estimate $\mu_{Y/x}$ within a prescribed accuracy

$$\varepsilon = \hat{\mu}_{Y/x} - \mu_{Y/x}$$

with (1 - α) percent confidence, we must again substitute

$$\sigma_x^2 = \frac{\Sigma (x_i - \bar{x})^2}{n}$$

in the t statistic. Solving for n we obtain

$$n = \left(1 + \frac{(x - \bar{x})^2}{\sigma_x^2}\right) \frac{s_\varepsilon^2 \; t_{d,1-\alpha/2}^2}{\varepsilon^2}$$

To demonstrate the use of this equation, consider the problem of sample size determination in our GPA-SAT score illustration. Say we would like to obtain a regression line that would enable us to predict $\mu_{Y/x=700}$ so as to have 95 percent assurance of having an error less than 0.10 GPA units. To evaluate the coefficients we have $\varepsilon = 0.10$ as specified and $x = 700$. We shall then agree that regardless of the size of n required we shall select the x's of our sample so that $\bar{x} = 1100$ and

$$\sigma_x^2 = \frac{\Sigma(x - \bar{x})^2}{n} = \frac{706602}{15} = 47107$$

In addition, we shall use

$$s_\varepsilon'^2 = 0.1351 = (0.37)^2$$

Since the $t_{d,1-\alpha/2}^2$ depends upon n, we must find our answer by an iterative process by first assuming a value for n. If we let $n = 50$, then $t_{48,.975} = 2.01$ and we obtain for n

$$n = \left(1 + \frac{(700 - 1100)^2}{47107}\right)\left(\frac{(0.1351)(4.04)}{(0.10)^2}\right)$$

$$= (1 + 3.3965)(54.58) = 239.96 = 240$$

We realize that n will be large, so we can replace $t_{d,1-\alpha/2}$ by $z_{1-\alpha/2} = 1.96$ and thus we no longer need to iterate. As the final value of n, we obtain

$$n = (4.3965)\frac{(0.1351)(1.96)^2}{(0.10)^2}$$

$$= (7.3965)(51.90) = 228.18 = 229$$

In the above inference considerations we use the standard approach for hypothesis testing, estimation, and sample size determination introduced in the earlier chapters. The student should appreciate the applicability of such approaches whenever a parameter is being estimated from sample data and a statistic that involves the parameter and its estimate is available, along with an associated probability density function.

13.5 The Generalized Regression Technique

As we remarked in the earlier section, we can either formulate a more appropriate regression model relating Y with the single independent variable X to improve the prediction of Y or obtain more information by including additional independent variables. Examples of different functional relationships between Y and a single independent variable X that might be used are

Functional Type	Analytical Form of $G(X,\underline{\theta})$
Linear	$\alpha + \beta x + \epsilon$
Quadratic-general	$\alpha + \beta x + \gamma x^2 + \epsilon$
Quadratic-pure	$\alpha + \beta x^2 + \epsilon$
Exponential	$\alpha e^{\beta x} + \epsilon$
Power	$\alpha x^\beta + \epsilon$
Generalized power	$\gamma + \alpha x^\beta + \epsilon$
General polynomial	$\alpha + \beta_1 x + \beta_2 x^2 + \cdots + \beta_p x^p + \epsilon$

If more independent variables are included, then a possible regression model is the multiple linear regression model given by

$$Y = \alpha + \beta_1 x_1 + \beta_2 x_2 + \beta_3 x_3 + \cdots + \beta_p x_p + \epsilon \qquad (13.22)$$

where x_1, x_2, $\ldots$, x_p are p different independent variables.

Although in the previous sections we considered only the simple linear regression model, the regression approach is a very general one. In all regression problems we must identify the dependent variable Y and the set of independent variables, denoted by the vector $\underline{X}$; select an appropriate regression model $Y = G(\underline{X},\underline{\theta})$, where the vector $\underline{\theta}$ represents the parameters of the model; and estimate the parameters $\underline{\theta}$, using the observed data.

As the regression model becomes more complex, difficulties arise in obtaining the parameter estimates and in determining the sampling distribution of the estimates. In addition, the computations become increasingly more complex. Since regression models, however, are used in so many disciplines, general-purpose computer programs have been written to perform the required computations for certain standard regression models. In this section we shall consider two standard regression models: multiple linear and polynomial regression. We shall not give the required computations but simply present the required data that are used as an input to a general-purpose computer program. We shall present only selected portions of the computer output and give their interpretation. The regression programs that we used are available in the Biomedical Statistical Package (BMD package[1]); this package is available at most computer centers, although there are also many other excellent statistical packages available. However, any other similar general regression program could have been used.

We shall continue to use the example of predicting the grade-point average of college students. Let us first consider what additional information might be available. It is apparent that the verbal and quantitative SAT scores could be used in place of their combined score. If, in addition, we knew each student's percentile rank in his high school graduating class, we would have three independent variables, denoted by x_1 = SAT verbal score, x_2 = SAT quantitative score, and x_3 = percentile rank in high school graduating class (91 to 100 percent indicates top 10 percent of class, 81 to 90 indicates the second 10 percent of the class). The augmented data that are now available on each of the 15 students is

Dixon, W. J. (ed)., Biomedical Computer Programs, Health Sciences Computing Facility, University of California at Los Angles.

Student Identification	Grade-point Average Y	Verbal SAT Score x_1	Quantitative SAT Score x_2	Percentile Rank x_3
1	1.44	377	366	79
2	2.34	381	421	86
3	1.95	420	431	70
4	2.18	450	456	33
5	2.56	451	491	90
6	1.95	512	492	2
7	2.84	475	575	38
8	2.11	500	597	5
9	2.67	584	563	49
10	2.48	564	644	12
11	3.61	574	674	75
12	2.73	671	631	68
13	2.85	654	693	34
14	3.44	603	803	99
15	3.11	726	725	88

Fig. 13.8 Augmented GPA regression data.

We shall use the multiple linear regression model

$$Y = \alpha + \beta_1 x_1 + \beta_2 x_2 + \beta_3 x_3 + \varepsilon$$

To fit the model to the data, the regression coefficients must be estimated; we use the same least squares criterion of minimizing the sums of squares of error:

$$SSE = \sum_{i=1}^{n} \varepsilon_i^2 = \sum_{i=1}^{n} \left(y_i - \hat{y}_i \right)^2$$

$$= \sum_{i=1}^{n} \left(y_i - \alpha - \beta_1 x_1 - \beta_2 x_2 - \beta_3 x_3 \right)^2 \qquad (13.23)$$

In this case, four equations that are generalizations of Eqs. (13.7) and (13.8) can be found; they are then solved multaneously to determine the four estimates a, b_1, b_2, and b_3. All the computations are generalizations of the single independent variable case; however, because any general purpose computer program performs these computations, we shall simply discuss how the computer program is applied.

We used the BMD02R program that performs a step-up variable selection analysis by first determining the best single variable to use in predicting Y, then including the second best variable. Thus additional variables are added to the regression model sequentially until either all the variables are included or an imposed stopping rule becomes effective. Punching the data given in Fig. 13.8 on cards according to the format described by the BMD02R program writeup, we obtained the output and show selected portions: (1) The best single variable is x_2 = SAT quantitative score, and the appropriate simple linear regression equation is

$$\hat{Y} = 0.33328 + 0.00388 \ x_2$$

This model has a s'_ϵ = 0.3306 and R^2 = 0.7028. The standard deviation of b_2 is s_{b_2} = 0.0007, and the degrees of freedom of the error are 13.

We note that the R^2 obtained using x_2 alone is larger than the R^2 = 0.63 we obtained by using the SAT combined score. In addition, the standard error of b_2 is only 0.0007, which indicates even without a formal t test that β_2 is significantly different from zero.
(2) The next variable to be included is x_3 = percentile rank in high school graduating class. The multiple linear regression equation is now

$$\hat{Y} = 0.08579 + 0.00380 \ x_2 - 0.00531 \ x_3$$

This model has a s'_ϵ = 0.2886 and R^2 = 0.7910. The standard deviation of the two β estimates are s_{b_2} = 0.000661 and s_{b_3} 0.00236, and the degrees of freedom of error are 12.

Through the addition of the variable x_3 we have decreased the standard error s'_ϵ from 0.3306 to 0.2886, with a corresponding increase in R^2. Both β_2 and β_3 are apparently significant in view of their standard deviations.

(3) The third variable to be included is x_1 = SAT verbal score. The regression model is now

$$\hat{Y} = 0.16751 - 0.00068\ x_1 + 0.00430\ x_2 - 0.00522\ x_3$$

The model now has an $s_\varepsilon' = 0.2986$ and an $R^2 = 0.7950$. The standard deviations of the β estimates are $s_{b_1} = 0.00147$, $s_{b_2} = 0.00125$, and $s_{b_3} = 0.00245$. The degrees of freedom of error are now 11.

An examination of the above output leads us to question the use of three independent variables in the prediction equation because the two variable regression model yields a smaller standard deviation of errors than the three variable regression model. The reason for the increase in s_ε' is that the divisor of SSE is decreased by 1 every time another independent variable is added to the model since

$$s_\varepsilon' = \sqrt{\frac{\Sigma \varepsilon_i^2}{n - p - 1}} = \sqrt{\frac{SSE}{n - p - 1}} \qquad (13.24)$$

where p is the number of independent variables used in the regression equation.

The goodness of fit measure R^2 is now generalized to

$$R^2 = \frac{SSR}{SSE} \qquad (13.25)$$

or

$$R^2 = \frac{b_1 \Sigma(x_{1i} - \bar{x}) \cdot y_i + b_2 \Sigma(x_{2i} - \bar{x}_2) \cdot y_i + b_3 \Sigma(x_{3i} - \bar{x}_3) \cdot y_i}{\Sigma(y_i - \bar{y})^2}$$

Included in the program output is a tabulation of the observed data, including the values of the predicted Y's using the regression model and the errors ε_i made in predicting each value. We give these results in Fig. 13.9.

Observed GPA Y	Predicted GPA $\hat{Y}$	Errors ε	SAT Quantitative Score X_2	Percentile Rank X_3	SAT Verbal Score X_1
1.44	1.90	-0.46	366	79	377
2.34	2.17	0.17	421	86	381
1.95	2.10	-0.15	431	70	420
2.18	2.00	0.19	456	33	450
2.56	2.44	0.12	491	90	451
1.95	1.95	0.00	492	2	512
2.84	2.52	0.32	575	38	475
2.11	2.42	-0.31	597	5	500
2.67	2.45	0.22	563	49	584
2.48	2.62	-0.14	644	12	564
3.61	3.07	0.54	674	75	574
2.73	2.78	-0.05	631	68	671
2.85	2.88	-0.03	693	34	654
3.44	3.73	-0.29	803	99	603
3.11	3.25	-0.14	725	88	726

Fig. 13.9 Comparison of observed and predicted Y's for the multiple regression model.

The combination SAT prediction equation yielded an $s'_\varepsilon = 0.37$ that is larger than the three variable equation value, $s'_\varepsilon = 0.30$. Thus the overall error is reduced when more variables are added, even though the error for a particular observation may increase when using the multiple linear regression model.

If the errors can be assumed to be normal with mean zero and variance σ^2_ε, statistical tests of hypothesis can be made or confidence intervals found for each of the parameters, using the techniques given in Sec. 13.4. Of course, the degrees of freedom associated with the t statistic are now $n - p - 1 = 15 - 3 - 1 = 11$.

Since the inclusion of all three variables did not greatly enhance the prediction of Y, we shall attempt to use only the quantitative SAT score in the polynomial regression model given by

$$Y = \alpha + \beta_1 x + \beta_2 x^2 + \beta_3 x^3 + \cdots + \beta_p x^p + \varepsilon \qquad (13.26)$$

We can transform the polynomial equation into multiple linear regression Eq. (13.22) by placing $x_2 = x^2$, $x_3 = x^3$ and so forth. This transformation would permit us to use the BMD02R program by modifying the input data to be the appropriate power of the variable x for each observation. To avoid these calculations and the repunching of cards, we shall use the polynomial regression program available in the BMD statistical package (BMD05R), which calculates the polynomial regression directly, with the require input being only the observations (x_i, y_i); the program stops when higher powers of X do not significantly increase the prediction ability.

We again present only selected parts of the output obtained from the BMD05R program. The program first obtains the simple linear regression equation, then proceeds to determine the quadratic equation, then the cubic. Thus we have

(1) The simple linear regression equation is

$$\hat{Y} = 0.33329 + 0.00388x$$

The standard deviation of the errors is 0.3306 and $R^2 = 0.7022$, and the standard deviation of the slope estimate s_b is 0.0007. This is, of course, the identical equation to the one obtained in the first step of the multiple linear regression program.

(2) The quadratic regression equation is

$$\hat{Y} = -0.08584 + 0.00541x + 0.0000x^2$$

We can easily see that a test as to whether the quadratic coefficient β_2 is equal to zero would be accepted; thus the simple linear model is adequate as a fit of the data. This can also be seen by plotting a scatter diagram of the quantitative SAT score and the associated grade-point average.

The technique used in changing the nonlinear polynomial regression model into a linear model can be used for many other linear regression models. For example, if we have two independent variables u and v, and we wish to predict Y using the model

$$Y = \alpha + \beta_1 u + \beta_2 v + \beta_3 \sqrt{u} + \beta_4 \cdot \log v + \beta_5 \cdot u \cdot v + \varepsilon$$

we can let $x_1 = u$, $x_2 = v$, $x_3 = \sqrt{u}$, $x_4 = \log v$, and $x_5 = u \cdot v$ and thus obtain the multiple linear model given by Eq. (13.22). The student should be cautious, however, in introducing such transformations because each new term in the model reduces the degrees of freedom for the error; thus models may fit the observed data simply because the number of model parameters is large.

13.6 The Use of Regression in Decision Making

Having shown how the regression approach can be used to develop prediction tools, we now consider how they might· be used for decision-making purposes. We shall restrict our considerations to a single actual example because it is felt that numerous applications which can be solved by the same approach will then become apparent.

We shall consider the college admission problem and assume that the admission officer must take one of the following three actions relative to each applicant: admit him to college, have a personal interview before any further decision, or reject the application. Since the admission officer knows the combined SAT score for each applicant, he can use a previously developed regression equation to predict the applicant's GPA at the end of the freshmen year.

The random variable in this decision problem is thus the predicted GPA for an individual, which is given by $\hat{Y}_x$. Since $\hat{Y}_x$ is a continuous random variable, we could consider the following class of strategies. Let $s[b_1, b_2]$ specify the following correspondence between the actions and the random variable:

If $\hat{Y}_x \leq b_1$, reject the application

If $b_1 < \hat{Y}_x \leq b_2$, invite the applicant for an interview

If $\hat{Y}_x > b_2$, admit the applicant

The values of b_1 and b_2 indicate different admission procedures; for example, any strategy for which b_1 is large represents a tough admission policy. A strategy having a large interval $[b_1, b_2]$ with b_1 being small reflects faith in the personal interview approach, and a small value of b_2 reflects an easy admission policy.

We shall once again step through the procedures used to determine the expected risk for a single strategy because once each step is appreciated, a computer program can be written to rapidly evaluate a large number of competitive strategies and thus enable us to pick a good strategy.

Step 1

Possible actions:

a_1: deny admission
a_2: have a personal interview
a_3: admit to college

Step 2

Possible outcomes of the random variable. The value of the predicted GPA, $\hat{Y}_x$, can theoretically fall between minus and plus infinity, although actually it is restricted to values between zero and 4.

Step 3

States of nature. We shall use the following four states of nature:

θ_1: failing the freshmen year
θ_2: on probation after freshmen year
θ_3: satisfactory record in freshmen year
θ_4: honor student during freshmen year

We must introduce a parameter whose value would identify each state of nature. For this purpose we use $\mu_{Y/x}$, the average GPA obtained by freshmen whose SAT combined score is x. In addition, we need to know the standard deviation for each state of nature. Since the standard deviation depends upon the particular value of x, we choose a representative x and assume that we have established that $\sigma_\varepsilon = 0.4$, $n = 15$, $\bar{x} = 1100.2$ and $\Sigma(x_i - \bar{x})^2 = 706602$. Using the fact that

$$\sigma_{Y/x} = \sigma_\varepsilon \sqrt{1 + \frac{1}{n} + \frac{(x - \bar{x})^2}{\Sigma(x_i - \bar{x})^2}}$$

and letting x be 105, 342, 1056, and 1532, respectively, for each state of nature, we obtain the following representation:

θ_1: $\mu_{Y/x} = 0.50$ $\qquad$ $\sigma_{Y/x} = 0.63$

θ_2: $\mu_{Y/x} = 1.00$ $\qquad$ $\sigma_{Y/x} = 0.55$

θ_3: $\mu_{Y/x} = 2.50$ $\qquad$ $\sigma_{Y/x} = 0.41$

θ_4: $\mu_{Y/x} = 3.50$ $\qquad$ $\sigma_{Y/x} = 0.46$

Step 4

Probabilities of occurrence. We know that the distribition of Y_x is

$$\hat{Y}_x \sim N(\mu_{Y/x}, \sigma_{Y/x})$$

and thus in this case we can again compute the action probabilities directly.

Step 5

Losses and regrets. The losses for each action and state of nature are

	a_1	a_2	a_3
θ_1	1	4	15
θ_2	4	3	10
θ_3	13	10	7
θ_4	20	8	5

However, we shall use the corresponding regrets given by

	a_1	a_2	a_3
θ_1	0	3	14
θ_2	1	0	7
θ_3	6	3	0
θ_4	15	3	0

Step 6

Designation of the competing strategies. We shall consider only one strategy: s[1.0,2.0].

Step 7

Evaluation of the action probabilities. To find the
action probability for action a_i and state of nature θ_k, the
following integral must be evaluated:

$$g_{ki}(m) = \int_{(b_{i-1}-\mu_{Y/x})/\sigma_{Y/x}}^{(b_i-\mu_{Y/x}/\sigma_{Y/x})} \frac{1}{\sqrt{2\pi}} e^{-z^2/2} \, dz$$

where $b_o = -\infty$ and $b_3 = \infty$. Thus we find that

$$g_{11}(1) = \int_{-\infty}^{(1.0-0.5)/0.63} \frac{1}{\sqrt{2\pi}} e^{-z^2/2} \, dz = \int_{-\infty}^{0.79} \frac{1}{\sqrt{2\pi}} e^{-z^2/2} \, dz$$

$$= 0.7852$$

$$g_{12}(1) = \int_{(1.0-0.5)/0.63}^{(2.0-0.5)/0.63} \frac{1}{\sqrt{2\pi}} e^{-z^2/2} \, dz = \int_{0.79}^{2.38} \frac{1}{\sqrt{2\pi}} e^{-z^2/2} \, dz$$

$$= 0.2061$$

432

$$g_{13}(1) = \int_{(2.0-0.5)/0.63}^{\infty} \frac{1}{\sqrt{2\pi}} e^{-z^2/2} \, dz = \int_{2.38}^{\infty} \frac{1}{\sqrt{2\pi}} e^{-z^2/2} \, dz$$

$$= 0.0087$$

Likewise, the action probabilities for θ_2, θ_3, and θ_4 can be found by using the appropriate values for $\mu_{Y/x}$ and $\sigma_{Y/x}$ and evaluating the integral. For example,

$$g_{21}(1) = \int_{-\infty}^{(1.0-1.0)/0.55} \frac{1}{\sqrt{2\pi}} e^{-z^2/2} \, dz = \int_{-\infty}^{0} \frac{1}{\sqrt{2\pi}} e^{-z^2/2} \, dz$$

$$= 0.5$$

$$g_{31}(1) = \int_{-\infty}^{(1.0-2.50)/0.41} \frac{1}{\sqrt{2\pi}} e^{-z^2/2} \, dz = \int_{-\infty}^{-3.66} \frac{1}{\sqrt{2\pi}} e^{-z^2/2} \, dz$$

$$= 0.0001$$

By evaluating the integrals for all actions and states of nature, we obtain the following action probabilities:

	a_1	a_2	a_3
θ_1	0.7852	0.2061	0.0087
θ_2	0.5000	0.4656	0.0344
θ_3	0.0001	0.1130	0.8869
θ_4	0.0000	0.0006	0.9994

Step 8

Computation of average risk. Using the equation

$$\bar{r}_k(m) = g_{11}(m) \cdot r_{11} + g_{12}(m) \cdot r_{12} + g_{13}(m) \cdot r_{13}$$

we can evaluate the average risk for the strategy. We find

$$\bar{r}_1(1) = (0.7852) \; 0 + (0.2061) \; 3 + (0.0087) \; 14$$

$$= 0.7401$$

$$\bar{r}_2(1) = (0.5000) \; 1 + (0.4656) \; 0 + (0.0344) \; 7$$

$$= 0.7408$$

$$\bar{r}_3(1) = (0.0001) \; 6 + (0.1130) \; 3 + (0.8869) \; 0$$

$$= 0.3390$$

$$\bar{r}_4(1) = (0.0000) \; 15 + (0.0006) \; 3 + (0.9994) \; 0$$

$$= 0.0018$$

Step 9

Selection of the strategy. To use the minimum risk criterion we need the prior probabilities for each state of nature. If these are given by $q_1 = 0.15$, $q_2 = 0.25$, $q_3 = 0.50$, and $q_4 = 0.10$, then the expected risk for the strategy $s[1.0, 2.0]$ is

$$R(1) = q_1 \bar{r}_1(1) + q_2 \bar{r}_2(1) + q_3 \bar{r}_3(1) + q_4 \bar{r}_4(1)$$

$$= (0.15)(0.7401) + (0.25)(0.7408)$$

$$+ (0.50)(0.3390) + (0.10)(0.0018)$$

$$= 0.4659$$

The expected risk of this strategy must then be compared to that of other strategies; the strategy having the smallest risk is the one chosen.

13.7 The Correlation Coefficient

Regression analysis is used for prediction purposes; correlation analysis is used to study the relationship that may exist between two random variables, where the values for the two variables are obtained on the same experimental unit. To develop such a measure of the relationship we need to know the distribution of the two random variables for the population being studied. Because this distribution must involve two variables X and Y, we need a bivariate distribution model, and it seems natural to generalize the normal distribution for this purpose. The generalized model is

$$f(X,Y) = \frac{1}{2\pi\sigma_x\sigma_y\sqrt{1-\rho^2}} \exp(D), \quad -\infty < X < +\infty, \quad -\infty < Y < +\infty$$

$$(13.27)$$

where

$$D = -\frac{1}{2(1-\rho^2)}\left[\frac{(X-\mu_x)^2}{\sigma_x^2} + \frac{(Y-\mu_y)^2}{\sigma_y^2} - 2\rho\frac{(X-\mu_x)}{\sigma_x}\frac{(Y-\mu_y)}{\sigma_y}\right]$$

We note that in the model, both the

$$\frac{(X-\mu_x)^2}{\sigma_x^2} \quad \text{and} \quad \frac{(Y-\mu_y)^2}{\sigma_y^2}$$

terms are present as they appear in their respective univariate normal models. In addition, a cross-product term and divisor are introduced with a new parameter ρ. This parameter ρ, the correlation coefficient, is a measure of the linear relationship between Y and X and is restricted to values between -1 and +1. If $\rho = 0$, then there is no relationship between X and Y in the population and the two variables are independent, and as ρ approaches ± 1, the strength of the linear relationship increases. When $\rho = \pm 1$, X and Y contain the same information about the experimental unit, and we can determine the value of one variable without error knowing the value of the other.

We wish to be able to estimate ρ, using observations of X's and Y's from a random sample of individuals assumed to have come from a bivariate normal population. The estimate we·use is the simple correlation coefficient r, which is determined by the formula

$$r = \frac{\Sigma x_i y_i - \Sigma x_i \Sigma y_i / n}{\sqrt{[\Sigma x_i^2 - (\Sigma x_i)^2/n][\Sigma y_i^2 - (\Sigma y_i)^2/n]}} \qquad (13.28)$$

There are many other formulas for computing r, but let us consider the following ungrouped sample data and demonstrate the use of Eq. (13.28) before discussing other available formulas.

Twelve patients with a specific disease are being studied, and the temperature (X) and the pulse rate (Y) of each patient on the third day after receiving a certain medication is measured. The data are

Observation i	Temperature X	Pulse Rate Y
1	99.4	78
2	100.1	83
3	98.6	70
4	97.9	76
5	101.4	75
6	102.6	89
7	99.0	80
8	98.6	77
9	101.1	86
10	98.0	79
11	100.5	84
12	99.3	87

Fig. 13.10 Sample data from a bivariate distribution.

The summations are:

$$n = 12$$
$$\Sigma x_i = 1196.5$$
$$\Sigma x_i^2 = 119324.37$$
$$\Sigma y_i = 964.0$$
$$\Sigma y_i^2 = 77786$$
$$\Sigma x_i y_i = 96170.2$$

We have

$$r = \frac{96170.2 - [(1196.5)(964)/12]}{\sqrt{[119324.37 - (1196.5)^2/12][77786 - (964)^2/12]}}$$

$$r = \frac{51.37}{\sqrt{(23.3492)(344.67)}} = \frac{51.37}{89.71} = 0.57$$

$$r^2 = 0.32$$

We note that the estimate of ρ is $+ 0.57$. It is tempting to therefore assume that the strength of the linear relationship between X and Y is a little better than halfway between none ($\rho = 0$) and perfect ($\rho = 1.0$). A better indicator of the relationship is given by ρ^2 because this measure indicates the percentage of elementary causation factors that are common to both X and Y.

Before considering the sampling problem associated with r we introduce other equivalent forms for the computation of r. They include

$$r = \frac{\Sigma(x_i - \bar{x})(y_i - \bar{y})}{\sqrt{\Sigma(x_i - \bar{x})^2 \cdot \Sigma(y_i - \bar{y})^2}} = \frac{\Sigma(x_i - \bar{x})(y_i - \bar{y})/n}{s_x s_y} \qquad (13.29)$$

and

$$r = \sum_i \frac{z_{xi} z_{yi}}{n} \qquad (13.30)$$

$$z_{xi} = \frac{x_i - \bar{x}}{s_x} \qquad \text{and} \qquad z_{yi} = \frac{y_i - \bar{y}}{s_y}$$

where s_x^2 and s_y^2 are the biased estimators of the variance (divisor: n). (The term sometimes used for r; the product moment correlation coefficient, derives from this latter form.)

Since both simple linear regression and correlation assume a linear relationship between the two variables of interest, it is to be expected that the correlation coefficient is related to the regression coefficient. One such relationship is

$$r^2 = \frac{SSR}{SST}$$

where SSR and SST are the sums of squares due to regression and total, respectively, for the regression of Y on X. The sign for r is the same as the sign of the regression coefficient b. The correlation coefficient can also be expressed as a function of the regression coefficient of Y on X, $b_{y/x}$, and the regression coefficient of X on Y, $b_{x/y}$:

$$r_{xy} = \sqrt{b_{y/x} \cdot b_{x/y}}$$

Thus the correlation coefficient is the geometric mean of the two linear regression coefficients.

The above relationship clearly illustrates that it is impossible to study cause and effect-type relationships through either regression or correlation analysis. In regression analysis the cause and effect relationship is assumed once the regression model is written, and correlation analysis does not specify the cause-effect role of the variable.

To make inferences about the unknown parameter ρ, using its estimator r, we need to have a test statistic whose sampling distribution is known. If ρ is assumed to be equal to zero, $\rho = 0$, then

$$t = \frac{r}{\sqrt{(1 - r^2)/(n - 2)}} = \frac{r\sqrt{n - 2}}{\sqrt{1 - r^2}} \qquad (13.31)$$

is known to have the Student's t distribution with $d = n - 2$ degrees of freedom.

When ρ is unequal to zero, $\rho \neq 0$, the above ratio no longer has the t distribution, and thus we cannot use it as the test statistic. However, we can apply a transformation of the statistic r such that the distribution of the transformed statistic rapidly approaches normality as the sample size increases.

The transformation is

$$z^*_\rho = (0.5) \cdot \ln \left(\frac{1 + \rho}{1 - \rho} \right) \quad \text{and} \quad z^*_r = (0.5) \cdot \ln \left(\frac{1 + r}{1 - r} \right) \quad (13.32)$$

The sampling distribution of z^*_r is approximately normally distributed, with mean z^*_ρ and variance $1/(n - 3)$. Thus standard error sampling theory can be used, and the three basic inference questions about r need only be changed into the corresponding questions about z^*_r.

We illustrate these techniques for making inferences about r by using the illustrative data on temperature and pulse rate.

1. Testing hypothesis

We might wish to test the hypothesis that $H_o : \rho = 0$ against the alternative $H_a : \rho \neq 0$. Since ρ is assumed to be equal to zero, we can use the t statistic. Using the example given in this section, where $r = 0.57$, we evaluate

$$t = \frac{0.57 \sqrt{10}}{\sqrt{1 - 0.57^2}} = 2.67$$

The degrees of freedom are $d = n - 2 = 10$, $\alpha = .05$ and the critical value of t is $t_{d, 1-\alpha/2} = t_{10, .975} = 2.228$. Since $2.67 > 2.228$, we reject the null hypothesis.

We could ask whether the true value ρ can be as large as 0.70 when r is equal to only 0.57; thus $H_o : \rho = 0.70$ and $H_a : \rho \neq 0.70$.

In this case we must first find

$$z_{0.70}^* = (0.5) \cdot \ln\left(\frac{1 + 0.70}{1 - 0.70}\right) = 0.867$$

$$z_{0.57}^* = (0.5) \cdot \ln\left(\frac{1 + 0.57}{1 - 0.57}\right) = 0.648$$

and

$$\sigma_{z_r^*} = \frac{1}{\sqrt{n - 3}} = \frac{1}{\sqrt{9}} = 0.33$$

Thus the test statistic is

$$z = \frac{z_r^* - z_\rho^*}{1/\sqrt{n - 3}} = \frac{0.648 - 0.867}{1/\sqrt{9}} = -0.66$$

The critical value using the 5 percent significance level is $z_{.975} = 1.96$. Since $-1.96 < -0.66 < 1.96$, we would not discard the null hypothesis that $\rho = 0.70$.

2. Confidence intervals

If we are interested in finding the confidence interval for ρ, we first determine the confidence interval for the transformed variable z_r^* whose mean and standard deviation are known, which is given by

$$z_r^* - z_{1-\alpha/2} \frac{1}{\sqrt{n - 3}} \leq z_\rho^* \leq z_r^* + z_{1-\alpha/2} \frac{1}{\sqrt{n - 3}}$$

In our illustration we obtain the following 95 percent confidence interval since $z_{.975} = 1.96$:

$$0.648 - 1.96 \frac{1}{\sqrt{9}} \leq z_\rho^* \leq 0.648 + 1.96 \frac{1}{\sqrt{9}}$$

$$-0.05 \leq z_\rho^* \leq + 1.301$$

The limits of z_ρ^* can now be converted to limits on ρ by using the inverse transformation.

Since

440

$$-0.05 \leq 0.5 \cdot \ln\left(\frac{1+\rho}{1-\rho}\right) \leq 1.301$$

we obtain the following relationship:

$$e^{2(-0.05)} \leq \frac{1+\rho}{1-\rho} \leq e^{2(1.301)}$$

The value of ρ that satisfies each side of the above inequality can now be found by some iterative technique.

For the example problem we shall simply determine values of ρ that satisfy each end point. Thus

$$e^{2(-0.05)} = e^{-0.1} = 0.9048$$

and the function value for ρ = -0.05 is

$$\frac{1+(-0.05)}{1-(-0.05)} = 0.9048$$

Likewise

$$e^{2(1.301)} = 13.4772$$

and the function value for ρ = 0.862 is

$$\frac{1+(+0.862)}{1-(+0.862)} = 13.4928$$

Thus the confidence interval for ρ in the example is

$$-0.05 \leq \rho \leq 0.86$$

Another technique for finding the end points is to evaluate the function

$$(0.5) \cdot \ln \frac{1+\rho}{1-\rho}$$

for different values of ρ. The value of ρ that corresponds to a particular function value can then be obtained by table look-up:

$$-0.5 \leq \rho \leq 0.86$$

3. Sample size

To determine the sample size required to estimate ρ within a certain accuracy ε, the following equation needs to be solved for n:

$$z = \frac{z_r^* - z_\rho^*}{\sqrt{\frac{1}{n-3}}} = \frac{\frac{1}{2}\ln\left(\frac{1+r}{1-r}\right) - \frac{1}{2}\ln\left(\frac{1+\rho}{1-\rho}\right)}{\sqrt{\frac{1}{n-3}}} = \frac{\varepsilon^*}{\sqrt{\frac{1}{n-3}}}$$

Since the value of ε^* not only depends upon the accuracy required, ε, but also upon the value of ρ, the sample size is difficult to determine before an estimate of ρ is available.

Using the central limit theorem it can be shown that the ratio

$$z = \frac{r - \rho}{\sqrt{\frac{1 - r^2}{n - 2}}}$$

approaches a N(0,1) distribution as the sample size increases. Solving for n, we obtain

$$n = \frac{z_{1-\alpha/2}^2(1 - r^2)}{\varepsilon^2} + 2$$

The maximum value of n would be found when r = 0:

$$n = \frac{z_{1-\alpha/2}^2}{\varepsilon^2} + 2$$

Consider the sample size required to estimate ρ within $\varepsilon = 0.1$, using a 95 percent confidence level. We have

$$n = \frac{(1.96)^2}{(0.1)^2} + 2 = 384.16 + 2 = 386.16 = 387$$

If the determined value of n is too small, a more exact method to determine n must be used since the central limit theorem would not be too effective.

In this section we introduced the correlation coefficient ρ as an index of the strength of the relationship between two random variables. The estimator of this index is given by r, and we showed how r can be used to make the standard inferences about the parameter ρ. It should be emphasized that the use of the correlation coefficient as an index of the relationship between two variables is based upon the assumption that the joint distribution of the two variables is given by the bivariate normal distribution. The consequence of this assumption is that if there is any relationship between the two variables, it must be a linear one. Thus the correlation coefficient indicates only the strength of the linear relationship, and because of this linearity a correspondence exists between simple linear regression and correlation analysis. Thus even if X and Y are both random variables, we can think of correlation as measuring how well X predicts Y or vice versa. This relationship should be appreciated because the sums of squares due to regression (SSR) and total sums of squares (SST) can be used to determine both the index of goodness of fit R^2 and the correlation coefficient since

$$r = \pm \sqrt{\frac{SSR}{SST}} = \pm \sqrt{R^2} \qquad (13.33)$$

The sign of the slope b is used to determine the sign of the correlation coefficient.

Just as the simple linear regression can be generalized to include more than one independent variable, correlation can also be generalized to study the linear relationship between several random variables $y_1, y_2, \ldots, y_j, \ldots, y_p$. We wish to know the strength of the linear relationship of one of the variables, say, y_p, with the remaining set of variables $y_1, y_2, \ldots, y_{p-1}$; this index is called the multiple correlation coefficient and is denoted by

$$\rho_{y_p \cdot y_1 y_2 \cdots y_{p-1}}$$

The estimator of this multiple correlation coefficient can be calculated knowing only the simple correlation of each variable with each of the others. However, using the relationship between regression and correlation, we can also determine the multiple correlation by fitting the regression line

$$y_p = \alpha + \beta_1 y_1 + \beta_2 y_2 + \cdots + \beta_{p-1} y_{p-1}$$

and then computing

$$r_{y_p \cdot y_1 y_2 \cdots y_{p-1}} = \sqrt{\frac{SSR}{SST}} = \sqrt{R^2} \qquad (13.34)$$

The multiple correlation coefficient is always positive because the concept of a direct or inverse relationship that was used to determine the sign of the simple correlation coefficient does not generalize when there are more than two variables.

We shall not consider the different techniques of calculating the multiple correlation since either a regression program such as the BMD02R to determine R^2 or a general-purpose program that calculates the multiple correlation coefficient directly can be used.

Another generalization of the simple correlation coefficient is the partial correlation coefficient denoted by $\rho_{y_1 y_2 / y_3}$. The concept behind this index is that two variables may exhibit a strong linear relationship simply because both are related to a third variable. Thus it is of interest to determine the strength of the relationship of the two variables when the effect of the third variable is removed. This partial correlation coefficient is given by

$$\rho_{x_1 x_2 / x_3} = \frac{\rho_{x_1 x_2} - \rho_{x_1 x_3} \rho_{x_2 x_3}}{\sqrt{\left(1 - \rho^2_{x_1 x_3}\right)\left(1 - \rho^2_{x_2 x_3}\right)}} \qquad (13.35)$$

and its estimator is given by the equivalent equation

$$r_{x_1x_2/x_3} = \frac{r_{x_1x_2} - r_{x_1x_3}r_{x_2x_3}}{\sqrt{\left(1 - r_{x_1x_3}^2\right)\left(1 - r_{x_2x_3}^2\right)}} \qquad (13.36)$$

Thus, knowing the simple correlations of variables $r_{x_1x_2}$, $r_{x_1x_3}$, and $r_{x_2x_3}$, the partial correlation can be found. If more than one variable, say x_3 and x_4, are partialed out, the computations can be made sequentially. The first-order partial correlations using Eq. (13.36) are found, and then the second-order partial correlations are found, using the same equation with the simple correlations being replaced by the first-order partial correlations. Thus the partial correlation of x_1x_2 after x_3 and x_4 have been partialed out is given by

$$r_{x_1x_2/x_3x_4} = \frac{r_{x_1x_2/x_3} - r_{x_1x_4/x_3}r_{x_2x_4/x_3}}{\sqrt{\left(1 - r_{x_1x_4/x_3}^2\right)\left(1 - r_{x_2x_4/x_3}^2\right)}} \qquad (13.37)$$

There are, however, also general-purpose computer programs that can be used to calculate the patial correlations.

The concept of correlations can be generalized even further to canonical and partial canonical correlations, but we feel that these are beyond the scope of this text.

13.8 Regression and Correlation Simulation

If we consider the simulation power of the computer, there are many interesting factors concerning regression and correlation techniques that might be investigated. For example, we could investigate the validity of the mathematically determined property that if the model errors are normally distributed with a mean of zero and variance σ_ε^2, then the distribution of the parameter estimates and the predicted values of Y are also normally distributed. Since the sampling distribution of the estimates determines the appropriate statistics to be used in making inferences about the parameter values, it is worthwhile for us to investigate the problem in a more empirical fashion. Similarly, we wish to investigate the sampling distribution of the sample correlation eoefficient.

To investigate the sampling distributions of the regression estimates, we generate a finite number of samples from the regression model, determine the estimates, and look at the resulting empirical sampling distribution of each estimate.

We shall assume that the true regression model for Y given X is

$$Y = \alpha + \beta x + \varepsilon$$

Then if $\varepsilon \sim N(0, \sigma_\varepsilon^2)$, we want to show that

(1) The observed Y's have a normal distribution

$$Y \sim N(\alpha + \beta x, \sigma_\varepsilon)$$

(2) The estimate of α has a normal distribution

$$a \sim N\left(\alpha, \ \sigma_\varepsilon \sqrt{\frac{1}{n} + \frac{\bar{x}^2}{\Sigma(x_i - \bar{x})^2}}\right)$$

(3) The estimate of β has a normal distribution

$$b \sim N\left(\beta, \ \sigma_\varepsilon \sqrt{\frac{1}{\Sigma(x_i - \bar{x})^2}}\right)$$

(4) The predicted Y's have a normal distribution

$$\hat{\mu}_{Y/x} = \hat{Y} \sim N\left(\alpha + \beta x, \ \sigma_\varepsilon \sqrt{\frac{1}{n} + \frac{(x - \bar{x})^2}{\Sigma(x_i - \bar{x})^2}}\right)$$

The inputs needed to perform the computer simulation study are the

(1) Model parameters α, β, and σ_ε

(2) Number of (x_i, y_i) points to be generated for each sample n

(3) Specific set of x's given by $(x_1, x_2, \ldots, x_n)$

(4) Number of samples to be generated

Once these inputs are available, the distribution of the Y's for each x_i is known, $Y_x \sim N(\alpha + \beta x, \sigma_\varepsilon)$, and thus the Y values from the appropriate normal population can be generated.

The program generates the empirical sampling distribution for each of the three estimators a, b, and $\hat{Y}_x$ (for each x) and writes the four characteristics of these empirical distributions. (The output from the program was given in Sec. 13.3.)

The program inputs are

$$\alpha = \text{ALPHA}$$
$$\beta = \text{BETA}$$
$$\sigma_\varepsilon = \text{SIGMAE}$$
$$n = \text{NSAM, number of points } (x_i, y_i) \text{ in each sample}$$
$$m = \text{M, number of samples to be generated}$$
$$x_i, i = 1, 2, \ldots n = (X(I), I=1, N), \text{ the set of x values to be used}$$

```
C     REGRESIMUL - GENERATION OF SAMPLING DIST. OF
C     REGRESSION ESTIMATES
      DIMENSION X(1ØØ),Y(1ØØ),YMU(1ØØ),A(2ØØ),B(2ØØ),
     1YPRED(2ØØ,1ØØ),YTEMP(2ØØ),YPSTD(1ØØ)
      READ(5,1) M,NSAM,IODD
    1 FORMAT(3I5)
      READ(5,2) (X(I),I=1,NSAM)
    2 FORMAT(8F1Ø.2)
      READ(5,3)ALPHA,BETA,SIGMAE
    3 FORMAT(3F1Ø.Ø)
      WRITE(6,4)ALPHA,BETA,SIGMAE
    4 FORMAT(1H1,21HREGRESSION SIMULATION/
     11H  6HALPHA=,F1Ø.2,2X,5HBETA=,F1Ø.2,2X,7HSIGMAE=,F1Ø.2)
      SUMX=Ø.
      SUMX2=Ø.
      SAM=NSAM
      DO 1Ø J=1,NSAM
      SUMX=SUMX + X(J)
      SUMX2 = SUMX2 + X(J)**2
```

447

```
1Ø YMU(J) = ALPHA + BETA*X(J)
   CSUMX = SUMX2-SUMX*SUMX/SAM
   XMEAN = SUMX/SAM
   DO 31 J= 1,NSAM
31 YPSTD(J)=SIGMAE*SQRT(1./SAM + (X(J) - XMEAN)**2/CSUMX)
   STDA = SIGMAE*SQRT(1./SAM + XMEAN**2/CSUMX)
   STDB = SIGMAE*SQRT(1./CSUMX)
   DO 11 I = 1,M
   SUMXY =Ø.
   SUMY =Ø.
   SUMY2 =Ø.
   DO 12 J = 1,NSAM
   CALL RANNOR(IODD,RY,YMU(J),SIGMAE)
   SUMY = SUMY + RY
   SUMXY = SUMXY + RY * X(J)
12 SUMY2 = SUMY2 +RY**2
   CSUMXY = SUMXY - SUMX*SUMY/SAM
   CSUMY = SUMY2 - SUMY*SUMY/SAM
   B(I) = CSUMXY/CSUMX
   A(I) = SUMY/SAM -B(I)*SUMX/SAM
   DO 13 J=1, NSAM
13 YPRED(I,J)  = A(I) + B(I)*X(J)
11 CONTINUE
   CALL CHAR(A,M,YM,YSIG,A3,A4)
   WRITE(6,26)  ALPHA,STDA
26 FORMAT(1HØ,16HDIST FOR ALPHA =,F12.2,3X,5HSTDA=,F12.4)
   WRITE(6,2Ø) YM,YSIG,A3,A4
2Ø FORMAT(1H ,4MEAN,F12.4/1X,3HSTD,F12.4/
  11X, 3HA3,F12.4/1X, 2HA4,F12.4)
   WRITE(6,27)BETA,STDB
27 FORMAT(1HØ,15HDIST FOR BETA = ,F12.2,3X,5HSTDB=,F12.4)
   CALL CHAR(B,M,YM,YSIG,A3,A4)
   WRITE(6,2Ø)YM,YSIG,A3,A4
   DO 22 J=1,NSAM
   DO 23 I=1,M
23 YTEMP(I) = YPRED(I,J)
   CALL CHAR(YTEMP,M,YM,YSIG,A3,A4)
   WRITE(6,24)X(J),YMU(J),YPSTD(J)
24 FORMAT(1HØ,4ØHTHEORETICAL MEAN AND STD FOR Y GIVEN
  1 X =,F12.2,2HIS,1X,F12.4,1X,3HAND,1X,F12.4)
22 WRITE (6,2Ø)YM,YSIG,A3,A4
   STOP
   END
```

Program 13.1 REGRESIMUL — generation of empirical sampling distribution of regression estimates.

The program inputs used to generate the results presented in Sec. 13.3 are

```
200  15  119
    743.      802.      851.      906.      942.     1004.     1050.     1097.
   1147.     1208.     1248.     1301.     1347.     1406.     1451.
    .2821  .002062      .37
```

448

We would like to study the sampling distribution of the sample correlation coefficient r for different values of the parameter ρ. We shall assume that the variables X and Y have a bivariate normal distribution, so we need to be able to generate random bivariate normal numbers to perform simulation studies.

The technique to generate random bivariate normal numbers with means μ_x and μ_y and variances σ_x^2 and σ_y^2 and correlation coefficient ρ is as follows.

If $z_1 \sim N(0,1)$ and $z_2 \sim N(0,1)$ and z_1 and z_2 are independent, then the variables X and Y are given by

$$X = z_1 \sigma_x + \mu_x \qquad\qquad (13.38)$$

$$Y = \left(\rho z_1 + z_2 \sqrt{1 - \rho^2} \right) \sigma_y + \mu_y \qquad\qquad (13.39)$$

and they have a bivariate normal distribution with the indicated means, variances, and correlation.

We can mathematically verify that this transformation from independent normal (0,1) numbers produces dependent normal numbers with the appropriate means, variances, and correlation by evaluating

$$E(X) = E(z_1 \sigma_x + \mu_x) = E(z_1)\sigma_x + \mu_x = 0\sigma_x + \mu_x = \mu_x$$

and $\sigma_x^2 = E(X - \mu_x)^2$

$$= E(z_1 \sigma_x + \mu_x - \mu_x)^2 = E(z_1 \sigma_x)^2 = E(z_1^2)\sigma_x^2 = \sigma_x^2$$

since when $z_1 \sim N(0,1)$, then $E(z_1^2) = E(z_1 - 0)^2 = \sigma_z^2 = 1$.

Likewise, for Y

$$E(Y) = E\left[\left(\rho z_1 + z_2 \sqrt{1 - \rho^2} \right) \sigma_y + \mu_y \right]$$

$$= \rho\sigma_y E(z_1) + \sqrt{1 - \rho^2}\, \sigma_y E(z_2) + \mu_y$$

$$= 0 + 0 + \mu_y = \mu_y$$

449

and

$$\sigma_y^2 = E(Y - \mu_y)^2 = E\left[\left(\rho z_1 + z_2 \sqrt{1 - \rho^2}\right)\sigma_y + \mu_y - \mu_y\right]^2$$

$$= E\left(\rho z_1 \sigma_y + z_2 \sqrt{1 - \rho^2}\,\sigma_y\right)^2$$

$$= \rho^2 \sigma_y^2 E\left(z_1^2\right) + \sigma_y^2(1 - \rho^2)E\left(z_2^2\right)$$

$$+ 2(1 - \rho^2)\sigma_y^2 E(z_1 z_2)$$

$$= \rho^2 \sigma_y^2 + \sigma_y^2(1 - \rho^2) + 0$$

$$= \sigma_y^2$$

It can also be shown that $E(X - \mu_x)(Y - \mu_y) = \sigma_x \sigma_y \rho$, so that the correlation of X and Y is

$$\frac{E(X - \mu_x)(Y - \mu_y)}{\sqrt{\sigma_x^2 \sigma_y^2}} = \rho$$

The fact that X and Y have a bivariate normal distribution is ensured because the transformations used to generate X and Y, Eqs. (13.38) and (13.39), are linear in X and Y.

The subroutine to generate these bivariate random normal numbers is given by

```
SUBROUTINE BIVAR (X,Y,IODD,XMU,XSIG, YMU,YSIG,RHO)
CALL RANNOR(IODD, Z1, Ø., 1.)
CALL RANNOR(IODD,Z2, Ø., 1.)
X = Z1 * XSIG + XMU
Y = (RHO * Z1 + Z2*SQRT(1.-RHO**2))*YSIG + YMU
RETURN
END
```

Program 13.2 SUBROUTINE BIVAR — generation of random normal bivariate numbers.

450

We can now generate the empirical sampling distribution of r for different values of ρ by generating m bivariate samples each of size n, then computing r, and determining the characteristics of the m values of r. The inputs for the program are

$$m = M = \text{number of samples to be generated}$$
$$n = \text{NSAM} = \text{sample size}$$
$$\text{IODD} = \text{random starting digit (positive odd)}$$
$$\text{NINT} = \text{number of intervals for frequency distribution}$$
$$\text{XMU and XSIG} = \text{population parameters for X variable}$$
$$\text{YMU and YSIG} = \text{population parameters for Y variable}$$
$$\text{RHO} = \text{the correlation coefficient } \rho$$

```
C      RHOGENER - GENERATION OF SAMPLING DISTRIBUTION OF R
       DIMENSION R(2ØØ), F(3Ø)
       READ(5,1) M,NSAM,NINT,IODD
  1    FORMAT(4I5)
       SAM = NSAM
       READ(5,2) XMU,XSIG,YMU,YSIG,RHO
  2    FORMAT(5F1Ø.2)
       WRITE (6,12) RHO,XMU,XSIG,YMU,YSIG,NSAM
  6    FORMAT(1H1,28HSAMPLING DIST OF R WHEN RHO=, F1Ø.4
      1 /1H ,3HXMU,F1Ø.2,4HXSIG,F1Ø.2,3HYMU,F1Ø.2,4HYSIG,
      2 F1Ø.2 /1H ,11HSAMPLE SIZE, I5)
       DO  3 I=1,M
       SUMX = Ø.
       SUMX2 = Ø.
       SUMY = Ø.
       SUMY2 = Ø.
       SUMXY = Ø.
       DO  4 J=1,NSAM
       CALL BIVAR(X,Y,IODD,XMU,XSIG,YMU,YSIG,RHO)
       SUMX = SUMX + X
       SUMX2 = SUMX2 + X**2
       SUMY = SUMY + Y
       SUMY2 = SUMY2 + Y**2
  4    SUMXY = SUMXY + X*Y
       COXY = SUMXY - SUMX*SUMY/SAM
       DEN = (SUMX2-SUMX**2/SAM) * (SUMY2-SUMY**2/SAM)
  3    R(I) = COXY/SQRT(DEN)
       CALL FREQ(R,M,F,-1.,.1,NINT)
       WRITE(6,1Ø) (J,F(J),J=1,NINT)
  1Ø   FORMAT(1HØ, I5,2X,F1Ø.Ø)
       CALL CHAR(R,M,RMU,RSIG,A3,A4)
       WRITE(6,11) RMU,RSIG,A3,A4
  11   FORMAT(1HØ,4HMEAN,F12.4/ 1H ,3HSTD,F12.4/
      1  1H ,2HA3,F12.4/ 1H ,2HA4,F12.4)
       STOP
       END
```

Program 13.3 RHOGENER — generation of empirical sampling distribution of r.

Using a sample size of n = 30. and $\mu_x = \mu_y = 0$ and $\sigma_x = \sigma_y = 1$, we run the program for different values of ρ. The inputs for each value of ρ are given by

Run 1

200	30	20 4113			
	0.	1.	0.	1.	0.

Run 2

200	30	20 4113			
	0.	1.	0.	1.	.5

Run 3

200	30	20 4113			
	0.	1.	0.	1.	.9

The frequency distribution and the four characteristics that are generated for each set of input values are

Intervals	$\rho = 0$	$\rho = 0.5$	$\rho = 0.9$
-0.5 — -0.4	4	0	0
-0.4 — -0.3	10	0	0
-0.3 — -0.2	20	0	0
-0.2 — -0.1	25	0	0
-0.1 — 0.0	40	0	0
0.0 — 0.1	45	2	0
0.1 — 0.2	25	8	0
0.2 — 0.3	17	12	0
0.3 — 0.4	8	29	0
0.4 — 0.5	4	53	0
0.5 — 0.6	2	45	0
0.6 — 0.7	0	35	0
0.7 — 0.8	0	13	7
0.8 — 0.9	0	3	87
0.9 — 1.0	0	0	106
Mean	0.0028	0.4930	0.8965
Standard dev.	0.1975	0.1524	0.0410
a_3	0.0721	-0.4177	-1.1140
a_4	2.9840	2.9865	4.6382

It can be clearly seen from the four characteristics of the three frequency distributions that only when $\rho = 0$ does the distribution appear close to being normal. As the sample size increases, the distribution of r should become more normal. The student should consider checking the z* transformation used in the inferences when $\rho \neq 0$ to see whether the use of this transformation causes the observed frequency distribution to approach the normal distribution more rapidly. This could be accomplished by a few changes to the correlation generation program.

13.9 Summary

This chapter introduced regression techniques that can be used as a tool to predict the random variable Y, given a set of independent variables $\underline{X}$. Although the regression model studied in detail was the linear model with only one independent variable, the multiple linear model was also introduced, with the analysis being accomplished by a general-purpose computer program. The nonlinear model that could be analyzed by a transformation into the multiple linear model was also presented.

Correlation analysis was considered as a means of measuring the strength of the linear relationship between two or more random variables. The technique for computing the simple correlation coefficient for only two random variables was discussed, and computer programs were recommended when the multiple correlation or partial correlation coefficients were to be found.

The use of regression in decision making was demonstrated by considering a typical problem. The simulation capability of the computer was used to study the sampling distribution of the estimates used for regression and correlation analysis.

EXERCISES

1. Consider the observed points

i	1	2	3	4	5	6	7
X	10	12	14	16	18	20	22
Y	5.0	6.1	6.9	8.8	10.2	14.1	18.0

   (a) Use the observed point to find the simple linear
       regression equation $Y = \alpha + \beta x + \varepsilon$.
   (b) Graph the points and the fitted equation on the same
       set of axes.
   (c) Find the predicted value for each y.

2. Consider the observed data, where X represents the IQ
   and Y represents the reading speed:

i	1	2	3	4	5	6	7	8
X	102	95	115	112	86	102	121	108
Y	14.7	12.0	17.1	15.8	10.1	11.6	19.8	17.5

   (a) Fit the linear regression model to the data.
   (b) Compute the standard deviation of the errors, using
       the differences between the observed and predicted
       values of Y. Check the results by calculating the
       sums of squares due to the regression and the total
       sums of squares.
   (c) Compute the correlation coefficient.
   (d) Assuming normality, find the estimated reading speed
       for individuals with IQ equal to 100, and determine
       the 95 percent confidence interval for such an
       estimate.
   (e) Plot the 90 percent confidence interval for the
       regression line.

3. Write a program to calculate the linear regression
   estimates from an ungrouped set of y's and x's. Test
   the program using the data in the chapter. (The program
   should determine a, b, s_{ε}', and $\hat{y}_i$ for each i.)

4. Write a program to calculate the simple correlation be-
ween a vector of y's and x's. Test the program using
the data in the chapter.

5. Use the regression simulation program, Program 13.1, to
generate m = 200 samples from the following regression
line. Use the same set of x's and a small value for
n = NSAM.

$$Y = 4.0 + 8.0x + \varepsilon$$

where $\varepsilon \sim N(0, \sigma_\varepsilon = 5)$.

(a) Compare the empirical results with the theoretical
results; for sake of example, $\bar{b}$ with β.
(b) Take one sample from the above regression line and
test the hypothesis that $\beta = 0$ and $\alpha = 4.00$, using
the 5 percent level of significance. Also find the
confidence limit for $\hat{Y}_{x=5}$, $\hat{Y}_{x=1}$, and $\hat{Y}_{x=2.75}$. Compare your findings with those obtained by other mem-
bers of the class.

6. Run a multiple linear regression program obtained from
the computer center, using the following data for a sample of 15 accountants:

Y = annual salary
X_1 = number of years since graduation from college
X_2 = college grade-point average
X_3 = SAT combined verbal and quantitative score

i	X_1	X_2	X_3	Y
1	5	2.1	107	12,150
2	15	3.7	135	16,000
3	5	2.5	105	12,900
4	1	2.8	120	13,100
5	7	3.6	151	14,300
6	11	2.8	122	17,600
7	10	2.9	107	15,000
8	3	3.1	111	13,000
9	6	3.5	124	13,600
10	14	2.2	96	15,600
11	12	3.5	132	18,800
12	8	3.0	148	14,500
13	3	2.6	130	12,900
14	11	2.1	121	11,900
15	10	2.4	105	14,500

7. Find the multiple correlation coefficient and partial correlation coefficients for the data of Exercise 6, using a standard computer program. Interpret these results relative to the problem of studying the relationship between salary and the background variables and in the light of the results obtained in Exercise 6.

8. Study the distribution of the z* transformation for the correlation coefficient when $\rho \neq 0$ by obtaining information about its sampling distribution. Use a bivariate normal generator subroutine for a specified value of ρ (each member of the class could use a different value of ρ), using the same sample size. For each sample observed compute r and then z*. Print out frequency distribution of z* and determine its characteristics. Is the sampling distribution of z* distribution approximately normal for the particular sample size chosen?

9. Consider the variables

Y = freshman grade-point average (GPA)
X_1 = SAT verbal score
X_2 = SAT quantitative score
X_3 = high school class standing
X_4 = number of credits in mathematics and science

In an attempt to predict the freshman grade-point average based upon information available prior to college admission, the following data was obtained on students who have completed their freshman year. Use an available multiple regression program to obtain the appropriate regression equation. Use the regression equation to estimate the GPA of a student whose scores are x_1 = 450, x_2 = 600, x_3 = 26, and x_4 = 2.5. Give the 95 percent confidence interval for the estimate.

Y	X_1	X_2	X_3	X_4
1.81	300	375	60	2.4
2.60	600	410	41	2.5
2.35	510	380	52	3.0
3.77	700	800	11	4.0
2.91	410	630	28	2.5
2.31	600	425	45	2.0
2.98	550	415	22	3.5
2.41	380	610	30	2.5
3.51	620	530	15	4.0
2.28	480	530	44	2.0
2.44	380	410	63	2.5
3.05	600	520	30	3.0
3.89	650	680	9	4.0
2.02	505	380	38	2.0
3.40	705	630	25	2.5
2.15	605	310	61	2.0

10. A study is made to determine if any candidates for Congress have excessive campaign expenditures. Information about 20 candidates is collected. It is felt that the expenditures depend upon the following variables:

Y = total campaign expenses

X_1 = population in district

X_2 = median family income

X_3 = number of years in Congress

X_4 = incumbant (1 = yes, 0 = no)

X_5 = political party (1 = democrat, 0 = republican)

X_6 = percentage of urbanization in the district

Use a multiple regression program to determine a regression equation that can be used to predict the total campaign expenses. Use this regression equation to select the two expenses that have the greatest deviation from the predicted regression line. Try the transformation $\sqrt{X_6}$ in place of X_6. Does the resulting regression equation provide a better fit?

Y	X_1	X_2	X_3	X_4	X_5	X_6
126	102	70	8	1	1	92
128	73	51	0	1	0	62
141	85	81	2	1	0	73
28	43	47	16	1	1	32
80	126	89	0	0	0	95
69	64	77	4	1	1	61
340	88	50	0	1	1	57
105	63	70	8	1	1	80
140	104	53	2	1	0	80
160	66	82	0	0	1	51
154	79	43	0	0	0	63
158	61	67	0	0	1	78
133	95	75	6	1	1	94
78	83	64	16	1	0	75
105	63	43	2	1	1	40
95	73	65	10	1	0	65
185	87	93	0	0	0	98
98	74	48	6	1	1	41
50	51	35	12	1	0	28
118	75	50	4	1	1	55

CHAPTER 14

AN INTRODUCTION TO EXPERIMENTAL DESIGN

14.1 Introduction

Whenever we encountered a random variable in the preceding chapters we tried to determine an appropriate mathematical model that describes the probability distribution of the random variable. Since each mathematical model has a set of parameters, we were interested in making inferences about these parameters. We saw that an estimator of the parameter as well as a test statistic that links the parameter and statistic were needed to make such inferences. We also needed the mathematical model for the test statistic.

We followed this approach in developing the sample mean $\bar{y}$ as an estimator of the population mean μ. We introduced the z test statistic, which has a normal distribution, and the t test statistic, which has a Student's t distribution. Although the mean is a very important parameter, the variance of a population is also of prime importance, so in this chapter we shall use the same approach to investigate the sampling distribution of the variance estimator and its use in making inferences about the population variance.

Up to now we considered only inferences about a single population. In this chapter we shall extend these considerations and introduce methods of testing how two or more populations may differ. Since we have already considered the problem of finding an appropriate distribution model for a population, we shall confine ourselves to making inferences about the equality of the parameter values of two or more normal distributions.

14.2 The Sampling Distribution of s'^2

An unbiased sample estimate of the population variance σ^2 is

$$s'^2 = \frac{\Sigma (y_i - \bar{y})^2}{n - 1} = \frac{\Sigma y_i^2 - (\Sigma y_i)^2/n}{n - 1}$$

The test statistic that links σ^2 and s'^2 is the chi square ratio, which is given by

$$\chi_d^2 = \frac{(n - 1)s'^2}{\sigma^2} = \frac{\Sigma (y_i - \bar{y})^2}{\sigma^2} \tag{14.1}$$

where $d = n - 1$ represents the degrees of freedom

If the distribution of the observed random variable Y is normal, the sampling distribution of the test statistic is given the by the chi-square distribution:

$$f(\chi^2; d) = \frac{1}{2^{d/2} \ (d/2)} \ (\chi^2)^{(d-2)/2} \ e^{-\chi^2/2} \qquad 0 \le \chi^2 \le \infty \tag{14.2}$$
$$d > 0$$

Since the χ^2 distribution has but one parameter, d, we can investigate how the value of d influences the shape of the model. To do this we can generate the ordinates of $f(\chi_d^2)$ for different values of d and plot the points; the results are shown in Fig. 14.1.

Figure 14.1 shows that when the degrees of freedom d = 10, the χ^2 distribution is essentially normal. In addition, it can be seen that the characteristics of the χ^2 distribution depend upon the value of d. In fact, $\mu = E(\chi^2) = d$ and $\sigma^2 = E(\chi^2 - d)^2 = 2 \cdot d$; $\alpha_3 = 4/\sqrt{2d}$ and $\alpha_4 = 3 + 12/d$. We could also generate the ordinates for the normal distribution with mean d and variance 2d and see how closely these normal ordinates coincide with the ordinates of the χ^2 distribution for increasing values of d. This is left as an exercise for the students.

The student should notice that the χ^2 distribution is

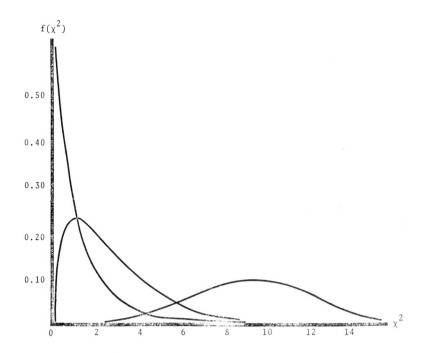

Fig. 14.1 χ^2 distribution for $d = 1$, 3, and 10
degrees of freedom.

asymmetrical, although the symmetry increases as the degrees
of freedom increase. Because of this lack of symmetry, the
critical values of the χ^2 distribution must be given for
both $\chi^2_{d,\alpha}$ and $\chi^2_{d,1-\alpha}$.

To investigate whether the χ^2 statistic has the χ^2 distribution we could generate samples from a normal distribu-
tion for a given μ and σ^2 and determine s'^2 and the χ ratio
for each sample. We can than tabulate the resulting χ^2
values into a frequency distribution and superimpose the
appropriate χ^2 distribution on the histogram for the generated frequency distribution.

14.3 The Use of the Chi-square Distribution

We shall use the χ^2 statistic and its distribution for
making inferences about σ^2. We shall again present each
type of inference separately.

1. Testing hypothesis

To demonstrate how to test a hypothesis about σ^2 we shall use an actual applied problem. A particular problem often encountered is that of determining the reliability of any measurement device or procedure. By reliability we mean the amount of variation that will occur if repeated measurements are made on the same experimental unit. The smaller the variation, the better the reliability of the measurement device or procedure. For example, the reliability question is encountered when measuring the pH value of blood specimens. Each morning the measurement instrument is checked to see if it is functioning properly by measuring the pH value for n specimens obtained from a standard homogeneous source. The variance of these readings is determined and compared with the allowable variation given by $\sigma_o^2 = 0.00013$ to determine whether the instrument is sufficiently reliable.

Let us suppose that on a particular day the sample variance obtained from n = 6 standard speciments is $s'^2 = 0.00019$. We then wish to know whether the instrument is functioning properly before processing the day's work. We test the hypothesis that this sample variance could have come from a population whose variance is σ_o^2. Thus the null hypothesis is

$$H_o : \sigma^2 = \sigma_o^2 = 0.00013$$

and in this case we are only interested in the one-sided alternative that

$$H_a : \sigma^2 > \sigma_o^2$$

To test the null hypothesis we compute the χ^2 ratio

$$\chi^2 = \frac{(n - 1)s'^2}{\sigma^2} = \frac{(5)(0.00019)}{0.00013} = 7.13$$

which has d = n - 1 = 5 degrees of freedom. We must now find the critical value of the χ^2 distribution. If the computed value is greater than the critical value, then the null hypothesis is rejected.

To find the critical value we can use the numerical integration subroutine or the tabulated values given in Appendix C-4. These tabulated values are given for different values of d and for selected values of α. If we wish to determine the value of χ_d^2 such that

$$\alpha = \Pr(\chi^2 < \chi_{d,\alpha}^2) = \int_0^{\chi_{d,\alpha}^2} f(\chi^2,d) \, d\chi^2$$

we look for the appropriate row having d degrees of freedom and the column that has the particular value of α.

For our particular example we wish to use the 5 percent level of significance and thus need to determine $\chi_{d,1-\alpha}^2 = \chi_{5,.95}^2 = 11.070$. Since the calculated value is less than 11.070, we do not reject the null hypothesis, and we would continue to use the instrument for measuring the pH values.

We can, of course, have different alternative hypothesis. For example, when the two-sided alternative hypothesis $H_a : \sigma^2 \neq \sigma_o^2$, is used, then the null hypothesis is rejected only if the computed χ^2 value falls outside the interval $\chi_{d,\alpha/2}^2 < \chi^2 < \chi_{d,1-\alpha/2}^2$. If the alternative hypothesis is $H_a : \sigma^2 < \sigma_o^2$, then the critical value is given by $\chi_{d,\alpha}^2$, and the null hypothesis is rejected if the computed value is less than the critical value.

2. Confidence intervals

To find the $(1 - \alpha)$ percent confidence interval for σ^2, we use the χ^2 distribution to determine the values of $\chi_{d,\alpha/2}^2$ and $\chi_{d,1-\alpha/2}^2$ so that

$$\Pr\left[\chi_{d,\alpha/2}^2 \leq \frac{(n-1)s'}{\sigma^2} \leq \chi_{d,1-\alpha/2}^2\right] = 1 - \alpha$$

By solving the inequality for σ^2, the required $(1 - \alpha)$percent confidence interval is found to be

$$\left[\frac{(n-1)s'^2}{\chi^2_{d,1-\alpha/2}} \le \sigma^2 \le \frac{(n-1)s'^2}{\chi^2_{d,\alpha/2}} \right]$$

To illustrate the approach we shall use the six measurements obtained on the particular day to get a 99 percent confidence interval for σ^2. Using the tabled values for the σ^2 distribution, we find that $\chi^2_{5,.005} = 0.412$ and $\chi^2_{5,.995} = 16.750$; thus the confidence interval is

$$\left[\frac{(5)(0.00019)}{16.75} \le \sigma^2 \le \frac{(5)(0.00019)}{0.412} \right]$$

or

$$[0.000057 \le \sigma^2 \le 0.002300]$$

The confidence interval for the population standard deviation can be found by taking the square root of the above confidence interval. Thus in our example we have

$$[0.0075 \le \sigma \le 0.0480]$$

3. Sample size

We are interested in estimating σ^2 within a prescribed accuracy ε with $(1-\alpha)$ percent confidence, but it should be recognized that the χ^2 statistic involves the sample size both explicitly through the $(n-1)$ factor and implicitly through the degrees of freedom as seen by

$$\left[\frac{(n-1)}{\chi^2_{d,(1-\alpha)/2}} \le \frac{\sigma^2}{s'^2} \le \frac{(n-1)}{\chi^2_{d,\alpha/2}} \right] \qquad (14.3)$$

Since we have a ratio relationship between σ^2 and s'^2, it is better to estimate s'^2 within a certain percentage of σ^2, say, ε. Thus we want s'^2 to fall within the interval

$$[\sigma^2 - \varepsilon \cdot \sigma^2 \le s'^2 \le \sigma^2 + \varepsilon \cdot \sigma^2] \qquad (14.4)$$

464

By inverting inequality (14.4) and multiplying by σ^2, we obtain

$$\left[\frac{1}{1+\varepsilon} \leq \frac{\sigma^2}{s'^2} \leq \frac{1}{1-\varepsilon}\right] \qquad (14.5)$$

We would like to find the value of n such that the left-hand side of inequality (14.3) equals $1/(1+\varepsilon)$ and the right-hand side equals $1/(1-\varepsilon)$. Usually it is not possible to find a single value of n that meets both sides of inequality (14.3). Thus each side is generally solved separately for n, and the larger of the two values of n is picked as the final value.

We shall show how n can be found to estimate σ^2 within an error bound of 20 percent with 90 percent confidence. Consider first the equation obtained using the left-hand side constraint of inequality (14.3), which is

$$\frac{n-1}{\chi^2_{d,1-\alpha/2}} = \frac{1}{1+\varepsilon} = 0.833$$

To determine the value of n that satisfies this equation, we must again use an iterative approach. Starting with an initial guess of n = 20, we obtain

$$\frac{(n-1)}{\chi^2_{19,.95}} = \frac{19}{30.1} = 0.631$$

which is less than the required 0.8333. Thus we shall need a larger value of n. We set up a table to record the successive trials as we attempt to find the correct value for n.

n	25	100	200	180	160	150	145
$\chi^2_{d,1-\alpha/2}$	36.4	123.1	232.8	211.1	189.3	178.4	172.9
$\dfrac{(n-1)}{\chi^2_{d,1-\alpha/2}}$	0.659	0.804	0.855	0.848	0.839	0.835	0.833

Thus a sample size of n = 145 will satisfy the first equation. We must now determine the value of n that satisfies the second equation, given by

$$\frac{n - 1}{\chi^2_{d,\alpha/2}} = \frac{1}{1 - \epsilon} = 1.250$$

and then pick the larger of the two values.

4. The decision problem

We now want to develop a decision rule that indicates when the pH measurement instrument should be readjusted before making any further readings. In determining this decision rule we need to consider the costs involved in servicing the equipment because in some cases it may be better to continue using the instrument even though its reliability has slightly decreased. The formulation of the decision rule is best accomplished by proceeding through our standard decision theory steps.

Step 1

Actions:

a_1: use the instrument
a_2: have the instrument serviced

Step 2

Possible outcomes of the random variable. The experiment consists of making readings on n samples from a standard specimen and determining the sample variance s'^2. The range of s'^2 is $0 \leq s'^2 < \infty$. We assume that the readings on the instrument come from a population that has a normal distribution $N(\mu, \sigma^2)$, and thus our parameter of interest is σ^2.

Step 3

The states of nature. Each state of nature represents a particular level of reliability of the instrument and thus can be identified by a value of σ^2. We shall consider only a finite number of possible states of nature by taking representative values of σ^2. We thus have

$$\theta_1: \sigma_1^2 = 0.00005 \qquad \theta_4: \sigma_4^2 = 0.00035$$
$$\theta_2: \sigma_2^2 = 0.00015 \qquad \theta_5: \sigma_5^2 = 0.00045$$
$$\theta_3: \sigma_3^2 = 0.00025 \qquad \theta_6: \sigma_6^2 = 0.00055$$

Step 4

Probabilities of occurrence. Under the assumption that the distribution of the readings is normal, we can find the probabilities of occurrence of s'^2 by using the χ^2 distribution since

$$Pr\left[s_j'^2 < \sigma_k^2\right] = Pr\left[\frac{(n-1)s_j'^2}{\sigma_k^2} < \chi_d^2\right]$$

Step 5

Regret functions. Since the parameter σ^2 used to represent the states of nature has an infinite range, we shall use the graphical approach to represent the regret function for each of the two actions. Representative regret functions are used to illustrate the technique.

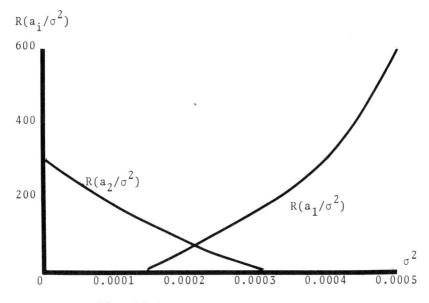

Fig. 14.2 Regret functions.

467

For the six states of nature, we can then read off the regrets; for example,

$$r_{11} = 0 \qquad r_{12} = 200$$
$$r_{21} = 0 \qquad r_{22} = 100$$

These regrets are given in the following list for each state of nature and action:

| | Actions | |
States of Nature	a_1	a_2
$\theta_1:\ \sigma_1^2 = 0.00005$	0	200
$\theta_2:\ \sigma_2^2 = 0.00015$	0	100
$\theta_3:\ \sigma_3^2 = 0.00025$	80	50
$\theta_4:\ \sigma_4^2 = 0.00035$	175	0
$\theta_5:\ \sigma_5^2 = 0.00045$	400	0
$\theta_6:\ \sigma_6^2 = 0.00055$	750	0

Step 6

The strategies. Since there are only two possible actions, a single value b_1 can be used to designate an entire class of strategies, which is

$$\text{If } s'^2 \le b_1, \text{ take action } a_1$$
$$\text{If } s'^2 > b_1, \text{ take action } a_2$$

In this presentation we shall consider only the following three strategies:

Strategy s_1: if $s'^2 > 0.0002$, take action a_2; otherwise take action a_1

Strategy s_2: if $s'^2 > 0.0003$, take action a_2; otherwise take action a_1

Strategy s_3:   if $s'^2 > 0.0004$, take action a_2;

otherwise take action a_1

Step 7

Determination of action probabilities. We can find the action probabilities directly because for each strategy we can determine

$$g_{k2}(m) = \Pr[a_2/\sigma_k^2] = \Pr\left[\chi_d^2 > \frac{(d)s'^2}{\sigma_k^2}\right]$$

and

$$g_{k1}(m) = \Pr[a_1/\sigma_k^2] = 1 - \Pr[a_2/\sigma_k^2]$$

If we have $n = 6$ readings to test the reliability of the instrument, then the action probabilities for strategy s_1 (take a_2 if $s'^2 > 0.0002$) for the state of nature θ_1 are

$$g_{12}(1) = \Pr[a_2/\sigma_1^2] = \Pr\left[\chi_5^2 > \frac{5(0.0002)}{0.00005}\right]$$

$$= \Pr[\chi_5^2 > 20] = 0.00125$$

$$g_{11}(1) = \Pr[a_1/\sigma_1^2] = 1 - \Pr[\chi_5^2 > 20] = 0.99875$$

For strategy s_2 (take a_2 if $s'^2 > 0.0003$), the action probabilities for each state of nature θ_k can be found by evaluating

$$g_{k2}(2) = \Pr[a_2/\sigma_k^2] = \Pr\left[\chi_5^2 > \frac{5(0.0003)}{\sigma_k^2}\right]$$

These action probabilities for the three strategies are

State of Nature	Strategy 1		Strategy 2		Strategy 3	
	a_1	a_2	a_1	a_2	a_1	a_2
$\theta_1 \; \sigma_1^2 = 0.00005$	0.00125	0.99875	0.00001	0.99999	0	1.0000
$\theta_2 : \sigma_2^2 = 0.00015$	0.24663	0.75337	0.07524	0.92476	0.02045	0.97955
$\theta_3 : \sigma_3^2 = 0.00025$	0.54941	0.45058	0.30622	0.69378	0.15624	0.84376
$\theta_4 : \sigma_4^2 = 0.00035$	0.72200	0.27800	0.50906	0.49094	0.33522	0.66478
$\theta_5 : \sigma_5^2 = 0.00045$	0.81762	0.12838	0.64838	0.35162	0.48735	0.51265
$\theta_6 : \sigma_6^2 = 0.00055$	0.87369	0.12631	0.74194	0.25806	0.60286	0.39714

Step 8

Determination of average regret. To determine the average regret, we evaluate

$$\bar{r}_k(m) = g_{k1}(m) \cdot r_{k1} + g_{k2}(m) \cdot r_{k2}$$

Thus

$$\bar{r}_1(1) = 0.00125(0) + 0.99875(200) = 199.75$$

By evaluating the average regret for each strategy and each state of nature, we obtain the following values:

State of Nature	Strategies		
	s_1	s_2	s_3
θ_1	199.75	200.00	200.00
θ_2	75.34	92.48	97.96
θ_3	66.48	54.68	54.69
θ_4	126.35	89.09	58.66
θ_5	327.05	259.35	194.94
θ_6	655.27	556.46	452.15

Step 9

Selection of minimum risk strategy. To find the risk for each strategy we need to have the prior probabilities.

Since we are presenting inputs that can be used for any value of the parameter σ^2, we can represent our prior probability information graphically, as shown in Fig. 14.3.

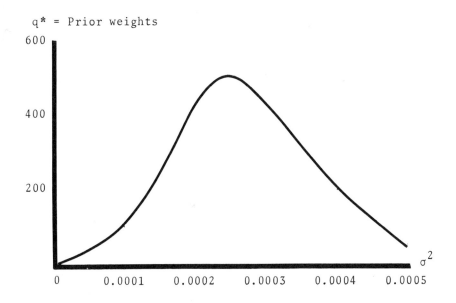

q* = Prior weights

Fig. 14.3 Prior weights for σ^2.

Because the curve shows only prior weights, and we must have probabilities that sum to 1, we must change the weights read from the curve into probabilities. For this purpose we read the q* values for each of the six states of nature and obtain

$$q_1^* = 50 \qquad q_2^* = 300 \qquad q_3^* = 500$$

$$q_4^* = 350 \qquad q_5^* = 150 \qquad q_6^* = 50$$

We determine $\Sigma q_i^* = 1400$, and thus find the required prior probabilities to be

$$q_1 = \frac{50}{1400} = 0.03571 \quad q_2 = \frac{300}{1400} = 0.21429 \quad q_3 = \frac{500}{1400} = 0.35742$$

$$q_4 = \frac{350}{1400} = 0.25000 \quad q_5 = \frac{150}{1400} = 0.10714 \quad q_6 = \frac{50}{1400} = 0.03571$$

Using these prior probabilities, we can now find the risk $R(m)$ for each strategy by evaluating

$$R(1) = \sum_{k=1}^{6} q_k \bar{r}_k(1) = 137.07$$

$$R(2) = \sum_{k=1}^{6} q_k \bar{r}_k(2) = 116.43$$

$$R(3) = \sum_{k=1}^{6} q_k \bar{r}_k(3) = 99.38$$

According to the minimum risk criterion, strategy s_3 should be used; it specifies that the instrument should be serviced if the observations yield a value of $s'^2 > 0.0004$. We note that the instrument would continue to be used even when the estimate of the reliability is greater than the standard of $\sigma^2 = 0.00016$ given earlier. This is brought about by the regrets associated with the problem as well as the prior prababilities.

We could further refine the decision rule by considering a set of strategies having a b_1 value closer to 0.0004 since some of these strategies might have a smaller risk. Of course, another refinement could be achieved by introducing additional states of nature.

14.4 Inferences About Two Normal Distributions

Two populations, both having normal distribution models, can differ only as to the value of their two parameters: μ and σ^2. The logical sequence in comparing the parameter values of the two normal distributions is to first test whether the variances are equal. The outcome of this test determines the appropriate test to use to check for the equality of the means.

Given a sample from each of the two populations of size n_1 and n_2, respectively, we can obtain the sample variances $s_1'^2$ and $s_2'^2$. We would like to use these two estimates to test the null hypothesis

$$H_o : \sigma_1^2 = \sigma_2^2$$

against any of the three possible alternatives: $\sigma_1^2 < \sigma_2^2$, $\sigma_1^2 > \sigma_2^2$, or $\sigma_1^2 \neq \sigma_2^2$. The test statistic used to test the null hypothesis is

$$F = \frac{s_1'^2}{s_2'^2} \qquad (14.6)$$

and its sampling distribution under the null hypothesis is

$$f(F; d_1, d_2) = \frac{\Gamma\left(\frac{d_1 + d_2}{2}\right)\left(\frac{d_1}{d_2}\right)^{d_1/d_2} F^{(d_1-2)/2}}{\Gamma\left(\frac{d_1}{2}\right)\Gamma\left(\frac{d_1}{2}\right)\left(1 + \frac{d_1 F}{d_2}\right)^{(d_1+d_2)/2}} \qquad (14.7)$$

$$F > 0 \qquad d_1, d_2 > 0$$

We recognize that the distribution of F has two parameters, d_1 and d_2, which in the two population case are given by $d_1 = n_1 - 1$ and $d_2 = n_2 - 1$. Since the F ratio given in Eq. (14.6) does not explicity involve d_1 or d_2, the question might be raised as to why they enter as parameters of the distribution function. We may more readily appreciate why d_1 and d_2 are parameters by looking at the more general form of the F ratio:

$$F = \frac{\chi_{d_1}^2 / d_1}{\chi_{d_2}^2 / d_2} \qquad (14.8)$$

which is the ratio of two independent χ^2's divided by their respective degrees of freedom. In this case the parameters d_1 and d_2 are explicitly involved. We can obtain the special form of F given by (14.6) by using the fact that when $H_o: \sigma_1^2 = \sigma_2^2$ is true, the F ratio reduces to

$$ F = \frac{\chi_{d_1}^2 / d_1}{\chi_{d_2}^2 / d_2} = \frac{\dfrac{(n_1 - 1)s_1'^2 / \sigma_1^2}{n_1 - 1}}{\dfrac{(n_2 - 1)s_2'^2 / \sigma_2^2}{n_2 - 1}} = \frac{s_1'^2}{s_2'^2} $$

We can again use the computer to study the shape of the distribution and the effect of the parameters d_1 and d_2 by plotting the ordinates of the model for different parameter values. In addition, we can study whether the ratio $s_1'^2 / s_2'^2$ does have the F distribution when $\sigma_1^2 = \sigma_2^2$ by writing a program that generates an empirical sampling distribution of the ratio.

Having introduced the F ratio and its distribution, let us illustrate its use in the test of hypothesis by using an actual problem. Suppose that the director of the laboratory is interested in determining whether his two laboratory technicians produce comparable readings on the pH measurement instrument. The director already knows that the readings obtained by each technician are normally distributed, and thus the comparability of the results obtained by the two technicians can be determined from their mean μ_1 and μ_2 and their variances σ_1^2 and σ_2^2. If $\mu_1 \neq \mu_2$, the director should appreciate that some technican bias exists; if $\sigma_1^2 \neq \sigma_2^2$, one technician is less reliable than the other.

To test for reliability and bias the director provides each technician with six specimens from a standard blood specimen and asks each to obtain a reading for each specimen. The mean and variance of the six readings for each technician are calculated and given below:

Technician #1	Technician #2
$n_1 = 6, \; d_1 = 5$	$n_2 = 6, \; d_2 = 5$
$\bar{y}_1 = 7.172$	$\bar{y}_2 = 7.153$
$s_1'^2 = 0.00024$	$s_2'^2 = 0.00013$

We first test to see if the two technicians are equally reliable; thus

$$H_o : \sigma_1^2 = \sigma_2^2$$

and

$$H_a : \sigma_1^2 \neq \sigma_2^2$$

The alternative hypothesis could also be the one-sided hypothesis $H_a : \sigma_1^2 > \sigma_2^2$ or $H_a : \sigma_1^2 < \sigma_2^2$. In all cases, however, the F ratio is determined by

$$F = \frac{\text{larger } s'^2}{\text{smaller } s'^2}$$

In our case we have

$$F = \frac{s_1'^2}{s_2'^2} = \frac{0.00024}{0.00013} = 1.85$$

The null hypothesis is then rejected at the α percent level of significance if the computed F value is greater than the appropriate critical F value. The appropriate critical F value for a two-sided test is $F_{d_1, d_2, 1-\alpha/2}$.

To find the critical value we use tabled values given in Appendix C-5. These tables give the F values for selected values of d_1 and d_2 that satisfy the following equation:

$$\alpha = \Pr \left[F < F_{d_1, d_2, \alpha} \right] = \int_0^{F_{d_1, d_2, \alpha}} f(F, d_1, d_2) \; dF$$

In our example, the critical F value for the 5 percent level of significance is $F_{5,5,.95} = 5.05$. Since the computed value of F is 1.85, which is less than 5.05, we do not reject the null hypothesis.

Since the director is now willing to assume that $\sigma_1^2 = \sigma_2^2 = \sigma^2$, his only concern is whether there is any bias in the readings of either technician. Because the director does not know the true value of μ for the standard specimen, he cannot test to see if the average reading for each technician is equal to μ. Thus he can only test to see if the difference $\mu_1 - \mu_2 = \delta_o$ is equal to zero.

To make inferences about δ_o, one uses the estimator

$$d_o = \bar{y}_1 - \bar{y}_2$$

and the director needs a test statistic that relates d_o with δ_o. The appropriate statistic (when it is assumed that $\sigma_1^2 = \sigma_2^2$) is

$$t = \frac{d_o - \delta_o}{\sqrt{s_p^2\left(\frac{1}{n_1} + \frac{1}{n_2}\right)}} \qquad (14.9)$$

where

$$s_p^2 = \frac{(n_1 - 1)s_1'^2 + (n_2 - 1)s_2'^2}{n_1 + n_2 - 2}$$

The t statistic has a Student's t distribution with $n_1 + n_2 - 2$ degrees of freedom. The "pooled variance estimate" s_p^2 in the denominator of the t statistic provides a single estimate of the common variance σ^2. We can now test the null hypothesis

$$H_o: \mu_1 - \mu_2 = \delta_o$$

against any of the standard three alternatives by comparing the computed t with the appropriate critical value obtained from the t table in Appendix C-3.

In our case, the director is interested in testing

$$H_o: \mu_1 - \mu_2 = \delta_o = 0$$

$$H_a: \mu_1 \neq \mu_2 \quad (\delta_o \neq 0)$$

To make this test we compute

$$t = \frac{7.192 - 7.153}{\sqrt{\left[\frac{5(0.00024) + 5(00013)}{5 + 5 - 2}\right]\left(\frac{1}{5} + \frac{1}{5}\right)}} = \frac{0.039}{\sqrt{0.0000925}}$$

$$= \frac{0.039}{0.0096} = 4.055$$

Since the critical value for t is $t_{8,.975} = 2.306$ (using the 5 percent level of significance), we reject the null hypothesis and thus conclude that μ_1 is different from μ_2.

If the common variance hypothesis had been rejected, we could not use the pooled t ratio but could use the ratio given by

$$t = \frac{d_o - \delta_o}{\sqrt{\frac{s'^2_1}{n_1} + \frac{s'^2_2}{n_2}}} \tag{14.10}$$

This statistic has an approximate Student's t distribution. The degrees of freedom must be found by evaluating the following expression:

$$d = \frac{s'^2_1/n_1 + s'^2_2/n_2}{\left(s'^2_1/n_1\right)^2/n_1 + \left(s'^2_2/n_2\right)^2/n_2} \tag{14.11}$$

When working with large samples, it can be assumed that $\sigma_1^2 \doteq s_1'^2$ and $\sigma_2^2 \doteq s_2'^2$; thus we can use the z statistic given by

$$z = \frac{d_o - \delta_o}{\sqrt{\sigma_1^2/n_1 + \sigma_2^2/n_2}} \qquad (14.12)$$

This statistic has a N(0,1) distribution, and the appropriate critical values of z can be obtained from the normal tables given in Appendix C-2.

There is still another way that the director could test for the bias. Instead of giving each technician a separate set of samples from the standard specimen, each technician is asked to make his readings, using the same set of samples. If Y represents the readings obtained by the first technician and X represents the readings obtained by the second technician, then y_i and x_i are related since they represent readings made on the ith sample. Neither the pooled t test nor the z test given above can be used because the y_i's and x_i's obtained are not independent. We can, however, consider the differences of the paired observations and use these differences to test for the bias.

Let us demonstrate this approach using a numerical illustration. If n = 6, the data obtained by the director using the paired observation approach are

Sample No. i	Technician # 1 Y	Technician # 2 X	Difference D
1	7.148	7.122	0.026
2	7.201	7.186	0.015
3	7.177	7.182	-0.005
4	7.153	7.144	0.009
5	7.181	7.153	0.028
6	7.166	7.150	0.016

Using this data, we obtain

$$\bar{d} = \frac{\Sigma d_i}{n} = \frac{0.099}{6} = 0.0165$$

$$s_d'^2 = \frac{\Sigma(d_i - \bar{d})^2}{n - 1} = \frac{\Sigma d_i - (\Sigma d_i)^2/n}{n - 1}$$

$$= \frac{0.002047 - (0.099)^2/6}{5} = \frac{0.0004135}{5} = 0.0000827$$

$$s_d' = \sqrt{0.0000827} = 0.00909$$

To test for bias we must have a parameter, its estimator, and a test statistic. In this case we assume that $d \sim N(\mu_d, \sigma_d)$ and thus consider the parameter μ_d and its estimator $\bar{d}$. The test statistic is

$$t = \frac{\bar{d} - \mu_d}{s_d'/\sqrt{n}}$$

which has the Student's t distribution with $n - 1$ degrees of freedom.

We can therefore test for bias as follows:

$$H_o: \mu_d = 0$$

$$H_a: \mu_d \neq 0$$

and

$$t = \frac{0.0165 - 0}{0.00909/\sqrt{6}} = \frac{0.0165}{0.0037} = 4.45$$

If 4.45 is greater than $t_{5, .975}$, we reject the null hypothesis at the 5 percent level of significance.

The use of pairing in an experimental design is recommended if there is a large variation between the experimental units. However, in using the paired t test, the degrees of freedom are reduced from $2n - 2$ to $n - 1$. Thus, if the experimental units are homogeneous, separate experimental units should be used, and the analysis should be performed using the pooled t test.

14.5 One-way Classification Designs

We often need to make inferences about the parameter values of more than two populations. For example, there may be more than two technicians who are making pH determinations, and we would like to check the comparability of their results. Let us assume that there are now four technicians and illustrate how we might compare them.

Each of the four technicians is given six separate samples from the standard specimen and obtains a reading for each sample. Let y_{ik} denote the reading obtained by the ith technician on his kth sample. The data obtained for this experimental design can then be represented by Fig. 14.4:

Technicians	Samples					
	1	2	3	4	5	6
T_1	$y_{11} = 7.174$	$y_{12} = 7.164$	$y_{13} = 7.162$	$y_{14} = 7.190$	$y_{15} = 7.172$	$y_{16} = 7.164$
T_2	$y_{21} = 7.187$	$y_{22} = 7.184$	$y_{23} = 7.163$	$y_{24} = 7.193$	$y_{25} = 7.186$	$y_{26} = 7.207$
T_3	$y_{31} = 7.173$	$y_{32} = 7.172$	$y_{33} = 7.158$	$y_{34} = 7.190$	$y_{35} = 7.151$	$y_{36} = 7.158$
T_4	$y_{41} = 7.161$	$y_{42} = 7.160$	$y_{43} = 7.126$	$y_{44} = 7.167$	$y_{45} = 7.161$	$y_{46} = 7.121$

Fig. 14.4 One-way classification experimental data.

We let n_i represent the number of readings made by the ith technician and a represent the number of technicians. Thus in our example we have $n_i = 6$ and $a = 4$.

If we assume that the distribution of the readings for each technician is given by $N(\mu_i, \sigma_i)$, we than want to compare μ_1, μ_2, μ_3, and μ_4 as well as σ_1^2, σ_2^2, σ_3^2, and σ_4^2. Paralleling the approach used in the two technician problem, we first test whether the distributions have a common variance σ^2 using the null hypothesis

$$H_o: \sigma_1^2 = \sigma_2^2 = \sigma_3^2 = \sigma_4^2 = \sigma^2$$

We can summarize these data for each technician separately to obtain

Technician	n_i	$\sum\limits_{k=1}^{n_i} y_{ik}$	$\bar{y}_i = \dfrac{\sum\limits_{k=1}^{n_i} y_{ik}}{n_i}$	$s_i'^2 = \dfrac{\sum\limits_{k=1}^{n_i} (y_{ik} - \bar{y}_i)^2}{n_i - 1}$
T_1	6	43.026	7.171	$\dfrac{0.00055}{5} = 0.00011$
T_2	6	43.120	7.187	$\dfrac{0.00102}{5} = 0.00020$
T_3	6	43.002	7.167	$\dfrac{0.00101}{5} = 0.00020$
T_4	6	42.896	7.149	$\dfrac{0.00205}{5} = 0.00041$
Totals	$\sum\limits_{i=1}^{a} n_i$	$\sum\limits_{i=1}^{a}\sum\limits_{k=1}^{n_i} y_{ik}$	$\bar{y} = \dfrac{\sum\limits_{i=1}^{a} n_i \bar{y}_i}{\sum\limits_{i=1}^{a} n_i}$	
	24	172.044	7.169	

(Procedures such as Bartlett's test are available for making such a test, but we consider them beyond the scope of this introductory text.) Instead of testing the hypothesis of equal variances, we shall simply assume them to be equal and test whether the means are all equal. The null hypothesis is that all technicians have a common mean:

$$H_o: \mu_1 = \mu_2 = \mu_3 = \mu_4 = \mu_o$$

The alternative hypothesis is:

$$H_a: \text{ at least one } \mu_i \neq \mu_o$$

We need a test statistic that reflects this null hypothesis and considers the general F ratio, which can also be expressed as the ratio of variance estimates and their expected values:

$$F = \frac{\chi_1^2/d_1}{\chi_2^2/d_2} = \frac{s_1'^2/E\left(s_1'^2\right)}{s_2'^2/E\left(s_2'^2\right)} \tag{14.13}$$

To use this F ratio we must obtain two independent variance estimates that when used in the ratio reflect the null hypothesis.

One variance estimate can be obtained by determining the variance of the sample means, which is given by

$$s'^2_{\bar{y}_i} = \frac{\sum\limits_{i=1}^{a} n_i(\bar{y}_i - \bar{y})^2}{a - 1} \tag{14.14}$$

The expected value of $s'^2_{\bar{y}_i}$ is

$$E\left(s'^2_{\bar{y}_i}\right) = \sigma^2 + \frac{\sum\limits_{i=1}^{a} n_i(\mu_i - \mu_o)^2}{a - 1} \tag{14.15}$$

where μ_o is the weighted average of the population means. We note that when the null hypothesis is true this expected value reduces to σ^2, and thus $s'^2_{\bar{y}_i}$ is one estimate that can be used in the F ratio.

In addition, the variances for each technician s'^2_i can be used to obtain a pooled estimate of σ^2 (since it is assumed that $\sigma^2_1 = \sigma^2_2 = \sigma^2_3 = \sigma^2_4 = \sigma^2$). This pooled estimate is given by

$$s'^2_w = \frac{\sum\limits_{i=1}^{a} (n_i - 1)s'^2_i}{\sum\limits_{i=1}^{a} (n_i - 1)} \tag{14.16}$$

$$= \frac{(n_1 - 1)s'^2_1 + (n_2 - 1)s'^2_2 + \cdots + (n_a - 1)s'^2_a}{n_1 + n_2 + \cdots + n_a - a}$$

This is simply an extension of pooled estimate for two populations. The expected value of this pooled estimate is $E(s'^2_w) = \sigma^2$.

It can also be shown that the s'^2_w is independent of

$s_{\bar{y}}'^{2}$. Thus, we can form a valid F ratio of independent χ^2 variables

$$F = \frac{s_{\bar{y}}'^{2}/E(s_{\bar{y}}'^{2})}{s_{w}'^{2}/E(s_{w}'^{2})} = \frac{s_{\bar{y}}'^{2}/[\sigma^2 + \Sigma n_i(\mu_i - \mu_o)^2/(a-1)]}{s_{w}'^{2}/\sigma^2} \qquad (14.17)$$

Note that when the null hypothesis is true,

$$\frac{\Sigma n_i(\mu_i - \mu_o)^2}{a - 1} = 0$$

the F ratio reduces to the ratio of the two variance estimates

$$F = \frac{s_{\bar{y}}'^{2}}{s_{w}'^{2}} \qquad (14.18)$$

This F ratio has the F distribution with $d_1 = a - 1$ and $d_2 = \Sigma(n_i - 1)$ degrees of freedom. Since the numerator under the alternative hypothesis is expected to be greater than the denominator, we treat this test as if it were one-sided and thus use the critical value $F_{d_1,d_2,1-\alpha}$.

The null hypothesis can now be tested since a computed F which is significantly greater than 1 indicates that

$$\frac{\Sigma n_i(\mu_i - \mu_o)^2}{a - 1} \neq 0$$

and for this to be true at least one of the μ_i's must differ from μ_o.

In our particular example, we test the null hypothesis using the 1 percent level of significance by computing

$$s_{\bar{y}}'^{2} = \frac{0.004392}{3} = 0.001464$$

$$s_w'^2 = \frac{0.004630}{20} = 0.000231$$

so that

$$F = \frac{0.001464}{0.000231} = 6.38$$

and this computed F value is compared to $F_{3,20,.99} = 4.94$; since $6.38 > 4.94$, the null hypothesis is rejected, and there is a difference between the technicians.

A more formal approach to the problem of testing for the equality of means is to first develop a model that explains the relationship between the observations and the model parameters similar to the approach we used in developing the regression model.

One possible model for the one-way classification experimental design is that each observation Y_{ik} is determined by the following relationship

$$Y_{ik} = \mu + \alpha_i + \varepsilon_{ik} \qquad \begin{array}{l} i = 1, \ldots, a \\ k = 1, \ldots, n_i \end{array}$$

This model states that there is an overall mean μ about which the mean for each technician given by μ_i varies. Thus the μ_i can be represented by $\mu + \alpha_i$, and it follows that

$$\sum_{i=1}^{a} \alpha_i = 0$$

We also assume that the ε_{ik}'s have a normal distribution with mean 0 and variance σ_ε^2; this implies that

$$Y_{ik} \sim N(\mu + \alpha_i, \sigma_\varepsilon)$$

The null hypothesis for this model is given by

$$H_o: \alpha_1 = \alpha_2 \overset{\cdots}{=} \alpha_a = 0$$

and the alternative hypothesis is

$$H_a: \text{at least one } \alpha_i \neq 0$$

The required computations for this design are given by the following algorithm.

(1) Find C, the correction term:

$$C = \frac{\left(\sum\limits_{i=1}^{a} \sum\limits_{j=1}^{n_i} y_{ij} \right)^2}{n} = \frac{(172.044)^2}{24} = 1233.297414$$

where n is total number of observations $n = \sum\limits_{i=1}^{a} n_i = 24$.

(2) Find SST, the total corrected sums of squares:

$$SST = \sum\limits_{i=1}^{a} \sum\limits_{j=1}^{n_i} y_{ij}^2 - C = 1233.306274 - 1233.297414$$

$$= 0.008860$$

(3) Find SSA, the sums of squares due to technician differences:

$$SSA = \sum\limits_{i=1}^{a} \frac{\left(\sum\limits_{j=1}^{n_i} y_{ij} \right)^2}{n_i} - C$$

$$= \frac{(43.026)^2}{6} + \frac{(43.120)^2}{6} + \frac{(43.002)^2}{6} + \frac{(42.896)^2}{6}$$

$$- 1233.297414$$

$$= 1233.301649 - 1233.297414 = 0.004235$$

(4) Find SSE, the sums of squares due to error:

$$SSE = SST - SSA = 0.008860 - 0.004235 = 0.004625$$

(5) Summarize the results in an analysis of variance table:

Source	Degrees of Freedom	Sums of Squares	Mean Square	F Ratio
Technician differences	$a - 1 = 3$	$SSA = 0.004235$	$MSA = 0.001411$	$F = \dfrac{0.001411}{0.000231}$ $= 6.10$
Error	$\sum_{i=1}^{a} (n_i - 1) = 20$	$SSE = 0.004625$	$MSE = 0.000231$	
Total	$n - 1 = 23$	$SST = 0.008860$		

The mean squares are the variance estimates ($MSA = s_y'^2$ and $MSE = s_w'^2$) and are obtained by dividing the sums of squares by the appropriate degrees of freedom. Note that there is a small difference between these results and the previous ones due to round-off error. Notice that we have partitioned the total sums of squares of deviations SST into

$$SST = SSA + SSE$$

where SSA represents the variation due to the differences between the technicians and SSE represents the variation of the readings due to the differences between the readings made by the same technician. Since these differences represent factors that are not controlled by the experimental design, they are called the experimental error.

(6) Determine the critical F value. This critical value is given by $F_{a-1, \Sigma(n_i-1), 1-\alpha}$. In the example, the critical value is $F_{3, 20, .99} = 4.94$, using the 1 percent level of significance.

(7) If the computed F value is greater than the critical F value, the null hypothesis should be rejected. In the example, 6.10 > 4.94, and therefore we reject the null hypothesis.
Since the null hypothesis has been rejected, we know that at least one $\alpha_i \neq 0$, and we now wish to determine which technician(s) caused the null hypothesis to be rejected.

Since sampling variation would cause the sample means $\bar{y}_i$ to differ even when the μ_i's are equal, it cannot simply be concluded that because the sample means are different the populations means are also different.

There are many techniques for determining which population means are different, but we shall consider only the technique of least significant differences. The least significant difference (LSD) between any two sample means $\bar{y}_i$ and $\bar{y}_j$ is given by

$$LSD = t_{d,1-\alpha/2} \sqrt{s_w'^2\left(\frac{1}{n_i} + \frac{1}{n_j}\right)}$$

where $s_w'^2$ is the mean square error that has $d = \Sigma(n_i - 1)$ degrees of freedom, α has the same value that was used in the F test, and n_i and n_j are the number of observations used in determining the sample mean $\bar{y}_i$ and $\bar{y}_j$, respectively. If for any i and j, $|\bar{y}_i - \bar{y}_j| > LSD$, we conclude that there is a difference between μ_i ($\mu + \alpha_i$) and μ_j ($\mu + \alpha_j$).

To compare any pair of means in our illustrative example, we need only compute one LSD since the n_i's are all equal. Because $t_{20,.995} = 2.845$ and $s_w'^2 = 0.000321$, we obtain

$$LSD = (2.845) \sqrt{0.000231\left(\frac{1}{6} + \frac{1}{6}\right)} = 0.0245$$

Thus any two sample means that differ by more than 0.0245 are considered to have come from populations whose means are different. We can order the sample means

$$\bar{y}_4 = 7.149 \quad \bar{y}_3 = 7.167 \quad \bar{y}_1 = 7.171 \quad \bar{y}_2 = 7.187$$

and then simply compare the differences of these ordered means. We find that only $|\bar{y}_4 - \bar{y}_2| > LSD$, and therefore only μ_4 and μ_2 are different, while all the other means are not.

The experimental design considered in this section is called a one-way classification design since the homogeneous experimental units are randomly assigned to different classes (in this case, technicians), and it is assumed that the differences in the observations on the experimental units are due to the particular class in which the unit is placed and the experimental error. Such a one-way classification design is often used to study the effect of a single factor. In this case each class is a particular level of the factor that is chosen in advance of the experimentation. For example, the factor of interest may be the effect of temperature on the random variable, and the different levels chosen in advance might be 200°, 210°, 220°.

The necessary computations to find the analysis of variance table for a one-way classification design do not require that the observations for each class or level be equal, although the illustration given did have an equal number of observations for each class.

14.6 Two-way Classification Designs

In the previous section we considered how an experiment could be designed to test whether the different technicians obtained significantly different readings. The experimental design can be generalized to accommodate more than one factor. In this section we shall consider two-factor experimental designs, but the student should appreciate that there are many different types of designs other than the ones presented here.

The purpose of a two-way classification design is to study the simultaneous effect of two factors upon the random variable of interest. Let us again consider a particular problem. Suppose the personnel officer in an organization wished to study the reading time of individuals having different ages (factor A) and different educational backgrounds (factor B). He establishes the four levels of factor A he wishes to consider, which are

$$A_1: \text{ less than 30 years}$$
$$A_2: \text{ 30 to 40 years}$$
$$A_3: \text{ 40 to 50 years}$$
$$A_4: \text{ over 50 years}$$

In addition, he decides to use three levels of factor B, which are

B_1: high school diploma but no college
B_2: some college but no degree
B_3: college degree

The personnel officer will have to provide employees (the experimental units) having the appropriate ages and educational levels for each of the $3 \cdot 4 = 12$ combinations or cells. The random variable of interest is the reading time Y and must be obtained on each experimental unit in each cell. Let y_{ijk} denote the reading time for the kth individual having level i of factor A and level j of factor B. Let n_{ij} denote the number of experimental units in each cell. The simplest two-way classification design would be to use the same number of experimental units in each cell, say n $n_{ij} = r$, and the methods presented in this section depend upon this assumption.

The data obtained for such an experimental design is given in Fig. 14.5:

	B_1	B_2	B_3
A_1	y_{111} $\vdots$ y_{11r}	y_{121} $\vdots$ y_{12r}	y_{131} $\vdots$ y_{13r}
A_2	y_{211} $\vdots$ y_{21r}	y_{221} $\vdots$ y_{22r}	y_{231} $\vdots$ y_{23r}
A_3	y_{311} $\vdots$ y_{31r}	y_{321} $\vdots$ y_{32r}	y_{331} $\vdots$ y_{33r}
A_4	y_{411} $\vdots$ y_{41r}	y_{421} $\vdots$ y_{42r}	y_{431} $\vdots$ y_{43r}

Fig. 14.5 Two-way experimental design.

489

When two factors are used in an experimental design, the model should indicate the effect of each factor on the random variable Y. One such model is

$$Y_{ijk} = \mu + \alpha_i + \beta_j + \varepsilon_{ijk} \qquad \begin{array}{l} i = 1,2, \ldots, a \\ j = 1,2, \ldots, b \\ k = 1,2, \ldots, r \end{array} \qquad (14.19)$$

$$\sum_{i=1}^{a} \alpha_i = \sum_{j=1}^{b} \beta_j = 0$$

In this model μ represents the overall mean, α_i represents the additive effect due to the age factor, β_j represents the additive effect due to the educational factor, and ε_{ijk} is the experimental error for the kth individual in the (i,j)th cell.

The model just given ignores any interaction between the two factors. We say that there is an interaction between two factors when a particular combination of A_i and B_j causes the observed values y_{ijk} to be higher or lower than would be expected by considering the two factors independently. We extend the above model to include the interaction effect and obtain

$$Y_{ijk} = \mu + \alpha_i + \beta_j + (\alpha\beta)_{ij} + \varepsilon_{ijk}, \begin{array}{l} i = 1,2, \ldots, a \\ j = 1,2, \ldots, b \\ k = 1,2, \ldots, r \end{array} \quad (14.20)$$

where $(\alpha\beta)_{ij}$ is the additional effect due to the ith level of A and the jth level of B. We assume that

$$\sum_{i=1}^{a} (\alpha\beta)_{ij} = \sum_{j=1}^{b} (\alpha\beta)_{ij} = 0$$

for all i and j. To determine whether there is a significant difference between the levels of each factor, we wish to test the following null hypotheses:

$H_o: \alpha_i = 0$ for all i vs. H_a: at least one $\alpha_i \neq 0$

$H_o: \beta_j = 0$ for all j vs. H_a: at least one $\beta_j \neq 0$

and to see whether there is a significant interaction between the two factors, we test

$$H_o: (\alpha\beta)_{ij} = 0 \text{ for all } i \text{ and } j$$

vs.

$$H_a: \text{at least one } (\alpha\beta)_{ij} \neq 0$$

To illustrate the procedure for analyzing a two-way classification design and testing the above hypothesis, we shall again use actual data. However, this time we shall let the computer generate the observations.

To generate the experimental data, we use the model

$$Y_{ijk} = \mu + \alpha_i + \beta_j + (\alpha\beta)_{ij} + \varepsilon_{ijk}$$

and specify the values of μ, α_i, β_j, and $(\alpha\beta)_{ij}$. The experimental error associated with each experimental unit must also be given, but by assuming that the errors are normally distributed with mean 0 and variance σ_ε^2, we need only specify σ_ε^2.

This information is sufficient to determine the distribution of the random variable Y. Since the errors have a normal distribution, the random variable also has a normal distribution. The mean and variance can be found by the standard procedure given below. The mean is given by

$$E(Y_{ijk}) = E[\mu + \alpha_i + \beta_j + (\alpha\beta)_{ij} + \varepsilon_{ijk}] \quad (14.21)$$

$$= \mu + \alpha_i + \beta_j + (\alpha\beta)_{ij} = \mu_{ij}$$

since $E(\varepsilon_{ijk}) = 0$, and the variance is given by

$$E(Y_{ijk} - \mu_{ij})^2 = E\Big([\mu + \alpha_i + \beta_j + (\alpha\beta)_{ij} + \varepsilon_{ijk}] \quad (14.22)$$

$$- [\mu + \alpha_i + \beta_j + (\alpha\beta)_{ij}]\Big)^2$$

$$= E(\varepsilon_{ijk})^2 = \sigma_\varepsilon^2$$

Thus the observations in the (i,j)th cell have a normal distribution with mean μ_{ij} and variance σ_ε^2

The parameter values used to generate the data are given in Fig. 14.6. The cell means determined by these parameter values are given in Fig. 14.6. For example,

$$\mu_{41} = \mu + \alpha_4 + \beta_1 + (\alpha\beta)_{41}$$

$$= 100 + 6 - 1 + 3 = 108$$

	B_1	B_2	B_3	
A_1	$\mu_{11} = 95$ $(\alpha\beta)_{11} = -1$	$\mu_{12} = 96$ $(\alpha\beta)_{12} = 0$	$\mu_{13} = 100$ $(\alpha\beta)_{13} = 1$	$\alpha_1 = -3$
A_2	$\mu_{21} = 95$ $(\alpha\beta)_{21} = -1$	$\mu_{22} = 96$ $(\alpha\beta)_{22} = 0$	$\mu_{23} = 100$ $(\alpha\beta)_{23} = 1$	$\alpha_2 = -3$
A_3	$\mu_{31} = 98$ $(\alpha\beta)_{31} = -1$	$\mu_{32} = 99$ $(\alpha\beta)_{32} = 0$	$\mu_{33} = 103$ $(\alpha\beta)_{33} = 1$	$\alpha_3 = 0$
A_4	$\mu_{41} = 108$ $(\alpha\beta)_{41} = 3$	$\mu_{42} = 105$ $(\alpha\beta)_{42} = 0$	$\mu_{43} = 105$ $(\alpha\beta)_{43} = -3$	$\alpha_4 = 6$
	$\beta_1 = -1$	$\beta_2 = -1$	$\beta_3 = 2$	$\mu = 100$ $\sigma_\varepsilon = 25$

Fig. 14.6 Parameter values for two-way design.

We can now generate the observations for each cell, using the random normal number generator given by Program 9.7 and the appropriate cell mean and variance. These observations, together with the totals and means, are given in Fig. 14.7, using the following notation. Let

n_{ij} = r = number of observations in (i,j)th cell

$n_i.$ = b · r = number of observations in ith level of A

$n._j$ = a · r = number of observations in jth level of B

n = n.. = a · b · r = total number of observations

In our example, n_{ij} = r = 4, $n_i.$ = 3 · 4, $n._j$ = 4 · 4 = 16, and n.. = 4 · 3 · 4 = 48. Note that we use a dot to indicate that a summation has been performed over the missing index. Similarly, we denote the observation totals for each cell, row, and column by T_{ij}, $T_i.$, and $T._j$, respectively, and the observation means by $\bar{y}_{ij}$, $\bar{y}_i.$, and $\bar{y}._j$.

To test the three null hypotheses we set up the analysis of variance table for the two-way classification design. The algorithm used to find the necessary sums of squares and mean squares is similar to the one used for the one-way classification design.

(1) Find the correction term C:

$$C = \frac{\left(\sum\limits_{i}^{a} \sum\limits_{j}^{b} \sum\limits_{k}^{r} y_{ijk} \right)^2}{n} = \frac{T_{..}^2}{48} = \frac{(4798)^2}{48} = 479600$$

(2) Find the corrected total sums of squares SST:

$$SST = \sum\limits_{i}^{a} \sum\limits_{j}^{b} \sum\limits_{k}^{r} y_{ijk}^2 - C$$

$$= 99^2 + 93^2 + 89^2 + 88^2 + 97^2 + \cdots + 103^2 - 479600$$

$$= 481476 - 479600 = 1876$$

	B_1	B_2	B_3	Total
A_1	99 \newline 93 \newline 89 \newline 88 \newline $T_{11} = 369$	99 \newline 94 \newline 94 \newline 100 \newline $T_{12} = 387$	93 \newline 97 \newline 99 \newline 86 \newline $T_{13} = 375$	$T_{1.} = 1131$ \newline $\bar{y}_{1.} = \dfrac{1131}{12} = 94.25$
A_2	97 \newline 98 \newline 102 \newline 104 \newline $T_{21} = 401$	82 \newline 92 \newline 99 \newline 102 \newline $T_{22} = 375$	98 \newline 108 \newline 101 \newline 111 \newline $T_{23} = 418$	$T_{2.} = 1194$ \newline $\bar{y}_{2.} = \dfrac{1194}{12} = 99.50$
A_3	96 \newline 107 \newline 95 \newline 100 \newline $T_{31} = 398$	99 \newline 95 \newline 106 \newline 100 \newline $T_{32} = 400$	103 \newline 112 \newline 105 \newline 104 \newline $T_{33} = 424$	$T_{3.} = 1222$ \newline $\bar{y}_{3.} = \dfrac{1222}{12} = 101.83$
A_4	107 \newline 107 \newline 102 \newline 105 \newline $T_{41} = 421$	105 \newline 104 \newline 102 \newline 105 \newline $T_{42} = 416$	104 \newline 107 \newline 100 \newline 103 \newline $T_{43} = 414$	$T_{4.} = 1251$ \newline $\bar{y}_{4.} = \dfrac{1251}{12} = 104.75$
Total	$T_{.1} = 1589$ \newline $\bar{y}_{.1} = \dfrac{1589}{16}$ \newline $= 99.31$	$T_{.2} = 1578$ \newline $\bar{y}_{.2} = \dfrac{1578}{16}$ \newline $= 98.63$	$T_{.3} = 1631$ \newline $\bar{y}_{.3} = \dfrac{1631}{16}$ \newline $= 101.93$	$T_{..} = 4798$ \newline $\bar{y}_{..} = \dfrac{4798}{48} = 99.96$

Fig. 14.7 Generated two-way classification experimental data.

(3) The corrected total sums of squares must now be partitioned into the sums of squares due to

    (a) age effect (factor A)
    (b) educational background effect (factor B)
    (c) interaction of factors A and B

(a) Find the sums of squares due to factor A (age effect) SSA:

$$SSA = \frac{\sum\limits_{i}^{a}\left(\sum\limits_{j}^{b}\sum\limits_{k}^{r} y_{ijk}\right)^2}{n_{i\cdot}} - C = \sum_{i=1}^{a} \frac{T_{i\cdot}^2}{b\cdot r} - C$$

$$= \frac{1}{4\cdot 3}\ (1131^2 + 1194^2 + 1222^2 + 1251^2) - 479600$$

$$= \frac{1}{12}\ (5763082) - 479600 = 656$$

(b) Find the sums of squares due to factor B, SSB:

$$SSB = \frac{\sum\limits_{j}^{b}\left(\sum\limits_{i}^{a}\sum\limits_{k}^{r} y_{ijk}\right)^2}{n_{\cdot j}} - C = \sum_{j=1}^{b} \frac{T_{\cdot j}^2}{a\cdot r} - C$$

$$= \frac{1}{4\cdot 4}\ (1589^2 + 1578^2 + 1631^2) - 479600$$

$$= \frac{1}{16}\ (7675166) - 479698 = 98$$

(c) Find the sums of squares due to cells SSST:

$$SSST = \frac{\sum\limits_{i}^{a}\sum\limits_{j}^{b}\left(\sum\limits_{k=1}^{r} y_{ijk}\right)^2}{r} - C = \sum_{i}^{a}\sum_{j}^{b} \frac{T_{ij}^2}{r} - C$$

$$= \frac{1}{4}\ (369^2 + 401^2 + \cdots + 414^2) - 479600$$

$$= \frac{1}{4}\ (1922578) - 479600 = 1044$$

(d) Find the sums of squares due to interaction SSAB:

$$SSAB = SSST - SSA - SSB = 1044 - 656 \quad - 98 = 290$$

(4) Find the sums of squares due to error SSE:

$$SSE = SST - SSST = 1876 - 1044 = 832$$

(5) Record degrees of freedom and sums of squares in analysis of variance table, and compute mean squares, which are the sums of squares divided by the degrees of freedom:

Source	Degrees of Freedom	Sums of Squares	Mean Squares
Factor A	$3 = a - 1$	656	219
Factor B	$2 = b - 1$	98	49
Interaction	$6 = (a - 1)(b - 1)$	290	48
Experimental error	$36 = ab(r - 1)$	832	23
Total	$47 = arb - 1$	1876	

(6) To test the null hypotheses, compute the F ratio and compare it with the critical F value, using $\alpha = 0.05$ level of significance:

Null Hypothesis	Computed F	Critical F	Conclusion
$H_o: \alpha_i = 0$	$\frac{219}{23} = 9.6$	$F_{3,36,.95} = 2.88$	Reject H_o
$H_o: \beta_j = 0$	$\frac{49}{23} = 2.1$	$F_{2,36,.95} = 3.28$	Do not reject H_o
$H_o: (\alpha\beta)_{ij} = 0$	$\frac{48}{23} = 2.1$	$F_{6,36,.95} = 2.38$	Do not reject H_o

We would therefore conclude that age does influence the reading speed of individuals but that educational background and the interaction of the two factors are not significant. Since we used nonzero values of α_i, β_j, and $(\alpha\beta)_{ij}$ in generating the data, our first conclusion is proper, but we failed to note the existence of the nonzero β's and $(\alpha\beta)_{ij}$'s. This type of error can always occur, particularly when the parameter values are close to zero and the sample size is small.

Since we rejected the first hypothesis, we should use the LSD test to determine which levels of each factor differ significantly from each other. The procedure is identical to the one given in the last section, so we shall not repeat it; instead, we shall use the observed data to estimate each parameter value and compare the estimate with the true value.

The estimate of μ is $\bar{y}.. = 99.96$ and this is close to the true value of 100. This closeness should be a little surprising since the standard deviation of the estimate is

$$s_{\bar{y}..} = \sqrt{\frac{\sigma_\epsilon^2}{n}} = \sqrt{\frac{25}{48}} = \sqrt{0.521} = 0.722$$

The estimate of the error variance is $s'^2 = 23.11$ and is relatively close to the true value of $\sigma_\epsilon^2 = 25$.

The estimates for the additive effect for each level of factor A are given by

$$a_1 = \hat{\alpha}_1 = \bar{y}_1. - \bar{y}.. = 94.25 - 99.96 = -5.71$$

$$a_2 = \hat{\alpha}_2 = \bar{y}_2. - \bar{y}.. = 99.50 - 99.96 = -0.46$$

$$a_3 = \hat{\alpha}_3 = \bar{y}_3. - \bar{y}.. = 101.83 - 99.96 = +1.87$$

$$a_4 = \hat{\alpha}_4 = \bar{y}_4. - \bar{y}.. = 104.25 - 99.96 = +4.29$$

These can be compared with the true values of $\alpha_1 = -3$, $\alpha_2 = -3$, $\alpha_3 = 0$, and $\alpha_4 = 6$. The size of the discrepancies can be explained by considering the standard deviation of these estimates, which is

$$\sigma_{a_i} = \frac{\sigma_\epsilon}{\sqrt{n_i.}} = \frac{5}{\sqrt{12}} = 1.44$$

If the hypothesis $H_o: \beta_j = 0$ had been discarded, we would also have estimated the β_j's which are given by

$$b_1 = \hat{\beta}_1 = \bar{y}_{.1} - \bar{y}.. = 99.31 - 99.96 = -0.65$$

$$b_2 = \hat{\beta}_2 = \bar{y}_{.2} - \bar{y}.. = 98.63 - 99.96 = -1.33$$

$$b_3 = \hat{\beta}_3 = \bar{y}_{.3} - \bar{y}.. = 101.93 - 99.96 = 1.97$$

as compared with the true values, which are $\beta_1 = -1$, $\beta_2 = -1$, and $\beta_3 = 2$. The standard deviation of these estimates is

$$\sigma_{b_j} = \frac{\sigma_\varepsilon}{\sqrt{n_{.j}}} = \frac{5}{\sqrt{16}} = 1.25$$

If the hypothesis of no interaction had been discarded, we would also want to estimate $(\alpha\beta)_{ij}$, which is given by

$$(ab)_{ij} = (\hat{\alpha\beta})_{ij} = \bar{y}_{ij} - \bar{y}.. - a_i - b_j$$

so that, for example,

$$(ab)_{11} = (\hat{\alpha\beta})_{11} = \frac{369}{4} - 99.96 + 5.71 + 0.65$$

$$= -1.35$$

The standard deviation of these estimates is

$$\sigma_{(ab)_{ij}} = \frac{\sigma_\varepsilon}{\sqrt{r}} = \frac{5}{\sqrt{4}} = 2.5$$

We were not able to pick up the influence of the $(\alpha\beta)_{ij}$'s because they are close to zero, and the standard deviation of the estimate is large relative to their true values.

In performing an experiment we control α, the probability of discarding the null hypothesis when it is true, by selecting the appropriate critical region. We also wish to design the experiment so that β, the probability of not dis-

carding the null hypothesis when it is false, is small.
This latter probability, however, is not easy to obtain for
a complex experimental design.

We can also use this computer simulation capability to
determine the power of the F test for different model parameter values. This power is defined to be the probability of
rejecting the null hypothesis when it is not true. In other
words, power is the relative frequency with which the F test
detects the nonzero values of the model parameters α_i and/or
β_j. The magnitude of nonzero values of α_i and β_j depend
upon the particular experimental problem and the factor be-
ing studied. Since experimentation is costly in both time
and money, it would be worthwhile if it could be determined
in advance whether a particular experimental design has a
high probability (power) of determining the particular differences of interest in the factor levels. The simulation
technique introduced above provides a tool that can be used
to make such feasibility studies before actually running the
experiments.

14.7 Computer Programs

In this chapter we introduced two new distributions,
the χ^2 and the F, and gave the relationship between the two.
If we are interested in studying these distributions empirically, we need to be able to generate random χ^2 and F numbers. In this section such computer routines are discussed.

To generate random χ^2 numbers, we use the fact that if
$z \sim N(0,1)$, then $z^2 = \chi_1^2$ with 1 degree of freedom. To
obtain a random χ^2 number with d degrees of freedom, we
need only add d independent χ_1^2 numbers. Thus a random χ^2
number is given by

$$\chi_d^2 = \sum_{i=1}^{d} (\chi_1^2)_i = \sum_{i=1}^{d} z_i^2 \quad \text{where } z_i \sim N(0,1)$$

The subroutine in Program 14.1 uses this relationship to
generate one random χ_d^2 number with d degrees of freedom,
CHI2:

```
      SUBROUTINE CHISQ(IODD,IDF,CHI2)
      CHI2 = Ø.
      DO  1   I= 1,  IDF
      CALL RANNOR(IODD, Y, Ø., 1.)
    1 CHI2 = CHI2 + Y**2
      RETURN
      END
```

Program 14.1 SUBROUTINE CHISQ — generation of one
chi-square number.

Random F numbers can be generated using the fact that F_{d_1,d_2}
is the ratio of two independent χ^2 numbers, each divided by
their respective degrees of freedom. This routine would
call SUBROUTINE CHISQ and is left as an exercise.

   To illustrate the use of SUBROUTINE CHISQ we shall
write a main program that generates a random sample of χ_d^2's,
computes the characteristics of the empirical distribution,
and forms its frequency distribution. The input for the
program is

$$d = IDF = \text{degrees of freedom}$$
$$m = M = \text{number of random numbers to be generated}$$
$$n = NINT = \text{number of intervals for frequency table}$$
$$IODD = \text{random starting digit}$$

Notice that the program uses SUBROUTINE CHAR to calculate
the sample characteristics. Using this program with any
given value of d, we can compare the sample characteristics
with their true values, which are

$$\mu = E[\chi_d^2] = d$$

$$\sigma^2 = E[\chi_d^2 - d]^2 = \nu_2 = 2d$$

$$\alpha_3 = \frac{\nu_3}{\nu_2^{3/2}} = \frac{8d}{(2d)^{3/2}} = \frac{4}{\sqrt{2d}}$$

$$\alpha_4 = \frac{\nu_4}{\nu_2^2} = \frac{(12d^2 + 48d)}{(2d)^2} = 3 + \frac{12}{d}$$

500

The main program is

```
C       CHISQUARE GENERATION PROGRAM FOR IDF DEGREES OF
C       FREEDOM
        DIMENSION CHI2(2ØØ),F(3Ø)
        READ(5,1) IDF,M,IODD,NINT
   1    FORMAT(4I5)
        DO 2 I=1,M
        CALL CHISQ(IODD,IDF,CHI2(I))
   2    CONTINUE
        CALL FREQ(CHI2,M,F,Ø., .5, NINT)
        WRITE(6,1Ø) IDF,M
   6    FORMAT(1HØ, 28HCHI-SQUARE DISTRIBUTION WITH, I5,
       1 24HDEGREES OF FREEDOM USING,  I5,  7HSAMPLES)
        WRITE(6,11) (J,F(J),J=1,NINT)
  11    FORMAT(1H , I5,2X,F1Ø.Ø)
        CALL CHAR(CHI2,M,YM,YSIG,A3,A4)
        WRITE(6,12) YM,YSIG,A3,A4
  12    FORMAT(1H , 4HMEAN, F12.2, 3X,  3HSTD, F12.2,3X,
       1  2HA3,F12.4, 3X, 2HA4,F12.4)
        STOP
        END
```

{Subroutine CHISQ}
{Subroutine FREQ}
{Subroutine CHAR}
{Subroutine RANNOR}
{Subroutine URANDN}

14.2 CHISQUARE — generation of χ^2 distribution.

We used the program to generate an empirical sample of χ^2 numbers with 3 degrees of freedom. The input used in this case is

3	200	13	30

and we obtained the following frequency table:

Interval	F
0.0 -0.5	15
0.5 -1.0	23
1.0 - 1.5	23
1.5 - 2.0	21
2.0 - 2.5	15
2.5 - 3.0	20
3.0 - 3.5	12
3.5 - 4.0	19

4.0 - 4.5	13
4.5 - 5.0	6
5.0 - 5.5	5
5.5 - 6.0	4
6.0 - 6.5	4
6.5 - 7.0	4
7.0 - 7.5	4
7.5 - 8.0	1
8.0 - 8.5	3
8.5 - 9.0	3
9.0 - 9.5	2
9.5 -10.0	0
10.0 -10.5	0
10.5 -11.0	0
11.0 -11.5	0
11.5 -12.0	0
12.0 -12.5	1
12.5 -13.0	0
13.0 -13.5	0
13.5 -14.0	1
14.0 -14.5	0
14.5 -15.0	1

We also obtained a sample mean of 3.08, a sample standard deviation of 2.50, an a_3 of 1.6585, and an a_4 of 6.9375. These sample values are close to the true values because for 3 degrees of freedom, $\mu = 3$, $\sigma = \sqrt{6} = 2.4494$, $\alpha_3 = 4/\sqrt{6} = 1.6333$, and $\alpha_4 = 7$.

We can also use the program results to verify that the distribution model given by the χ^2 distribution provides a good fit of the empirically generated histogram. We generated the ordinates for the χ^2_3 distribution and plotted these values on the histogram, as seen in Fig. 14.8:

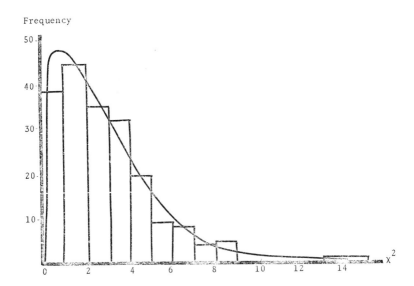

Fig. 14.8 Random χ_3^2 numbers and superimposed χ_3^2 ordinates.

A simple modification of Program 14.2 would enable us to verify that as d becomes large, the ratio

$$z = \frac{\chi_d^2 - d}{\sqrt{2d}}$$

approaches the standard normal distribution.

14.8 Summary

This chapter introduced the techniques that can be used to make inferences about a single population variance. The test statistic that links the variance estimate and the parameter value is the χ^2 ratio, which is given by

$$\chi_d^2 = \frac{(n - 1)s'^2}{\sigma^2}$$

This ratio has a χ^2 distribution with d = n - 1 degrees of freedom when the observations used to calculate s'^2 are a random sample from a normal distribution.

We also considered tests about the mean and variance of two populations. We introduced the F test statistic, which can be used to test the hypothesis that two normal distributions have the same variance. This test statistic has the form

$$F_{d_1,d_2} = \frac{s_1'^2}{s_2'^2}$$

and has the F distribution with $d_1 = n_1 - 1$ and $d_2 = n_2 - 1$ degrees of freedom.

In addition, we considered testing the hypothesis that the means of the two populations are equal. The test statistic used to test this hypothesis depends upon the results obtained from the equal variance test. Three tests were considered:

(1) If the variances are assumed to be equal, the pooled t test is used.
(2) If the variances are assumed to be unequal, an approximate t test is used.
(3) If the observations can be matched, a paired t test is used.

These considerations about the equality of population means were extended to more than two populations. In this case an appropriate model must be assumed, and the analysis of variance procedure is used to partition the total corrected sums of squares into the sums of squares attributed to each of the model effects. The F test provides a means of testing the null hypothesis that none of the model effects is significant. Two typical experimental designs were analyzed: the one-way and two-way classification designs.

The computer can be used to study the sampling properties of the χ^2 and F ratios. Most of these studies are left as exercises since they are very similar to the ones given in the previous chapters. The subroutine to generate a single random χ^2 number was given and its use was demonstrated.

1. The variable of interest is Y = problem-solving time for
   a particular test. Studies have shown that as individuals grow older, the variation in the time needed to
   solve a problem increases. Assume that it is known that
   at 70 years of age the general population has a variance
   of σ_y^2 = 17.5 sec. A theory explains that this increase
   in variance is caused by some people "aging" too rapidly,
   so a special diet has been developed to overcome this
   effect. Twenty-five randomly selected 70-year-old
   individuals who have been on this diet for a period of
   time were measured and yielded the following data:

   45.1, 51.3, 52.0, 46.3, 36.8, 57.1, 50.6, 72.8,
   34.5, 77.1, 41.3, 56.6, 37.2, 38.3., 47.1, 36.1,
   26.3, 51.1, 74.0, 6.4, 67.2, 68.1, 37.4, 54.2, 50.5

   On the basis of these results, would you be willing to
   conclude that the special diet reduced the variance?

2. Use the above data to find the 95 percent confidence
   interval for the variance of Y.

3. If you were required to obtain an estimate of σ^2 for
   Exercise 1 so there would be 90 percent assurance
   that the estimate would be in error by no more than 10
   percent, how large a sample of 70-year-olds would you
   use in your study?

4. If a special diet can be achieved by fortifying the salt
   used by individuals, then a doctor might take one of
   four actions:

   a_1 = recommend against fortification
   a_2 = do nothing relative to such a salt
         fortification
   a_3 = recommend that salt be fortified
   a_4 = require that all salt be fortified

Consider that the underlying state of nature can be represented by

$$\theta_1: \sigma_1^2 = 12.6$$

$$\theta_2: \sigma_2^2 = 15.0$$

$$\theta_3: \sigma_3^2 = 17.5$$

$$\theta_4: \sigma_4^2 = 20.0$$

Thus the parameter of interest is σ^2, and the observed random variable is s'^2. The losses are given by

	a_1	a_2	a_3	a_4
θ_1	15	10	3	1
θ_2	10	6	2	2
θ_3	5	1	5	10
θ_4	1	5	25	50

Develop some reasonable competing strategies that might be followed by the doctor. Use the prior probabilities $q_1 = 0.10$, $q_2 = 0.15$, $q_3 = 0.50$, $q_4 = 0.25$ to determine which strategy would have the minimum expected loss.

5. Two teaching methods are known to produce about the same average results, but there is some question as to which is more consistent. If they are of the same consistency, then the less expensive method would be adopted. To test the hypothesis that $\sigma_1^2 = \sigma_2^2$ against the alternative hypothesis that $\sigma_1^2 \neq \sigma_2^2$, each method was used on a separate sample of students, with the following results:

$$n_1 = 70 \qquad s_1'^2 = 95.3$$
$$n_2 = 50 \qquad s_2'^2 = 121.3$$

Using the 95 percent level of significance, would you reject the equal consistency hypothesis?

From a decision-making viewpoint, can you envision a situation where the α_3 of the two distributions of

506

outcomes would be of importance? Write a discussion of such a situation and how you would propose to consider it if you believe it to be possible.

6. Consider the problem of how to test which form of advertising is best to promote the sale of a product:

$$A_1 = \text{newspaper}$$
$$A_2 = \text{radio}$$
$$A_3 = \text{TV}$$

A 15-week test period is selected for the test, with each form of advertising being used for five different weeks and the increase in the total weekly sales compared. The particular weeks used for each type of advertising have been selected randomly from the 15 possible weeks. The gain of sales is determined by subtracting the sales from the expected sales for the week if there had been no advertising with the actual sales. This difference is the response variable Y of our experiment. We can tabulate the results of the experiment in the following one-way table:

A_1	A_2	A_3
790	640	721
853	832	834
940	734	737
820	969	931
830	742	923

Using this data, test the hypothesis that the average response is the same for each form of advertising. If you discard this null hypothesis, use the least significant difference technique to determine if one form of advertising is superior to either of the other two.

7. The personnel officer of a company is interested in what type of relationship exists between the educational

background of employees and their reading time. In making a study of this problem he uses the following educational classes: A_1:high school not completed; A_2:high school completed, no college; A_3:some college but no degree; A_4:a college degree. His experiment consists of selecting samples of 10 employees from each educational class and measuring their reading time.

The data obtained are given below. Each observation is represented by y_{ij}, where i denotes the educational class and j denotes the kth observation in that class.

A_1 High School Not Completed	A_2 Completed High School	A_3 Some College	A_4 College Degree
7.2	5.7	5.8	3.7
6.7	5.7	4.1	6.3
6.5	7.9	4.2	5.6
7.4	5.2	5.0	5.7
6.4	6.6	7.6	5.1
8.4	4.6	4.6	3.6
7.5	6.3	6.0	4.4
7.0	7.0	6.0	5.1
5.5	4.8	5.2	6.4
5.7	5.7	5.4	6.2

Test the hypothesis that the reading speeds are the same for all levels of education. If this hypothesis is discarded, divide the four levels into as many different groups as the LSD approach might indicate.

8. The curriculum committee is considering whether students in a college should be required to take a speed-reading course. It was felt that the benefit, as measured by Y = gain in reading speed 6 months after completing the course, would be affected by the year in school of the student as well as by the type of course load he is carrying. An experiment is performed in which five stu-

dents from each of the four undergraduate years and for
each of three colleges were given the course. Six
months later their reading speed is tested. The results
are recorded in the following two-way table:

College	Freshman	Sophomore	Junior	Senior
Arts & Sciences	110, 105, 101, 88, 101	115, 133, 117, 121, 113	112, 93 93, 108, 114	76, 81, 84, 90, 75
Engineering	142, 129, 135, 139, 137	121, 90, 120, 105, 103	117, 108, 86, 102, 106	103, 86, 80, 106, 94
Business	161, 148, 140, 143, 160	100, 97, 117, 101, 111	122, 125, 119, 137, 120	118, 108, 109, 123, 122

Test the null hypothesis relative to the differences in
the two main effects and their interaction.

9. Write a program to generate the ordinates of χ_d^2 for any
value of d. Plot the ordinates for d = 1, 10, 30, and
50 on the same graph, and determine how the value of d
affects the shape of the model.

10. Write a program using FUNCTION DEFINT to determine
specific probabilities for the χ^2 distribution

$$Pr(a \leq \chi^2 \leq b) = \int_a^b f(\chi^2) \, d\chi^2$$

Test the program by checking the results with the tabled
values of χ^2.

11. Generate m = 200 samples from the normal distribution,
calculate s'^2 and χ^2 and tabulate the resulting χ^2
values into a frequency distribution for the following
normal distributions:

(a) d = 10 μ = 10 σ^2 = 25
(b) d = 10 μ = 50 σ^2 = 100
(c) d = 50 μ = 0 σ^2 = 1
(d) d = 50 μ = 100 σ^2 = 100

Do μ and σ^2 affect the resulting χ^2 values?

12. Adapt the program written in Exercise 10 to generate the expected χ^2 frequencies for the frequency tables generated in Exercise 11. Perform a goodness of fit test.

13. Write a program to determine the ordinates for the F distribution for the following values of d_1 and d_2:

  (a) d_1 = 3 d_2 = 3

  (b) d_1 = 10 d_2 = 4

  (c) d_1 = 30 d_2 = 30

  (d) d_1 = 1 d_2 = 20

Plot the distributions on the same graph. There is a relationship between the t distribution and the F distribution when d_1 = 1, $t_d^2 = F_{1,d}$. Verify the relationship.

14. Generate two samples from each of the following normal distributions. Determine $s_1'^2$ and $s_2'^2$ and the F ratio. Repeat the process m = 200 times. Tabulate the resulting m F values into a frequency table for the following values:

  (a) d_1 = 3 d_2 = 3 μ = 0 σ^2 = 1

  (b) d_1 = 10 d_2 = 4 μ = 10 σ^2 = 25

  (c) d_1 = 30 d_2 = 30 μ = 50 σ^2 = 100

  (d) d_1 = 30 d_2 = 30 μ = 0 σ^2 = 1

Do μ and σ^2 affect the F distribution? Why or why not?

15. Write a program using FUNCTION DEFINT to generate the expected frequencies for the frequency tables generated in Exercise 6. Perform a goodness of fit test.

16. Calculate the total sums of squares and s_y if there are three samples with four observations each, and $\bar{y}_1 = 10$, $\bar{y}_2 = 15$, $\bar{y}_3 = 20$, and $s_1'^2 = 4$, $s_2'^2 = 5$, and $s_3'^2 = 6$. Also find the pooled variance estimate and between variance estimate.

17. Write a program to perform a one-way analysis of variance.

    (a) Use the data given in Sec. 14.5 to test the program.
    (b) Consider an experiment with three treatments: A_1, A_2, and A_3. The model is

$$y_{ij} = \mu + \alpha_i + \epsilon_{ij} \qquad \begin{array}{l} i = 1, 2, 3 \\ j = 1, 2, 3, 4, 5 \end{array}$$

Generate random samples from each treatment using the following parameter values.

$$\mu = 50 \qquad \sigma^2 = 10 \qquad \alpha_1 = -3 \qquad \alpha_2 = 4 \qquad \text{and} \qquad \alpha_3 = -1$$

In this case

$$A_1 \sim N(\mu + \alpha_1, \sigma)$$
$$A_2 \sim N(\mu + \alpha_2, \sigma)$$
$$A_3 \sim N(\mu + \alpha_3, \sigma)$$

Use the computer program to calculate the appropriate entries for the analysis of variance table to perform the appropriate tests of hypothesis.

18. Consider the following two-factor experiment:

$$Y_{ijk} = \mu + \alpha_i + \beta_j + (\alpha\beta)_{ij} + \varepsilon_{ijk} \qquad \begin{aligned} i &= 1, 2, 3 \\ j &= 1, 2 \\ k &= 1, 2, 3, 4 \end{aligned}$$

Also,

$$
\begin{aligned}
\alpha_1 &= 2 & (\alpha\beta)_{11} &= 10 \\
\alpha_2 &= -2 & (\alpha\beta)_{12} &= -10 \\
\alpha_3 &= 0 & (\alpha\beta)_{21} &= -6 \\
\beta_1 &= 0 & (\alpha\beta)_{22} &= 6 \\
\beta_2 &= 0 & (\alpha\beta)_{31} &= -4 \\
& & (\alpha\beta)_{32} &= 4
\end{aligned}
$$

Verify that the appropriate sums are equal to zero according to the model assumptions. Then generate the $r = 4$ random samples for each cell. Set up the analysis of variance table and perform the appropriate tests.

CHAPTER 15

OTHER USES OF THE CHI-SQUARE DISTRIBUTION

15.1 Introduction

Until now we have emphasized the analysis of either discrete or continuous data. However, variables whose values
are simply code names indicating the class to which an individual belongs are often encountered, for example, sex, race,
and personality traits; these types of variables are often
called nominal variables.

An experimental design involving a numerical response
variable can be analyzed by the techniques we have already
considered. However, if the response variable is a nominal
one, some new method of analysis is required. The chi-square
statistic is used in analyzing such data; we shall consider
some of these methods in this chapter.

15.2 Goodness of Fit Tests

When we considered the problem of obtaining a mathematical model to represent the distribution of a random variable, we encountered the question of whether the model adequately fitted the observed distribution. Using the test of
hypothesis concept with the χ^2 statistic, we can now develop
a more objective solution to this problem than by simply
making graphic comparisons.

We wish to test the null hypothesis that the probabilities of an observation being in a given class are those
specified by the model. The alternative hypothesis is that
some other set of probabilities governs the occurrence of
the observations. To test this null hypothesis we assume
that the observed data have been tabulated into classes and

513

that observed frequencies are thus available for each class. The corresponding expected frequencies can then be obtained using the model probabilities assigned to each class. The appropriate test statistic to test the null hypothesis is given by

$$\chi^2 = \sum_{j=1}^{m} \frac{(f_{oj} - f_{ej})^2}{f_{ej}}$$

where f_{oj} is the observed frequency for the jth class and f_{ej} is the expected frequency for the same class. If all $f_{ej} > 3$, then this test statistic has an approximate χ^2 distribution with d = m - 1 - k degrees of freedom. The k represents the number of model parameters that had to be estimated from the observed data. If $f_{ej} < 3$ for any j, we can combine that particular class with an adjacent class until all expected frequencies are greater than 3. If the computed χ^2 statistic is greater than the critical χ^2 value for a given significance level, the null hypothesis is rejected, and thus the model would not be used.

Let us illustrate the technique by testing whether a discrete uniform model adequately fits the observed distribution given in Sec. 7.4, Chap. 7. The observed distribution was

Y	F
1	30
2	35
3	33
4	36
5	29
6	37

The expected frequencies using the model f(Y) = 1/6 for each value of Y are equal to 200 (1/6) = 33.33. Thus the value of the critical ratio is given by

$$\chi^2 = \frac{(30 - 33.33)^2}{33.33} + \frac{(35 - 33.33)^2}{33.33} + \cdots + \frac{(37 - 33.33)^2}{33.33}$$

$$= 1.6$$

Since in this example no parameter had to be estimated, the degrees of freedom are given by $d = m - 1 = 6 - 1 = 5$. Using the 5 percent level of significance, we find that $\chi^2_{5,.95} = 11.07$, and thus we would not reject the null hypothesis and would be willing to use the uniform distribution as the mathematical model.

As another illustration we shall test whether the binomial model provided a satisfactory fit for the true-false test data as given in Sec. 7.5, Chap. 7. The observed and expected frequencies are

Y	f_{oj}	f_{ej}
0	35	24.88
1	45	74.72
2	120	105.18
3	112	92.12
4	42	56.16
5	23	25.31
6	10	8.70
7	2	2.34
8	1	.51
9	0	.08
10	0	0
11	0	0
12	0	0
13	0	0
14	0	0
15	0	0
16	0	0

(rows 6–16 for f_{oj} bracketed as 13; rows 6–16 for f_{ej} bracketed as 11.63)

The test statistic is

$$\chi^2 = \frac{(35 - 24.88)^2}{24.88} + \frac{(45 - 74.72)^2}{74.72} + \cdots + \frac{(23 - 25.31)^2}{25.31}$$

$$+ \frac{(13 - 11.63)^2}{11.63}$$

$$= 26.11$$

The critical χ^2 value now has $d = 7 - 1 - 1 = 5$ degrees of freedom since the parameter θ has to be estimated. Using the 5 percent level of significance, the critical value is $\chi^2_{5,.95} = 11.070$, and thus we reject the null hypothesis that the binomial model provides a good fit and would have to seek a different model.

To illustrate the χ^2 goodness of fit test for a continuous model, let us test whether the normal model provides an adequate representation for the reading-time data given in Fig. 8.3, Chap. 8.

Interval	f_{oj}	f_{ej}	
2.0 - 2.4	21	22.17	
2.5 - 2.9	43	41.45	
3.0 - 3.4	68	72.53	
3.5 - 3.9	97	91.38	
4.0 - 4.4	72	82.92	
4.5 - 4.9	53	54.18	
5.0 - 5.4	21	25.49	
5.5 - 5.9	11 } 15	8.63 }	11.12
6.0 - 6.4	4 }	2.49 }	
	390	401.24	

$$\chi^2 = \frac{(21 - 22.17)^2}{22.17} + \cdots + \frac{(15 - 11.12)^2}{11.12}$$

$$= 4.36$$

Note that the expected frequency for the last interval was less than 3, and thus the last two intervals were combined into one. Both μ and σ had to be estimated in fitting the normal model; thus the degrees of freedom are given by $d = m - 1 - 2 = 8 - 1 - 2 = 5$. Using the 5 percent level of significance, we obtain $\chi^2_{5,.95} = 11.07$, and thus we would not discard the null hypotnesis.

Through the use of the χ^2 test we are actually evaluating the probability that the differences between the observed and expected frequencies could have arisen by chance when the model being tested is the true model. If this probability is too small (less than our level of significance), we choose to discard the model because the data showed evidence that it may have come from some other distribution. The use of the χ^2 distribution for the test statistic could be mathematically justified, but this is beyond the scope of this text. We can, however, perform an empirical study of the adequacy of the χ^2 distribution by generating sets of observed frequencies from a known distribution model, computing the χ^2 ratio and comparing its observed sampling distribution with the tabulated values of the χ^2 distribution with $d = m - 1 - k$ degrees of freedom.

15.3 One-way Contingency Tables

Since the χ^2 test provides a critical ratio that can be used in comparing expected frequencies with observed frequencies, we can consider how it might be used when data are tabulated into what are called one-way contingency tables. One-way contingency tables are obtained when random observations from a population are classified into one of several classes according to a particular characteristic of the observation.

Let us consider an actual illustration. Our personnel manager is concerned about the educational background of new employees since a new employment program has recently been put into effect. He fears that the program may adversely effect the educational background of new employees, although he has been assured that the new program will produce new

employees with the same educational distribution as the old
program produced. The manager decides to test this assertion by taking a random sample of 100 new employees, determining their educational background, and classifying each
new employee in the sample into one of five classes:

C_1: no high school diploma
C_2: high school diploma but no college
C_3: some college, no degree
C_4: bachelor degree, no graduate work
C_5: some graduate work

When the old recruiting program was in effect, experience indicated that the following proportion might be expected in each of the classes: $p_1 = 0.17$, $p_2 = 0.39$, $p_3 =$
0.27, $p_4 = 0.14$, and $p_5 = 0.04$; thus the null hypothesis
(which assumes that the new program produces the same frequencies as the old one) is

$$H_o: p_1 = 0.17$$
$$p_2 = 0.34$$
$$p_3 = 0.27$$
$$p_4 = 0.14$$
$$p_5 = 0.04$$

and the alternative hypothesis is that at least two p_j's are
unequal to their specified values.

These probabilities (proportions) can be used to find
the expected frequencies for each class since $f_{ej} = n \cdot p_j$;
it is these expected frequencies that are then compared with
the observed frequencies obtained from the sample of 100 new
employees. The data are summarized in the following frequency table:

Educational Class	C_1	C_2	C_3	C_4	C_5
Obs. frequency f_{oj}	30	40	20	8	2
Exp. frequency f_{ej}	17	38	27	14	4
Discrepancy	+13	+2	-7	-6	-2

These frequencies can be converted into the χ^2 test statistic, which yields

$$\chi^2 = \frac{(13)^2}{17} + \frac{(2)^2}{38} + \frac{(-7)^2}{27} + \frac{(-6)^2}{14} + \frac{(-2)^2}{4}$$

$$= 9.94 + 0.11 + 1.81 + 2.57 + 1.00 = 15.43$$

The degrees of freedom are $5 - 1 = 4$ since no parameter values need to be estimated. To test the null hypothesis at the 1 percent level we compare the $\chi^2 = 15.43$ with the $\chi^2_{4,.99} = 13.28$ value. Since $15.43 > 13.28$, the null hypothesis is rejected at the 1 percent level of significance, and the personnel manager would be inclined to bring this problem to the attention of top management.

We note that one-way contingency tables are similar to frequency distributions, but the classes need not relate to the numerical value of a random variable. In general, expected frequencies for each class may come from theoretical hypotheses or prior experience.

15.4 Two-way Contingency Tables

The techniques presented in the preceding section are applicable when there is a single characteristic that is used to assign individuals to classes. Let us now consider the case when two characteristics are available on each observation.

We again use an example to illustrate the approach. Consider a study in which both the reading time and the socioeconomic class for each individual in the population of interest can be determined. Assume that there are four socioeconomic classes, denoted by A_1, A_2, A_3, A_4, and five classes of reading times, B_1, B_2, B_3, B_4, and B_5. A sample of 1455 individuals was taken, and each observation was classified into a two-way contingency table according to his reading time and socioeconomic class:

Socioeconomic class	Reading Time B_1	B_2	B_3	B_4	B_5	Total
A_1	20	30	41	46	55	192
A_2	41	54	61	63	71	290
A_3	63	75	90	94	102	424
A_4	72	90	105	125	157	549
Total	196	249	297	328	385	1455

Although the χ^2 test could be used to test any null hypothesis that assigns probabilities to each of the 20 classes, the usual null hypothesis of interest in such two-way classifications is whether the two characteristics used in making the classification are independent. In our example this null hypothesis states that the socioeconomic class of an individual does not influence his reading time. This hypothesis can formally be represented by

$$H_o: \Pr[A_1 \cap B_j] = \Pr[A_i] \cdot \Pr[B_j] \quad \text{for all } i = 1,2, \ldots ,k$$
$$\text{and } j = 1,2, \ldots ,m$$

Note that k represents the number of classes for one variable and m the number of classes for the other. The alternative hypothesis is

$$H_a: \Pr[A_i \cap B_j] \neq \Pr[A_i] \cdot \Pr[B_j] \quad \text{for some } i \text{ and } j$$

Thus the null hypothesis states that the expected probabilities for each cell can be found by multiplying the appropriate marginal probabilities. This is a direct consequence of the assumption that the two characteristics are independent, as already discussed in Chap. 11.

If we wish to test the null hypothesis of independence,

the marginal probabilities must first be estimated. Let R_i denote the frequency total for the ith row and C_j denote the frequency total for the jth column. The estimates of the marginal probabilities are

$$Pr[A_i] = \frac{R_i}{n} \quad \text{and} \quad Pr[B_j] = \frac{C_j}{n}$$

where n represents the total number of observations. The expected probability for each cell under the null hypothesis can then be found by multiplying the appropriate marginal probabilities. Thus in the example,

$$Pr[A_1] = \frac{192}{1455} \quad Pr[B_1] = \frac{196}{1455}$$

and

$$Pr[A_1 \cap B_1] = (0.13196) \cdot (0.13471) = 0.1778$$

The expected frequency for the (i,j)th cell can then be obtained by multiplying the expected probability by the total number of observations n or by using the formula

$$E_{ij} = n \cdot Pr[A_i \cap B_j] = \frac{R_i \cdot C_j}{n}$$

Thus the expected frequency for the A_1B_1 cell is

$$E_{11} = (0.01778) \cdot (1455) = \frac{(192) \cdot (196)}{1455} = 25.86$$

Since the observed frequency for this cell was 20, we note that we had fewer observations in this cell than expected under the null hypothesis.

The observed and expected frequencies for the illustration are

	B_1	B_2	B_3	B_4	B_5	Marginal Total
A_1	20	30	41	46	55	192
	25.86	32.86	39.19	43.28	50.80	
A_2	41	54	61	63	71	290
	39.09	49.63	59.20	65.37	76.74	
A_3	63	75	90	94	102	424
	57.12	72.56	86.55	95.58	112.19	
A_4	72	90	105	125	157	549
	73.95	93.95	112.06	123.76	145.27	
Marginal Totals	196	249	297	328	385	1455

The computed χ^2 statistic for the null hypothesis is given by

$$\chi^2 = \sum_{i=1}^{k} \sum_{j=1}^{m} \frac{(O_{ij} - E_{ij})^2}{E_{ij}}$$

The degrees of freedom are given by $d = (k - 1) \cdot (m - 1)$ and the observed frequency for the (i,j)th cell is denoted by O_{ij}. The student should note that the total degrees of freedom $m \cdot k - 1$ are reduced to $(k - 1) \cdot (m - 1)$. This reduction in the degrees of freedom is due to the fact that $m + k - 2$ marginal probabilities had to be estimated.

For the two-way contingency example, the computed test statistic is obtained by evaluating

$$\chi^2 = \frac{(20 - 25.86)^2}{25.86} + \frac{(30 - 32.86)^2}{32.86} + \cdots + \frac{(157 - 145.27)^2}{145.27}$$

$$= 6.625$$

Since the critical value is $\chi^2_{12, .95} = 21.03$, we would not reject at the 5 percent level of significance the null hypothesis that the two characteristics are independent.

The use of contingency tables and the χ^2 test of significance is an effective way of testing whether two or more characteristics are independent. The procedure can be extended to accommodate three-way contingency tables or even higher-order ones since the computation of the critical χ^2 value is readily generalized. The restriction that all E_{ij} be greater than 3 still holds, and, in addition, the number of classes should be less than 30.

15.5 Introduction to the FORTRAN A Format

When dealing with frequency and contingency tables, the identification of classes is often given by "words" defined by a set of alphanumeric characters. In data acquisition, alphanumeric words are also used, and often such words are actually punched into data cards. Thus it is useful to be able to write computer programs that can read and print such alphanumerical identifications.

Whenever alphanumeric characters are to be processed by a FORTRAN program, a variable name must be assigned to each set of characters. These variable names can be either a real or integer name or subscripted variable names can also be used. The format specification used to read and write alphanumeric variables is

nAw: repeat the Aw format n times and read each w
       columns as one alphanumeric field

For example, the statements

```
        READ(5,2) MSG1, MSG2
2       FORMAT(2A3)
        WRITE(6,3) MSG1,MSG2
3       FORMAT(1X,2A3)
```

would cause two alphanumeric fields to be read, one in columns 1 to 3, the other in columns 4 to 6. The two sets of characters would then be stored in the locations designated by MSG1 and MSG2, respectively. The WRITE statement then causes the same characters to be printed out.

The same rules that apply to the F, I, and X format specifications discussed in Chap. 5 apply to the A format. In addition, the maximum number of alphanumeric characters than can be stored in a storage location must be known. Let v designate the maximum number of characters that can be stored under one variable name. If the specification Aw is used to read, then the following rules apply:

If w > v: then leftmost w-v characters in data field are skipped and remaining v characters are read and stored.

If w ≤ v: then w characters are read and v-w blanks are inserted on the right.

When printing alphanumeric variables using the specification Aw, the following rules apply:

If w > v: then printed field will contain v characters right-justified in the field and preceded by w-v blanks.

If w ≤ v: then only leftmost w characters of variable are printed.

It should be appreciated that long words may have to be split into parts, and each part must be stored under a different designation. Although the value of v depends upon the computer being used, we shall assume in the examples that v = 4. For example, if the following characters were punched in columns 1 to 15

THISISANEXAMPLE

and are read using

```
          READ(5,1) A,B,C,D.
     1    FORMAT(A4,A2,A2,A7)
```

the following characters will be stored in the respective
locations:

     A: THIS
     B: IS␢␢
     C: AN␢␢
     D: MPLE

Although these statements fail to store the 15 characters,
the following modified statements would store the entire
message by words:

```
        READ(5,1) A,B,C,D,E
    1   FORMAT(A4,A2,A2,A4,A3)
```

When writing general-purpose programs, identifying
titles are often desirable. We can use the A format to read
and write titles that are part of the input data without
changing the program each time. Suppose that a title should
have at the most 80 characters. Since we assume that the
maximum number of characters stored in one location is $v = 4$, we must have $80/4 = 20$ variable names to store the entire
title, and we would therefore use a subscripted variable to
store the title. The reading and writing is accomplished by
the following FORTRAN statements:

```
        DIMENSION TITLE(20)
        READ(5,1) (TITLE(I),I=1,20)
    1   FORMAT(20A4)
        WRITE(6,2) (TITLE(I),I=1,20)
    2   FORMAT(1X,20A4)
```

A data card containing the title anywhere in columns 1 to 80
would be needed, and the above program would write out exactly the 80 characters punched in the title card.

    Alphanumeric variables can be compared using a logical
IF statement. For example, let us assume that we have inventory data cards for 100 articles in a store that has an
alphanumeric code punched indicating whether the article was

old or new. We wish to read in the data cards, count the number of old and new articles, and print out the numbers. We could accomplish this by reading in one data card containing the alphanumeric codes for the two types of articles that will be encountered and then reading in the remaining 100 data cards and comparing them with the codes. The following performs these steps:

```
        FTYPE1 = Ø.
        FTYPE2 = Ø.
        READ(5,1) TYPE1,TYPE2
   1    FORMAT(2A3)
        DO 2 I=1,1ØØ
        READ(5,3) DATA
   3    FORMAT(A3)
        IF(DATA .EQ. TYPE1) GO TO 1Ø
        FTYPE2 = FTYPE2 + 1.
        GO TO 2
  1Ø    FTYPE1 = FTYPE1 + 1.
   2    CONTINUE
        WRITE(6,11) TYPE1, FTYPE1, TYPE2, FTYPE2
  11    FORMAT(1H1,5HNØ OF,A3,9HITEMS ARE, F1Ø.Ø,
       1 1HØ,5HNØ OF, A3,9HITEMS ARE, F1Ø.Ø)
        STOP
        END
```

The first data card should contain the alphanumeric identification for the two types of items. If these designations are OLD and NEW, the first data card would have the following characters in columns 1 to 6:

OLDNEW

The remaining 100 data cards would have either the identification OLD or NEW punched in columns 1 to 3. If the designations for the items are different, the first data card need simply give the appropriate code. Although the designation involves more than three characters, the appropriate FORMAT changes would be needed.

The student should now be able to write a program that creates a two-way contingency table when the characters are given as alphanumeric codes. This will be given as an exercise.

15.6 Summary

In this chapter we considered the uses of the χ^2 statistic as given by

$$\chi^2 = \sum_{j=1}^{m} \frac{(f_{oj} - f_{ej})}{f_{ej}}$$

This test statistic has an approximate χ^2 distribution, with degrees of freedom given by d = m - 1 - (number of parameters used in determining the expected frequencies). The observed frequencies for each of the m classes is given by f_{oj}; the expected frequencies under the null hypothesis are given by f_{ej}. The use of the χ^2 distribution requires that each f_{ej} be greater than 3.

We gave illustrations to show how the χ^2 statistic could be used to test whether a particular continuous or discrete model provided an adequate fit of the observed data, whether the observed proportions for a nominal variable agreed with the theoretically determined values, and whether two variables were independent.

We introduced the A format, which enables us to read, write, and compare alphanumeric data. We presented a program to illustrate the use of the A format both in reading in titles and in making comparisons.

EXERCISES

1. Using the 5 percent level of significance, test the adequacy of the type III model fit given in Sec. 8.6, Chap. 8.

2. In an attitute study, the following contingency table summarized the results of an random sample of the population of interest:

Status	Attitude		
	Approved	Disapproved	No Opinion
Undergraduate	128	64	15
Graduate	41	19	4
Faculty	28	12	5

Using the 95 percent level of significance, test whether attitude depends upon status.

3. A sample of 60 graduate students enrolled in a statistics course were asked to list their major area in which they were enrolled. The results are

LA, BA, AG, AG, LA, LA, ENG, ENG, ENG, LA, BA, AG, LA,
AG, AG, BA, ENG, LA, BA, BA, LA, LA, AG, AG, ENG, ENG,
LA, LA, ENG, AG, LA, LA, BA, AG, LA, LA, BA, ENG, ENG,
LA, AG, LA, LA, AG, BA, AG, LA, LA, AG, ENG, AG, LA,
LA, BA, BA, LA, LA, ENG, LA, LA

The coded characters represent

LA = liberal arts
BA = business administration
AG = agriculture
ENG = engineering

If the total enrollment in the graduate school had 50 percent in liberal arts, 30 percent in business administration, and 10 percent each in agriculture and engineer-

528

ing, use the above data to test the hypothesis at the 5 percent level of significance that the students are equally likely to enroll in the course regardless of their college.

4. If the first 30 observations given in Exercise 3 were obtained from the first semester enrollement and the second 30 observations from the second semester enroll-ment, write a computer program that would

   (a) Read in the 60 observations in alphanumerical form.
   (b) Determine the observed frequencies for each class in the two-way table (semester vs. college) and print out the table with coded headings.
   (c) Print out an interpretation for each code as given in Exercise 3.
   (d) Determine the expected frequency for each class under the independence assumption and print out these frequencies.
   (e) Check to see that each expected frequency is greater than 3. If H is not > 3, then print out a warning message.
   (f) Read in the appropriate 95 percent critical value of χ^2.
   (g) Compute the χ^2 value and compare this with the actual tial value. Print out the appropriate conclusion along with the critical χ^2 value and the computed χ^2 value.

5. Write a more general computer program that will handle a general contingency table analysis for testing indepen-dence. Read in each observed cell frequency.

CHAPTER 16

SAMPLE SURVEY DESIGNS

16.1 Introduction

Data that are needed for decision making, planning, or
analysis purposes can be obtained from several different
sources. The system being studied can be simulated in a
laboratory under different controlled conditions, and results can then be analyzed to obtain information relative
to system behavior. Chapter 14 illustrated methods of
analyzing the data obtained from such controlled experiments. However, such an approach may be either too expensive or infeasible, or the simulation may be of questionable
validity. In such cases, one alternative to the experimental approach is to survey actual units from the population
of interest to obtain the data needed for such analyses. To
use the sample survey approach, two major problems must be
resolved: (1) a method of acquiring the sample of units
from the population, and (2) a method of obtaining the required measurements once the units are available must be
developed. In spite of the difficulties in trying to
solve these problems, the survey approach is often the only
method to which one has recourse. For example, to measure
the attitudes or opinions of a population, an opinion poll, in
which individuals are contacted and questioned relative to
their opinion on some subject, must be used. Public health,
wage, population, and market surveys are other examples of
this type of investigation.

A person could of course perform the survey by contact-
ing every unit in the population. However, since the costs
and time needed for such a census approach are often prohib-
itive, sample surveys are usually made. Sampling theory

must then be used to determine the best design for such sample surveys and to indicate how the survey data should be analyzed. In this chapter we shall introduce some of the more basic ideas associated with such sample surveys.

Many surveys are made to estimate the average value of a variable for some finite population. Sometimes, however, interest centers upon population totals or percentages. Since the basic techniques, other than the particular estimator to be used, are essentially the same for all types of estimation, we shall consider only the population average in this chapter.

When a sample survey approach rather than a census approach is used, an appropriate design for the survey must be made. The design should specify the particular estimator to be used, the appropriate sample size to meet some precision requirement for the estimator, and the method of sample selection. In the following sections we shall introduce some useful survey designs that are associated with the problem of estimating the population mean.

To aid in the critical problem of selecting a representative sample, we shall assume that a listing of all the units of the entire population is available. This listing is often called the population frame. Once such a population frame is available, the survey design to be used depends upon the information that is available for each unit of the population.

To illustrate the techniques throughout the chapter we shall use as our example the problem of estimating the average rent paid by families in single-family dwellings presently renting houses in a particular urban renewal area.

16.2 The Simple Random Sample Design

If the only information available for the population frame prior to the survey is the identification of the individual units, then the only sampling technique that can be used is simple random sampling, and the best estimator of the population mean is the sample mean. In terms of our illustrative example, this would occur when only the identifications of the individual single family dwellings, i.e., addresses, are known.

Let y_i denote the rent paid by the ith unit in the sample of size n, which has been selected at random from the population of N units. The sample mean estimator of the population mean μ_y (the average rent paid in the area) is given by

$$\hat{\mu}_y = \bar{y} = \frac{\sum\limits_{i=1}^{n} y_i}{n} \qquad (16.1)$$

The variance of the sample mean estimator for a finite population is

$$\sigma_{\bar{y}}^2 = \frac{N - n}{n(N - 1)} \sigma_y^2 \qquad (16.2)$$

where σ_y^2 is the variance of the population.

Assuming that the sample is large enough so that the central limit theorem can effectively be applied, the sample size which should be used to obtain an estimate that falls within ε units of the population mean with a $100(1 - \alpha)$ percent level of confidence is given by

$$n = \frac{z_{1-\alpha/2}^2 \cdot \sigma_y^2 \cdot N}{z_{1-\alpha/2}^2 \cdot \sigma_y^2 + (N - 1) \cdot \varepsilon^2} \qquad (16.3)$$

To apply this simple random sampling design formula to our example, let us assume that the population frame consists of N = 210 housing units. It is specified that the sampling error is to be less than \$5.00, that is, $|\varepsilon| = |\bar{y} - \mu_y| \leq \$5.00$, with a 95 percent level of assurance. To apply the formula, we need a value of σ_y^2; we shall use the value of σ_y^2 = \$115.00 which has previously been obtained from similar studies.

We would then compute

$$n = \frac{(1.96)^2 \ (115) \ (210)}{(1.96)^2 \ (115) + (209) \ (5.0)^2} = 16.4 = 17$$

We thus need to randomly select 17 units from the list of 210 rental dwellings to obtain the units that are to be surveyed.

To select a random sample, we can use the uniform random number generator given in Appendix B to obtain a random
number u_i. We let each u_i identify the unit to be surveyed
by determining $i = [N \cdot u_i + 1]$, where $[N \cdot u_i + 1]$ means
the greater integer less than the value of $N \cdot u_i + 1$. If
duplicate i's are obtained, we ignore the second one and
obtain another random number until a total of 17 separate
units are identified to be included in the sample.

A survey of the 17 units must now be made to obtain the
rent paid by each tenant. Although the problem of actually
obtaining the rent being paid involves many considerations,
we shall not attempt any discussion of this problem but
simply assume that we can obtain the 17 values and summarize
then to find

$$\bar{y} = \$59.12 \quad s_y' = \$12.53 \quad a_3 = 0.0367 \quad a_4 = 2.3372$$

We would probably not publish the latter two characteristics
because most persons would not appreciate their implications
but would simple report

$$\mu_y = \$59.12 \pm \$2.92$$

where the 2.92 is the standard deviation of the estimator
that is obtained using Eq. (16.2):

$$s_{\bar{y}} = \sqrt{\frac{210 - 17}{(17) \cdot (209)}} \, (12.53) = 2.92$$

We may elect to report that the 95 percent confidence interval for the mean rental for the population is

$$\mu_y = 59.12 \pm (1.96)(2.92) = \$59.12 \pm \$5.72$$

The sample mean is the most efficient estimator of the
population mean (it has the smallest standard deviation)
when no information is available about the population frame.
However, if additional information is available for each

533

unit in the populations, then the approach outlined above
can be improved upon by using a different estimator, by
changing the sample design, or both. In the next section we
consider how to evolve other estimators using such additional information, and in Sec. 16.4 we shall present different
sample designs.

16.3 Other Estimators for the Population Mean

It is often possible that the population mean of another variable X is available and that the value of this variable is either already known for each unit of the population
or its value can be obtained along with the value of the Y
variable when the units in the sample are surveyed. If X is
linearly related to the variable of interest Y, then the information about Y contained in X can readily be used to obtain a more efficient estimator of μ_y.

In our illustrative example such a variable might be
the assessed evaluation of each house because such an evaluation can often be obtained from official sources for each
house in the entire city. Thus the population average μ_x
for the 210 rental units easily can be obtained. Let us
assume that for our example, μ_x = \$2073.00. We look up the
assessed evaluation for each house in the sample and augment
the rental survey results with these values as shown below:

| | Y | X |
Identification	Monthly Rental	Assessed Evaluation
1	80.00	2100
2	72.50	2450
3	60.00	1800
4	50.00	2000
5	52.50	2350
6	45.00	1800
7	60.00	2600
8	80.00	3000
9	62.50	1500
10	55.00	1800
11	67.50	2500
12	35.00	1250
13	55.00	1900
14	70.00	2500
15	65.00	2000
16	45.00	1500
17	50.00	1900

Fig. 16.1 The sample survey data.

The sums and sums of squares and cross products for these data are:

$$n = 17$$
$$\Sigma x_i = 34950 \qquad \Sigma y_i = 1005.00$$
$$\bar{x} = 2055.88 \qquad \bar{y} = 59.12$$
$$\Sigma x_i^2 = 75197500 \qquad \Sigma y_i^2 = 61925.00$$

$$\Sigma x_i y_i = 21312.50$$

The new estimators that we shall introduce are most effective if the relationship between Y and X is linear. Thus a check of the linearity of Y on X should be made by plotting the observed sample points on a scatter diagram. If this is done for our 17 points, we observe that Y and X for our data are linearly related. On the other hand, if the relationship appears to be nonlinear, transformations such as the log x, $\sqrt{x}$, should be tried in an attempt to improve the linearity of the relationship and then the new transformed variable used as the X in our analyses.

The two new estimators we shall consider are the ratio and the linear regression. The formulas for these are

$$\text{Ratio estimator:} \quad \hat{\mu}_y = \bar{y}_r = \frac{\bar{y}}{\bar{x}} \mu_x \qquad (16.4)$$

and

$$\text{Linear regression estimator:}$$

$$\hat{\mu}_y = \bar{y}_L = \bar{y} + b(\mu_x - \bar{x}) \qquad (16.5)$$

where b is the regression coefficient for the linear regression of Y on X.

Using the estimators for the illustrative example, we find that the ratio estimator gives the estimate

$$\hat{\mu}_y = \bar{y}_r = \frac{59.12}{2055.88} (2073.00) = \$59.61$$

and the linear regression estimate yields

$$\hat{\mu}_y = \bar{y}_L = 59.12 + 0.01946 \quad (2073.00 - 2055.88)$$

$$= 59.12 + 0.33 = 59.45$$

since the estimate of the regression coefficient is given by

$$b = \frac{\sum_{i=1}^{n} x_i y_i - \sum_{i=1}^{n} x_i \cdot \sum_{i=1}^{n} y_i / n}{\sum_{i=1}^{n} x_i^2 - \left(\sum_{i=1}^{n} x_i\right)^2 / n} \qquad (16.6)$$

For the 17 observations we obtain

$$b = \frac{2131250 - (34950)(1005)/17}{75197500 - (34950)^2/17}$$

$$= \frac{65088.24}{3344412} = 0.01946$$

Since the sample mean estimate was 59.12, we now have three different values that could be used to estimate the same unknown parameter. We would surely like to know which estimator is best. For example, the bias of an estimator could be used as the criterion for estimator selection. Since it can be shown that the sample mean is always unbiased and the other two estimators are not, this criterion would always lead one to use the sample mean. In many cases, however, if the bias of an estimator is small and if it has a smaller variance than the unbiased estimator, the biased estimator may be more desirable. Since the bias of the ratio and regression estimator is generally small, we shall use the variance as a means of comparing the three estimators.

For ease of notation, we shall introduce the "finite correction term"

$$1 - f = 1 - \frac{n}{N} = \frac{N - n}{n}$$

which is encountered when dealing with finite populations. Using this notiation, we can rewrite the variance of the sample mean estimator as

$$\sigma^2_{\bar{y}_s} = \frac{N - n}{n(N - 1)} \cdot \sigma^2_y = \frac{N - n}{n(N - 1)} \cdot \frac{\Sigma(y_i - \mu_y)^2}{N} \tag{16.7}$$

$$= \frac{1}{n} \cdot \frac{N - n}{N} \cdot \frac{\Sigma(y_i - \mu_y)^2}{N - 1} = (1 - f) \frac{\sigma'^2_y}{n}$$

Note that we have introduced the expression

$$\sigma'^2_y = \frac{\displaystyle\sum_{i=1}^{N} (y_i - \mu_y)^2}{N - 1} \tag{16.8}$$

to simplify the formula. We shall use this notation throughout the remainder of the chapter.

The variance of the ratio estimator is approximated by

$$\sigma^2_{\bar{y}_r} \doteq (1 - f) \frac{\mu^2_y}{n} \left(\frac{\sigma'^2_y}{\mu^2_y} + \frac{\sigma'^2_x}{\mu^2_x} - \frac{2\sigma'_{yx}}{\mu_y \mu_x} \right) \tag{16.9}$$

where

$$\sigma'^2_x = \frac{\displaystyle\sum_{i=1}^{N} (x_i - \mu_x)^2}{N - 1} \tag{16.10}$$

and

$$\sigma'_{yx} = \frac{\displaystyle\sum_{i=1}^{N} (x_i - \mu_x) \cdot (y_i - \mu_y)}{N - 1} \tag{16.11}$$

The variance of the linear regression estimator when b is assumed to be known is

$$\sigma^2_{\bar{y}_L} = \frac{(1 - f)}{n} \left(\sigma'^2_y - 2b \cdot \sigma'_{yx} + b^2 \cdot \sigma'^2_x \right) \qquad (16.12)$$

If b must be computed from the sample data, the variance of the linear regression estimator is approximated by

$$\sigma^2_{\bar{y}_L} \doteq \frac{(1 - f)}{n} \sigma'^2_y (1 - \rho^2) \qquad (16.13)$$

$$\doteq \frac{(1 - f)}{n} \left(\sigma'^2_y - \frac{\sigma'^2_{xy}}{\sigma'^2_x} \right) \qquad (16.14)$$

For our illustrative rental example let us compute the variances for the three estimates we obtained earlier to see which one has the smaller estimated variance. From the sample data in Fig. 16.1 we obtain the following values:

$$\hat{\sigma}'^2_x = s'^2_x = \frac{\sum_{i=1}^{n} x^2_i - \left(\sum_{i=1}^{n} x_i \right)^2 / n}{n - 1} = \frac{75197500 - (34950)^2 / 17}{16}$$

$$= 209025.74$$

$$\hat{\sigma}'^2_y = s'^2_y = \frac{\sum_{i=1}^{n} y^2_i - \left(\sum_{i=1}^{n} y_i \right)^2 / n}{n - 1} = \frac{61925.00 - (1005)^2 / 17}{16}$$

$$= 156.99$$

$$\hat{\sigma}'_{yx} = s'_{yx} = \frac{\sum_{i=1}^{n} x_i y_i - \left(\sum_{i=1}^{n} x_i \sum_{i=1}^{n} y_i \right) / n}{n - 1}$$

$$= \frac{213.250 - (34950)(1005) / 17}{16}$$

$$= 4068.01$$

Since $1 - f = 1 - 17/210 = 0.9190$, we find that the variance estimates are

$$\hat{\sigma}_{\bar{y}}^2 = s_{\bar{y}}^2 = (0.9190)\left(\frac{156.99}{17}\right) = 8.49$$

$$\hat{\sigma}_{\bar{y}_r}^2 = s_{\bar{y}_r}^2 = (0.9190)\left(\frac{(59.12)^2}{17}\right)\left(\frac{156.99}{(59.12)^2} + \frac{209025.74}{(2055.88)^2}\right.$$

$$\left. - \frac{2(4068.01)}{(59.12)(2055.88)}\right)$$

$$= (0.9190)(205.5985)(0.027431) = 5.18$$

$$\hat{\sigma}_{\bar{y}_L}^2 = s_{\bar{y}_L}^2 = \frac{(0.9190)}{17}\left(156.99 - \frac{(4068.01)^2}{209025.74}\right) = 4.21$$

Note that in this case the linear regression estimate has the smallest estimated variance, so we would be inclined to use it in place of the sample mean or the ratio estimator. In general, the ratio estimator is most effective when the regression of Y on X goes through the origin. If the regression has, however, a nonzero intercept, the linear regression estimator should be used. In fact, when there is a perfect linear relationship between Y and X, the ratio (for a zero intercept) and the linear regression estimator (for a nonzero intercept) will estimate μ_y perfectly with a sample of size 1. If the linear relationship is, however, not strong enough, then the sample mean estimator will have a smaller variance.

We can develop more quantitative guides to the estimator selection problem by comparing the variances. The three rules that can be evolved are:

(1) If

$$\frac{R\sigma_x'^2}{2} < \sigma_{xy}'$$

where $R = \mu_y/\mu_x$, then the ratio estimator will have a smaller variance than the sample mean estimator.

(2) If

$$|2b\sigma'_{yx}| > b^2\sigma'_x$$

then the linear regression estimator has a smaller variance than the sample mean estimator.

(3) If

$$R^2\sigma'^2_x - 2R\sigma'_{xy} + \sigma'^2_{xy}/\sigma'^2_x > 0$$

then the linear regression estimator will have a smaller variance than the ratio estimator.

Let us use these selection rules to confirm the conclusion that we made earlier when we actually computed the sample variances and picked the linear regression estimator. We evaluate each of the expressions given by the three rules and find that for (1):

$$\frac{(59.12)(209026)}{(2055.88)(2)} = 3005.42 < 4068.01$$

so that the ratio estimator is preferred to the sample mean estimator. For (2),

$$|2(0.01946)(4068.01)| > (0.01946)^2(209026)$$

$$158.73 > 79.16$$

and thus the regression estimator is preferred to the sample mean estimator. Finally, the decision between the ratio and the regression estimator is made by evaluating (3):

$$(0.02876)^2(209026) - 2(0.02876)(4068.01) + (4068.0)^2/209026$$

$$= 18.07$$

and since the result is greater than zero, the regression

estimator should be used. The advantage of the rules is that we can often apply them before making the more detailed variance computations.

16.4 Stratified Designs

When it is possible to obtain the value of the variable X for each population unit prior to the actual survey, it may be possible to divide the entire population frame into separate subpopulations so that the variance of Y within each subpopulation is less than its variance over the entire population. Such a division of the population is called stratification and it can be used to evolve more efficient sample designs.

To evolve an optimal stratification rule, we need to know the Y value for each unit of the population. Since this information about Y is, of course, unknown, sample design stratification procedures use the X variable information rather than Y.

Regardless of how the strata are determined, the stratified design approach generally calls for a random sample to be selected from each stratum. An appropriate estimator is calculated for each stratum from the sampled units. These stratum estimators are then combined into an estimator of the population mean.

For notational purposes, we need to be able to identify each individual unit and the subpopulation (stratum) to which the unit belongs. To do this we shall use a double subscript notation; y_{ik} will represent the Y value for the ith unit from the kth stratum. The same notation is used for the X variable; thus x_{ik} is the X value of the ith unit in the kth stratum.

There are a total of K strata; the number of units in the kth stratum is denoted by N_k, and the sample size for a stratum is denoted by n_k. Each stratum has a mean and variance for both the X and Y variable; they are denoted by μ_{xk}, σ'^2_{xk}, μ_{yk}, σ'^2_{yk}, respectively. The sample estimates of these parameters are $\bar{x}_k$, s'^2_{xk}, $\bar{y}_k$, and s'^2_{yk}. The population mean for X and Y is μ_x and μ_y and the respective sample means are $\bar{x}$ and $\bar{y}$.

To evolve a stratified design based upon known X values, we follow the six steps listed below:

(1) Order the units of the population according to their X values.
(2) Divide the ordered list into strata.
(3) Determine the estimator to be used.
(4) Determine the total sample size n.
(5) Determine the strata sample sizes n_k, and randomly select this number of units from each strata.
(6) Perform the survey and find the population estimates and its variance.

We shall illustrate a technique appropriate for each of these steps using the problem of estimating the average rent paid in a particular urban renewal area. In the previous sections the population frame consisted N = 210 units, but in this section we shall use a population frame of only N = 50 units for ease of representation.

1. Order the units of the population according to their X values

In our illustration we shall again use the assessed evaluation for each rental unit in the population area obtained from official records prior to the survey for the X variable. The units in the population are ordered by the X value and are given in Fig. 16.2:

Strata 1	Strata 2	Strata 3
$x_{11} = 4100$	$x_{21} = 2350$	$x_{31} = 1710$
$x_{12} = 3800$	$x_{22} = 2300$	$x_{32} = 1700$
$x_{13} = 3750$	$x_{23} = 2150$	$x_{33} = 1650$
$x_{14} = 3400$	$x_{24} = 2150$	$x_{34} = 1650$
$x_{15} = 3200$	$x_{25} = 2100$	$x_{35} = 1630$
$x_{16} = 3000$	$x_{26} = 2000$	$x_{36} = 1625$
$x_{17} = 3000$	$x_{27} = 1950$	$x_{37} = 1620$
$x_{18} = 2800$	$x_{28} = 1925$	$x_{38} = 1600$
$x_{19} = 2700$	$x_{29} = 1910$	$x_{39} = 1600$
$x_{1,10} = 2650$	$x_{2,10} = 1875$	$x_{3,10} = 1575$
$x_{1,11} = 2500$	$x_{2,11} = 1850$	$x_{3,11} = 1550$
	$x_{2,12} = 1800$	$x_{3,12} = 1500$
	$x_{2,13} = 1775$	$x_{3,13} = 1500$
	$x_{2,14} = 1775$	$x_{3,14} = 1500$
	$x_{2,15} = 1770$	$x_{3,15} = 1450$
		$x_{3,16} = 1430$
		$x_{3,17} = 1425$
		$x_{3,18} = 1400$
		$x_{3,19} = 1400$
		$x_{3,20} = 1350$
		$x_{3,21} = 1350$
		$x_{3,22} = 1325$
		$x_{3,23} = 1325$
		$x_{3,24} = 1250$

Fig. 16.2 The ordered listing of N = 50 units according to X value.

2. Determination of the strata

Since for most populations the variance decreases with the range, we can often effectively stratify the population by dividing the ordered listing into strata. Although there is the problem of how many strata to use and where to place the strata division marks, we shall not consider these problems in this introductory textbook but simply illustrate the procedure by selecting three strata, with $N_1 = 11$, $N_2 = 15$, and $N_3 = 24$.

To illustrate the effectiveness of the above stratification, we have computed the sums, sums of squares, and the standard deviations for each stratum and for the total population. The sums and sums of squares for the strata are:

$$N_1 = 11 \qquad \Sigma x_{i1} = 34,900 \qquad \Sigma x_{i1}^2 = 113,515,000$$

$$N_2 = 15 \qquad \Sigma x_{i2} = 29,660 \qquad \Sigma x_{i2}^2 = 59,165,600$$

$$N_3 = 24 \qquad \Sigma x_{i3} = 36,165 \qquad \Sigma x_{i3}^2 = 54,878,425$$

$$N = 50 \qquad \Sigma\Sigma x_{ik} = 100,725 \qquad \Sigma\Sigma x_{ik}^2 = 227,559,025$$

The within strata standard deviations are:

$$\sigma'_{x1} = 527.90 \qquad \sigma'_{x2} = 192.33 \qquad \sigma'_{x3} = 128.92$$

and the population standard deviation is

$$\sigma'_x = 709.25$$

Thus it can be seen that the stratification has effectively reduced the total variation to smaller within strata variances.

3. Determination of the estimator

In this first example we shall assume that the sample mean has been chosen as the estimator and postpone the discussion of the use of other types of estimators until later. The sample mean estimator for a stratified population is given by

$$\hat{\mu}_y = \bar{y}_{ss} = \frac{\displaystyle\sum_{k=1}^{K} N_k \cdot \bar{y}_k}{N} \qquad (16.15)$$

where $\bar{y}_k$ = mean of sample units obtain from kth stratum

$$= \sum_{i=1}^{n_k} y_{ik}/n_k, \text{ and its variance is estimated by}$$

$$s_{\bar{y}_{ss}}^2 = \frac{1}{N^2}\left(\frac{\sum_{k=1}^{K} N_k(N_k - n_k)s_{yk}'^2}{n_k}\right) \tag{16.16}$$

where

$$s_{yk}'^2 = \frac{\sum_{i=1}^{n_k} y_{ik}^2 - \left(\sum_{i=1}^{n_k} y_{ik}'\right)^2/n_k}{n_k - 1}$$

4. Determination of the total sample size n

The total sample size can be determined by using any one of the three following methods:

(a) The total sample size n can be given.

(b) The minimum sample size n can be determined by using the constraint that the variance of the estimator has a certain prescribed value:

$$n = \frac{\left(\sum_{k=1}^{K} N_k\sigma_k'\right)^2}{N^2\sigma_{\bar{y}_{ss}}^2 + \sum_{k=1}^{K} N_k^2\sigma_k'^2} \tag{16.17}$$

To illustrate the use of this formula, suppose it is required that $\sigma_{\bar{y}_{ss}} \leq \$1.50$. Since we only know the within stratum variance for the X variable, we must convert these variances to the corresponding variances for the Y variable. To find this relationship we use the regression of Y or X and recognize that

$$Y = bX + \varepsilon$$
$$\sigma_y^2 = b^2\sigma_x^2 + \sigma_\varepsilon^2$$

We need a prior estimate of b and σ_ε to determine that variance of Y. To obtain an estimator of b, we consider that the monthly rent charged for a house is approximately 1 percent of the actual value (V) and that the property is assessed at one-third the actual value. Thus $Y = (0.01)V$ and $X = V/3$ so $Y = (0.01)(3X) = 0.03X$ and $b = 0.03$. To estimate σ_ε we use the fact that $\sigma_\varepsilon = \sigma_y$ and estimate σ_y by using the range. Since the houses in the area have about the same value, the range of rents may be about \$20: $\sigma_\varepsilon = 20/5 = \$4.00$. We can now apply the relationship $\sigma_{yk}^2 = (0.03)^2 \sigma_{xk}^2 + (4)^2$ and find that

$$\sigma_{y1}^2 = (0.0009)(527.90)^2 + 16 = 266.81$$
$$\sigma_{y2}^2 = (0.0009)(192.33)^2 + 16 = 49.29$$
$$\sigma_{y3}^2 = (0.0009)(128.92)^2 + 16 = 30.94$$

Using these values we obtain:

$$n = \frac{[11(16.33) + 15(7.02) + 24(5.56)]^2}{(50)^2(\$1.50)^2 + [11(266.81) + 15(49.29) + 24(30.94]}$$

$$= \frac{17503}{10042} = 17.43 \quad = 18$$

(c) The sample size can also be determined to minimize the survey cost while still meeting the requirement that the estimator have a prescribed variance. If we assume that there is a cost c_{ik} associated with surveying the ith unit in the kth stratum, we can convert these individual costs into average stratum costs c_k, by simply averaging the unit costs over each stratum. The required sample size is then given in terms of the average stratum costs c_k and is

$$n = \frac{\left(\sum_{k=1}^{K} N_k \sigma_k' \sqrt{c_k}\right)\left(\sum_{k=1}^{K} N_k \sigma_k'/\sqrt{c_k}\right)}{N^2 \sigma_{y_{ss}}^2 + \sum_k N_k \sigma_k'^2} \qquad (16.18)$$

To apply the formula, we need the average survey cost for each stratum. For the example, let us assume that they are given by $c_1 = 4.0$, $c_2 = 2.5$, and $c_3 = 1.0$. Using Eq. (16.18), we then obtain the sample size by evaluating

$$n = \left(\frac{11(16.33)\sqrt{4.0} + 15(7.02)\sqrt{2.5} + 24(5.56)\sqrt{1.0}}{848106} \right).$$

$$\left(\frac{11(16.33)}{\sqrt{2.0}} + \frac{15(7.02)}{\sqrt{2.5}} + \frac{24(5.56)}{\sqrt{1.0}} \right)$$

$$= \frac{(659.19)(289.85)}{10042} = 18.94 = 19$$

5. Determination of sample sizes in the strata

The three methods generally used to obtain the n_k's are given below. In each case the total sample size is assumed to be given.

(a) Proportional allocation can be used. In this case the total sample size is divided among the strata according to their sizes by using the following formula:

$$n_k = n \frac{N_k}{N} \tag{16.19}$$

If a total sample size of $n = 18$ is used, then the proportional allocation is

$$n_1 = \frac{18(11)}{50} = 3.96 = 4$$

$$n_2 = \frac{18(15)}{50} = 5.40 = 5$$

$$n_3 = \frac{18(24)}{50} = 8.64 = 9$$

Note the round-off problem encountered so as to keep $n = \sum n_k$ $= 18$. It should be noted that though $n = 18$ was determined in 4(b), the variance constraint will not be met if proportional allocation is used.

(b) Optimum allocation determines the sample sizes in the strata so the variance of the estimator is minimized. The formula is

$$n_k = n \frac{N_k \sigma'_k}{\Sigma N_k \sigma'_k} \qquad (16.20)$$

Using a total sample size of n = 18, we obtain the following allocation:

$$n_1 = 18 \left(\frac{11(16.33)}{418.37} \right) = 7.73 = 8$$

$$n_2 = 18 \left(\frac{15(7.02)}{418.37} \right) = 4.53 = 4$$

$$n_3 = 18 \left(\frac{24(5.56)}{418.37} \right) = 5.74 = 6$$

When the total sample size is determined to meet a prescribed variance constraint, then the strata sample size should be determined by 4(b) should be used.

$$n_1 = 19 \left(\frac{11(16.63)}{418.37} \right) = 8.31 = 8$$

$$n_2 = 19 \left(\frac{15(7.02)}{418.37} \right) = 4.78 = 5$$

$$n_3 = 19 \left(\frac{24(5.56)}{418.37} \right) = 6.06 = 6$$

Since the sample sizes must be integers and thus the computed value must be rounded off, the resulting integer sample sizes should be substituted into the equation for the variance of the estimator as given by Eq. (16.16) to be sure that the total variance constraint is met by the rounded-off n_k's. Using the above values of n_k in the equation yields $\sigma_{\bar{y}_{ss}} = \$1.39$, which indicates that the design satisfies the contraint, which is $\sigma_{\bar{y}_{ss}} \leq \$1.50$. If the constraint is not met, some n_k's will need to be increased or the rounding-off accomplished in a somewhat different manner.

(c) The optimum allocation for minimizing the survey cost can also be used to determine the strata sample sizes. In this case the values for n_k are found directly by using

$$n_k = n \frac{N_k \sigma'_k / \sqrt{c_k}}{\sum_k N_k \sigma'_k / \sqrt{c_k}} \tag{16.21}$$

Using Eq. (16.21), together with the total sample size obtained by 4(c), we obtain the following allocation:

$$n_1 = 19 \left(\frac{11(16.33)/\sqrt{4.0}}{295.51} \right) = 5.77 = 6$$

$$n_2 = 19 \left(\frac{15(7.02)/\sqrt{2.5}}{295.51} \right) = 4.28 = 4$$

$$n_3 = 19 \left(\frac{24(5.56)/\sqrt{1.0}}{295.51} \right) = 8.58 = 9$$

When this allocation is checked to see if the variance constraint was met, it is found that $\sigma_{\bar{y}_{ss}} = \$1.51$, and thus the design just fails to meet the constraint. However, it was determined to keep $n = 19$ rather than to use $n = 20$ since the difference between the required standard deviation and the actual one is only 0.01.

It is of interest to compare these possible allocations. Figure 16.3 gives the allocation, the standard deviation of the estimator, and the expected cost of the survey

$$\text{Expected survey cost} = \sum_{k=1}^{K} n_k c_k$$

Different Total Sample Sizes
for Variance Constraint of 1.50

Allocation	Minimum Sample Size $n = 18$		Minimum Survey Cost Sample Size $n = 19$	
	Proportional	Optimum	Optimum	Optimum Cost
n_1	4	8	8	6
n_2	5	4	5	4
n_3	9	6	6	9
Standard deviation	1.77	1.46	1.39	1.51
Expected cost	37.5	48.0	50.5	43.0

Fig. 16.3 Comparison of different strata allocations.

The table in Fig. 16.3 displays the results of the several possible designs but does not enable one to make direct comparisons. In each design the three design factors (sample size, standard deviation of the mean estimate, and expected cost) all vary. Thus the optimum design for sample size $n = 18$ has a higher expected cost but a smaller standard deviation than the design that proportionally assigns the values to the n_k's. As expected the proportional allocation does not meet the variance constraint of 1.50, while the optimum allocation does meet the constraint. To minimize the survey cost, a sample of $n = 19$ and the optimum cost allocation should be used.

6. In selecting the sample and performing the survey we decided to use the sample size and strata sample sizes obtained from 4(b) and 5(b). Thus $n = 19$, $n_1 = 8$, $n_2 = 5$, and $n_3 = 6$. Eight random numbers between 1 and 11 were obtained, five between 1 and 15, and six between 1 and 25. The survey was made; the sampled observations are:

Strata 1	Strata 2	Strata 3
n = 8	n = 5	n = 6

Strata 1	Strata 2	Strata 3
$y_{11} = 110.00$	$y_{22} = 65.00$	$y_{33} = 50.00$
$y_{12} = 1C7.00$	$y_{25} = 65.00$	$y_{36} = 47.50$
$y_{13} = 112.50$	$y_{27} = 50.00$	$y_{37} = 42.50$
$y_{15} = 95.00$	$y_{2,10} = 56.00$	$y_{3,10} = 39.00$
$y_{16} = 80.00$	$y_{2,13} = 48.00$	$y_{3,19} = 40.00$
$y_{17} = 100.00$		$y_{3,23} = 37.50$
$y_{19} = 75.00$		
$y_{1,11} = 70.00$		

Fig. 16.4 Sample survey results.

We summarize the results by obtaining an estimate of μ_y and its standard deviation. To accomplish this, we first find the strata estimates, which are

$$\bar{y}_1 = \frac{749.50}{8} \qquad \bar{y}_2 = \frac{284.00}{5} \qquad \bar{y}_3 = \frac{256.50}{6}$$

$$= 93.69 \qquad = 56.80 \qquad = 42.75$$

$$s_1' = \sqrt{\frac{1936.47}{7}} \qquad s_2' = \sqrt{\frac{258}{4}} \qquad s_3' = \sqrt{\frac{124.375}{5}}$$

$$= \sqrt{276.64} \qquad = \sqrt{64.7} \qquad = \sqrt{24.875}$$

$$= \$16.63 \qquad = \$8.04 \qquad = \$4.99$$

Thus the estimated average rent and its standard deviation is: is:

$$\bar{y}_{ss} = \frac{\Sigma N_k \bar{y}_k}{N} = \frac{11(93.69) + 15(56.80) + 24(42.75)}{50} = \$58.17$$

$$s_{\bar{y}_{ss}} = \sqrt{\frac{1}{2500}\left[11(3)\left(\frac{276.64}{8}\right) + 15(10)\left(\frac{64.7}{5}\right) + 24(18)\left(\frac{24.875}{6}\right)\right]}$$

$$= \sqrt{1.9492} = 1.40$$

In this section we introduced the concept of stratification of the population into subpopulations to improve the
efficiency of the sample estimate. The goal of such strati-
fications is to make each stratum as homogeneous relative to
the response variable to be studied as possible. We showed
that one method of accomplishing such a stratification is to
use a single descriptive variable X, which is linearly correlated with the response variable. However, there are many
other methods that can be used. We have also simply introduced some problems without giving techniques to resolve
them; for example, the number of strata to be used must be
determined. These additional considerations are beyond the
scope of this introductory chapter.

16.5 The Use of Other Estimators in
Stratified Designs

The ratio and regression estimators can also be used in
stratified designs. Since there are K subpopulations, the
ratio or regression coefficient used in determining the K
estimators can be found for each stratum separately, or a
single value can be obtained for the entire population.
Thus for each type of estimator there are different formulas
that can be used; we can identify seven different estimators
that can be used in stratified designs:

(1) A sample mean estimator (as used in the previous
    section)
(2) A ratio estimator, with a separate ratio for each
    stratum
(3) A ratio estimator, with the same ratio used for all
    strata
(4) A linear regression estimator, with separate re-
    gression coefficients known in advance for each
    stratum
(5) A linear regression estimator, with the same re-
    gression coefficient known in advance used for all
    strata
(6) A linear regression estimator, with separate re-

gression coefficients that must be estimated for each stratum

(7) A linear regression estimator, with a single regression coefficient that must be estimated used for all strata

We shall give the equation for each of these estimators and their estimated standard deviation, using the same notation that was introduced in the preceding sections. Once a particular estimator is chosen, the sample size question must be considered. Formulas similar to the one used to determine the sample size for the sample mean estimator, Eq. (16.3), are available, but we shall not consider them. We therefore restrict our considerations to giving some general guidelines as to how to use the estimators and then illustrate the use of the combined ratio estimator for the illustrative example.

1. Sample mean

$$\bar{y}_{ss} = \frac{\sum\limits_{k=1}^{K} N_k \bar{y}_k}{N} \tag{16.22}$$

$$s_{\bar{y}_{ss}} = \frac{1}{N} \sqrt{\sum_{k=1}^{K} N_k (N_k - n_k) \frac{s'^2_{yk}}{n_k}} \tag{16.23}$$

2. Separate ratio

$$\bar{y}_{rs} = \frac{1}{N} \left(\sum_{k=1}^{K} N_k \frac{\bar{y}_k}{\bar{x}_k} \mu_{xk} \right) \tag{16.24}$$

$$s_{\bar{y}_{rs}} = \frac{1}{N} \sqrt{\sum_{k=1}^{K} \left(\frac{1 - f_k}{n_k} \right) N_k^2 \bar{y}_k^2 \left(\frac{s'^2_{yk}}{\bar{y}_k^2} + \frac{s'^2_{xk}}{\bar{x}_k^2} - 2 \frac{s'_{yxk}}{\bar{x}_k \bar{y}_k} \right)} \tag{16.25}$$

where

$$1 - f_k = 1 - \frac{n_k}{N_k} \qquad .$$

553

3. Combined ratio $\quad \bar{y}_{rc} = \dfrac{\bar{y}}{\bar{x}} \mu_x$ $\qquad\qquad\qquad$ (16.26)

where

$$\bar{y} = \frac{\displaystyle\sum_{k=1}^{K} N_k \bar{y}_k}{N} \qquad \bar{x} = \frac{\displaystyle\sum_{k=1}^{K} N_k \bar{x}_k}{N}$$

and

$$s_{\bar{y}_{rc}} = \frac{1}{N} \sqrt{\sum_{k=1}^{K} \left(\frac{1 - f_k}{n_k}\right) N_k^2 \bar{y}^2 \left(\frac{s'^2_{yk}}{\bar{y}^2} + \frac{s'^2_{xk}}{\bar{x}^2} - 2\,\frac{s'_{yxk}}{\bar{y}\bar{x}}\right)} \quad (16.27)$$

4. Separate regression — predetermined b_k, $k = 1, \ldots, K$:

$$\bar{y}_{Ls_1} = \sum_{k=1}^{K} \frac{N_k}{N} \left[\bar{y}_k + b_k\!\left(\mu_{xk} - \bar{x}_k\right)\right] \qquad\qquad (16.28)$$

$$s_{\bar{y}_{Ls_1}} = \frac{1}{N} \sqrt{\sum_{k=1}^{K} \frac{N_k^2(1 - f_k)}{n_k} \left(s'^2_{yk} - 2b_k s'_{yxk} + b_k^2 s'^2_{xk}\right)} \quad (16.29)$$

5. Combined regression — predetermined b, $k = 1,2, \ldots, K$

$$\bar{y}_{Lc_1} = \bar{y} + b(\mu_x - \bar{x})$$

where

$$\bar{y} = \frac{\displaystyle\sum_{k=1}^{K} N_k \bar{y}_k}{N} \qquad\qquad \bar{x} = \frac{\displaystyle\sum_{k=1}^{K} N_k \bar{x}_k}{N} \qquad (16.30)$$

$$s_{\bar{y}_{Lc_1}} = \frac{1}{N} \sqrt{\sum_{k=1}^{K} \frac{N_k^2(1 - f_k)}{n_k} \left(s'^2_{yk} - 2b s'_{yxk} + b^2 s'_{xk}\right)} \quad (16.31)$$

6. Separate regression — calculated b_k in each stratum:

$$\bar{y}_{Ls_2} = \sum_{k=1}^{K} \frac{N_k}{N} \left[\bar{y}_k + b_k (\mu_{xk} - \bar{x}_k) \right] \qquad (16.32)$$

where

$$b_k = \frac{\displaystyle\sum_{i=1}^{n_k} (y_{ik} - \bar{y}_k)(x_{ik} - \bar{x}_k)}{\displaystyle\sum_{i=1}^{n_k} (x_{ik} - \bar{x}_k)^2}$$

$$s_{\bar{y}_{Ls_2}} = \frac{1}{N} \sqrt{\sum_{k=1}^{K} \frac{N_k^2 (1 - f_k)}{n_k} \left[s_{yk}'^2 - \frac{s_{xyk}'^2}{s_{xk}'^2} \right]} \qquad (16.33)$$

7. Combined regression — calculated b used for all strata:

$$\bar{y}_{Lc_2} = \bar{y} + b(\mu_x - \bar{x}) \qquad \bar{y} = \Sigma N_k \bar{y}_k / N \atop \bar{x} = \Sigma N_k \bar{x}_k / N \qquad (16.34)$$

where

$$b = \frac{\displaystyle\sum_{k=1}^{K} N_k^2 (1 - f_k)/n_k (n_k - 1) \sum_{i=1}^{n_k} (y_{ik} - \bar{y}_k)(x_{ki} - \bar{x}_k)}{\displaystyle\sum_{k=1}^{K} N_k^2 (1 - f_k)/n_k (n_k - 1) \sum_{i=1}^{n_k} (x_{ik} - \bar{x}_k)^2}$$

$$s_{\bar{y}_{Lc_2}} = \frac{1}{N} \sqrt{\sum_{k=1}^{K} \frac{N_k^2 (1 - f_k)}{n_k (n_k - 2)} \left(\sum_{i=1}^{n_k} (y_{ik} - \bar{y}_k)^2 - b^2 (x_{ik} - \bar{x}_k)^2 \right)}$$

The introduction of the concept of stratification in-creased the number of possible estimators for the population mean to 7. The general guidelines given previously for choosing between the sample mean, the ratio, and the linear regression estimator still apply.

The stratification procedure itself should be used whenever such a procedure will yield relatively homogeneous strata so that most of the variation is between the strata. If such a stratification procedure yields different ratios or regression coefficients in each stratum, then the separate ratio or regression estimators should be used.

To illustrate the application of these additional estimators for stratified designs, let us again use the urban renewal rental problem. In the previous section we used a three-strata design and picked a random sample from each strata (Fig. 16.4). Now, instead of using the stratified sample mean estimator we shall take advantage of the information given about Y by the descriptive variable X, whose value is known for each unit in the population. Although we could use more than one of the estimators given above, we shall show only the actual calculations for the combined regression estimator when the regression coefficient b has a predetermined value. This estimator is given by Eq. (16.30) and its variance is given by Eq. (16.31).

From the population frame given in Fig. 16.2 we know that

$$N = 50 \quad N_1 = 11 \quad N_2 = 15 \quad N_3 = 24$$

and thus

$$f_1 = \frac{8}{11} = 0.7273 \quad f_2 = \frac{5}{15} = 0.3333 \quad f_3 = \frac{6}{24} = 0.2500$$

and

$$\mu_x = 2014.50$$

Summarizing the sample results given in Fig. 16.4 for the Y variable, we find that

$$\bar{y}_1 = \frac{749.50}{8} = 93.69$$
$$\bar{y}_2 = \frac{284.00}{5} - 56.80$$

$$\bar{y}_3 = \frac{256.50}{6} = 42.75$$

and thus the overall sample mean is

$$\bar{y} = \frac{11(93.69) + 15(56.80) + 24(42.75)}{50} = 58.17$$

For the X variable, we refer to Fig. 16.2 to obtain the corresponding x values for the sampled units

$$\bar{x}_1 = \frac{26050}{8} = 3256.25$$

$$\bar{x}_2 = \frac{10000}{5} = 2000.00$$

$$\bar{x}_3 = \frac{9195}{6} = 1532.50$$

and thus

$$\bar{x} = \frac{11(3256.25) + 15(2000.00) + 24(1532.50)}{50}$$

$$= 2051.975$$

The sample variances for the Y variable are

$$s_{y1}'^2 = \frac{\sum_i y_{1i}^2 - \left(\sum_i y_{1i}\right)^2/n_1}{n_1 - 1} = \frac{(72156.25) - (749.5)^2/8}{7}$$

$$= 276.64$$

$$s_{y2}'^2 = \frac{\sum_i y_{2i}^2 - \left(\sum_i y_{2i}\right)^2/n_2}{n_2 - 1} = 64.7$$

$$s_{y3}'^2 = \frac{\sum_i y_{3i}^2 - \left(\sum_i y_{3i}\right)^2/n_3}{n_3 - 1} = 24.875$$

For the X variable, we have

$$s'^2_{x1} = 323884$$

$$s'^2_{x2} = 42187$$

$$s'^2_{x3} = 18487$$

and the sample covariance for X and Y are

$$s'_{yx1} = \frac{\sum\limits_{i} y_{1i}x_{1i} - \left(\sum\limits_{i} y_{1i}x_{1i}\right)^2/n_1}{n_1 - 1}$$

$$= \frac{2500975 - (26050)(749.5)/8}{7}$$

$$= 8630.80$$

Likewise,

$$s'_{yx2} = 1425.00 \qquad s'_{yx3} = 512.75$$

The predetermined value of the regression coefficient is assumed to be b = 0.0333. This value is based on the assumption that the monthly rents are about 1 percent of the value of the property, and appraised evaluation is about one-third of the actual value. The combined linear regression estimator of μ_y is then

$$\hat{\mu}_y = \bar{y}_{Ls_1} = \bar{y} + b(\mu_x - \bar{x})$$

$$= 58.17 + 0.0333(2014.500 - 2051.975)$$

$$= 58.17 - 1.2479 = 56.92$$

We note that the regression estimate adjusts the estimate to compensate for the difference between the population and sample mean for the X variable.

To illustrate the computation of the estimated variance, we expand Eq. (16.31) to give terms for each of the three stratum:

$$s^2_{\bar{y}_{Ls_1}} = \frac{1}{N^2} \left\{ \left[\frac{N_1^2(1 - f_1)}{n_1} \left(s'^2_{y1} - 2b \cdot s'_{yx1} + b^2 \cdot s'^2_{x1} \right) \right] \right.$$

$$+ \left[\frac{N_2^2(1 - f_2)}{n_2} \left(s'^2_{y2} - 2b \cdot s'_{yx2} + b^2 \cdot s'^2_{x2} \right) \right]$$

$$\left. + \left[\frac{N_3^2(1 - f_3)}{n_3} \left(s'^2_{y3} - 2b \cdot s'_{yx3} + b \cdot s'^2_{x3} \right) \right] \right\}$$

$$= \frac{1}{50^2} \left\{ \left[\frac{11^2(1 - 0.7273)}{8} \left(276.64 - 2(0.0333)(8630.80) \right. \right. \right.$$

$$\left. \left. + (0.0333)^2 (323884) \right) \right]$$

$$+ \left[\frac{15^2(1 - 0.3333)}{5} \left(64.7 - 2(0.0333)(1425.00) \right. \right.$$

$$\left. \left. + (0.0333)^2(42187) \right) \right]$$

$$+ \left[\frac{24^2(1 - 0.2500)}{6} \left(24.875 - 2(0.0333)(512.75) \right. \right.$$

$$\left. \left. \left. + (0.0333)^2(18487) \right) \right] \right\}$$

$$= \frac{1}{2500} \left[(4.1246)(60.98) + (30.0015)(16.58) \right.$$

$$\left. + (72.000)(11.23) \right]$$

$$= \frac{1}{2500} (1557.50) = 0.6230$$

and thus

$$s_{\bar{y}_{Ls_1}} = \sqrt{0.6230} = 0.79$$

We note that the standard deviation of the regression estimator is smaller than that for the sample mean estimator whose standard deviation was 1.39. Although these variance estimators are subject to large sampling errors because the samples are so small, the improvement (if it exists) is caused by the strong linear relationship of X and Y.

Although the algebraic complexity of the numerous formulas presented in this section, when considered initially, may leave the student with a feeling of confusion, the straightforward nature of any particular application should soon make such a feeling vanish. The illustrative example of the use of the combined regression estimated with predetermined regression coefficient surely demonstrated this point.

16.6 Summary

This chapter introduced some of the basic techniques for designing a sample survey. Although the design of a survey is based on the concepts developed in sampling theory, the sample survey approaches presented in this chapter stress the finite nature of the population and as such have certain unique characteristics.

A survey design should specify the particular estimator to be used and the appropriate sample size and the method of selecting the sample. In determining the design, we always assume that a complete listing representing all units of the population is available.

We introduced different types of estimators that can be used to estimate the population mean, depending upon the additional information available for the population frame prior to the actual survey. Some guidelines were also given as to when particular estimators would have a smaller variance. There are many additional problems that must be considered in sample survey designs, but these are considered beyond the scope of this introductory chapter.

We wish to estimate the average income of a population of 60 faculty members. The population frame is given below. Both a descriptive variable X and the random variable Y to be surveyed are also given below, where X = years of experience and Y = income.

Id	X	Y	Id	X	Y	Id	X	Y
1	22	38,000	21	12	20,500	41	8	17,750
2	18	36,400	22	11	20,500	42	9	17,750
3	16	35,000	23	10	20,000	43	7	17,500
4	17	31,000	24	14	20,000	44	10	17,500
5	15	30,000	25	10	20,000	45	8	17,500
6	12	30,000	26	8	19,500	46	8	17,500
7	14	29,000	27	6	19,250	47	9	17,250
8	13	27,500	28	9	19,000	48	7	17,250
9	18	27,000	29	10	19,000	49	6	17,250
10	10	26,400	30	11	19,000	50	7	17,000
11	13	26,000	31	9	18,750	51	10	17,000
12	12	25,000	32	11	18,750	52	8	17,000
13	16	25,000	33	12	18,500	53	7	16,750
14	12	25,000	34	9	18,500	54	6	16,750
15	15	24,500	35	7	18,500	55	6	16,750
16	20	22,750	36	14	18,500	56	7	16,750
17	19	22,000	37	9	18,500	57	6	16,750
18	7	22,000	38	10	18,000	58	10	16,500
19	10	21,500	39	6	18,000	59	6	16,500
20	10	20,000	40	10	18,000	60	7	16,500

1. Using the descriptive variable X, determine the sample size required to have 95 percent confidence that the sample mean will not differ from the population mean by more than $1,000. Note that Y = 2000 X + 8, and thus $\sigma_{\bar{x}} = 1/2000\ \sigma_{\bar{y}}$.

2. Use the actual Y values to determine the appropriate sample size.

3. Select a random sample from the population, with the sample size as determined in Exercise 1. Determine $\bar{y}$ and $s_{\bar{y}}$. Compare your $\bar{y}$ with the results obtained by other members of the class and with the population mean μ_y.

4. Write a computer program that will read in the survey results and determine the estimate and its standard deviation as found in Exercise 3.

5. Use the ratio estimator to obtain an estimate of μ_y and its estimated and actual standard deviation, using the sample results obtained in Exercise 3.

6. Use the sample results obtained in Exercise 3 to find the regression estimate of μ_y and its estimated standard deviation.

7. Check to see which of the above three estimates should be more efficient, using the three rules given in the chapter.

8. Stratify the population according to their X value by dividing the population into two strata of equal size. Select random samples of the same size from each stratum so that the total sample size is the same as that determined in Exercise 1. Find the stratified sample mean estimate and its estimated standard deviation.

9. Use the samples selected in Exercise 8 to find one of the other type of estimators that can be used for stratified designs. Estimate the standard deviation of the estimator that you use. Also compute the standard deviation of the estimator, using the known Y population values and compare the two results. Compare your results with those obtained by other members of the class.

CHAPTER 17

STATISTICAL CLASSIFICATION

17.1 Introduction

Many decisions that must be made about simple day-to-day activities or more complex problems are of a classification nature. A classification decision is being made when the doctor diagnoses a patient's illness, when the admissions officer of a college accepts or rejects an applicant, when the quality control engineer accepts or rejects a manufactured item, and even when you decide whether or not it is going to rain.

Statistical classification deals with the problems of assigning an observation to one of several possible groups or populations on the basis of past data, using information obtained on one or more observable variables. To more clearly define the classification problem, let us discuss an actual illustration. For ease we shall restrict the size and dimension of the problem. Let us consider the decision that must be made by a personnel manager when considering whether to employ an applicant as a salesman. We restrict the information available to the manager about the applicant to two measurements: Y_1—the age of the applicant, and Y_2—the individual's minimum monthly living expenses.

The manager must decide whether to hire the individual on the basis of these two measures. For example, if the individual's descriptive measures are $(y_1, y_2) = (42$ years, \$750), should he be hired?

In general any decision rule relative to classification is based upon a vector of observed values $(y_1, y_2, \ldots, y_k)$, which is used to divide the sample space into nonoverlapping classification regions. In our example any observation

(y_1, y_2) can be represented by a point in a two-dimensional sample space, so a two-population classification decision rule will divide the two-dimensional sample space into two regions. Such a division of the sample space would be obtained if the company had the policy of not hiring anyone who is over 40 years old. In addition, the personnel manager feels that a person hired for a sales position should be in the socioeconomic class represented by a minimum monthly expense of $600.00. An applicant must conform both to the company's hiring policy and the personnel manager's feeling and thus the candidate whose measurements were given above would not be hired. The regions for this particular classification rule are represented in Fig. 17.1. Thus, if the candidate's measurements fall in region R_1 he would be hired, but if they fall into R_2 he would not be hired.

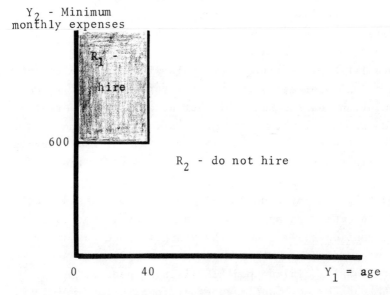

Fig. 17.1 Regions of classification.

Although classification decision rules can be determined by such policy statements that are often based upon one's judgement, we would like to consider how to evolve a classification decision rule from a statistical point of view. This means that we want to evolve a classification rule that

564

is based upon data and has certain optimum operational characteristics which can at least be estimated in advance. We
shall restrict ourselves in this initial consideration to
the two-population classification problem, although many of
the techniques generalize to the case where there are more
than two populations.

17.2 The One-variable Classification Problem

In our development of the basic concepts we shall first
consider the case when there is but one descriptive variable
and only two populations. To statistically formulate the
univariate two-population statistical classification rule,
we assume that the single descriptive variable Y has a different distribution for each of the two populations. We a
assume that each distribution is unimodal, and denote one
population by π_1 and the other population by π_2. A graphic
representation of the two distributions of Y is shown in
Fig. 17.2:

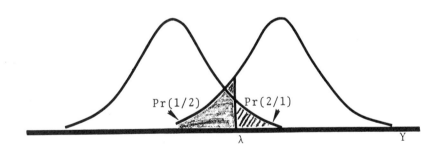

Fig. 17.2 The distribution of Y for π_1 and π_2.

For the two-population representation given in Fig.
17.2, a classification rule might be

If $y > \lambda$, classify the individual into π_2,
otherwise classify him into π_1.

565

This rule seems reasonable because the larger the value of Y, the more likely it is that the individual came from π_2. In fact, if the two distributions are normal with equal variances, this rule has optimum properties.

The effectiveness of the classification rule given above depends upon the value of λ used and the probabilities of making the two types of classification errors. In the two-population cases there are only two types of classification errors:

Pr(1/2): probability of classifying individual into π_1 when he really belongs to π_2

and

Pr(2/1): probability of classifying individual into π_2 when he really belongs to π_1

These two probabilities are represented by the shaded areas in Fig. 17.2. It can be seen that the probability values that would be obtained depend upon the value of the classification constant λ. If λ were increased, Pr(2/1) would become smaller and Pr(1/2) would become larger.

If the distribution for the populations are normal, with equal variances, these probabilities are given by

$$Pr(2/1) = \int_{\lambda}^{\infty} f_{\pi_1}(y) \, dy = \int_{\lambda}^{\infty} \frac{1}{\sqrt{2\pi}\sigma} e^{-\frac{1}{2}(y-\mu_1)^2/\sigma^2} \, dy$$

$$Pr(1/2) = \int_{-\infty}^{\lambda} f_{\pi_2}(y) \, dy = \int_{-\infty}^{\lambda} \frac{1}{\sqrt{2\pi}\sigma} e^{-\frac{1}{2}(y-\mu_2)^2/\sigma^2} \, dy$$

where

μ_1 = mean for population π_1
μ_2 = mean for population π_2
σ = common standard deviation

To determine the value of λ, we can apply any of the following criteria:

(1) Determine λ to make one of the probabilities of mis-classification take on a given value. For example, if we want $Pr(2/1) = \alpha_0$, we select λ so that

$$Pr(2/1) = \int_\lambda^\infty f_{\pi_1}(y) \, dy = \int_\lambda^\infty \frac{1}{\sqrt{2\pi}\sigma} e^{-\frac{1}{2}(y-\mu_1)^2/\sigma^2} \, dy = \alpha_0$$

We may wish that $Pr(1/2) = \beta_0$. Then we select λ so that

$$Pr(1/2) = \int_{-\infty}^\lambda f_{\pi_2}(y) \, dy = \int_{-\infty}^\lambda \frac{1}{\sqrt{2\pi}\sigma} e^{-\frac{1}{2}(y-\mu_2)^2/\sigma^2} \, dy = \beta_0$$

(2) Determine λ to make the two probabilities of misclassi-fication equal. In the equal variance situation, this is accomplished by letting

$$\lambda = \frac{\mu_1 + \mu_2}{2}$$

(3) If losses have been assigned to each type of incorrect classification decision, and the prior probabilities for an individual belonging to each population are available, then we can determine λ to minimize the expected loss EL where

$$EL = q_1 \, Pr(2/1) \, L(2/1) + q_2 \, Pr(1/2) \, L(1/2)$$

where q_1 and q_2 are the prior probabilities that an individual comes from π_1 or π_2, respectively, and $L(2/1)$ and $L(1/2)$ are the losses associated with misclassifying an individual from π_1 into π_2 and with misclassifying an individual from π_2 into π_1, respectively. The value of λ in this case is given by

$$\lambda = \frac{\frac{1}{2}(\mu_1^2 - \mu_2^2) + \sigma^2 \ln[q_2 \cdot L(1/2)/q_1 \cdot L(2/1)]}{\mu_1 - \mu_2}$$

Although any of these different criteria can be used to evolve a classification rule, we usually compare the effectiveness of any classification rule by either considering the probabilities of misclassification errors or by determining the expected loss. In the above consideration we assumed that the mean and variance of the two normal distributions are known. Since we usually do not know these population parameters, we shall use data obtained from past experience to estimate them. To be practical, we must also consider how to take advantage of information contained in more than one descriptive variable.

17.3 The Multivariate Classification Problem

If there is more than one descriptive variable, and we still wish to apply the statistical classification technique given in the last section, we can follow a procedure that reduces the multivariate observation into an univariate classification statistic that retains all the classification information. We reduce the multivariate observation by using a linear function of the several observed variables. Let the linear function be represented by

$$w = b_1 y_1 + b_2 y_2 + \cdots + b_p y_p$$

where $y_1, \ldots, y_p$ represent the set of available descriptive variables. Through the use of this function, each multivariate observation is reduced to a single classification statistic w, which we use in determining the classification rule.

Our principle requirement in accomplishing this reduction is to have a sample of n_ℓ observations, $y_{1i}, y_{2i}, \ldots, y_{pi}$, $i = 1, 2, \ldots, n_\ell$, from each of the two possible populations ($\ell = 1, 2$). These data are then used to determine the coefficients of the linear function and to estimate the mean and standard deviation of the resulting w variable for each population. We shall present the procedure through our employment classification illustration. To better illustrate the general approach, we shall add another descriptive variable, Y_3 = the expected monthly income, to the two variables Y_1 = age and Y_2 = monthly living expenses, introduced earlier. Thus we first want to determine the coefficients b_1, b_2, and b_3 for the function

$$w = b_1 y_1 + b_2 y_2 + b_3 y_3$$

Step 1

Acquisition of samples from the two populations. Consider that we have the following data from past experience:

	π_1 - Successful Salesman				π_2 - Unsuccessful Salesmen		
i	Age y_1	Expenses y_2	Expected Income y_3	i	Age y_1	Expenses y_2	Expected Income y_3
1	42	600	800	1	41	300	350
2	48	300	500	2	40	500	300
3	42	350	500	3	50	500	700
4	42	500	600	4	41	100	500
5	39	250	450	5	37	300	600
6	42	500	600	6	33	500	500
7	53	1200	1500	7	42	650	650
8	49	500	600	8	52	750	700
9	43	600	600	9	45	500	500
10	41	700	700				
11	49	350	500				
12	31	300	550				

We restrict our data for ease of presentation to n_1 = 12 successful salesmen and n_2 = 9 unsuccessful salesmen and assume that each sample is representative of its population. Let y_{ki} represent the ith observation for variable k.

Determination of the coefficient, b_1, b_2, and b_3.
(a) Determine the sums and sums of the squares and cross products for the y's for each sample:

Population π_1: $n_1 = 12$

$$\sum_{\pi_1} y_{1i} = 521 \quad \sum_{\pi_1} y_{2i} = 6150 \quad \sum_{\pi_1} y_{3i} = 7900$$

$$\sum_{\pi_1} y_{1i}^2 = 22983 \quad \sum_{\pi_1} y_{2i}^2 = 3887500 \quad \sum_{\pi_1} y_{3i}^2 = 607500$$

$$\sum_{\pi_1} y_{1i}y_{2i} = 275100 \qquad \sum_{\pi_1} y_{1i}y_{3i} = 351500$$

$$\sum_{\pi_1} y_{2i}y_{3i} = 4807500$$

Population π_2: $n_2 = 9$

$$\sum_{\pi_2} y_{1i} = 381 \quad \sum_{\pi_2} y_{2i} = 4100 \quad \sum_{\pi_2} y_{3i} = 4800$$

$$\sum_{\pi_2} y_{1i}^2 = 16413 \quad \sum_{\pi_2} y_{2i}^2 = 217500 \quad \sum_{\pi_2} y_{3i}^2 = 2725000$$

$$\sum_{\pi_2} y_{1i}y_{2i} = 177800 \qquad \sum_{\pi_2} y_{1i}y_{3i} = 206750$$

$$\sum_{\pi_2} y_{2i}y_{3i} = 228250$$

(b) Compute the vector of sample means and their differences. Let $\bar{y}_k^{(\ell)}$ denote the mean for variable k in population ℓ, (ℓ = 1,2). Thus the difference between the populations for each

570

variable is given by

$$d_k = \bar{y}_{k1}^{(1)} - \bar{y}_{k2}^{(2)} \qquad k = 1,2,3$$

Variable	Mean π_1	π_2	Difference d_k
1	43.42	42.33	1.09
2	512.50	455.55	56.95
3	658.33	533.33	125.00

(c) Determine the corrected sums of squares and cross products for each population.

For the ith population, the corrected sum of cross products for variables j and k is given by

$$SS_{jk}^{(\ell)} = \sum_{\pi_\ell} \left(y_{ji} - \bar{y}_j^{(\ell)} \right) \left(y_{ki} - \bar{y}_k^{(\ell)} \right)$$

$$= \sum_{\pi_\ell} y_{ji} y_{ki} - \frac{\sum_{\pi_\ell} y_{1i} \sum_{\pi_\ell} y_{ki}}{n_\ell}$$

where n_ℓ is the member of observations in the ith population.

Thus for population π_1 we obtain the following values:

$$SS_{11}^{(1)} = \sum_{\pi_1} y_{1i}^2 - \frac{\left(\sum_{\pi_1} y_{1i} \right)^2}{n_1}$$

$$= 22983 - \frac{(521)^2}{12} = 362.92$$

$$SS_{12}^{(1)} = \sum_{\pi_1} y_{1i}y_{2i} - \frac{\sum_{\pi_1} y_{1i} \sum_{\pi_1} y_{2i}}{n_1}$$

$$= 275100 - \frac{(521)(6150)}{12} = 8087.50$$

$$SS_{13}^{(1)} = 351500 - \frac{(521)(7900)}{12} = 8508.33$$

$$SS_{22}^{(1)} = 3887500 - \frac{(6150)^2}{12} = 735625.00$$

$$SS_{23}^{(1)} = 4807500 - \frac{(6150)(7900)}{12} = 758750.00$$

$$SS_{33}^{(1)} = 21607500 - \frac{(7900)^2}{12} = 874166.67$$

For population π_2 we obtain

$$SS_{11}^{(2)} = 16413 - \frac{(381)^2}{9} = 284.00$$

$$SS_{12}^{(2)} = 177800 - \frac{(381)(4100)}{9} = 4233.33$$

$$SS_{13}^{(2)} = 206750 - \frac{(381)(4800)}{9} = 3550.00$$

$$SS_{22}^{(2)} = 2175000 - \frac{(4100)^2}{9} = 37222.22$$

$$SS_{23}^{(2)} = 2282500 - \frac{(4100)(4800)}{9} = 95833.33$$

$$SS_{33}^{(2)} = 27285000 - \frac{(4800)^2}{9} = 165000.00$$

(d) Compute the variance-covariance matrix for the combined sample. The elements of the matrix can be found by evaluating

$$s_{jk} = \frac{1}{n-2} \left[SS_{jk}^{(1)} + SS_{jk}^{(2)} \right]$$

where the subscripts j and k denote the variables being used. Thus we obtain the variances and covariances for each combination of the three variables:

$$s_{11} = 34.04 \qquad s_{12} = 648.46 \qquad s_{13} = 634.65$$

$$s_{22} = 54886.70 \qquad s_{23} = 44978.07$$

$$s_{33} = 54692.98$$

These variances and covariances can be summarized as a symmetric matrix

$$S = \begin{bmatrix} 34.04 & 648.46 & 634.65 \\ 648.46 & 54886.70 & 44978.07 \\ 634.65 & 44978.07 & 54692.98 \end{bmatrix}$$

(e) The coefficients b_1, b_2, and b_3 can then be found by solving the simultaneous linear equations given by

$$S\ \underline{b} = \underline{d}$$

or

$$s_{11}\ b_1 + s_{12}\ b_2 + s_{13}\ b_3 = d_1$$
$$s_{12}\ b_1 + s_{22}\ b_2 + s_{23}\ b_3 = d_2$$
$$s_{13}\ b_1 + s_{23}\ b_2 + s_{33}\ b_3 = d_3$$

The simultaneous equations are

$$34.04 \, b_1 + 648.46 \, b_2 + 634.65 \, b_3 = 1.09$$

$$648.46 \, b_1 + 54886.70 \, b_2 + 44978.07 \, b_3 = 56.95$$

$$634.65 \, b_1 + 44978.07 \, b_2 + 54692.98 \, b_3 = 125.00$$

A computer program that can be used to solve simultaneous linear equations and obtain b_1, b_2, and b_3 is given in Appendix B. To solve the equations without the computer, we can use the method of elimination. The computations are given below. First divide each equation by the coefficient of b_1:

$$b_1 + 19.0499 \, b_2 + 18.6442 \, b_3 = 0.032021 \qquad (17.1)$$

$$b_1 + 84.6416 \, b_2 + 69.3614 \, b_3 = 0.08782 \qquad (17.2)$$

$$b_1 + 70.8707 \, b_2 + 86.1782 \, b_3 = 0.19696 \qquad (17.3)$$

Then subtract (17.1) from (17.2) and (17.3) from (17.2), which eliminates b_1:

$$65.5917 \, b_2 + 50.7172 \, b_3 = 0.05580 \qquad (17.4)$$

$$13.7709 \, b_2 - 16.8168 \, b_3 = -0.10914 \qquad (17.5)$$

Divide the resulting equations by the coefficient of b_2:

$$b_2 + 0.7732 \, b_3 = 0.0008507 \qquad (17.6)$$

$$b_2 - 1.2212 \, b_3 = -0.0079254 \qquad (17.7)$$

and subtract Eq. (17.7) from Eq. (17.6) to obtain

$$1.9944 \, b_3 = 0.008776 \qquad (17.8)$$

$$b_3 = 0.004400 \qquad (17.9)$$

The other values b_1 and b_2 are obtained by substituting (17.9) into (17.6) to obtain

$$b_2 = 0.008507 - .7732(0.004400) \qquad (17.10)$$

$$b_2 = -0.003317$$

and (17.9) and (17.10) into (17.1) to obtain

$$b_1 = 0.032021 - 19.0499(-0.003317)$$

$$- (18.6442)(0.004400) \qquad (17.11)$$

$$b_1 = 0.013175$$

We can check these solutions by using them to evaluate whether each of the original equations is satisfied. For example,

$$34.04(0.013175) + 648.46(-0.003317) + 634.65(0.004400)$$

$$= 1.08982 = 1.09$$

The classification statistic is then

$$w = 0.01317 \; y_1 - 0.003317 \; y_2 + 0.004400 \; y_3$$

Step 3

Estimation of mean, variance, and standard deviation of w. (a) The mean for population π_1 is

$$\bar{w}_{\pi_1} = b_1 \bar{y}_{11} + b_2 \bar{y}_{21} + b_3 \bar{y}_{31}$$

$$= 0.013175(43.42) - 0.003317(512.50) + 0.004400(658.33)$$

$$= 1.769$$

(b) The mean for population π_2 is

$$\bar{w}_{\pi_2} = b_1\bar{y}_{12} + b_2\bar{y}_{22} + b_3\bar{y}_{32}$$

$$= 0.013175(42.33) - 0.003317(455.55) + 0.004400(533.33)$$

$$= 1.393$$

and (c) the common variance is

$$s_w^2 = b_1^2 s_{11} + b_2^2 s_{22} + b_3^2 s_{33} + 2b_1 b_2 s_{12} + 2b_1 b_3 s_{13} + 2b_2 b_3 s_{23}$$

$$= (0.013175)^2(34.04) + (-0.003317)^2(54887.76)$$

$$+ (0.004400)^2(54692.98) + 2(0.013175)(-0.003317)$$

$$(648.46) + 2(0.013175)(0.004400)(634.65)$$

$$+ 2(-0.003317)(0.004400)(44978.98)$$

$$= 0.3703$$

and the standard deviation is

$$s_w = \sqrt{s_w^2} = 0.6085$$

Step 4

Determination of classification rule and the appropriate classification constant. (a) Determine the appropriate rule to be used. Since $\bar{w}_{\pi_1} < \bar{w}_{\pi_2}$, we use the rule

If $w > \lambda$, then classify the observation into π_1, otherwise classify it into π_2.

If, however $\bar{w}_{\pi_1} > \bar{w}_{\pi_2}$, we would have used the rule

If $w > \lambda$, then classify the observation into π_2, otherwise classify it into π_1.

(b) The classification constant λ can be determined in many ways; we shall present three of the more standard approaches. Since the distribution of the variable w must be assumed, we shall assume that the variables y_1, y_2, y_3 have a multivariate normal distribution because w then has a normal distribution.

(i) Choose λ so that one of the probabilities of misclassification is fixed. Assume that we want to find λ such that $Pr(2/1) = 0.2$. Since we assume that w has a normal distribution, this probability can be expressed as

$$Pr(2/1) = \int_{-\infty}^{\lambda} f_{\pi_1}(w) \; dw = \int_{-\infty}^{z_1} f(z) \; dz = 0.2$$

Using the normal tables, we can determine z_1 and then solve for λ by using the relationship

$$z_1 = \frac{\lambda - \bar{w}_{\pi_1}}{s_w} = -0.84$$

Thus

$$\lambda = s_w z_1 + \bar{w}_{\pi_1} = 0.6085(-0.84) + 1.769$$

$$\lambda = 1.26$$

(ii) Choose λ so that the two probabilities of misclassification are equal. Thus we would select

$$\lambda = \frac{\bar{w}_{\pi_1} + \bar{w}_{\pi_2}}{2} = \frac{1.769 + 1.393}{2} = 1.581$$

(iii) Choose λ so that the classification rule has a minimum expected loss. For this procedure we require that the prior probabilities and losses are known. Let us use q_1 = 0.35, q_2 = 0.65, $L(2/1)$ = 1, and $L(1/2)$ = 5. Then

$$\lambda = \frac{0.5(\mu_1^2 - \mu_2^2) + \sigma^2 \ln[q_2 \cdot L(1/2)/q_1 \cdot L(2/1)]}{\mu_1 - \mu_2}$$

$$= \frac{0.5(1.769^2 - 1.393^2) + (0.3703)\ln[(0.35)5/(0.65)1]}{1.769 - 1.393}$$

$$= \frac{0.5946 + (0.3703)\ln(2.69)}{0.376} = \frac{0.5946 + 0.3662}{0.376}$$

$$= 2.555$$

Step 5

Evaluation of operational effectiveness of the classification rule. We can estimate the operational effectiveness for each of the classification constants and compare them by determining the two probabilities of error and the associated expected loss of each rule.

(a) For rule (i) we set $Pr(2/1)$ = 0.2 and obtain λ = 1.26. We thus need only to determine

$$Pr(1/2) = \int_{\lambda}^{\infty} f_{\pi_2}(w)\ dw = \int_{z_2}^{\infty} f(z)\ dz = 0.4129$$

since $z_2 = \dfrac{1.26 - 1.393}{0.6085} = -0.22$

(b) For rule (ii) we obtain λ = 1.581 and thus need to find

$$Pr(2/1) = \int_{-\infty}^{\lambda} f_{\pi_1}(w)\ dw = \int_{-\infty}^{z_1} f(z)\ dz = 0.3783$$

since $z_1 = \dfrac{1.581 - 1.769}{0.6085} = -0.31$

and we can verify that $Pr(1/2) = Pr(2/1)$ by determining

$$Pr(1/2) = \int_{\lambda}^{\infty} f_{\pi_2}(w) \, dw = \int_{z_2}^{\infty} f(z) \, dz = 0.3783$$

since $z_1 = \dfrac{1.581 - 1.393}{0.6085} = 0.31$

(c) for rule (iii) $\lambda = 2.555$, and we find that

$$Pr(1/2) = \int_{2.555}^{\infty} f_{\pi_2}(w) \, dw = \int_{\frac{2.555-1.393}{0.6085}}^{\infty} f(z) \, dz = 0.0281.$$

and

$$Pr(2/1) = \int_{-\infty}^{2.555} f_{\pi_1}(w) \, dw = \int_{-\infty}^{\frac{2.555-1.769}{0.6085}} f(z) \, dz = 0.9015$$

To compare the three rules, we evaluate the expected loss for each rule by using the relationship

$$EL = q_1 \, Pr(2/1) \, L(2/1) + q_2 \, Pr(1/2) \, L(1/2)$$

For rule (i) we have

$$EL = (0.35)(0.2)(1) + (0.65)(0.4129)(5) = 1.48$$

For rule (ii) we have

$$EL = (0.35)(0.3783)(1) = (0.65)(0.3983)(5) = 1.23$$

For rule (iii) we have

$$El = (0.35)(0.9015)(1) + (0.65)(0.0281)95) = 0.41$$

which of course is the minimum of the three.

If the personnel manager is concerned about the size of the probabilities of misclassification and/or the expected loss, the operational effectiveness of the classification decision rule can be improved by either measuring the variables used with more precision or by adding more discriminatory variables to the descriptive vector. Of course, in the illustration the sample sizes are small, and thus the estimate of the mean and standard deviation of the w variable are not very reliable. In addition, the evaluation of the effectiveness of the technique assumes that the variables have a normal distribution, and if these assumptions are not met, the probability estimates will not be correct.

17.4 Summary

In this chapter we considered how to evolve a statistical classification rule that would classify an individual into one of two populations on the basis of a set of measurements on the individual.

We have illustrated the multivariate classification technique using an employment decision example; we shall now summarize the steps:

(1) The first requirement is to define the vector of observable variables that will be used in making the classification. Statistical criteria can be used to determine whether variables are useful for classification purposes. For example, the sample means can be obtained for the two populations, and the differences between the means relative to their standard deviations can be checked to select variables having large relative mean differences. In addition, one can also look at the correlation between the variables being considered and eliminate those that are too highly correlated with variables already selected. The cost of measuring the variable should also be compared with the additional classification power that the variable provides. Variable selection requires good judgement and should be tested by actually determining the operational effectiveness of the resulting classification rule.

(2) Once the set of variables is determined, their distribution should be checked to see how close the joint dis-

tribution of the variable is to being a multivariate normal distribution. The efficiency of our linear classification rule depends upon how good the normal assumption is. If a significant lack of normality is apparent, or if the covariance matrices are not close to being the same, we can use transformations on the variables to try to improve the situation.

(3) After the variables to be used have been determined, we need a classification statistic based on the variables to be used in the classification rule. We introduced the linear classification statistic

$$w = b_1 y_1 + b_2 y_2 + \cdots + b_p y_p$$

and justified its use from the fact that if the variables are normally distributed with equal covariance matrices and the coefficients are properly determined, the linear function given above minimizes the probability of making a misclassification. If the normality assumptions are not met, then some other function of the observable variables may be required.

(4) Given the linear classification function, one must determine the coefficients by solving the system of simultaneous equations as given in our illustration.

(5) The final step in developing the classification rule is to determine the classification constant. We introduced three criteria for determining this constant.

(6) Finally, one should make an evaluation of the operational effectiveness of the resulting classification rule. The statistician does this by considering the probabilities of making different types of errors and determining the expected loss if losses and prior probabilities are available.

Although we restricted ourselves to the two-population case, the approach presented can be generalized by considering many populations. One approach to the multipopulation problem is to perform sequential classifications by first aggregating the populations into two groups. Each group is then considered as a super population, and the two-population technique can then be used to obtain a rule for classifying an individual into one of the super populations. Each super population is then broken down into two smaller aggregates, another classification rule evolved, and a classification made. This is repeated until the final classification rule places an individual into a single population.

If one knows the distribution of the vector of variables for each population $f_k(\underline{y})$, the likelihood ratio classification rule can be used to classify an individual into one of the populations. This rule states that

If $f_k(\underline{y}) > f_j(\underline{y})$ for all $j \neq k$, classify the individual into population π_k.

To apply this rule we need to assume an appropriate distribution model for each $f_k(\underline{y})$, estimate the parameters from the available samples, compute each $f_k(\underline{y})$ for the observed $\underline{y}$, and classify the observation into the population associated with the maximum $f_k(\underline{y})$.

These multiple population approaches are, however, beyond the scope of this chapter, and thus we presented only a two-population technique.

EXERCISES

1. Develop a classification rule for the employment decision, using only the age variable Y, whose sampled values are given in the chapter. Find the classification constant λ to make the two probabilities of misclassification equal, and estimate the probabilities using the normality assumption. Check this probability by observing how many of the individuals in the two samples would have been incorrectly classified if the rule had been applied to them.

2. Work Exercise 1, using the two variables Y_1 and Y_2.

3. On the basis of the work done in the text and in exercise 2, how much would you be willing to pay to obtain Y_3 for an individual being considered for employment?

4. Write a program to develop the three variable classification rules. Test the program using the data given in this chapter.

The problem of determining the accuracy of a numerical observation arises in dealing with the recorded values of a continuous variable or when using results obtained by computations. There are two types of accuracy.

(1) The significant digit accuracy of any number is the number of digits that are measured when the value is originally obtained. A measured digit is any digit recorded other than a zero that is inserted to simply locate the decimal point. Thus we say that 27.063 has five significant digits and 12.0047 has six, but 0.00146 and 277000. have only three. In the last number it is assumed that the last seven was measured but none of the zeros were measured.

(2) The decimal-place accuracy of any number is obtained by counting the number of digits including zeros in the number that are to the right of the decimal point. Thus 2.65 has two decimal place accuracy, 0.07641 has five, and 0.0103 has four places.

When recording a number obtained by a measurement it is necessary to use a round-off rule in order to eliminate the digits that are not to be recorded. One round-off rule keeps the last digit to be retained at its measured value if the digits to the right are less than 5000..., and the last digit to be retained is increased by 1 if the digits to the right of it are greater than 5000.... If we wish to have two decimal place accuracy using this technique, we would record the value of 28.27314 as 28.27; 19.23651 would become 19.24. In case the digits to the right of the last recorded digit are exactly 5000...; the "odd-add" convention can be used. This rule states that odd digits are to be increased by 1 and even digits remain the same. Thus 17.635 becomes 17.64 while 0.425 becomes 0.42.

Another round-off rule would be to simply truncate the digits to be dropped. Thus the last digit to be recorded is always retained at its measured value. Still another rule is to always round-up; thus the last digit retained is always increased by 1.

When we perform arithmetic operations using measured values we must be concerned with the accuracy of the results since the results that are obtained are affected by the error in the original numbers. The following rule can be used:

In an arithmetic operation, the accuracy of the result is no more accurate than the least accurate of the numbers used in the operation.

To apply this rule in the case of additions or subtractions, the accuracy of the numbers used is determined by the decimal place accuracy criterion; in multiplications and divisions (and square roots), the significant digit criterion is used. Thus

$$26.141 + 32.2 - 16.264 = 42.057 = 42.1$$

since 32.2 has but one decimal place, and

$$\frac{(14.18) \cdot (0.143)}{267.1} = .0076289 = .00763$$

since 0.143 has only three significant digits.

When we consider computer programming it will be interesting to consider how the computer rounds off under different circumstances. This is an important consideration when using the computer to perform a large number of computations since the round-off errors can sometimes accumulate and thus become very significant. Often the choice of the formula will affect the accuracy of the result obtained. Thus we may choose to evaluate products by using logarithms to keep the magnitude of the intermediate results under control. A technical study of round-off error in computer evaluations is, however, beyond the scope of this course.

In most data analyses the intermediate computations can be carried out with as much accuracy as the computational system allows. At the end of the numerical evaluations, however, the final value should be rounded off in line with the accuracy that is really present in the original observations. In determining the final accuracy of the result, the recording accuracy and the inaccuracies introduced by the intermediate computations must be considered. For example, the mean or other such averages can be expressed with one decimal place beyond that of the recorded data used in determining the mean. This apparent enhancement of the accuracy comes about through the tendency of the recording errors to compensate each other upon addition since some will be positive while the others will be negative.

This subroutine sorts an array of N values. After
execution of the subroutine is completed, the smallest value
in the array will be stored in Y(1), the second smallest
value in Y(2),..., and the largest value will be stored in
Y(N).

The arguments of the subroutine that must be passed
from the calling program are

(Y(I),I=1,N) = array to be sorted
N = number of elements in array

```
      SUBROUTINE SORT (Y,N)
      DIMENSION Y(1)
C     Y IS THE ARRAY TO BE SORTED, I.E. AT COMPLETION Y(1)
C     WILL BE THE SMALLEST VALUE AND Y(N) WILL BE THE
C     LARGEST
      N1 = N -1
      DO 1 I=1,N1
      J = I + 1
      DO 2 K = J,N
      IF (Y(I) .LE. Y(K)) GO TO 2
      TEMP = Y(I)
      Y(I) = Y(K)
      Y(K) = TEMP
    2 CONTINUE
    1 CONTINUE
      RETURN
      END
```

The following function uses Simpson's rule to find the integral of any integrable function f(y). The function must be given by an ARITHMETIC STATEMENT FUNCTION. The calling arguments are the lower limit of the integral Y0, the upper limit of integration YM, and the number of intervals M, <u>which must be even</u>. Thus

$$\text{DEFINT(Y0,YM,M)} = \int_{YM}^{YO} f(y)\ dy$$

$$= \frac{1}{3}\ \Delta y [f(y_0) + 4f(y_1) + 2f(y_2) + \cdots + 2f(y_{m-2}) +$$

$$4f(y_{m-1}) + f(y_m)]$$

where

$$\Delta y = \frac{YM - YO}{M} \quad \text{and} \quad y_i = y_{i-1} + \Delta y$$

```
      FUNCTION DEFINT (YØ,YM,M)
      F(Y) = (MUST BE DEFINED BY AN ARITHMETIC STATEMENT
             FUNCTION)
C     CHECK IF M IS AN EVEN NUMBER
      IF ((M/2)*2.NE.M) GO TO 1Ø
      DELTAY = (YM - YØ)/ FLOAT(M)
      DEFINT = F(YØ) + F(YM) + 4.*F(YM-DELTAY)
      IF (M.EQ.2) GO TO 2
      Y = YØ+DELTAY
      M3 = (M -2)/2
      DEL = 2.*DELTAY
      DO 1  I=1,M3
      DEFINT = DEFINT + 4.*F(Y) + 2.*F(Y+DELTAY)
    1 Y = Y + DELTAY
    2 DEFINT = DEFINT * DELTAY/3
      RETURN
   1Ø WRITE(6,3) N
    3 FORMAT(1X,13HN IS NOT EVEN, I5)
      RETURN
      END
```

The algorithm for the numerical evaluation of the natural logarithm of $\Gamma(x)$ is:

$$\ln \Gamma(x) = \frac{1}{2} \ln (2\pi) + (x - \frac{1}{2}) \ln x + \frac{1}{12x} - \frac{1}{360x^3}$$

$$+ \frac{1}{1260x^5} - \frac{1}{1680x^7} \; , \; x > 15.$$

Since

$$\Gamma(x + 1) = x \, \Gamma(x)$$

any value of $x \leq 15$ is evaluated by incrementing $x = 16$ and using the relationship

$$\Gamma(x) = \frac{\Gamma(x + a)}{x(x + 1)(x + 2) \; ... \; (x + a - 1)}$$

so that

$$\ln \Gamma(x) = \ln \Gamma(x + a) - \ln [x(x + 1)(x + 2) \; ...$$

$$(x + a - 1)].$$

The program is:

```
      FUNCTION ALGAMA(A)
      ALGAMA=Ø.
      IF(A.LE.Ø.) TO TO 1ØØ
      DETERM=1.
      DB=A
    7 IF(DB.GT.15.) GO TO 1Ø
      DTERM=DTERM*DB
      DB=DB+1.
      GO TO 7
   1Ø DQ=1./DB**2
      DR=1./ DB
      ALGAMA=Ø.91893853 - DB + (DB-Ø.5)*ALOG(DB)+
     1(((-Ø.5952380̸6E-3*DQ +.79365Ø8ØE-3*DQ)-.27777778E-2)
     2 *DQ + Ø.83333333E-1)*DR-ALOG(DTERM)
      GO TO 99
  1ØØ WRITE(6,98) A
   98 FORMAT (1H ,1ØX,8HARGUMENT,E14.8,15HIS UNACCEPTABLE)
   99 RETURN
      END
```

The multiplicative congruential procedure is used to generate each random number u_{i+1} from the previous one, u_i, using the following relationship:

$$u_i = \alpha u_{i-1} \ (\text{Mod } m)$$

The values of α and m depend upon the computer. If the product is greater than the maximum integer m allowed by the computer, it is repeatedly reduced by m until the result is between zero and m. For example, if a decimal computer with p digits per number, is used, the

$$\alpha = 10^{[p/2]+1} + 3$$

where the notation $[p/2]$ represents the greatest integer less than $p/2$. If the computer is a binary machine with p bits per number, then

$$\alpha = 2^{[p/2]+1} + 3$$

For example, a binary computer with 36 bits per word (1 for sign, 35 for values) would use $p = 35$, $\alpha = 2^{18} + 3$, $m = 2^{36} - 1$. A binary computer with 32 bits per word (1 for sign, 31 for values) would use $p = 31$, $\alpha = 2^{16} + 3$, $m = 2^{32}$, while a decimal computer with 11 digits per word would use $p = 11$, $\alpha = 10^6 + 3$, $m = 10^{11}$.

The FORTRAN subroutine to generate uniform random numbers using this technique is given below.

$P = p$ = number of bits used for a number (excluding sign bit)

$B = b$ = base: $b = 2$ if binary computer; $b = 10$ if decimal computer

$MAXINT$ = maximum integer number for particular computer
$= m^P - 1$

IODD = any odd integer number less than MAXINT (must be given at first entry into subroutine)

  IY = resulting integer number to be used for next entry into subroutine

   Y = uniform random number between zero and 1.

```
SUBROUTINE URANDN(IODD,IY,Y)
P = NUMBER OF BITS PER INTEGER WORD - 1 ) must be
B = BASE                                 } specified
MAXINT = MAXIMUM INTEGER NUMBER          ) for par-
JP = P                                     ticular
IP = P/2                                   computer
IP = IP + 1                                system
IALFA = B**IP + 3.
IY = IODD*IALFA
IF (IY .LT. Ø) IY = IY + MAXINT + 1
Y = IY
YM = B**(-JP)
Y = Y * YM
RETURN
END
```

B-5 A routine to solve simultaneous   *MATINV(A,N,B,M,DET)
    linear equations or invert a
    matrix

This subroutine solves a simultaneous system of n
linear equations represented by

$$a_{11} x_1 + a_{12} x_2 + \ldots + a_{1n} x_n = b_1$$
$$a_{21} x_1 + a_{22} x_2 + \ldots + a_{2n} x_n = b_2$$
$$\vdots \qquad \vdots \qquad \qquad \vdots \qquad \vdots$$
$$a_{i1} x_1 + a_{i2} x_2 + \ldots + a_{in} x_n = b_i$$
$$\vdots \qquad \vdots \qquad \qquad \vdots \qquad \vdots$$
$$a_{n1} x_1 + a_{n2} x_2 + \ldots + a_{nn} x_n = b_n$$

where the a_{ij} are the coefficients that are known, the b_i
are also known, and x_j are the unknown values to be determined. Note that the number of equations is equal to the
number of unknowns.

The subroutine can also be used to find the inverse of
a matrix A, where A is represented by

$$A = \begin{bmatrix} a_{11} & a_{12} & \cdots & a_{1n} \\ a_{21} & a_{22} & \cdots & a_{2n} \\ \vdots & \vdots & & \vdots \\ a_{i1} & a_{i2} & \cdots & a_{in} \\ \vdots & \vdots & & \vdots \\ a_{n1} & a_{n2} & \cdots & a_{nm} \end{bmatrix}$$

The arguments of the subroutine are

---

*Program Library package, administered by Share, Inc., Suite
750, 25 Broadway, New York, N.Y. 10004.

Input:

    $(A(I,J),J=1,N),I=1,N)$ = matrix of coefficients

    N = number of rows of matrix

    $(B(I),I=1,N)$ = vector of coefficients on right-
                          hand side of system of equations
                          (not necessary if only inverse is
                          to be found).

  M = indicates whether there are equations to be solved
      or only an inverse to be found.

      = 0 when inverse is to be found (no input
      needed for B)

      = 1, when system of simultaneous equations to be
      solved

Output:

    $(A(I,J),J=1,N),I=1,N)$ = inverse of matrix A
    $(B(I),I=1,N)$ = vector of solutions $(x_1, x_2, \ldots ,$
                           $x_n)$ if M = 1
  DET = determinant of matrix


```
      SUBROUTINE MATINV(A,N,B,M,DET)
C     TAKEN FROM SHARE F4Ø2
C     THIS SUBROUTINE COMPUTES THE INVERSE AND DETERMINANT
C     OF THE MATRIX A, OF ORDER N, BY THE GAUSS- JORDAN
C     METHOD.  A INVERSE REPLACES A, THE DETERMINANT OF A
C     IS PLACED IN DET.IF M=1 THE VECTOR B CONTAINS THE
C     CONSTANT VECTOR WHEN MATINV IS CALLED, AND THIS IS
C     REPLACED BY THE SOLUTION VECTOR.  IF M = Ø NO
C     SIMULTANEOUS EQUATIONS ARE CALLED FOR, AND B IS NOT
C     PERTINENT. N IS NOT TO EXCEED 12Ø.
C     A,N,B,M AND DET ARE IN THE ARGUMENT LIST AND ARE
C     DUMMY VARIABLES, A IS MATRIX TO BE INVERTED, N IS
C     ORDER, B IS CONSTANT VECTOR, AND DET IS WHERE
C     DETERMINANT IS PLACED, N IS THE RANK OF THE SUBSET
C     TO BE INVERTED
      DIMENSION IPIVOT(5Ø),INDEX(5Ø,2)
      DOUBLE PRECISION A(N,N),B(N,1),DET,PIVOT(5Ø),SWAP,
     1AMAX,T
      EQUIVALENCE (IROW,JROW),(ICOLUM,JCOLUM),(AMAX,T,SWAP)
C     INITIALIZATION
      DET = 1.ØØØØØØØD+ØØ
      DO 2Ø J = 1,N
   2Ø IPIVOT(J) = Ø
      DO 55Ø I = 1,N
C     SEARCH FOR PIVOT ELEMENT
```

```
      AMAX = Ø
      DO 1Ø5 J = 1,N
      IF(IPIVOT(J) - 1) 6Ø,1Ø5,6Ø
  6Ø  DO 1ØØ K = 1,N
      IF (IPIVOT(K) - 1) 8Ø,1ØØ,74Ø
  8Ø  IF (DABS(AMAX) - DABS(A(J,K))) 85,1ØØ,1ØØ
  85  IROW = J
      ICOLUM = K
      AMAX = A(J,K)
 1ØØ  CONTINUE
 1Ø5  CONTINUE
      IPIVOT(LCOLUM) = IPIVOT(LCOLUM) + 1
C        INTERCHANGE ROWS TO PUT PIVOT ELEMENT ON DIAGONAL
      IF (IROW - ICOLUM) 14Ø,26Ø,14Ø
 14Ø  DET = - DET
      DO 2ØØ L = 1,N
      SWAP = A(IROW,L)
      A(IROW,L) = A(ICOLUM,L)
 2ØØ  A(ICOLUM,L) = SWAP
      IF (M) 26Ø,26Ø,21Ø
 21Ø  DO 25Ø L = 1,M
      SWAP = B(IROW,L)
      B(IROW,L) = B(ICOLUM,L)
 25Ø  B(ICOLUM,L) = SWAP
 26Ø  INDEX(I,1) = IROW
      INDEX(I,2) = ICOLUM
      PIVOT(I) = A(ICOLUM,ICOLUM)
      DET = DET * PIVOT(I)
C        DIVIDE PIVOT ROW BY PIVOT ELEMENT
      A(ICOLUM,ICOLUM) = 1.Ø
      DO 35Ø L = 1,N
 35Ø  A(ICOLUM,L) = A(ICOLUM,L)/PIVOT(I)
      IF(M) 38Ø,38Ø,36Ø
 36Ø  DO 37Ø L = 1,M
 37Ø  B(ICOLUM,L) = B(ICOLUM,L)/PIVOT(I)
C        REDUCE NON PIVOT ROWS
 38Ø  DO 55Ø L1 = 1,N
      IF (L1 - ICOLUM) 4ØØ,55Ø,4ØØ
 4ØØ  T = A(L1,ICOLUM)
      A(L1,ICOLUM) = Ø
      DO 45Ø L = 1,N
 45Ø  A(L1,L) = A(L1,L) - A(ICOLUM,L)*T
      IF(M) 55Ø,55Ø,46Ø
 46Ø  DO 5ØØ L = 1,M
 5ØØ  B(L1,L) = B(L1,L) - B(ICOLUM,L)*T
 55Ø  CONTINUE
C        INTERCHANGE COLUMNS
      DO 71Ø I = 1,N
      L = N+1 - I
      IF(INDEX(L,1) - INDEX(L,2)) 63Ø,71Ø,63Ø
 63Ø  JROW = INDEX(L,1)
      JCOLUM = INDEX(L,2)
      DO 7Ø5 K = 1,N
      SWAP = A(K,JROW)
```

```
      A(K,JROW) = A(K,JCOLUM)
      A(K,JCOLUM) = SWAP
705 CONTINUE
710 CONTINUE
740 RETURN
      END
```

$$\Pr[Y \leq k] = \sum_{y=0}^{k} C_y^n \, \theta^y (1-\theta)^{n-y}$$

1. n = 5

θ

k	0.01	0.05	0.10	0.20	0.30	0.40	0.50	0.60	0.70	0.80	0.90	0.95	0.99	k
0	.951	.774	.590	.328	.168	.078	.031	.010	.002	.000	.000	.000	.000	0
1	.999	.977	.919	.737	.528	.337	.188	.087	.031	.007	.000	.000	.000	1
2	1.000	.999	.991	.942	.837	.683	.500	.317	.163	.058	.009	.001	.000	2
3	1.000	1.000	1.000	.993	.969	.913	.812	.663	.472	.263	.081	.023	.001	3
4	1.000	1.000	1.000	1.000	.998	.990	.969	.922	.832	.672	.410	.226	.049	4

2. n = 10

θ

k	0.01	0.05	0.10	0.20	0.30	0.40	0.50	0.60	0.70	0.80	0.90	0.95	0.99	k
0	.904	.599	.349	.107	.028	.006	.001	.000	.000	.000	.000	.000	.000	0
1	.996	.914	.736	.376	.149	.046	.011	.002	.000	.000	.000	.000	.000	1
2	1.000	.988	.930	.678	.383	.167	.055	.012	.002	.000	.000	.000	.000	2
3	1.000	.999	.987	.879	.650	.382	.172	.055	.011	.001	.000	.000	.000	3
4	1.000	1.000	.998	.967	.850	.633	.377	.166	.047	.006	.000	.000	.000	4
5	1.000	1.000	1.000	.994	.953	.834	.623	.367	.150	.033	.002	.000	.000	5
6	1.000	1.000	1.000	.999	.989	.945	.828	.618	.350	.121	.013	.001	.000	6
7	1.000	1.000	1.000	1.000	.998	.988	.945	.833	.617	.322	.070	.012	.000	7
8	1.000	1.000	1.000	1.000	1.000	.998	.989	.954	.851	.624	.264	.086	.004	8
9	1.000	1.000	1.000	1.000	1.000	1.000	.999	.994	.972	.893	.651	.401	.096	9

3. n = 15

θ

k	0.01	0.05	0.10	0.20	0.30	0.40	0.50	0.60	0.70	0.80	0.90	0.95	0.99	k
0	.860	.463	.206	.035	.005	.000	.000	.000	.000	.000	.000	.000	.000	0
1	.990	.829	.549	.167	.035	.005	.000	.000	.000	.000	.000	.000	.000	1
2	1.000	.964	.816	.398	.127	.027	.004	.000	.000	.000	.000	.000	.000	2
3	1.000	.995	.944	.648	.297	.091	.018	.002	.000	.000	.000	.000	.000	3
4	1.000	.999	.987	.836	.515	.217	.059	.009	.001	.000	.000	.000	.000	4
5	1.000	1.000	.998	.939	.722	.403	.151	.034	.004	.000	.000	.000	.000	5
6	1.000	1.000	1.000	.982	.869	.610	.304	.095	.015	.001	.000	.000	.000	6
7	1.000	1.000	1.000	.996	.950	.787	.500	.213	.050	.004	.000	.000	.000	7
8	1.000	1.000	1.000	.999	.985	.905	.696	.390	.131	.018	.000	.000	.000	8
9	1.000	1.000	1.000	1.000	.996	.966	.849	.597	.278	.061	.002	.000	.000	9
10	1.000	1.000	1.000	1.000	.999	.991	.941	.783	.485	.164	.013	.001	.000	10
11	1.000	1.000	1.000	1.000	1.000	.998	.982	.909	.703	.352	.056	.005	.000	11
12	1.000	1.000	1.000	1.000	1.000	1.000	.996	.973	.873	.602	.184	.036	.000	12
13	1.000	1.000	1.000	1.000	1.000	1.000	1.000	.995	.965	.833	.451	.171	.010	13
14	1.000	1.000	1.000	1.000	1.000	1.000	1.000	1.000	.995	.965	.794	.537	.140	14

4. n = 20

θ

k	0.01	0.05	0.10	0.20	0.30	0.40	0.50	0.60	0.70	0.80	0.90	0.95	0.99	k
0	.818	.358	.122	.002	.001	.000	.000	.000	.000	.000	.000	.000	.000	0
1	.983	.736	.392	.069	.008	.001	.000	.000	.000	.000	.000	.000	.000	1
2	.999	.925	.677	.206	.035	.004	.000	.000	.000	.000	.000	.000	.000	2
3	1.000	.984	.867	.411	.107	.016	.001	.000	.000	.000	.000	.000	.000	3
4	1.000	.997	.957	.630	.238	.051	.006	.000	.000	.000	.000	.000	.000	4
5	1.000	1.000	.989	.804	.416	.126	.021	.002	.000	.000	.000	.000	.000	5
6	1.000	1.000	.998	.913	.608	.250	.058	.006	.000	.000	.000	.000	.000	6
7	1.000	1.000	1.000	.968	.772	.416	.132	.021	.001	.000	.000	.000	.000	7
8	1.000	1.000	1.000	.990	.887	.596	.252	.057	.005	.000	.000	.000	.000	8
9	1.000	1.000	1.000	.997	.952	.755	.412	.128	.017	.001	.000	.000	.000	9
10	1.000	1.000	1.000	.999	.983	.872	.588	.245	.048	.003	.000	.000	.000	10
11	1.000	1.000	1.000	1.000	.995	.943	.748	.404	.113	.010	.000	.000	.000	11
12	1.000	1.000	1.000	1.000	.999	.979	.868	.584	.228	.032	.000	.000	.000	12
13	1.000	1.000	1.000	1.000	1.000	.994	.942	.750	.392	.087	.002	.000	.000	13
14	1.000	1.000	1.000	1.000	1.000	.998	.979	.874	.584	.196	.011	.000	.000	14
15	1.000	1.000	1.000	1.000	1.000	1.000	.994	.949	.762	.370	.043	.003	.000	15
16	1.000	1.000	1.000	1.000	1.000	1.000	.999	.984	.893	.589	.133	.016	.000	16
17	1.000	1.000	1.000	1.000	1.000	1.000	1.000	.996	.965	.794	.323	.075	.001	17
18	1.000	1.000	1.000	1.000	1.000	1.000	1.000	.999	.992	.931	.608	.264	.017	18
19	1.000	1.000	1.000	1.000	1.000	1.000	1.000	1.000	.999	.988	.878	.642	.182	19

5. n = 25

θ

k	0.01	0.05	0.10	0.20	0.30	0.40	0.50	0.60	0.70	0.80	0.90	0.95	0.99	k
0	.778	.277	.072	.004	.000	.000	.000	.000	.000	.000	.000	.000	.000	0
1	.974	.642	.271	.027	.002	.000	.000	.000	.000	.000	.000	.000	.000	1
2	.998	.873	.537	.092	.009	.000	.000	.000	.000	.000	.000	.000	.000	2
3	1.000	.966	.764	.234	.033	.002	.000	.000	.000	.000	.000	.000	.000	3
4	1.000	.993	.902	.421	.090	.009	.000	.000	.000	.000	.000	.000	.000	4
5	1.000	.999	.967	.617	.193	.029	.002	.000	.000	.000	.000	.000	.000	5
6	1.000	1.000	.991	.780	.341	.074	.007	.000	.000	.000	.000	.000	.006	6
7	1.000	1.000	.998	.891	.512	.154	.022	.001	.000	.000	.000	.000	.000	7
8	1.000	1.000	1.000	.953	.677	.274	.054	.004	.000	.000	.000	.000	.000	8
9	1.000	1.000	1.000	.983	.811	.425	.115	.013	.000	.000	.000	.000	.000	9
10	1.000	1.000	1.000	.994	.902	.586	.212	.034	.002	.000	.000	.000	.000	10
11	1.000	1.000	1.000	.998	.956	.732	.345	.078	.006	.000	.000	.000	.000	11
12	1.000	1.000	1.000	1.000	.983	.846	.500	.154	.017	.000	.000	.000	.000	12
13	1.000	1.000	1.000	1.000	.994	.922	.655	.268	.044	.002	.000	.000	.000	13
14	1.000	1.000	1.000	1.000	.998	.966	.788	.414	.098	.006	.000	.000	.000	14
15	1.000	1.000	1.000	1.000	1.000	.987	.885	.575	.189	.017	.000	.000	.000	15
16	11.000	1.000	1.000	1.000	1.000	.996	.946	.726	.323	.047	.000	.000	.000	16
17	1.000	1.000	1.000	1.000	1.000	.999	.978	.846	.488	.109	.002	.000	.000	17
18	1.000	1.000	1.000	1.000	1.000	1.000	.993	.926	.659	.220	.009	.000	.000	18
19	1.000	1.000	1.000	1.000	1.000	1.000	.998	.971	.807	.383	.033	.001	.000	19
20	1.000	1.000	1.000	1.000	1.000	1.000	1.000	.991	.910	.579	.098	.007	.000	20
21	1.000	1.000	1.000	1.000	1.000	1.000	1.000	.998	.967	.766	.236	.034	.000	21
22	1.000	1.000	1.000	1.000	1.000	1.000	1.000	1.000	.991	.902	.463	.127	.002	22
23	1.000	1.000	1.000	1.000	1.000	1.000	1.000	1.000	.998	.973	.729	.358	.026	23
24	1.000	1.000	1.000	1.000	1.000	1.000	1.000	1.000	1.000	.996	.928	.723	.222	24

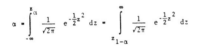

$$\alpha = \int_{-\infty}^{z_\alpha} \frac{1}{\sqrt{2\pi}}\, e^{-\frac{1}{2}z^2}\, dz = \int_{z_{1-\alpha}}^{\infty} \frac{1}{\sqrt{2\pi}}\, e^{-\frac{1}{2}z^2}\, dz$$

and

$$z_\alpha = -z_{1-\alpha}$$

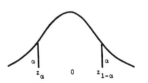

z	.00	.01	.02	.03	.04	.05	.06	.07	.08	.09	z
0.0	.5000	.5040	.5080	.5120	.5160	.5199	.5239	.5279	.5319	.5359	0.0
0.1	.5398	.5438	.5478	.5517	.5557	.5596	.5636	.5675	.5714	.5753	0.1
0.2	.5793	.5832	.5871	.5910	.5948	.5987	.6026	.6064	.6103	.6141	0.2
0.3	.6179	.6217	.6255	.6293	.6331	.6368	.6406	.6443	.6480	.6517	0.3
0.4	.6554	.6591	.6628	.6664	.6700	.6736	.6772	.6808	.6844	.6879	0.4
0.5	.6915	.6950	.6985	.7019	.7054	.7088	.7123	.7157	.7190	.7224	0.5
0.6	.7257	.7291	.7324	.7357	.7389	.7422	.7454	.7486	.7517	.7549	0.6
0.7	.7580	.7611	.7642	.7673	.7703	.7734	.7764	.7794	.7823	.7852	0.7
0.8	.7881	.7910	.7939	.7967	.7995	.8023	.8051	.8078	.8106	.8133	0.8
0.9	.8159	.8186	.8212	.8238	.8264	.8289	.8315	.8340	.8365	.8389	0.9
1.0	.8413	.8438	.8461	.8485	.8508	.8531	.8554	.8577	.8599	.8621	1.0
1.1	.8643	.8665	.8686	.8708	.8729	.8749	.8770	.8790	.8810	.8830	1.1
1.2	.8849	.8869	.8888	.8907	.8925	.8944	.8962	.8980	.8997	.9015	1.2
1.3	.9032	.9049	.9066	.9082	.9099	.9115	.9131	.9147	.9162	.9177	1.3
1.4	.9192	.9207	.9222	.9236	.9251	.9265	.9279	.9292	.9306	.9319	1.4
1.5	.9332	.9345	.9357	.9370	.9382	.9394	.9406	.9418	.9429	.9441	1.5
1.6	.9452	.9463	.9474	.9484	.9495	.9505	.9515	.9525	.9535	.9545	1.6
1.7	.9554	.9564	.9573	.9582	.9591	.9599	.9608	.9616	.9625	.9633	1.7
1.8	.9641	.9649	.9656	.9664	.9671	.9678	.9686	.9693	.9699	.9706	1.8
1.9	.9713	.9719	.9726	.9732	.9738	.9744	.9750	.9756	.9761	.9767	1.9
2.0	.9772	.9778	.9783	.9788	.9793	.9798	.9803	.9808	.9812	.9817	2.0
2.1	.9821	.9826	.9830	.9838	.9838	.9842	.9846	.9850	.9854	.9857	2.1
2.2	.9861	.9864	.9868	.9871	.9875	.9878	.9881	.9884	.9887	.9890	2.2
2.3	.9893	.9896	.9898	.9901	.9904	.9906	.9909	.9911	.9913	.9916	2.3
2.4	.9918	.9920	.9922	.9925	.9927	.9929	.9931	.9932	.9934	.9936	2.4
2.5	.9938	.9940	.9941	.9943	.9945	.9946	.9948	.9949	.9951	.9952	2.5
2.6	.9953	.9955	.9956	.9957	.9959	.9960	.9961	.9962	.9963	.9964	2.6
2.7	.9965	.9966	.9967	.9968	.9969	.9970	.9971	.9972	.9973	.9974	2.7
2.8	.9974	.9975	.9976	.9977	.9977	.9978	.9979	.9979	.9980	.9981	2.8
2.9	.9981	.9982	.9982	.9983	.9984	.9984	.9985	.9985	.9986	.9986	2.9
3.0	.9987	.9987	.9987	.9988	.9988	.9989	.9989	.9989	.9990	.9990	3.0
3.1	.9090	.9991	.9991	.9991	.9992	.9992	.9992	.9992	.9993	.9993	3.1
3.2	.9993	.9993	.9994	.9994	.9994	.9994	.9994	.9995	.9995	.9995	3.2
3.3	.9995	.9995	.9996	.9996	.9996	.9996	.9996	.9996	.9996	.9997	3.3
3.4	.9997	.9997	.9997	.9997	.9997	.9997	.9997	.9997	.9998	.9998	3.4
3.5	.9998	.9998	.9998	.9998	.9998	.9998	.9998	.9998	.9998	.9998	3.5

$$\alpha = \int_{-\infty}^{t_\alpha} f(t,d)dt = \int_{t_{1-\alpha}}^{\infty} f(t,d)dt$$

and

$$t_\alpha = -t_{1-\alpha}$$

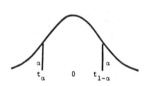

d	$t_{.60}$	$t_{.70}$	$t_{.80}$	$t_{.90}$	$t_{.95}$	$t_{.975}$	$t_{.99}$	$t_{.995}$	$t_{.999}$	d
1	0.325	0.727	1.376	3.078	6.314	12.706	31.821	63.657	318.309	1
2	0.289	0.617	1.061	1.886	2.920	4.303	6.965	9.925	22.327	2
3	0.277	0.584	0.978	1.638	2.353	3.182	4.541	5.841	10.215	3
4	0.271	0.569	0.941	1.533	2.132	2.776	3.747	4.604	7.173	4
5	0.267	0.559	0.920	1.476	2.015	2.571	3.365	4.032	5.893	5
6	0.265	0.553	0.906	1.440	1.943	2.447	3.143	3.707	5.208	6
7	0.263	0.549	0.896	1.415	1.895	2.365	2.998	3.499	4.785	7
8	0.262	0.546	0.889	1.397	1.860	2.306	2.896	3.355	4.501	8
9	0.261	0.543	0.883	1.383	1.833	2.262	2.821	3.250	4.297	9
10	0.260	0.542	0.879	1.372	1.812	2.228	2.764	3.169	4.144	10
11	0.260	0.540	0.876	1.363	1.796	2.201	2.718	3.106	4.025	11
12	0.259	0.539	0.873	1.356	1.782	2.179	2.681	3.055	3.930	12
13	0.259	0.538	0.870	1.350	1.771	2.160	2.650	3.012	3.852	13
14	0.258	0.537	0.868	1.345	1.761	2.145	2.624	2.977	3.787	14
15	0.258	0.536	0.866	1.341	1.753	2.131	2.602	2.947	3.733	15
16	0.258	0.535	0.865	1.337	1.746	2.120	2.583	2.921	3.686	16
17	0.257	0.534	0.863	1.333	1.740	2.110	2.567	2.898	3.646	17
18	0.257	0.534	0.862	1.330	1.734	2.101	2.552	2.878	3.610	18
19	0.257	0.533	0.861	1.328	1.729	2.093	2.539	2.861	3.579	19
20	0.257	0.533	0.860	1.325	1.725	2.086	2.528	2.845	3.552	20
21	0.257	0.532	0.859	1.323	1.721	2.080	2.518	2.831	3.527	21
22	0.256	0.532	0.858	1.321	1.717	2.074	2.508	2.819	3.505	22
23	0.256	0.532	0.858	1.319	1.714	2.069	2.500	2.807	3.485	23
24	0.256	0.531	0.857	1.318	1.711	2.064	2.492	2.797	3.467	24
25	0.256	0.531	0.856	1.316	1.708	2.060	2.485	2.787	3.450	25
26	0.256	0.531	0.856	1.315	1.706	2.056	2.479	2.779	3.435	26
27	0.256	0.531	0.855	1.314	1.703	2.052	2.473	2.771	3.421	27
28	0.256	0.530	0.855	1.313	1.701	2.048	2.467	2.763	3.408	28
29	0.056	0.530	0.854	1.311	1.699	2.045	2.462	2.756	3.396	29
30	0.256	0.530	0.854	1.310	1.697	2.042	2.457	2.750	3.385	30
31	0.256	0.530	0.853	1.309	1.696	2.040	2.453	2.744	3.375	31
32	0.255	0.530	0.853	1.309	1.694	2.037	2.449	2.738	3.365	32
33	0.255	0.530	0.853	1.308	1.692	2.035	2.445	2.733	3.356	33
34	0.255	0.529	0.852	1.307	1.691	2.032	2.441	2.728	3.348	34
35	0.255	0.529	0.852	1.306	1.690	2.030	2.438	2.724	3.340	35
36	0.255	0.529	0.852	1.306	1.688	2.028	2.434	2.719	3.333	36
37	0.255	0.529	0.851	1.305	1.687	2.026	2.431	2.715	3.326	37
38	0.255	0.529	0.851	1.304	1.686	2.024	2.429	2.712	3.319	38
39	0.255	0.529	0.851	1.304	1.685	2.023	2.426	2.708	3.313	39
40	0.255	0.529	0.851	1.303	1.684	2.021	2.423	2.704	3.307	40
41	0.255	0.529	0.850	1.303	1.683	2.020	2.421	2.701	3.301	41
42	0.255	0.528	0.850	1.302	1.682	2.018	2.418	2.698	3.296	42
43	0.255	0.528	0.850	1.302	1.681	2.017	2.416	2.695	3.291	43
44	0.255	0.528	0.850	1.301	1.680	2.015	2.414	2.692	3.286	44
45	0.255	0.528	0.850	1.301	1.679	2.014	2.412	2.690	3.281	45
46	0.255	0.528	0.850	1.300	1.679	2.013	2.410	2.687	3.277	46
47	0.255	0.528	0.849	1.300	1.678	2.012	2.408	2.685	3.273	47
48	0.255	0.528	0.849	1.299	1.677	2.011	2.407	2.682	3.269	48
49	0.255	0.528	0.849	1.299	1.677	2.010	2.405	2.680	3.265	49
50	0.255	0.528	0.849	1.299	1.676	2.009	2.403	2.678	3.261	50

CRITICAL VALUES OF THE t DISTRIBUTION, (continued)

d	$t_{.60}$	$t_{.70}$	$t_{.80}$	$t_{.90}$	$t_{.95}$	$t_{.975}$	$t_{.99}$	$t_{.995}$	$t_{.999}$	d
51	0.255	0.528	0.849	1.298	1.675	2.008	2.402	2.676	3.258	51
52	0.255	0.528	0.849	1.298	1.675	2.007	2.400	2.674	3.255	52
53	0.255	0.528	0.848	1.298	1.674	2.006	2.399	2.672	3.251	53
54	0.255	0.528	0.848	1.297	1.674	2.005	2.397	2.670	3.248	54
55	0.255	0.527	0.848	1.297	1.673	2.004	2.396	2.668	3.245	55
56	0.255	0.527	0.848	1.297	1.673	2.003	2.395	2.667	3.242	56
57	0.255	0.527	0.848	1.297	1.672	2.002	2.394	2.665	3.239	57
58	0.255	0.527	0.848	1.296	1.672	2.002	2.392	2.663	3.237	58
59	0.254	0.527	0.848	1.296	1.671	2.001	2.391	2.662	3.234	59
60	0.254	0.527	0.848	1.296	1.671	2.000	2.390	2.660	3.232	60
61	0.254	0.527	0.848	1.296	1.670	2.000	2.389	2.659	3.229	61
62	0.254	0.527	0.847	1.295	1.670	1.999	2.388	2.657	3.227	62
63	0.254	0.527	0.847	1.295	1.669	1.998	2.387	2.656	3.225	63
64	0.254	0.527	0.847	1.295	1.669	1.998	2.386	2.655	3.223	64
65	0.254	0.527	0.847	1.295	1.669	1.997	2.385	2.654	3.220	65
66	0.254	0.527	0.847	1.295	1.668	1.997	2.384	2.652	3.218	66
67	0.254	0.527	0.847	1.294	1.668	1.996	2.383	2.651	3.216	67
68	0.254	0.527	0.847	1.294	1.668	1.995	2.382	2.650	3.214	68
69	0.254	0.527	0.847	1.294	1.667	1.995	2.382	2.649	3.213	69
70	0.254	0.527	0.847	1.294	1.667	1.994	2.381	2.648	3.211	70
71	0.254	0.527	0.847	1.294	1.667	1.994	2.380	2.647	3.209	71
72	0.254	0.527	0.847	1.293	1.666	1.993	2.379	2.646	3.207	72
73	0.254	0.527	0.847	1.293	1.666	1.993	2.379	2.645	3.206	73
74	0.254	0.527	0.847	1.293	1.666	1.993	2.378	2.644	3.204	74
75	0.254	0.527	0.846	1.293	1.665	1.992	2.377	2.643	3.202	75
76	0.254	0.527	0.846	1.293	1.665	1.992	2.376	2.642	3.201	76
77	0.254	0.527	0.846	1.293	1.665	1.991	2.376	2.641	3.199	77
78	0.254	0.527	0.846	1.292	1.665	1.991	2.375	2.640	3.198	78
79	0.254	0.527	0.846	1.292	1.664	1.990	2.374	2.640	3.197	79
80	0.254	0.526	0.846	1.292	1.664	1.990	2.374	2.639	3.195	80
81	0.254	0.526	0.846	1.292	1.664	1.990	2.373	2.638	3.194	81
82	0.254	0.526	0.846	1.292	1.664	1.989	2.373	2.637	3.193	82
83	0.254	0.526	0.846	1.292	1.663	1.989	2.372	2.636	3.191	83
84	0.254	0.526	0.846	1.292	1.663	1.989	2.372	2.636	3.190	84
85	0.254	0.526	0.846	1.292	1.663	1.988	2.371	2.635	3.189	85
86	0.254	0.526	0.846	1.291	1.663	1.988	2.370	2.634	3.188	86
87	0.254	0.526	0.846	1.291	1.663	1.988	2.370	2.634	3.187	87
88	0.254	0.526	0.846	1.291	1.662	1.987	2.369	2.633	3.185	88
89	0.254	0.526	0.846	1.291	1.662	1.987	2.369	2.632	3.184	89
90	0.254	0.526	0.846	1.291	1.662	1.987	2.368	2.632	3.183	90
91	0.254	0.526	0.846	1.291	1.662	1.986	2.368	2.631	3.182	91
92	0.254	0.526	0.846	1.291	1.662	1.986	2.368	2.630	3.181	92
93	0.254	0.526	0.846	1.291	1.661	1.986	2.367	2.630	3.180	93
94	0.254	0.526	0.845	1.291	1.661	1.986	2.367	2.629	3.179	94
95	0.254	0.526	0.845	1.291	1.661	1.985	2.366	2.629	3.178	95
96	0.254	0.526	0.845	1.290	1.661	1.985	2.366	2.628	3.177	96
97	0.254	0.526	0.845	1.290	1.661	1.985	2.365	2.627	3.176	97
98	0.254	0.526	0.845	1.290	1.661	1.984	2.365	2.627	3.175	98
99	0.254	0.526	0.845	1.290	1.660	1.984	2.365	2.626	3.175	99
100	0.254	0.526	0.845	1.290	1.660	1.984	2.364	2.626	3.174	100
∞	0.253	0.524	0.842	1.282	1.645	1.960	2.326	2.576	3.090	∞

$$\alpha = \int_{0}^{\chi_{\alpha}^{2}} f(\chi^2, d)\, d\chi^2$$

$$= \int_{\chi_{1-\alpha}^{2}}^{\infty} f(\chi^2, d)\, d\chi^2$$

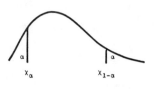

d	$\chi^2_{.001}$	$\chi^2_{.005}$	$\chi^2_{.01}$	$\chi^2_{.025}$	$\chi^2_{.05}$	$\chi^2_{.10}$	$\chi^2_{.20}$	$\chi^2_{.30}$	$\chi^2_{.40}$	$\chi^2_{.50}$	d
1	0.000	0.000	0.000	0.001	0.004	0.016	0.064	0.148	0.275	0.455	1
2	0.002	0.010	0.020	0.051	0.103	0.211	0.446	0.713	1.022	1.386	2
3	0.024	0.072	0.115	0.216	0.352	0.584	1.005	1.424	1.869	2.366	3
4	0.091	0.207	0.297	0.484	0.711	1.064	1.649	2.195	2.753	3.357	4
5	0.210	0.412	0.554	0.831	1.145	1.610	2.343	3.000	3.655	4.351	5
6	0.381	0.676	0.872	1.237	1.635	2.204	3.070	3.828	4.570	5.348	6
7	0.598	0.989	1.239	1.690	2.167	2.833	3.822	4.671	5.493	6.346	7
8	0.857	1.344	1.646	2.180	2.733	3.490	4.594	5.527	6.423	7.344	8
9	1.152	1.735	2.088	2.700	3.325	4.168	5.380	6.393	7.357	8.343	9
10	1.479	2.156	2.558	3.247	3.940	4.865	6.179	7.267	8.295	9.342	10
11	1.834	2.603	3.053	3.816	4.575	5.578	6.989	8.148	9.237	10.341	11
12	2.214	3.074	3.571	4.404	5.226	6.304	7.807	9.034	10.182	11.340	12
13	2.617	3.565	4.107	5.009	5.892	7.042	8.634	9.926	11.129	12.340	13
14	3.041	4.075	4.660	5.629	6.571	7.790	9.467	10.821	12.078	13.339	14
15	3.483	4.601	5.229	6.262	7.261	8.547	10.307	11.721	13.030	14.339	15
16	3.942	5.142	5.812	6.908	7.962	9.312	11.152	12.624	13.983	15.338	16
17	4.416	5.697	6.408	7.564	8.672	10.085	12.002	13.531	14.937	16.338	17
18	4.905	6.265	7.015	8.231	9.390	10.865	12.857	14.440	15.893	17.338	18
19	5.407	6.844	7.633	8.907	10.117	11.651	13.716	15.352	16.850	18.338	19
20	5.921	7.434	8.260	9.591	10.851	12.443	14.578	16.266	17.809	19.337	20
21	6.447	8.034	8.897	10.283	11.591	13.240	15.445	17.182	18.768	20.337	21
22	6.983	8.643	9.542	10.982	12.338	14.041	16.341	18.101	19.729	21.337	22
23	7.529	9.260	10.196	11.689	13.091	14.848	17.186	19.021	20.690	22.337	23
24	8.085	9.886	10.856	12.401	13.848	15.659	18.062	19.943	21.652	23.337	24
25	8.649	10.520	11.524	13.120	14.611	16.473	18.940	20.867	22.616	24.337	25
26	9.222	11.160	12.198	13.844	15.379	17.292	19.820	21.792	23.579	25.336	26
27	9.803	11.808	12.879	14.573	16.151	18.114	20.703	22.719	24.544	26.336	27
28	10.391	12.461	13.565	15.308	16.928	18.939	21.588	23.647	25.509	27.336	28
29	10.986	13.121	14.256	16.047	17.708	19.768	22.475	24.577	26.475	28.336	29
30	11.588	13.787	14.953	16.791	18.493	20.599	23.364	25.508	27.442	29.336	30
31	12.196	14.458	15.655	17.539	19.281	21.434	24.255	26.440	28.409	30.336	31
32	12.811	15.134	16.362	18.291	20.072	22.271	25.148	27.373	29.376	31.336	32
33	13.431	15.815	17.074	19.047	20.867	23.110	26.042	28.307	30.344	32.336	33
34	14.057	16.501	17.789	19.806	21.664	23.952	26.938	29.242	31.313	33.336	34
35	14.688	17.192	18.509	20.569	22.465	24.797	27.836	30.178	32.282	34.336	35
36	15.324	17.887	19.233	21.336	23.269	25.643	28.735	31.115	33.252	35.336	36
37	15.965	18.586	19.960	22.106	24.075	26.492	29.635	32.053	34.222	36.336	37
38	16.611	19.289	20.691	22.878	24.884	27.343	30.537	32.992	35.192	37.335	38
39	17.262	19.996	21.426	23.654	25.695	28.196	31.441	33.932	36.163	38.335	39
40	17.916	20.707	22.164	24.433	26.509	29.051	32.345	34.872	37.134	39.335	40
41	18.575	21.421	22.906	25.215	27.326	29.907	33.251	35.813	38.105	40.335	41
42	19.239	22.138	23.650	25.999	28.144	30.765	34.157	36.755	39.077	41.335	42
43	19.906	22.859	24.398	26.785	28.965	31.625	35.065	37.698	40.050	42.335	43
44	20.576	23.584	25.148	27.575	29.787	32.487	35.974	38.641	41.022	43.335	44
45	21.251	24.311	25.901	28.366	30.612	33.350	36.884	39.585	41.995	44.335	45
46	21.929	25.041	26.657	29.160	31.439	34.215	37.795	40.529	42.968	45.335	46
47	22.610	25.775	27.416	29.956	32.268	35.081	38.708	41.474	43.942	46.335	47
48	23.295	26.511	28.177	30.755	33.098	35.949	39.620	42.420	44.915	47.335	48
49	23.983	27.249	28.941	31.555	33.930	36.818	40.534	43.366	45.889	48.335	49
50	24.674	27.991	29.707	32.357	34.764	37.689	41.449	44.313	46.864	49.335	50

d	$\chi^2_{.001}$	$\chi^2_{.005}$	$\chi^2_{.01}$	$\chi^2_{.025}$	$\chi^2_{.05}$	$\chi^2_{.10}$	$\chi^2_{.20}$	$\chi^2_{.30}$	$\chi^2_{.40}$	$\chi^2_{.50}$	d
51	25.368	28.735	30.475	33.162	35.600	38.560	42.365	45.261	47.838	50.335	51
52	26.065	29.481	31.246	33.968	36.437	39.433	43.281	46.209	48.813	51.335	52
53	26.765	30.230	32.018	34.776	37.276	40.308	44.199	47.157	49.788	52.335	53
54	27.468	30.981	32.793	35.586	38.116	41.183	45.117	48.106	50.764	53.335	54
55	28.173	31.735	33.570	36.398	38.958	42.060	46.036	49.055	51.739	54.335	55
56	28.881	32.490	34.350	37.212	39.801	42.937	46.955	50.005	52.715	55.335	56
57	29.592	33.248	35.131	38.027	40.646	43.816	47.875	50.956	53.691	56.335	57
58	30.305	34.008	35.913	38.844	41.492	44.696	48.797	51.906	54.667	57.335	58
59	31.020	34.770	36.698	39.662	42.339	45.577	49.718	52.858	55.643	58.335	59
60	31.738	35.534	37.485	40.482	43.188	46.459	50.641	53.809	56.620	59.335	60
61	32.459	36.300	38.273	41.303	44.038	47.342	51.564	54.761	57.597	60.335	61
62	33.181	37.068	39.063	42.126	44.889	48.226	52.487	55.714	58.574	61.335	62
63	33.906	37.838	39.855	42.950	45.741	49.111	53.412	56.666	59.551	62.335	63
64	34.633	38.610	40.649	43.776	46.595	49.996	54.336	57.620	60.528	63.335	64
65	35.362	39.383	41.444	44.603	47.450	50.883	55.262	58.573	61.506	64.335	65
66	36.093	40.158	42.240	45.431	48.305	51.770	56.188	59.527	62.484	65.335	66
67	36.826	40.935	43.038	46.261	49.162	52.659	57.115	60.481	63.461	66.335	67
68	37.561	41.713	43.838	47.092	50.020	53.548	58.042	61.436	64.440	67.335	68
69	38.298	42.494	44.639	47.924	50.879	54.438	58.970	62.391	65.418	68.334	69
70	39.036	43.275	45.442	48.758	51.739	55.329	59.898	63.346	66.396	69.334	70
71	39.777	44.058	46.246	49.592	52.600	56.221	60.827	64.302	67.375	70.334	71
72	40.519	44.843	47.051	50.428	53.462	57.113	61.756	65.258	68.353	71.334	72
73	41.264	45.629	47.858	51.265	54.325	58.006	62.686	66.214	69.332	72.334	73
74	42.010	46.417	48.666	52.103	55.189	58.900	63.616	67.170	70.311	73.334	74
75	42.757	47.206	49.475	52.942	56.054	59.795	64.547	68.127	71.290	74.334	75
76	43.507	47.997	50.286	53.782	56.920	60.690	65.478	69.084	72.270	75.334	76
77	44.258	48.788	51.097	54.623	57.786	61.586	66.409	70.042	73.249	76.334	77
78	45.010	49.582	51.910	55.466	58.654	62.483	67.431	70.999	74.228	77.334	78
79	45.764	50.376	52.725	56.309	59.522	63.380	68.274	71.957	75.208	78.334	79
80	46.520	51.172	53.540	57.153	60.391	64.278	69.207	72.915	76.188	79.334	80
81	47.277	51.969	54.357	57.998	61.261	65.176	70.140	73.874	77.168	80.334	81
82	48.036	52.767	55.174	58.845	62.132	66.076	71.074	74.833	78.148	81.344	82
83	48.796	53.567	55.993	59.692	63.004	66.976	72.008	75.792	79.128	82.334	83
84	49.557	54.368	56.813	60.540	63.876	67.876	72.943	76.751	80.108	83.334	84
85	50.320	55.170	57.634	61.389	64.749	68.777	73.878	77.710	81.089	84.334	85
86	51.085	55.973	58.456	62.239	65.623	69.679	74.813	78.670	82.069	85.334	86
87	51.850	56.777	59.279	63.089	66.498	70.581	75.749	79.630	83.050	86.334	87
88	52.617	57.582	60.103	63.941	67.373	71.484	76.685	80.590	84.031	87.334	88
89	53.386	58.389	60.928	64.793	68.249	72.387	77.622	81.550	85.012	88.334	89
90	54.155	59.196	61.754	65.647	69.126	73.291	78.558	82.511	85.993	89.334	90
91	54.926	60.005	62.581	66.501	70.003	74.196	79.496	83.472	86.974	90.334	91
92	55.698	60.815	63.409	67.356	70.882	75.100	80.433	84.433	87.955	91.334	92
93	56.472	61.625	64.238	68.211	71.760	76.006	81.371	85.394	88.936	92.334	93
94	57.246	62.437	65.068	69.068	72.640	76.912	82.309	86.356	89.917	93.334	94
95	58.022	63.250	65.898	69.925	73.520	77.818	83.248	87.317	90.899	94.334	95
96	58.799	64.063	66.730	70.783	74.401	78.725	84.187	88.279	91.881	95.334	96
97	59.577	64.878	67.562	71.642	75.282	79.633	85.126	89.241	92.862	96.334	97
98	60.356	65.694	68.396	72.501	76.164	80.541	86.065	90.204	93.844	97.344	98
99	61.137	66.510	69.230	73.361	77.046	81.449	87.005	91.166	94.826	98.334	99
100	61.918	67.328	70.065	74.222	77.929	82.358	87.945	92.129	95.808	99.334	100

d	$\chi^2_{.50}$	$\chi^2_{.60}$	$\chi^2_{.70}$	$\chi^2_{.80}$	$\chi^2_{.90}$	$\chi^2_{.95}$	$\chi^2_{.975}$	$\chi^2_{.99}$	$\chi^2_{.995}$	$\chi^2_{.999}$	d
1	0.455	0.708	1.074	1.642	2.706	3.841	5.024	6.635	7.879	10.828	1
2	1.386	1.833	2.408	3.219	4.605	5.991	7.378	9.210	10.597	13.815	2
3	2.366	2.946	3.665	4.642	6.251	7.815	9.348	11.345	12.838	16.266	3
4	3.357	4.045	4.878	5.989	7.779	9.488	11.143	13.277	14.860	18.467	4
5	4.351	5.132	6.064	7.289	9.236	11.070	12.832	15.086	16.750	20.515	5
6	5.348	6.211	7.231	8.558	10.645	12.592	14.449	16.812	18.548	22.458	6
7	6.346	7.283	8.383	9.803	12.017	14.067	16.013	18.475	20.278	24.322	7
8	7.344	8.351	9.524	11.030	13.362	15.507	17.535	20.090	21.955	26.124	8
9	8.343	9.414	10.656	12.242	14.684	16.919	19.023	21.666	23.589	27.877	9
10	9.342	10.473	11.781	13.442	15.987	18.307	20.483	23.209	25.188	29.588	10
11	10.341	11.530	12.899	14.631	17.275	19.675	21.920	24.725	26.757	31.264	11
12	11.340	12.584	14.011	15.812	18.549	21.026	23.337	26.217	28.299	32.909	12
13	12.340	13.636	15.119	16.985	19.812	22.362	24.736	27.688	29.819	34.528	13
14	13.339	14.685	16.222	18.151	21.064	23.685	26.119	29.141	31.319	36.123	14
15	14.339	15.733	17.322	19.311	22.307	24.996	27.488	30.578	32.801	37.697	15
16	15.338	16.780	18.418	20.465	23.542	26.296	28.845	32.000	34.267	39.252	16
17	16.338	17.824	19.511	21.615	24.769	27.587	30.191	33.409	35.718	40.790	17
18	17.388	18.868	20.601	22.760	25.989	28.869	31.526	34.805	37.156	42.312	18
19	18.338	19.910	21.689	23.900	27.204	30.144	32.852	36.191	38.582	43.820	19
20	19.337	20.951	22.775	25.037	28.412	31.410	34.170	37.566	39.997	45.315	20
21	20.337	21.991	23.858	26.171	29.615	32.671	35.479	38.932	41.401	46.797	21
22	21.337	23.031	24.939	27.301	30.813	33.924	36.781	40.289	42.796	48.268	22
23	22.337	24.069	26.018	28.429	32.007	35.172	38.076	41.638	44.181	49.728	23
24	23.337	25.106	27.096	29.553	33.196	36.415	39.364	42.980	45.558	51.178	24
25	24.337	26.143	28.172	30.675	34.382	37.652	40.646	44.314	46.928	52.620	25
26	25.336	27.179	29.246	31.795	35.563	38.885	41.923	45.642	48.290	54.052	26
27	26.336	28.214	30.319	32.912	36.741	40.113	43.195	46.963	49.645	55.476	27
28	27.336	29.249	31.391	34.027	37.916	41.337	44.461	48.278	50.993	56.892	28
29	28.336	30.283	32.461	35.139	39.087	42.557	45.722	49.588	52.336	58.301	29
30	29.336	31.316	33.530	36.250	40.256	43.773	46.979	50.892	53.672	59.703	30
31	30.336	32.349	34.598	37.359	41.422	44.985	48.232	52.191	55.003	61.098	31
32	31.336	33.381	35.665	38.466	42.585	46.194	49.480	53.486	56.328	62.487	32
33	32.336	34.413	36.731	39.572	43.745	47.400	50.725	54.776	57.648	63.870	33
34	33.336	35.444	37.795	40.676	44.903	48.602	51.966	56.061	58.964	65.247	34
35	34.366	36.475	38.859	41.778	46.059	49.802	53.203	57.342	60.275	66.619	35
36	35.336	37.505	39.922	42.879	47.212	50.998	54.437	58.619	61.581	67.985	36
37	36.336	38.535	40.984	43.978	48.363	52.192	55.668	59.892	62.883	69.346	37
38	37.335	39.564	42.045	45.076	49.513	53.384	56.896	61.162	64.181	70.703	38
39	38.335	40.593	43.105	46.173	50.660	54.572	58.120	62.428	65.476	72.055	39
40	39.335	41.622	44.165	47.269	51.805	55.758	59.342	63.691	66.766	73.402	40
41	40.355	42.651	45.224	48.363	52.949	56.942	60.561	64.950	68.053	74.745	41
42	41.335	43.679	46.282	49.456	54.090	58.124	61.777	66.206	69.336	76.084	42
43	42.335	44.706	47.339	50.548	55.230	59.303	62.990	67.459	70.616	77.418	43
44	43.335	45.734	48.396	51.639	56.369	60.481	64.201	68.709	71.893	78.749	44
45	44.335	46.761	49.452	52.729	57.505	61.656	65.410	69.957	73.166	80.077	45
46	45.335	47.787	50.507	53.818	58.641	62.830	66.617	71.201	74.436	81.400	46
47	46.335	48.814	51.562	54.906	59.774	64.001	67.821	72.443	75.704	82.720	47
48	47.335	49.840	52.616	55.993	60.907	65.171	69.023	73.683	76.969	84.037	48
49	48.335	50.866	53.670	57.079	62.038	66.339	70.222	74.919	78.231	85.350	49
50	49.335	51.892	54.723	58.164	63.167	67.505	71.420	76.154	79.490	86.661	50

CRITICAL VALUES OF THE CHI-SQUARE DISTRIBUTION, (continued)

d	$\chi^2_{.50}$	$\chi^2_{.60}$	$\chi^2_{.70}$	$\chi^2_{.80}$	$\chi^2_{.90}$	$\chi^2_{.95}$	$\chi^2_{.975}$	$\chi^2_{.99}$	$\chi^2_{.995}$	$\chi^2_{.999}$	d
51	50.335	52.917	55.775	59.248	64.295	68.669	72.616	77.386	80.747	87.968	51
52	51.335	53.942	56.827	60.332	65.422	69.832	73.810	78.616	82.001	89.272	52
53	52.335	54.967	57.879	61.414	66.548	70.993	75.002	79.843	83.253	90.573	53
54	53.335	55.992	58.930	62.496	67.673	72.153	76.192	81.069	84.502	91.872	54
55	54.335	57.016	59.980	63.577	68.796	73.311	77.380	82.292	85.749	93.167	55
56	55.335	58.040	61.031	64.658	69.919	74.468	78.567	83.513	86.994	94.460	56
57	56.335	59.064	62.080	65.737	71.040	75.624	79.752	84.733	88.236	95.751	57
58	57.335	60.088	63.129	66.816	72.160	76.778	80.936	85.950	89.477	97.039	58
59	58.335	61.111	64.178	67.894	73.279	77.931	82.117	87.166	90.715	98.324	59
60	59.335	62.135	65.227	68.972	74.397	79.082	83.298	88.379	91.952	99.607	60
61	60.335	63.158	66.274	70.049	75.514	80.232	84.476	89.591	93.186	100.888	61
62	61.335	64.181	67.322	71.125	76.630	81.381	85.654	90.801	94.419	102.166	62
63	62.335	65.204	68.369	72.201	77.745	82.529	86.830	92.010	95.649	103.442	63
64	63.335	66.226	69.416	73.276	78.860	83.675	88.004	93.217	96.878	104.716	64
65	64.335	67.249	70.462	74.351	79.973	84.821	89.177	94.422	98.105	105.988	65
66	65.335	68.271	71.508	75.424	81.085	85.965	90.349	95.626	99.330	107.258	66
67	66.335	69.293	72.554	76.498	82.197	87.108	91.519	96.828	100.554	108.525	67
68	67.335	70.315	73.600	77.571	83.308	88.250	92.689	98.028	101.776	109.791	68
69	68.334	71.337	74.645	78.643	84.418	89.391	93.856	99.227	102.996	111.055	69
70	69.334	72.358	75.689	79.715	85.527	90.531	95.023	100.425	104.215	112.317	70
71	70.334	73.380	76.734	80.786	86.635	91.670	96.189	101.621	105.432	113.577	71
72	71.334	74.401	77.778	81.857	87.743	92.808	97.353	102.816	106.648	114.835	72
73	72.334	75.422	78.821	82.927	88.850	93.945	98.516	104.010	107.862	116.091	73
74	73.334	76.443	79.865	83.997	89.956	95.081	99.678	105.202	109.074	117.364	74
75	74.334	77.464	80.908	85.066	91.061	96.217	100.839	106.393	110.286	118.599	75
76	75.334	78.485	81.951	86.135	92.166	97.351	101.999	107.583	111.495	119.850	76
77	76.334	79.505	82.994	87.203	93.270	98.484	103.158	108.771	112.704	121.100	77
78	77.334	80.526	84.036	88.271	94.374	99.617	104.316	109.958	113.911	122.348	78
79	78.334	81.546	85.078	89.338	95.476	100.749	105.473	111.144	115.117	123.594	79
80	79.334	82.566	86.120	90.405	96.578	101.879	106.629	112.329	116.321	124.839	80
81	80.334	83.586	87.161	91.472	97.680	103.010	107.783	113.512	117.524	126.082	81
82	81.334	84.606	88.202	92.538	98.780	104.139	108.937	114.695	118.726	127.324	82
83	82.334	85.626	89.243	93.604	99.880	105.267	110.090	115.876	119.927	128.565	83
84	83.334	86.646	90.284	94.669	100.980	106.395	111.242	117.057	121.126	129.804	84
85	84.334	87.665	91.325	95.734	102.079	107.522	112.393	118.236	122.325	131.041	85
86	85.334	88.685	92.365	96.799	103.177	108.648	113.544	119.414	123.522	132.277	86
87	86.334	89.704	93.405	97.863	104.275	109.773	114.693	120.591	124.718	133.512	87
88	87.334	90.723	94.445	98.927	105.372	110.898	115.841	121.767	125.912	134.745	88
89	88.334	91.742	95.484	99.991	106.469	112.022	116.989	122.942	127.106	135.977	89
90	89.334	92.761	96.524	101.054	107.565	113.145	118.136	124.116	128.299	137.208	90
91	90.334	93.780	97.563	102.117	108.661	114.268	119.282	125.289	129.490	138.438	91
92	91.334	94.799	98.602	103.179	109.756	115.390	120.427	126.462	130.681	139.666	92
93	92.334	95.818	99.641	104.241	110.850	116.511	121.571	127.633	131.871	140.893	93
94	93.334	96.836	100.679	105.303	111.944	117.632	122.715	128.803	133.059	142.119	94
95	94.334	97.855	101.717	106.364	113.038	118.752	123.858	129.973	134.246	143.343	95
96	95.334	98.873	102.755	107.425	114.131	119.871	125.000	131.141	135.433	144.567	96
97	96.334	99.892	103.793	108.486	115.223	120.990	126.141	132.309	136.619	145.789	97
98	97.334	100.910	104.831	109.547	116.315	122.108	127.282	133.476	137.803	147.010	98
99	98.334	101.928	105.868	110.607	117.407	123.225	128.422	134.642	138.987	148.230	99
100	99.334	102.946	106.906	111.667	118.498	124.342	129.561	135.807	140.169	149.449	100

$$\alpha = \int_{0}^{F_{\alpha}} f(F,d_1,d_2)\, dF$$

$$= \int_{F_{1-\alpha}}^{\infty} f(F,d_1,d_2)\, dF$$

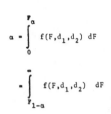
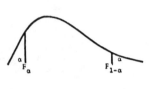

ν_1, DEGREES OF FREEDOM FOR NUMERATOR

ν_2	Cum. prop.	1	2	3	4	5	6	7	8	9	10	11	12	Cum. prop.
1	.0005	$.0^6 62$	$.0^5 50$	$.0^5 38$	$.0^2 94$	.016	.022	.027	.032	.036	.039	.042	.045	.0005
	.001	$.0^5 25$	$.0^2 10$	$.0^5 60$	.013	.021	.028	.034	.039	.044	.048	.051	.054	.001
	.005	$.0^6 62$	$.0^5 51$	.018	.032	.044	.054	.062	.068	.073	.078	.082	.085	.005
	.010	$.0^3 25$	.010	.029	.047	.062	.073	.082	.089	.095	.100	.104	.107	.010
	.025	$.0^5 15$	.026	.057	.082	.100	.113	.124	.132	.139	.144	.149	.153	.025
	.05	$.0^6 62$	.054	.099	.130	.151	.167	.179	.188	.195	.201	.207	.211	.05
	.10	.025	.117	.181	.220	.246	.265	.279	.289	.298	.304	.310	.315	.10
	.25	.172	.389	.494	.553	.591	.617	.637	.650	.661	.670	.680	.684	.25
	.50	1.00	1.50	1.71	1.82	1.89	1.94	1.98	2.00	2.03	2.04	2.05	2.07	.50
	.75	5.83	7.50	8.20	8.58	8.82	8.98	9.10	9.19	9.26	9.32	9.36	9.41	.75
	.90	39.9	49.5	53.6	55.8	57.2	58.2	58.9	59.4	59.9	60.2	60.5	60.7	.90
	.95	161	200	216	225	230	234	237	239	241	242	243	244	.95
	.975	648	800	864	900	922	937	948	957	963	969	973	977	.975
	.99	405^1	500^1	540^1	562^1	576^1	586^1	593^1	598^1	602^1	606^1	608^1	611^1	.99
	.995	162^2	200^2	216^2	225^2	231^2	234^2	237^2	239^2	241^2	242^2	243^2	244^2	.995
	.999	406^3	500^3	540^3	562^3	576^3	586^3	593^3	598^3	602^3	606^3	609^3	611^3	.999
	.9995	162^4	200^4	216^4	225^4	231^4	234^4	237^4	239^4	241^4	242^4	243^4	244^4	.9995
2	.0005	$.0^5 50$	$.0^3 50$	$.0^4 42$	.011	.020	.029	.037	.044	.050	.056	.061	.065	.0005
	.001	$.0^5 20$	$.0^2 10$	$.0^4 68$	.016	.027	.037	.046	.054	.061	.067	.072	.077	.001
	.005	$.0^5 50$	$.0^2 50$	.020	.038	.055	.069	.081	.091	.099	.106	.112	.118	.005
	.01	$.0^5 20$	.010	.032	.056	.075	.092	.105	.116	.125	.132	.139	.144	.01
	.025	$.0^2 13$	.026	.062	.094	.119	.138	.153	.165	.175	.183	.190	.196	.025
	.05	$.0^5 50$	.053	.105	.144	.173	.194	.211	.224	.235	.244	.251	.257	.05
	.10	.020	.111	.183	.231	.265	.289	.307	.321	.333	.342	.350	.356	.10
	.25	.133	.333	.439	.500	.540	.568	.588	.604	.616	.626	.633	.641	.25
	.50	.667	1.00	1.13	1.21	1.25	1.28	1.30	1.32	1.33	1.34	1.35	1.36	.50
	.75	2.57	3.00	3.15	3.23	3.28	3.31	3.34	3.35	3.37	3.38	3.39	3.39	.75
	.90	8.53	9.00	9.16	9.24	9.29	9.33	9.35	9.37	9.38	9.39	9.40	9.41	.90
	.95	18.5	19.0	19.2	19.2	19.3	19.3	19.4	19.4	19.4	19.4	19.4	19.4	.95
	.975	38.5	39.0	39.2	39.2	39.3	39.3	39.4	39.4	39.4	39.4	39.4	39.4	.975
	.99	98.5	99.0	99.2	99.2	99.3	99.3	99.4	99.4	99.4	99.4	99.4	99.4	.99
	.995	198	199	199	199	199	199	199	199	199	199	199	199	.995
	.999	998	999	999	999	999	999	999	999	999	999	999	999	.999
	.9995	200^1	200^1	200^1	200^1	200^1	200^1	200^1	200^1	200^1	200^1	200^1	200^1	.9995
3	.0005	$.0^5 46$	$.0^3 50$	$.0^4 44$	.012	.023	.033	.043	.052	.060	.067	.074	.079	.0005
	.001	$.0^5 19$	$.0^2 10$	$.0^4 71$	.018	.030	.042	.053	.063	.072	.079	.086	.093	.001
	.005	$.0^5 46$	$.0^2 50$	.021	.041	.060	.077	.092	.104	.115	.124	.132	.138	.005
	.01	$.0^5 19$	.010	.034	.060	.083	.102	.118	.132	.143	.153	.161	.168	.01
	.025	$.0^2 12$	.026	.065	.100	.129	.152	.170	.185	.197	.207	.216	.224	.025
	.05	$.0^5 46$	.052	.108	.152	.185	.210	.230	.246	.259	.270	.279	.287	.05
	.10	.019	.109	.185	.239	.276	.304	.325	.342	.356	.367	.376	.384	.10
	.25	.122	.317	.424	.489	.531	.561	.582	.600	.613	.624	.633	.641	.25
	.50	.585	.881	1.00	1.06	1.10	1.13	1.15	1.16	1.17	1.18	1.19	1.20	.50
	.75	2.02	2.28	2.36	2.39	2.41	2.42	2.43	2.44	2.44	2.44	2.45	2.45	.75
	.90	5.54	5.46	5.39	5.34	5.31	5.28	5.27	5.25	5.24	5.23	5.22	5.22	.90
	.95	10.1	9.55	9.28	9.12	9.01	8.94	8.89	8.85	8.81	8.79	8.76	8.74	.95
	.075	17.4	16.0	15.4	15.1	14.9	14.7	14.6	14.5	14.5	14.4	14.4	14.3	.975
	.99	34.1	30.8	29.5	28.7	28.2	27.9	27.7	27.5	27.3	27.2	27.1	27.1	.99
	.995	55.6	49.8	47.5	46.2	45.4	44.8	44.4	44.1	43.9	43.7	43.5	43.4	.995
	.999	167	149	141	137	135	133	132	131	130	129	129	128	.999
	.9995	266	237	225	218	214	211	209	208	207	206	204	204	.9995

ν_2, DEGREES OF FREEDOM FOR DENOMINATOR

Read $.0^3 56$ as .00056, 200^1 as 2,000, 162^4 as 1,620,000, and so on.

ν_1, DEGREES OF FREEDOM FOR NUMERATOR

Cum. prop.	15	20	24	30	40	50	60	100	120	200	500	∞	Cum. prop.	
.0005	.051	.058	062	.066	.069	.072	.074	.077	.078	.080	.081	.083	.0005	**1**
.001	.060	.067	.071	.075	.079	.082	.084	.088	.089	.091	.092		.001	
.005	.093	.101	.105	.109	.113	.116	.118	.121	.122	.124	.126	.127	.005	
.01	.115	.124	.128	.132	.137	.139	.141	.145	.146	.148	.150	.151	.01	
.025	.161	.170	.175	.180	.184	.187	.189	.193	.194	.196	.198	.199	.025	
.05	.220	.230	.235	.240	.245	.248	.250	.254	.255	.257	.259	.261	.05	
.10	.325	.336	.342	.347	.353	.356	.358	.362	.364	.366	.368	.370	.10	
.25	.698	.712	.719	.727	.734	.738	.741	.747	.749	.752	.754	.756	.25	
.50	2.09	2.12	2.13	2.15	2.16	2.17	2.17	2.18	2.18	2.19	2.19	2.20	.50	
.75	9.49	9.58	9.63	9.67	9.71	9.74	9.76	9.78	9.80	9.82	9.84	9.85	.75	
.90	61.2	61.7	62.0	62.3	62.5	62.7	62.8	63.0	63.1	63.2	63.3	63.3	.90	
.95	246	248	249	250	251	252	252	253	253	254	254	254	.95	
.975	985	993	997	100^1	101^1	101^1	101^1	101^1	101^1	102^1	102^1	102^1	.975	
.99	616^1	621^1	623^1	626^1	629^1	630^1	631^1	633^1	634^1	635^1	636^1	637^1	.99	
.995	246^2	248^2	249^2	250^2	251^2	252^2	253^2	253^2	254^2	254^2	254^2	255^2	.995	
.999	616^3	621^3	623^3	626^3	629^3	630^3	631^3	633^3	634^3	635^3	636^3	637^3	.999	
.9995	246^4	248^4	249^4	250^4	251^4	252^4	252^4	253^4	253^4	253^4	254^4	254^4	.9995	
.0005	.076	.088	.094	.101	.108	.113	.116	.122	.124	.127	.130	.132	.0005	**2**
.001	.088	.100	.107	.114	.121	.126	.129	.135	.137	.140	.143	.145	.001	
.005	.130	.142	.150	.157	.165	.169	.173	.179	.181	.184	.187	.189	.005	
.01	.157	.171	.178	.186	.193	.198	.201	.207	.209	.212	.215	.217	.01	
.025	.210	.224	.232	.239	.247	.251	.255	.261	.263	.266	.269	.271	.025	
.05	.272	.286	.294	.302	.309	.314	.317	.324	.326	.329	.332	.334	.05	
.10	.371	.386	.394	.402	.410	.415	.418	.424	.426	.429	.433	.434	.10	
.25	.657	.672	.680	.689	.697	.702	.705	.711	.713	.716	.719	.721	.25	
.50	1.38	1.39	1.40	1.41	1.42	1.42	1.43	1.43	1.43	1.44	1.44	1.44	.50	
.75	3.41	3.43	3.43	3.44	3.45	3.45	3.46	3.47	3.47	3.48	3.48	3.48	.75	
.90	9.42	9.44	9.45	9.46	9.47	9.47	9.47	9.48	9.48	9.49	9.49	9.49	.90	
.95	19.4	19.4	19.5	19.5	19.5	19.5	19.5	19.5	19.5	19.5	19.5	19.5	.95	
.975	39.4	39.4	39.5	39.5	39.5	39.5	39.5	39.5	39.5	39.5	39.5	39.5	.975	
.99	99.4	99.4	99.5	99.5	99.5	99.5	99.5	99.5	99.5	99.5	99.5	99.5	.99	
.995	199	199	199	199	199	199	199	199	199	199	199	200	.995	
.999	999	999	999	999	999	999	999	999	999	999	999	999	.999	
.9995	200^1	200^1	200^1	200^1	200^1	200^1	200^1	200^1	200^1	200^1	200^1	200^1	.9995	
.0005	.093	.109	.117	.127	.136	.143	.147	.156	.158	.162	.166	.169	.0005	**3**
.001	.107	.123	.132	.142	.152	.158	.162	.171	.173	.177	.181	.184	.001	
.005	.154	.172	.181	.191	.201	.207	.211	.220	.222	.227	.231	.234	.005	
.01	.185	.203	.212	.222	.232	.238	.242	.251	.253	.258	.262	.264	.01	
.025	.241	.259	.269	.279	.289	.295	.299	.308	.310	.314	.318	.321	.025	
.05	.304	.323	.332	.342	.352	.358	.363	.370	.373	.377	.382	.384	.05	
.10	.402	.420	.430	.439	.449	.455	.459	.467	.469	.474	.476	.480	.10	
.25	.658	.675	.684	.693	.702	.708	.711	.719	.721	.724	.728	.730	.25	
.50	1.21	1.23	1.23	1.24	1.25	1.25	1.25	1.26	1.26	1.26	1.27	1.27	.50	
.75	2.46	2.46	2.46	2.47	2.47	2.47	2.47	2.47	2.47	2.47	2.47	2.47	.75	
.90	5.20	5.18	5.18	5.17	5.16	5.15	5.15	5.14	5.14	5.14	5.14	5.13	.90	
.95	8.70	8.66	8.63	8.62	8.59	8.58	8.57	8.55	8.55	8.54	8.53	8.53	.95	
.975	14.3	14.2	14.1	14.1	14.0	14.0	14.0	13.9	13.9	13.9	13.9	13.9	.975	
.99	26.9	26.7	26.6	26.5	26.4	26.4	26.3	26.2	26.2	26.2	26.1	26.1	.99	
.995	43.1	42.8	42.6	42.5	42.3	42.2	42.1	42.0	42.0	41.9	41.9	41.8	.995	
.999	127	126	126	125	125	125	124	124	124	124	124	123	.999	
.9995	203	201	200	199	199	198	198	197	197	197	196	196	.9995	

ν_2, DEGREES OF FREEDOM FOR DENOMINATOR

ν_1, DEGREES OF FREEDOM FOR NUMERATOR

	Cum. prop.	1	2	3	4	5	6	7	8	9	10	11	12	Cum. prop.
4	.0005	$.0^844$	$.0^550$	$.0^446$	.013	.024	.036	.047	.057	.066	.075	.082	.089	.0005
	.001	$.0^518$	$.0^410$	$.0^373$	.019	.032	.046	.058	.069	.079	.089	.097	.104	.001
	.005	$.0^444$	$.0^550$	.022	.043	.064	.083	.100	.114	.126	.137	.145	.153	.005
	.01	$.0^418$	.010	.035	.063	.088	.109	.127	.143	.156	.167	.176	.185	.01
	.025	$.0^311$	.026	.066	.104	.135	.161	.181	.198	.212	.224	.234	.243	.025
	.05	$.0^444$	.052	.110	.157	.193	.221	.243	.261	.275	.288	.298	.307	.05
	.10	.018	.108	.187	.243	.284	.314	.338	.356	.371	.384	.394	.403	.10
	.25	.117	.309	.418	.484	.528	.560	.583	.601	.615	.627	.637	.645	.25
	.50	.549	.828	.941	1.00	1.04	1.06	1.08	1.09	1.10	1.11	1.12	1.13	.50
	.75	1.81	2.00	2.05	2.06	2.07	2.08	2.08	2.08	2.08	2.08	2.08	2.08	.75
	.90	4.54	4.32	4.19	4.11	4.05	4.01	3.98	3.95	3.94	3.92	3.91	3.90	.90
	.95	7.71	6.94	6.59	6.39	6.26	6.16	6.09	6.04	6.00	5.96	5.94	5.91	.95
	.975	12.2	10.6	9.98	9.60	9.36	9.20	9.07	8.98	8.90	8.84	8.79	8.75	.975
	.99	21.2	18.0	16.7	16.0	15.5	15.2	15.0	14.8	14.7	14.5	14.4	14.4	.99
	.995	31.3	26.3	24.3	23.2	22.5	22.0	21.6	21.4	21.1	21.0	20.8	20.7	.995
	.999	74.1	61.2	56.2	53.4	51.7	50.5	49.7	49.0	48.5	48.0	47.7	47.4	.999
	.9995	106	87.4	80.1	76.1	73.6	71.9	70.6	69.7	68.9	68.3	67.8	67.4	.9995
5	.0005	$.0^843$	$.0^550$	$.0^447$	.014	.025	.038	.050	.061	.070	.081	.089	.096	.0005
	.001	$.0^517$	$.0^410$	$.0^375$	.019	.034	.048	.062	.074	.085	.095	.104	.112	.001
	.005	$.0^443$	$.0^550$	.022	.045	.067	.087	.105	.120	.134	.146	.156	.165	.005
	.01	$.0^417$	.010	.035	.064	.091	.114	.134	.151	.165	.177	.188	.197	.01
	.025	$.0^311$	.025	.067	.107	.140	.167	.189	.208	.223	.236	.248	.257	.025
	.05	$.0^443$	.052	.111	.160	.198	.228	.252	.271	.287	.301	.313	.322	.05
	.10	.017	.108	.188	.247	.290	.322	.347	.367	.383	.397	.408	.418	.10
	.25	.113	.305	.415	.483	.528	.560	.584	.604	.618	.631	.641	.650	.25
	.50	.528	.799	.907	.965	1.00	1.02	1.04	1.05	1.06	1.07	1.08	1.09	.50
	.75	1.69	1.85	1.88	1.89	1.89	1.89	1.89	1.89	1.89	1.89	1.89	1.89	.75
	.90	4.06	3.78	3.62	3.52	3.45	3.40	3.37	3.34	3.32	3.30	3.28	3.27	.90
	.95	6.61	5.79	5.41	5.19	5.05	4.95	4.88	4.82	4.77	4.74	4.71	4.68	.95
	.975	10.0	8.43	7.76	7.39	7.15	6.98	6.85	6.76	6.68	6.62	6.57	6.52	.975
	.99	16.3	13.3	12.1	11.4	11.0	10.7	10.5	10.3	10.2	10.1	9.96	9.89	.99
	.995	22.8	18.3	16.5	15.6	14.9	14.5	14.2	14.0	13.8	13.6	13.5	13.4	.995
	.999	47.2	37.1	33.2	31.1	29.7	28.8	28.2	27.6	27.2	26.9	26.6	26.4	.999
	.9995	63.6	49.8	44.4	41.5	39.7	38.5	37.6	36.9	36.4	35.9	35.6	35.2	.9995
6	.0005	$.0^843$	$.0^550$	$.0^447$	.014	.026	.039	.052	.064	.075	.085	.094	.103	.0005
	.001	$.0^517$	$.0^410$	$.0^375$	.020	.035	.050	.064	.078	.090	.101	.111	.119	.001
	.005	$.0^443$	$.0^550$	.022	.045	.069	.090	.109	.126	.140	.153	.164	.174	.005
	.01	$.0^417$	.010	.036	.066	.094	.118	.139	.157	.172	.186	.197	.207	.01
	.025	$.0^311$	.025	.068	.109	.143	.172	.195	.215	.231	.246	.258	.268	.025
	.05	$.0^443$	.052	.112	.162	.202	.233	.259	.279	.296	.311	.324	.334	.05
	.10	.017	.107	.189	.249	.294	.327	.354	.375	.392	.406	.418	.429	.10
	.25	.111	.302	.413	.481	.524	.561	.586	.606	.622	.635	.645	.654	.25
	.50	.515	.780	.886	.942	.977	1.00	1.02	1.03	1.04	1.05	1.05	1.06	.50
	.75	1.62	1.76	1.78	1.79	1.79	1.78	1.78	1.78	1.77	1.77	1.77	1.77	.75
	.90	3.78	3.46	3.29	3.18	3.11	3.05	3.01	2.98	2.96	2.94	2.92	2.90	.90
	.95	5.99	5.14	4.76	4.53	4.39	4.28	4.21	4.15	4.10	4.06	4.03	4.00	.95
	.975	8.81	7.26	6.60	6.23	5.99	5.82	5.70	5.60	5.52	5.46	5.41	5.37	.975
	.99	13.7	10.9	9.78	9.15	8.75	8.47	8.26	8.10	7.98	7.87	7.79	7.72	.99
	.995	18.6	14.5	12.9	12.0	11.5	11.1	10.8	10.6	10.4	10.2	10.1	10.0	.995
	.999	35.5	27.0	23.7	21.9	20.8	20.0	19.5	19.0	18.7	18.4	18.2	18.0	.999
	.9995	46.1	34.8	30.4	28.1	26.6	25.6	24.9	24.3	23.9	23.5	23.2	23.0	.9995

ν_2, DEGREES OF FREEDOM FOR DENOMINATOR

v_1, DEGREES OF FREEDOM FOR NUMERATOR

Cum. prop.	15	20	24	30	40	50	60	100	120	200	500	∞	Cum. prop.	v_2
.0005	.105	.125	.135	.147	.159	.166	.172	.183	.186	.191	.196	.200	.0005	**4**
.001	.121	.141	.152	.163	.176	.183	.188	.200	.202	.208	.213	.217	.001	
.005	.172	.193	.204	.216	.229	.237	.242	.253	.255	.260	.266	.269	.005	
.01	.204	.226	.237	.249	.261	.269	.274	.285	.287	.293	.298	.301	.01	
.025	.263	.284	.296	.308	.320	.327	.332	.342	.346	.351	.356	.359	.025	
.05	.327	.349	.360	.372	.384	.391	.396	.407	.409	.413	.418	.422	.05	
.10	.424	.445	.456	.467	.478	.485	.490	.500	.502	.508	.510	.514	.10	
.25	.664	.683	.692	.702	.712	.718	.722	.731	.733	.737	.740	.743	.25	
.50	1.14	1.15	1.16	1.16	1.17	1.18	1.18	1.18	1.18	1.19	1.19	1.19	.50	
.75	2.08	2.08	2.08	2.08	2.08	2.08	2.08	2.08	2.08	2.08	2.08	2.08	.75	
.90	3.87	3.84	3.83	3.82	3.80	3.80	3.79	3.78	3.78	3.77	3.76	3.76	.90	
.95	5.86	5.80	5.77	5.75	5.72	5.70	5.69	5.66	5.66	5.65	5.64	5.63	.95	
.975	8.66	8.56	8.51	8.46	8.41	8.38	8.36	8.32	8.31	8.29	8.27	8.26	.975	
.99	14.2	14.0	13.9	13.8	13.7	13.7	13.7	13.6	13.6	13.5	13.5	13.5	.99	
.995	20.4	20.2	20.0	19.9	19.8	19.7	19.6	19.5	19.5	19.4	19.4	19.3	.995	
.999	46.8	46.1	45.8	45.4	45.1	44.9	44.7	44.5	44.4	44.3	44.1	44.0	.999	
.9995	66.5	65.5	65.1	64.6	64.1	63.8	63.6	63.2	63.1	62.9	62.7	62.6	.9995	
.0005	.115	.137	.150	.163	.177	.186	.192	.205	.209	.216	.222	.226	.0005	**5**
.001	.132	.155	.167	.181	.195	.204	.210	.223	.227	.233	.239	.244	.001	
.005	.186	.210	.223	.237	.251	.260	.266	.279	.282	.288	.294	.299	.005	
.01	.219	.244	.257	.270	.285	.293	.299	.315	.322	.328	.331		.01	
.025	.280	.304	.317	.330	.344	.353	.359	.370	.374	.380	.386	.390	.025	
.05	.345	.369	.382	.395	.408	.417	.422	.432	.437	.442	.448	.452	.05	
.10	.440	.463	.476	.488	.501	.508	.514	.524	.527	.532	.538	.541	.10	
.25	.669	.690	.700	.711	.722	.728	.732	.741	.743	.748	.752	.755	.25	
.50	1.10	1.11	1.12	1.12	1.13	1.13	1.14	1.14	1.14	1.15	1.15	1.15	.50	
.75	1.89	1.88	1.88	1.88	1.88	1.88	1.87	1.87	1.87	1.87	1.87	1.87	.75	
.90	3.24	3.21	3.19	3.17	3.16	3.15	3.14	3.13	3.12	3.12	3.11	3.10	.90	
.95	4.62	4.56	4.53	4.50	4.46	4.44	4.43	4.41	4.40	4.39	4.37	4.36	.95	
.975	6.43	6.33	6.28	6.23	6.18	6.14	6.12	6.08	6.07	6.05	6.03	6.02	.975	
.99	9.72	9.55	9.47	9.38	9.29	9.24	9.20	9.13	9.11	9.08	9.04	9.02	.99	
.995	13.1	12.9	12.8	12.7	12.5	12.5	12.4	12.3	12.3	12.2	12.2	12.1	.995	
.999	25.9	25.4	25.1	24.9	24.6	24.4	24.3	24.1	24.1	23.9	23.8	23.8	.999	
.9995	34.6	33.9	33.5	33.1	32.7	32.5	32.3	32.1	32.0	31.8	31.7	31.6	.9995	
.0005	.123	.148	.162	.177	.193	.203	.210	.225	.229	.236	.244	.249	.0005	**6**
.001	.141	.166	.180	.195	.211	.222	.229	.243	.247	.255	.262	.267	.001	
.005	.197	.224	.238	.253	.269	.279	.286	.301	.304	.312	.318	.324	.005	
.01	.232	.258	.273	.288	.304	.313	.321	.334	.338	.346	.352	.357	.01	
.025	.293	.320	.334	.349	.364	.375	.381	.394	.398	.405	.412	.415	.025	
.05	.358	.385	.399	.413	.428	.437	.444	.457	.460	.467	.472	.476	.05	
.10	.453	.478	.491	.505	.519	.526	.533	.546	.548	.556	.559	.564	.10	
.25	.675	.696	.707	.718	.729	.736	.741	.751	.753	.758	.762	.765	.25	
.50	1.07	1.08	1.09	1.10	1.10	1.11	1.11	1.11	1.12	1.12	1.12	1.12	.50	
.75	1.76	1.76	1.75	1.75	1.75	1.75	1.74	1.74	1.74	1.74	1.74	1.74	.75	
.90	2.87	2.84	2.82	2.80	2.78	2.77	2.76	2.75	2.74	2.73	2.73	2.72	.90	
.95	3.94	3.87	3.84	3.81	3.77	3.75	3.74	3.71	3.70	3.69	3.68	3.67	.95	
.975	5.27	5.17	5.12	5.07	5.01	4.98	4.96	4.92	4.90	4.88	4.86	4.85	.975	
.99	7.56	7.40	7.31	7.23	7.14	7.09	7.06	6.99	6.97	6.93	6.90	6.88	.99	
.995	9.81	9.59	9.47	9.36	9.24	9.17	9.12	9.03	9.00	8.95	8.91	8.88	.995	
.999	17.6	17.1	16.9	16.7	16.4	16.3	16.2	16.0	16.0	15.9	15.8	15.7	.999	
.9995	22.4	21.9	21.7	21.4	21.1	20.9	20.7	20.5	20.4	20.3	20.2	20.1	.9995	

v_2, DEGREES OF FREEDOM FOR DENOMINATOR

ν_1, DEGREES OF FREEDOM FOR NUMERATOR

ν_2, DEGREES OF FREEDOM FOR DENOMINATOR

7

Cum. prop.	1	2	3	4	5	6	7	8	9	10	11	12	Cum. prop.
.0005	$.0^6 42$	$.0^5 50$	$.0^4 48$	.014	.027	.040	.053	.066	.078	.088	.099	.108	.0005
.001	$.0^5 17$	$.0^4 10$	$.0^3 76$	.020	.035	.051	.067	.081	.093	.105	.115	.125	.001
.005	$.0^4 42$	$.0^3 50$	.023	.046	.070	.093	.113	.130	.145	.159	.171	.181	.005
.01	$.0^3 17$	.010	.036	.067	.096	.121	.143	.162	.178	.192	.205	.216	.01
.025	$.0^2 10$	.025	.068	.110	.146	.176	.200	.221	.238	.253	.266	.277	.025
.05	$.0^2 42$	.052	.113	.164	.205	.238	.264	.286	.304	.319	.332	.343	.05
.10	.017	.107	.190	.251	.297	.332	.359	.381	.399	.414	.427	.438	.10
.25	.110	.300	.412	.481	.528	.562	.588	.608	.624	.637	.649	.658	.25
.50	.506	.767	.871	.926	.960	.983	1.00	1.01	1.02	1.03	1.04	1.04	.50
.75	1.57	1.70	1.72	1.72	1.71	1.71	1.70	1.70	1.69	1.69	1.69	1.68	.75
.90	3.59	3.26	3.07	2.96	2.88	2.83	2.78	2.75	2.72	2.70	2.68	2.67	.90
.95	5.59	4.74	4.35	4.12	3.97	3.87	3.79	3.73	3.68	3.64	3.60	3.57	.95
.975	8.07	6.54	5.89	5.52	5.29	5.12	4.99	4.90	4.82	4.76	4.71	4.67	.975
.99	12.2	9.55	8.45	7.85	7.46	7.19	6.99	6.84	6.72	6.62	6.54	6.47	.99
.995	16.2	12.4	10.9	10.0	9.52	9.16	8.89	8.68	8.51	8.38	8.27	8.18	.995
.999	29.2	21.7	18.8	17.2	16.2	15.5	15.0	14.6	14.3	14.1	13.9	13.7	.999
.9995	37.0	27.2	23.5	21.4	20.2	19.3	18.7	18.2	17.8	17.5	17.2	17.0	.9995

8

Cum. prop.	1	2	3	4	5	6	7	8	9	10	11	12	Cum. prop.
.0005	$.0^6 42$	$.0^5 50$	$.0^4 48$	.014	.027	.041	.055	.068	.081	.092	.102	.112	.0005
.001	$.0^5 17$	$.0^4 10$	$.0^3 76$	.020	.036	.053	.068	.083	.096	.109	.120	.130	.001
.005	$.0^4 42$	$.0^3 50$	.027	.047	.072	.095	.115	.133	.149	.164	.176	.187	.005
.01	$.0^3 17$	.010	.036	.068	.097	.123	.146	.166	.183	.198	.211	.222	.01
.025	$.0^2 10$	.025	.069	.111	.148	.179	.204	.226	.244	.259	.273	.285	.025
.05	$.0^2 42$	.052	.113	.166	.208	.241	.268	.291	.310	.326	.339	.351	.05
.10	.017	.107	.190	.253	.299	.335	.363	.386	.405	.421	.435	.445	.10
.25	.109	.298	.411	.481	.529	.563	.589	.610	.627	.640	.654	.661	.25
.50	.499	.757	.860	.915	.948	.971	.988	1.00	1.01	1.02	1.02	1.03	.50
.75	1.54	1.66	1.67	1.66	1.66	1.65	1.64	1.64	1.64	1.63	1.63	1.62	.75
.90	3.46	3.11	2.92	2.81	2.73	2.67	2.62	2.59	2.56	2.54	2.52	2.50	.90
.95	5.32	4.46	4.07	3.84	3.69	3.58	3.50	3.44	3.39	3.35	3.31	3.28	.95
.975	7.57	6.06	5.42	5.05	4.82	4.65	4.53	4.43	4.36	4.30	4.24	4.20	.975
.99	11.3	8.65	7.59	7.01	6.63	6.37	6.18	6.03	5.91	5.81	5.73	5.67	.99
.995	14.7	11.0	9.60	8.81	8.30	7.95	7.69	7.50	7.34	7.21	7.10	7.01	.995
.999	25.4	18.5	15.8	14.4	13.5	12.9	12.4	12.0	11.8	11.5	11.4	11.2	.999
.9995	31.6	22.8	19.4	17.6	16.4	15.7	15.1	14.6	14.3	14.0	13.8	13.6	.9995

9

Cum. prop.	1	2	3	4	5	6	7	8	9	10	11	12	Cum. prop.
.0005	$.0^6 41$	$.0^5 50$	$.0^4 48$	.015	.027	.042	.056	.070	.083	.094	.105	.115	.0005
.001	$.0^5 17$	$.0^4 10$	$.0^3 77$	.021	.037	.054	.070	.085	.099	.112	.123	.134	.001
.005	$.0^4 42$	$.0^3 50$	.023	.047	.073	.096	.117	.136	.153	.168	.181	.192	.005
.01	$.0^3 17$	.010	.037	.068	.098	.125	.149	.169	.187	.202	.216	.228	.01
.025	$.0^2 10$	.025	.069	.112	.150	.181	.207	.230	.248	.265	.279	.291	.025
.05	$.0^2 40$	.052	.113	.167	.210	.244	.272	.296	.315	.331	.345	.358	.05
.10	.017	.107	.191	.254	.302	.338	.367	.390	.410	.426	.441	.452	.10
.25	.108	.297	.410	.480	.529	.564	.591	.612	.629	.643	.654	.664	.25
.50	.494	.749	.852	.906	.939	.962	.978	.990	1.00	1.01	1.01	1.02	.50
.75	1.51	1.62	1.63	1.63	1.62	1.61	1.60	1.60	1.59	1.59	1.58	1.58	.75
.90	3.36	3.01	2.81	2.69	2.61	2.55	2.51	2.47	2.44	2.42	2.40	2.38	.90
.95	5.12	4.26	3.86	3.63	3.48	3.37	3.29	3.23	3.18	3.14	3.10	3.07	.95
.975	7.21	5.71	5.08	4.72	4.48	4.32	4.20	4.10	4.03	3.96	3.91	3.87	.975
.99	10.6	8.02	6.99	6.42	6.06	5.80	5.61	5.47	5.35	5.26	5.18	5.11	.99
.995	13.6	10.1	8.72	7.96	7.47	7.13	6.88	6.69	6.54	6.42	6.31	6.23	.995
.999	22.9	16.4	13.9	12.6	11.7	11.1	10.7	10.4	10.1	9.89	9.71	9.57	.999
.9995	28.0	19.9	16.8	15.1	14.1	13.3	12.8	12.4	12.1	11.8	11.6	11.4	.9995

THE CRITICAL VALUES OF THE F DISTRIBUTION, (continued)

ν₁, DEGREES OF FREEDOM FOR NUMERATOR

Cum. prop.	15	20	24	30	40	50	60	100	120	200	500	∞	Cum. prop.	ν₂
.0005	.130	.157	.172	.188	.206	.217	.225	.242	.246	.255	.263	.268	.0005	7
.001	.148	.176	.191	.208	.225	.237	.245	.261	.266	.274	.282	.288	.001	
.005	.206	.235	.251	.267	.285	.296	.304	.319	.324	.332	.340	.345	.005	
.01	.241	.270	.286	.303	.320	.331	.339	.355	.358	.366	.373	.379	.01	
.025	.304	.333	.348	.364	.381	.392	.399	.413	.418	.426	.433	.437	.025	
.05	.369	.398	.413	.428	.445	.455	.461	.476	.479	.485	.493	.498	.05	
.10	.463	.491	.504	.519	.534	.543	.550	.562	.566	.571	.578	.582	.10	
.25	.679	.702	.713	.725	.737	.745	.749	.760	.762	.767	.772	.775	.25	
.50	1.05	1.07	1.07	1.08	1.08	1.09	1.09	1.10	1.10	1.10	1.10	1.10	.50	
.75	1.68	1.67	1.67	1.66	1.66	1.66	1.66	1.65	1.65	1.65	1.65	1.65	.75	
.90	2.63	2.59	2.58	2.56	2.54	2.52	2.51	2.50	2.49	2.48	2.48	2.47	.90	
.95	3.51	3.44	3.41	3.38	3.34	3.32	3.30	3.27	3.27	3.25	3.24	3.23	.95	
.975	4.57	4.47	4.42	4.36	4.31	4.28	4.25	4.21	4.20	4.18	4.16	4.14	.975	
.99	6.31	6.16	6.07	5.99	5.91	5.86	5.82	5.75	5.74	5.70	5.67	5.65	.99	
.995	7.97	7.75	7.65	7.53	7.42	7.35	7.31	7.22	7.19	7.15	7.10	7.08	.995	
.999	13.3	12.9	12.7	12.5	12.3	12.2	12.1	11.9	11.9	11.8	11.7	11.7	.999	
.9995	16.5	16.0	15.7	15.5	15.2	15.1	15.0	14.7	14.7	14.6	14.5	14.4	.9995	
.0005	.136	.164	.181	.198	.218	.230	.239	.257	.262	.271	.281	.287	.0005	8
.001	.155	.184	.200	.218	.238	.250	.259	.277	.282	.292	.306	.306	.001	
.005	.214	.244	.261	.279	.299	.311	.319	.337	.341	.351	.358	.364	.005	
.01	.250	.281	.297	.315	.334	.346	.354	.372	.376	.385	.392	.398	.01	
.025	.313	.343	.360	.377	.395	.407	.415	.431	.435	.442	.450	.456	.025	
.05	.379	.409	.425	.441	.459	.469	.477	.493	.496	.505	.510	.516	.05	
.10	.472	.500	.515	.531	.547	.556	.563	.578	.581	.588	.595	.599	.10	
.25	.684	.707	.718	.730	.743	.751	.756	.767	.769	.775	.780	.783	.25	
.50	1.04	1.05	1.06	1.07	1.07	1.07	1.08	1.08	1.08	1.09	1.09	1.09	.50	
.75	1.62	1.61	1.60	1.60	1.59	1.59	1.59	1.58	1.58	1.58	1.58	1.58	.75	
.90	2.46	2.42	2.40	2.38	2.36	2.35	2.34	2.32	2.32	2.31	2.30	2.29	.90	
.95	3.22	3.15	3.12	3.08	3.04	3.02	3.01	2.97	2.97	2.95	2.94	2.93	.95	
.975	4.10	4.00	3.95	3.89	3.84	3.81	3.78	3.74	3.73	3.70	3.68	3.67	.975	
.99	5.52	5.36	5.28	5.20	5.12	5.07	5.03	4.96	4.95	4.91	4.88	4.86	.99	
.995	6.81	6.61	6.50	6.40	6.29	6.22	6.18	6.09	6.06	6.02	5.98	5.95	.995	
.999	10.8	10.5	10.3	10.1	9.92	9.80	9.73	9.57	9.54	9.46	9.39	9.34	.999	
.9995	13.1	12.7	12.5	12.2	12.0	11.8	11.8	11.6	11.5	11.4	11.4	11.3	.9995	
.0005	.141	.171	.188	.207	.228	.242	.251	.270	.276	.287	.297	.303	.0005	9
.001	.160	.191	.208	.228	.249	.262	.271	.291	.296	.307	.316	.323	.001	
.005	.220	.253	.271	.290	.310	.324	.332	.351	.356	.366	.376	.382	.005	
.01	.257	.289	.307	.326	.346	.358	.368	.386	.391	.400	.410	.415	.01	
.025	.320	.352	.370	.388	.408	.420	.428	.446	.450	.459	.467	.473	.025	
.05	.386	.418	.435	.452	.471	.483	.490	.508	.510	.518	.526	.532	.05	
.10	.479	.509	.525	.541	.558	.568	.575	.588	.594	.602	.610	.613	.10	
.25	.687	.711	.723	.736	.749	.757	.762	.773	.776	.782	.787	.791	.25	
.50	1.03	1.04	1.05	1.05	1.06	1.06	1.07	1.07	1.07	1.08	1.08	1.08	.50	
.75	1.57	1.56	1.56	1.55	1.55	1.54	1.54	1.53	1.53	1.53	1.53	1.53	.75	
.90	2.34	2.30	2.28	2.25	2.23	2.22	2.21	2.19	2.18	2.17	2.17	2.16	.90	
.95	3.01	2.94	2.90	2.86	2.83	2.80	2.79	2.76	2.75	2.73	2.72	2.71	.95	
.975	3.77	3.67	3.61	3.56	3.51	3.47	3.45	3.40	3.39	3.37	3.35	3.33	.975	
.99	4.96	4.81	4.73	4.65	4.57	4.52	4.48	4.42	4.40	4.36	4.33	4.31	.99	
.995	6.03	5.83	5.73	5.62	5.52	5.45	5.41	5.32	5.30	5.26	5.21	5.19	.995	
.999	9.24	8.90	8.72	8.55	8.37	8.26	8.19	8.04	8.00	7.93	7.86	7.81	.999	
.9995	11.0	10.6	10.4	10.2	9.94	9.80	9.71	9.53	9.49	9.40	9.32	9.26	.9995	

ν₂, DEGREES OF FREEDOM FOR DENOMINATOR

ν_1, DEGREES OF FREEDOM FOR NUMERATOR

ν_2, DEGREES OF FREEDOM FOR DENOMINATOR

	Cum. prop.	1	2	3	4	5	6	7	8	9	10	11	12	Cum. prop.
10	.0005	$.0^6 41$	$.0^5 50$	$.0^2 49$	.015	.028	.043	.057	.071	.085	.097	.108	.119	.0005
	.001	$.0^5 17$	$.0^2 10$	$.0^2 77$	.021	.037	.054	.071	.087	.101	.114	.126	.137	.001
	.005	$.0^4 41$	$.0^5 50$	.023	.048	.073	.098	.119	.139	.156	.171	.185	.197	.005
	.01	$.0^3 17$	.010	.037	.069	.100	.127	.151	.172	.190	.206	.220	.233	.01
	.025	$.0^2 10$	.025	.069	.113	.151	.183	.210	.233	.252	.269	.283	.296	.025
	.05	$.0^2 41$	.052	.114	.168	.211	.246	.275	.299	.319	.336	.351	.363	.05
	.10	.017	.106	.191	.255	.303	.340	.370	.394	.414	.430	.444	.457	.10
	.25	.107	.296	.409	.480	.529	.565	.592	.613	.631	.645	.657	.667	.25
	.50	.490	.743	.845	.899	.932	.954	.971	.983	.992	1.00	1.01	1.01	.50
	.75	1.49	1.60	1.60	1.59	1.59	1.58	1.57	1.56	1.56	1.55	1.55	1.54	.75
	.90	3.28	2.92	2.73	2.61	2.52	2.46	2.41	2.38	2.35	2.32	2.30	2.28	.90
	.95	4.96	4.10	3.71	3.48	3.33	3.22	3.14	3.07	3.02	2.98	2.94	2.91	.95
	.975	6.94	5.46	4.83	4.47	4.24	4.07	3.95	3.85	3.78	3.72	3.66	3.62	.975
	.99	10.0	7.56	6.55	5.99	5.64	5.39	5.20	5.06	4.94	4.85	4.77	4.71	.99
	.995	12.8	9.43	8.08	7.34	6.87	6.54	6.30	6.12	5.97	5.85	5.75	5.66	.995
	.999	21.0	14.9	12.6	11.3	10.5	9.92	9.52	9.20	8.96	8.75	8.58	8.44	.999
	.9995	25.5	17.9	15.0	13.4	12.4	11.8	11.3	10.9	10.6	10.3	10.1	9.93	.9995
11	.0005	$.0^6 41$	$.0^5 50$	$.0^2 49$	.015	.028	.043	.058	.072	.086	.099	.111	.121	.0005
	.001	$.0^5 16$	$.0^2 10$	$.0^2 78$	.021	.038	.055	.072	.088	.103	.116	.129	.140	.001
	.005	$.0^4 40$	$.0^5 50$	.023	.048	.074	.099	.121	.141	.158	.174	.188	.200	.005
	.01	$.0^3 16$	.010	.037	.069	.100	.128	.153	.175	.193	.210	.224	.237	.01
	.025	$.0^2 10$	.025	.069	.114	.152	.185	.212	.236	.256	.273	.288	.301	.025
	.05	$.0^2 41$	.052	.114	.168	.212	.248	.278	.302	.323	.340	.355	.368	.05
	.10	.017	.106	.192	.256	.305	.342	.373	.397	.417	.435	.448	.461	.10
	.25	.107	.295	.408	.481	.529	.565	.592	.614	.633	.645	.658	.667	.25
	.50	.486	.739	.840	.893	.926	.948	.964	.977	.986	.994	1.00	1.01	.50
	.75	1.47	1.58	1.58	1.57	1.56	1.55	1.54	1.53	1.53	1.52	1.52	1.51	.75
	.90	3.23	2.86	2.66	2.54	2.45	2.39	2.34	2.30	2.27	2.25	2.23	2.21	.90
	.95	4.84	3.98	3.59	3.36	3.20	3.09	3.01	2.95	2.90	2.85	2.82	2.79	.95
	.975	6.72	5.26	4.63	4.28	4.04	3.88	3.76	3.66	3.59	3.53	3.47	3.43	.975
	.99	9.65	7.21	6.22	5.67	5.32	5.07	4.89	4.74	4.63	4.54	4.46	4.40	.99
	.995	12.2	8.91	7.60	6.88	6.42	6.10	5.86	5.68	5.54	5.42	5.32	5.24	.995
	.999	19.7	13.8	11.6	10.3	9.58	9.05	8.66	8.35	8.12	7.92	7.76	7.62	.999
	.9995	23.6	16.4	13.6	12.2	11.2	10.6	10.1	9.76	9.48	9.24	9.04	8.88	.9995
12	.0005	$.0^6 41$	$.0^5 50$	$.0^2 49$	.015	.028	.044	.058	.073	.087	.101	.113	.124	.0005
	.001	$.0^5 16$	$.0^2 10$	$.0^2 78$	.021	.038	.056	.073	.089	.104	.118	.131	.143	.001
	.005	$.0^4 39$	$.0^5 50$	.023	.048	.075	.100	.122	.143	.161	.177	.191	.204	.005
	.01	$.0^3 16$	.010	.037	.070	.101	.130	.155	.176	.196	.212	.227	.241	.01
	.025	$.0^2 10$	.025	.070	.114	.153	.186	.214	.238	.259	.276	.292	.305	.025
	.05	$.0^2 41$	.052	.114	.169	.214	.250	.280	.305	.325	.343	.358	.372	.05
	.10	.016	.106	.192	.257	.306	.344	.375	.400	.420	.438	.452	.466	.10
	.25	.106	.295	.408	.480	.530	.566	.594	.616	.633	.649	.662	.671	.25
	.50	.484	.735	.835	.888	.921	.943	.959	.972	.981	.989	.995	1.00	.50
	.75	1.46	1.56	1.56	1.55	1.54	1.53	1.52	1.51	1.51	1.50	1.50	1.49	.75
	.90	3.18	2.81	2.61	2.48	2.39	2.33	2.28	2.24	2.21	2.19	2.17	2.15	.90
	.95	4.75	3.89	3.49	3.26	3.11	3.00	2.91	2.85	2.80	2.75	2.72	2.69	.95
	.975	6.55	5.10	4.47	4.12	3.89	3.73	3.61	3.51	3.44	3.37	3.32	3.28	.975
	.99	9.33	6.93	5.95	5.41	5.06	4.82	4.64	4.50	4.39	4.30	4.22	4.16	.99
	.995	11.8	8.51	7.23	6.52	6.07	5.76	5.52	5.35	5.20	5.09	4.99	4.91	.995
	.999	18.6	13.0	10.8	9.63	8.89	8.38	8.00	7.71	7.48	7.29	7.14	7.01	.999
	.9995	22.2	15.3	12.7	11.2	10.4	9.74	9.28	8.94	8.66	8.43	8.24	8.08	.9995

ν_1, DEGREES OF FREEDOM FOR NUMERATOR

Cum. prop.	15	20	24	30	40	50	60	100	120	200	500	∞	Cum. prop.	
.0005	.145	.177	.195	.215	.238	.251	.262	.282	.288	.299	.311	.319	.0005	10
.001	.164	.197	.216	.236	.258	.272	.282	.303	.309	.321	.331	.338	.001	
.005	.226	.260	.279	.299	.321	.334	.344	.365	.370	.380	.391	.397	.005	
.01	.263	.297	.316	.336	.357	.370	.380	.400	.405	.415	.424	.431	.01	
.025	.327	.360	.379	.398	.419	.431	.441	.459	.464	.474	.483	.488	.025	
.05	.393	.426	.444	.462	.481	.493	.502	.518	.523	.532	.541	.546	.05	
.10	.486	.516	.532	.549	.567	.578	.586	.602	.605	.614	.621	.625	.10	
.25	.691	.714	.727	.740	.754	.762	.767	.779	.782	.788	.793	.797	.25	
.50	1.02	1.03	1.04	1.05	1.05	1.06	1.06	1.06	1.06	1.07	1.07	1.07	.50	
.75	1.53	1.52	1.52	1.51	1.51	1.50	1.50	1.49	1.49	1.49	1.48	1.48	.75	
.90	2.24	2.20	2.18	2.16	2.13	2.12	2.11	2.09	2.08	2.07	2.06	2.06	.90	
.95	2.85	2.77	2.74	2.70	2.66	2.64	2.62	2.59	2.58	2.56	2.55	2.54	.95	
.975	3.52	3.42	3.37	3.31	3.26	3.22	3.20	3.15	3.14	3.12	3.09	3.08	.975	
.99	4.56	4.41	4.33	4.25	4.17	4.12	4.08	4.01	4.00	3.96	3.93	3.91	.99	
.995	5.47	5.27	5.17	5.07	4.97	4.90	4.86	4.77	4.75	4.71	4.67	4.64	.995	
.999	8.13	7.80	7.64	7.47	7.30	7.19	7.12	6.98	6.94	6.87	6.81	6.76	.999	
.9995	9.56	9.16	8.96	8.75	8.54	8.42	8.33	8.16	8.12	8.04	7.96	7.90	.9995	
.0005	.148	.182	.201	.222	.246	.261	.271	.293	.299	.312	.324	.331	.0005	11
.001	.168	.202	.222	.243	.266	.282	.292	.313	.320	.332	.343	.353	.001	
.005	.231	.266	.286	.308	.330	.345	.355	.376	.382	.394	.403	.412	.005	
.01	.268	.304	.324	.344	.366	.380	.391	.412	.417	.427	.439	.444	.01	
.025	.332	.368	.386	.407	.429	.442	.450	.472	.476	.485	.495	.503	.025	
.05	.398	.433	.452	.469	.490	.503	.513	.529	.535	.543	.552	.559	.05	
.10	.490	.524	.541	.559	.578	.588	.595	.614	.617	.625	.633	.637	.10	
.25	.694	.719	.730	.744	.758	.767	.773	.780	.788	.794	.799	.803	.25	
.50	1.02	1.03	1.03	1.04	1.05	1.05	1.05	1.06	1.06	1.06	1.06	1.06	.50	
.75	1.50	1.49	1.49	1.48	1.47	1.47	1.47	1.46	1.46	1.46	1.45	1.45	.75	
.90	2.17	2.12	2.10	2.08	2.05	2.04	2.03	2.00	2.00	1.99	1.98	1.97	.90	
.95	2.72	2.65	2.61	2.57	2.53	2.51	2.49	2.46	2.45	2.43	2.42	2.40	.95	
.975	3.33	3.23	3.17	3.12	3.06	3.03	3.00	2.96	2.94	2.92	2.90	2.88	.975	
.99	4.25	4.10	4.02	3.94	3.86	3.81	3.78	3.71	3.69	3.66	3.62	3.60	.99	
.995	5.05	4.86	4.76	4.65	4.55	4.49	4.45	4.36	4.34	4.29	4.25	4.23	.995	
.999	7.32	7.01	6.85	6.68	6.52	6.41	6.35	6.21	6.17	6.10	6.04	6.00	.999	
.9995	8.52	8.14	7.94	7.75	7.55	7.43	7.35	7.18	7.14	7.06	6.98	6.93	.9995	
.0005	.152	.186	.206	.228	.253	.269	.280	.305	.311	.323	.337	.345	.0005	12
.001	.172	.207	.228	.250	.275	.291	.302	.326	.332	.344	.357	.365	.001	
.005	.235	.272	.292	.315	.339	.355	.365	.388	.393	.405	.417	.424	.005	
.01	.273	.310	.330	.352	.375	.391	.401	.422	.428	.441	.450	.458	.01	
.025	.337	.374	.394	.416	.437	.450	.461	.481	.487	.498	.508	.514	.025	
.05	.404	.439	.458	.478	.499	.513	.522	.541	.545	.556	.565	.571	.05	
.10	.496	.528	.546	.564	.583	.595	.604	.621	.625	.633	.641	.647	.10	
.25	.695	.721	.734	.748	.762	.771	.777	.789	.792	.799	.804	.808	.25	
.50	1.01	1.02	1.03	1.03	1.04	1.04	1.05	1.05	1.05	1.05	1.06	1.06	.50	
.75	1.48	1.47	1.46	1.45	1.45	1.44	1.44	1.43	1.43	1.43	1.42	1.42	.75	
.90	2.11	2.06	2.04	2.01	1.99	1.97	1.96	1.94	1.93	1.92	1.91	1.90	.90	
.95	2.62	2.54	2.51	2.47	2.43	2.40	2.38	2.35	2.34	2.32	2.31	2.30	.95	
.975	3.18	3.07	3.02	2.96	2.91	2.87	2.85	2.80	2.79	2.76	2.74	2.72	.975	
.99	4.01	3.86	3.78	3.70	3.62	3.57	3.54	3.47	3.45	3.41	3.38	3.36	.99	
.995	4.72	4.53	4.43	4.33	4.23	4.17	4.12	4.04	4.01	3.97	3.93	3.90	.995	
.999	6.71	6.40	6.25	6.09	5.93	5.83	5.76	5.63	5.59	5.52	5.46	5.42	.999	
.9995	7.74	7.37	7.18	7.00	6.80	6.68	6.61	6.45	6.41	6.33	6.25	6.20	.9995	

ν_2, DEGREES OF FREEDOM FOR DENOMINATOR

THE CRITICAL VALUES OF THE F DISTRIBUTION, (continued)

n_1 DEGREES OF FREEDOM FOR NUMERATOR

n_2 DEGREES OF FREEDOM FOR DENOMINATOR

	Cum. prop.	1	2	3	4	5	6	7	8	9	10	11	12	Cum. prop.
15	.0005	$.0^4 41$	$.0^5 50$	$.0^2 49$	.015	.029	.045	.061	.076	.091	.105	.117	.129	.0005
	.001	$.0^3 16$	$.0^2 10$	$.0^2 79$	.021	.039	.057	.075	.092	.108	.123	.137	.149	.001
	.005	$.0^4 39$	$.0^2 50$	.023	.049	.076	.102	.125	.147	.166	.183	.198	.212	.005
	.01	$.0^4 16$	.010	.037	.070	.103	.132	.158	.181	.202	.219	.235	.249	.01
	.025	$.0^3 10$	.025	.070	.116	.156	.190	.219	.244	.265	.284	.300	.315	.025
	.05	$.0^2 41$	.051	.115	.170	.216	.254	.285	.311	.333	.351	.368	.382	.05
	.10	.016	.106	.192	.258	.309	.348	.380	.406	.427	.446	.461	.475	.10
	.25	.105	.293	.407	.480	.531	.568	.596	.618	.637	.652	.667	.676	.25
	.50	.478	.726	.826	.878	.911	.933	.948	.960	.970	.977	.984	.989	.50
	.75	1.43	1.52	1.52	1.51	1.49	1.48	1.47	1.46	1.46	1.45	1.44	1.44	.75
	.90	3.07	2.70	2.49	2.36	2.27	2.21	2.16	2.12	2.09	2.06	2.04	2.02	.90
	.95	4.54	3.68	3.29	3.06	2.90	2.79	2.71	2.64	2.59	2.54	2.51	2.48	.95
	.975	6.20	4.76	4.15	3.80	3.58	3.41	3.29	3.20	3.12	3.06	3.01	2.96	.975
	.99	8.68	6.36	5.42	4.89	4.56	4.32	4.14	4.00	3.89	3.80	3.73	3.67	.99
	.995	10.8	7.70	6.48	5.80	5.37	5.07	4.85	4.67	4.54	4.42	4.33	4.25	.995
	.999	16.6	11.3	9.34	8.25	7.57	7.09	6.74	6.47	6.26	6.08	5.93	5.81	.999
	.9995	19.5	13.2	10.8	9.48	8.66	8.10	7.68	7.36	7.11	6.91	6.75	6.60	.9995
20	.0005	$.0^4 40$	$.0^5 50$	$.0^5 50$	.015	.029	.046	.063	.079	.094	.109	.123	.136	.0005
	.001	$.0^3 16$	$.0^2 10$	$.0^2 79$	.022	.039	.058	.077	.095	.112	.128	.143	.156	.001
	.005	$.0^4 39$	$.0^2 50$	.023	.050	.077	.104	.129	.151	.171	.190	.206	.221	.005
	.01	$.0^3 16$	.010	.037	.071	.105	.135	.162	.187	.208	.227	.244	.259	.01
	.025	$.0^3 10$	.025	.071	.117	.158	.193	.224	.250	.273	.292	.310	.325	.025
	.05	$.0^2 40$	.051	.115	.172	.219	.258	.290	.318	.340	.360	.377	.393	.05
	.10	.016	.106	.193	.260	.312	.353	.385	.412	.435	.454	.472	.485	.10
	.25	.104	.292	.407	.480	.531	.569	.598	.622	.641	.656	.671	.681	.25
	.50	.472	.718	.816	.868	.900	.922	.938	.950	.959	.966	.972	.977	.50
	.75	1.40	1.49	1.48	1.47	1.45	1.44	1.43	1.42	1.41	1.40	1.39	1.39	.75
	.90	2.97	2.59	2.38	2.25	2.16	2.09	2.04	2.00	1.96	1.94	1.91	1.89	.90
	.95	4.35	3.49	3.10	2.87	2.71	2.60	2.51	2.45	2.39	2.35	2.31	2.28	.95
	.975	5.87	4.46	3.86	3.51	3.29	3.13	3.01	2.91	2.84	2.77	2.72	2.68	.975
	.99	8.10	5.85	4.94	4.43	4.10	3.87	3.70	3.56	3.46	3.37	3.29	3.23	.99
	.995	9.94	6.99	5.82	5.17	4.76	4.47	4.26	4.09	3.96	3.85	3.76	3.68	.995
	.999	14.8	9.95	8.10	7.10	6.46	6.02	5.69	5.44	5.24	5.08	4.94	4.82	.999
	.9995	17.2	11.4	9.20	8.02	7.28	6.76	6.38	6.08	5.85	5.66	5.51	5.38	.9995
24	.0005	$.0^4 40$	$.0^5 50$	$.0^5 50$	.015	.030	.046	.064	.080	.096	.112	.126	.139	.0005
	.001	$.0^3 16$	$.0^2 10$	$.0^2 79$	.022	.040	.059	.079	.097	.115	.131	.146	.160	.001
	.005	$.0^4 40$	$.0^2 50$	.023	.050	.078	.106	.131	.154	.175	.193	.210	.226	.005
	.01	$.0^3 16$	.010	.038	.072	.106	.137	.165	.189	.211	.231	.249	.264	.01
	.025	$.0^3 10$	.025	.071	.117	.159	.195	.227	.253	.277	.297	.315	.331	.025
	.05	$.0^2 40$	.051	.116	.173	.221	.260	.293	.321	.345	.365	.383	.399	.05
	.10	.016	.106	.193	.261	.313	.355	.388	.416	.439	.459	.476	.491	.10
	.25	.104	.291	.406	.480	.532	.570	.600	.623	.643	.659	.671	.684	.25
	.50	.469	.714	.812	.863	.895	.917	.932	.944	.953	.961	.967	.972	.50
	.75	1.39	1.47	1.46	1.44	1.43	1.41	1.40	1.39	1.38	1.38	1.37	1.36	.75
	.90	2.93	2.54	2.33	2.19	2.10	2.04	1.98	1.94	1.91	1.88	1.85	1.83	.90
	.95	4.26	3.40	3.01	2.78	2.62	2.51	2.42	2.36	2.30	2.25	2.21	2.18	.95
	.975	5.72	4.32	3.72	3.38	3.15	2.99	2.87	2.78	2.70	2.64	2.59	2.54	.975
	.99	7.82	5.61	4.72	4.22	3.90	3.67	3.50	3.36	3.26	3.17	3.09	3.03	.99
	.995	9.55	6.66	5.52	4.89	4.49	4.20	3.99	3.83	3.69	3.59	3.50	3.42	.995
	.999	14.0	9.34	7.55	6.59	5.98	5.55	5.23	4.99	4.80	4.64	4.50	4.39	.999
	.9995	16.2	10.6	8.52	7.39	6.68	6.18	5.82	5.54	5.31	5.13	4.98	4.85	.9995

ν_1, DEGREES OF FREEDOM FOR NUMERATOR

Cum. prop.	15	20	24	30	40	50	60	100	120	200	500	∞	Cum. prop.	
.0005	.159	.197	.220	.244	.272	.290	.303	.330	.339	.353	.368	.377	.0005	15
.001	.181	.219	.242	.266	.294	.313	.325	.352	.360	.375	.388	.398	.001	
.005	.246	.286	.308	.333	.360	.377	.389	.415	.422	.435	.448	.457	.005	
.01	.284	.324	.346	.370	.397	.413	.425	.450	.456	.469	.483	.490	.01	
.025	.349	.389	.410	.433	.458	.474	.485	.508	.514	.526	.538	.546	.025	
.05	.416	.454	.474	.496	.519	.535	.545	.565	.571	.581	.592	.600	.05	
.10	.507	.542	.561	.581	.602	.614	.624	.641	.647	.658	.667	.672	.10	
.25	.701	.728	.742	.757	.772	.782	.788	.802	.805	.812	.818	.822	.25	
.50	1.00	1.01	1.02	1.02	1.03	1.03	1.04	1.04	1.04	1.04	1.04	1.05	.50	
.75	1.43	1.41	1.41	1.40	1.39	1.39	1.38	1.38	1.37	1.37	1.36	1.36	.75	
.90	1.97	1.92	1.90	1.87	1.85	1.83	1.82	1.79	1.79	1.77	1.76	1.76	.90	
.95	2.40	2.33	2.29	2.25	2.20	2.18	2.16	2.12	2.11	2.10	2.08	2.07	.95	
.975	2.86	2.76	2.70	2.64	2.59	2.55	2.52	2.47	2.46	2.44	2.41	2.40	.975	
.99	3.52	3.37	3.29	3.21	3.13	3.08	3.05	2.98	2.96	2.92	2.89	2.87	.99	
.995	4.07	3.88	3.79	3.69	3.59	3.52	3.48	3.39	3.37	3.33	3.29	3.26	.995	
.999	5.54	5.25	5.10	4.95	4.80	4.70	4.64	4.51	4.47	4.41	4.35	4.31	.999	
.9995	6.27	5.93	5.75	5.58	5.40	5.29	5.21	5.06	5.02	4.94	4.87	4.83	.9995	
.0005	.169	.211	.235	.263	.295	.316	.331	.364	.375	.391	.408	.422	.0005	20
.001	.191	.233	.258	.286	.318	.339	.354	.386	.395	.413	.429	.441	.001	
.005	.258	.301	.327	.354	.385	.405	.419	.448	.457	.474	.490	.500	.005	
.01	.297	.340	.365	.392	.422	.441	.455	.483	.491	.508	.521	.532	.01	
.025	.363	.406	.430	.456	.484	.503	.514	.541	.548	.562	.575	.585	.025	
.05	.430	.471	.493	.518	.544	.562	.572	.595	.603	.617	.629	.637	.05	
.10	.520	.557	.578	.600	.623	.637	.648	.671	.675	.685	.694	.704	.10	
.25	.708	.736	.751	.767	.784	.794	.801	.816	.820	.827	.835	.840	.25	
.50	.989	1.00	1.01	1.01	1.02	1.02	1.02	1.03	1.03	1.03	1.03	1.03	.50	
.75	1.37	1.36	1.35	1.34	1.33	1.33	1.32	1.31	1.31	1.30	1.30	1.29	.75	
.90	1.84	1.79	1.77	1.74	1.71	1.69	1.68	1.65	1.64	1.63	1.62	1.61	.90	
.95	2.20	2.12	2.08	2.04	1.99	1.97	1.95	1.91	1.90	1.88	1.86	1.84	.95	
.975	2.57	2.46	2.41	2.35	2.29	2.25	2.22	2.17	2.16	2.13	2.10	2.09	.975	
.99	3.09	2.94	2.86	2.78	2.69	2.64	2.61	2.54	2.52	2.48	2.44	2.42	.99	
.995	3.50	3.32	3.22	3.12	3.02	2.96	2.92	2.83	2.81	2.76	2.72	2.69	.995	
.999	4.56	4.29	4.15	4.01	3.86	3.77	3.70	3.58	3.54	3.48	3.42	3.38	.999	
.9995	5.07	4.75	4.58	4.42	4.24	4.15	4.07	3.93	3.90	3.82	3.75	3.70	.9995	
.0005	.174	.218	.244	.274	.309	.331	.349	.384	.395	.416	.434	.449	.0005	24
.001	.196	.241	.268	.298	.332	.354	.371	.405	.417	.437	.455	.469	.001	
.005	.264	.310	.337	.367	.400	.422	.437	.469	.479	.498	.515	.527	.005	
.01	.304	.350	.376	.405	.437	.459	.473	.505	.513	.529	.546	.558	.01	
.025	.370	.415	.441	.468	.498	.518	.531	.562	.568	.585	.599	.610	.025	
.05	.437	.480	.504	.530	.558	.575	.588	.613	.622	.637	.649	.659	.05	
.10	.527	.566	.588	.611	.635	.651	.662	.685	.691	.704	.715	.723	.10	
.25	.712	.741	.757	.773	.791	.802	.809	.825	.829	.837	.844	.850	.25	
.50	.983	.994	1.00	1.01	1.01	1.02	1.02	1.02	1.02	1.02	1.03	1.03	.50	
.75	1.35	1.33	1.32	1.31	1.30	1.29	1.29	1.28	1.28	1.27	1.27	1.26	.75	
.90	1.78	1.73	1.70	1.67	1.64	1.62	1.61	1.58	1.57	1.56	1.54	1.53	.90	
.95	2.11	2.03	1.98	1.94	1.89	1.86	1.84	1.80	1.79	1.77	1.75	1.73	.95	
.975	2.44	2.33	2.27	2.21	2.15	2.11	2.08	2.02	2.01	1.98	1.95	1.94	.975	
.99	2.89	2.74	2.66	2.58	2.49	2.44	2.40	2.33	2.31	2.27	2.24	2.21	.99	
.995	3.25	3.06	2.97	2.87	2.77	2.70	2.66	2.57	2.55	2.50	2.46	2.43	.995	
.999	4.14	3.87	3.74	3.59	3.45	3.35	3.29	3.16	3.14	3.07	3.01	2.97	.999	
.9995	4.55	4.25	4.09	3.93	3.76	3.66	3.59	3.44	3.41	3.33	3.27	3.22	.9995	

ν_2, DEGREES OF FREEDOM FOR DENOMINATOR

DEGREES OF FREEDOM FOR NUMERATOR

	Cum. prop.	1	2	3	4	5	6	7	8	9	10	11	12	Cum. prop.
30	.0005	$.0^440$	$.0^550$	$.0^550$	.015	.030	.047	.065	.082	.098	.114	.129	.143	.0005
	.001	$.0^516$	$.0^210$	$.0^380$	.022	.040	.060	.080	.099	.117	.134	.150	.164	.001
	.005	$.0^440$	$.0^350$	.024	.050	.079	.107	.133	.156	.178	.197	.215	.231	.005
	.01	$.0^516$	.010	.038	.072	.107	.138	.167	.192	.215	.235	.254	.270	.01
	.025	$.0^210$	.025	.071	.118	.161	.197	.229	.257	.281	.302	.321	.337	.025
	.05	$.0^240$	.051	.116	.174	.222	.263	.296	.325	.349	.370	.389	.406	.05
	.10	.016	.106	.193	.262	.315	.357	.391	.420	.443	.464	.481	.497	.10
	.25	.103	.290	.406	.480	.532	.571	.601	.625	.645	.661	.676	.688	.25
	.50	.466	.709	.807	.858	.890	.912	.927	.939	.948	.955	.961	.966	.50
	.75	1.38	1.45	1.44	1.42	1.41	1.39	1.38	1.37	1.36	1.35	1.35	1.34	.75
	.90	2.88	2.49	2.28	2.14	2.05	1.98	1.93	1.88	1.85	1.82	1.79	1.77	.90
	.95	4.17	3.32	2.92	2.69	2.53	2.42	2.33	2.27	2.21	2.16	2.13	2.09	.95
	.975	5.57	4.18	3.59	3.25	3.03	2.87	2.75	2.65	2.57	2.51	2.46	2.41	.975
	.99	7.56	5.39	4.51	4.02	3.70	3.47	3.30	3.17	3.07	2.98	2.91	2.84	.99
	.995	9.18	6.35	5.24	4.62	4.23	3.95	3.74	3.58	3.45	3.34	3.25	3.18	.995
	.999	13.3	8.77	7.05	6.12	5.53	5.12	4.82	4.58	4.39	4.24	4.11	4.00	.999
	.9995	15.2	9.90	7.90	6.82	6.14	5.66	5.31	5.04	4.82	4.65	4.51	4.38	.9995
40	.0005	$.0^410$	$.0^550$	$.0^550$	.016	.030	.048	.066	.084	.100	.117	.132	.147	.0005
	.001	$.0^516$	$.0^210$	$.0^380$	.022	.042	.061	.081	.101	.119	.137	.153	.169	.001
	.005	$.0^440$	$.0^350$	.024	.051	.080	.108	.135	.159	.181	.201	.220	.237	.005
	.01	$.0^516$	.010	.038	.073	.108	.140	.169	.195	.219	.240	.259	.276	.01
	.025	$.0^999$	.025	.071	.119	.162	.199	.232	.260	.285	.307	.327	.344	.025
	.05	$.0^240$	.051	.116	.175	.224	.265	.299	.329	.354	.376	.395	.412	.05
	.10	.016	.106	.194	.263	.317	.360	.394	.424	.448	.469	.488	.504	.10
	.25	.103	.290	.405	.480	.533	.572	.603	.627	.647	.664	.680	.691	.25
	.50	.463	.705	.802	.854	.885	.907	.922	.934	.943	.950	.956	.961	.50
	.75	1.36	1.44	1.42	1.40	1.39	1.37	1.36	1.35	1.34	1.33	1.32	1.31	.75
	.90	2.84	2.44	2.23	2.09	2.00	1.93	1.87	1.83	1.79	1.76	1.73	1.71	.90
	.95	4.08	3.23	2.84	2.61	2.45	2.34	2.25	2.18	2.12	2.08	2.04	2.00	.95
	.975	5.42	4.05	3.46	3.13	2.90	2.74	2.62	2.53	2.45	2.39	2.33	2.29	.975
	.99	7.31	5.18	4.31	3.83	3.51	3.29	3.12	2.99	2.89	2.80	2.73	2.66	.99
	.995	8.83	6.07	4.98	4.37	3.99	3.71	3.51	3.35	3.22	3.12	3.03	2.95	.995
	.999	12.6	8.25	6.60	5.70	5.13	4.73	4.44	4.21	4.02	3.87	3.75	3.64	.999
	.9995	14.4	9.25	7.33	6.30	5.64	5.19	4.85	4.59	4.38	4.21	4.07	3.95	.9995
60	.0005	$.0^440$	$.0^550$	$.0^551$	.016	.031	.048	.067	.085	.103	.120	.136	.152	.0005
	.001	$.0^516$	$.0^210$	$.0^380$	.022	.041	.062	.083	.103	.122	.140	.157	.174	.001
	.005	$.0^440$	$.0^350$	.024	.051	.081	.110	.137	.162	.185	.206	.225	.243	.005
	.01	$.0^316$	.010	.038	.073	.109	.142	.172	.199	.223	.245	.265	.283	.01
	.025	$.0^999$	.025	.071	.120	.163	.202	.235	.264	.290	.313	.333	.351	.025
	.05	$.0^240$	.051	.116	.176	.226	.267	.303	.333	.359	.382	.402	.419	.05
	.10	.016	.106	.194	.264	.318	.362	.398	.428	.453	.475	.493	.510	.10
	.25	.102	.289	.405	.480	.534	.573	.604	.629	.650	.667	.680	.695	.25
	.50	.461	.701	.798	.849	.880	.901	.917	.928	.937	.945	.951	.956	.50
	.75	1.35	1.42	1.41	1.38	1.37	1.35	1.33	1.32	1.31	1.30	1.29	1.29	.75
	.90	2.79	2.39	2.18	2.04	1.95	1.87	1.82	1.77	1.74	1.71	1.68	1.66	.90
	.95	4.00	3.15	2.76	2.53	2.37	2.25	2.17	2.10	2.04	1.99	1.95	1.92	.95
	.975	5.29	3.93	3.34	3.01	2.79	2.63	2.51	2.41	2.33	2.27	2.22	2.17	.975
	.99	7.08	4.98	4.13	3.65	3.34	3.12	2.95	2.82	2.72	2.63	2.56	2.50	.99
	.995	8.49	5.80	4.73	4.14	3.76	3.49	3.29	3.13	3.01	2.90	2.82	2.74	.995
	.999	12.0	7.76	6.17	5.31	4.76	4.37	4.09	3.87	3.69	3.54	3.43	3.31	.999
	.9995	13.6	8.65	6.81	5.82	5.20	4.76	4.44	4.18	3.98	3.82	3.69	3.57	.9995

ν_2, DEGREES OF FREEDOM FOR DENOMINATOR

Cum. prop.	ν_1, DEGREES OF FREEDOM FOR NUMERATOR												Cum. prop.	
	15	20	24	30	40	50	60	100	120	200	500	∞		
.0005	.179	.226	.254	.287	.325	.350	.369	.410	.420	.444	.467	.483	.0005	30
.001	.202	.250	.278	.311	.348	.373	.391	.431	.442	.465	.488	.503	.001	
.005	.271	.320	.349	.381	.416	.441	.457	.495	.504	.524	.543	.559	.005	
.01	.311	.360	.388	.419	.454	.476	.493	.529	.538	.559	.575	.590	.01	
.025	.378	.426	.453	.482	.515	.535	.551	.585	.592	.610	.625	.639	.025	
.05	.445	.490	.516	.543	.573	.592	.606	.637	.644	.658	.676	.685	.05	
.10	.534	.575	.598	.623	.649	.667	.678	.704	.710	.725	.735	.746	.10	
.25	.716	.746	.763	.780	.798	.810	.818	.835	.839	.848	.856	.862	.25	
.50	.978	.989	.994	1.00	1.01	1.01	1.01	1.02	1.02	1.02	1.02	1.02	.50	
.75	1.32	1.30	1.29	1.28	1.27	1.26	1.26	1.25	1.24	1.24	1.23	1.23	.75	
.90	1.72	1.67	1.64	1.61	1.57	1.55	1.54	1.51	1.51	1.50	1.48	1.46	.90	
.95	2.01	1.93	1.89	1.84	1.79	1.76	1.74	1.70	1.68	1.66	1.64	1.62	.95	
.975	2.31	2.20	2.14	2.07	2.01	1.97	1.94	1.88	1.87	1.84	1.81	1.79	.975	
.99	2.70	2.55	2.47	2.39	2.30	2.25	2.21	2.13	2.11	2.07	2.03	2.01	.99	
.995	3.01	2.82	2.73	2.63	2.52	2.46	2.42	2.32	2.30	2.25	2.21	2.18	.995	
.999	3.75	3.49	3.36	3.22	3.07	2.98	2.92	2.79	2.76	2.69	2.63	2.59	.999	
.9995	4.10	3.80	3.65	3.48	3.32	3.22	3.15	3.00	2.97	2.89	2.82	2.78	.9995	
.0005	.185	.236	.266	.301	.343	.373	.393	.441	.453	.480	.504	.525	.0005	40
.001	.209	.259	.290	.326	.367	.396	.415	.461	.473	.500	.524	.545	.001	
.005	.279	.331	.362	.396	.436	.463	.481	.524	.534	.559	.581	.599	.005	
.01	.319	.371	.401	.435	.473	.498	.516	.556	.567	.592	.613	.628	.01	
.025	.387	.437	.466	.498	.533	.556	.573	.610	.621	.641	.662	.674	.025	
.05	.454	.502	.529	.558	.591	.613	.627	.658	.669	.685	.704	.717	.05	
.10	.542	.585	.609	.636	.664	.683	.696	.724	.731	.747	.762	.772	.10	
.25	.720	.752	.769	.787	.806	.819	.828	.846	.851	.861	.870	.877	.25	
.50	.972	.983	.989	.994	1.00	1.00	1.01	1.01	1.01	1.01	1.02	1.02	.50	
.75	1.30	1.28	1.26	1.25	1.24	1.23	1.22	1.21	1.21	1.20	1.19	1.19	.75	
.90	1.66	1.61	1.57	1.54	1.51	1.48	1.47	1.43	1.42	1.41	1.39	1.38	.90	
.95	1.92	1.84	1.79	1.74	1.69	1.66	1.64	1.59	1.58	1.55	1.53	1.51	.95	
.975	2.18	2.07	2.01	1.94	1.88	1.83	1.80	1.74	1.72	1.69	1.66	1.64	.975	
.99	2.52	2.37	2.29	2.20	2.11	2.06	2.02	1.94	1.92	1.87	1.83	1.80	.99	
.995	2.78	2.60	2.50	2.40	2.30	2.23	2.18	2.09	2.06	2.01	1.96	1.93	.995	
.999	3.40	3.15	3.01	2.87	2.73	2.64	2.57	2.44	2.41	2.34	2.28	2.23	.999	
.9995	3.68	3.39	3.24	3.08	2.92	2.82	2.74	2.60	2.57	2.49	2.41	2.37	.9995	
.0005	.192	.246	.278	.318	.365	.398	.421	.478	.493	.527	.561	.585	.0005	60
.001	.216	.270	.304	.343	.389	.421	.444	.497	.512	.545	.579	.602	.001	
.005	.287	.343	.376	.414	.458	.488	.510	.559	.572	.602	.633	.652	.005	
.01	.328	.383	.416	.453	.495	.524	.545	.592	.604	.633	.658	.679	.01	
.025	.396	.450	.481	.515	.555	.581	.600	.641	.654	.680	.704	.720	.025	
.05	.463	.514	.543	.575	.611	.633	.652	.690	.700	.719	.746	.759	.05	
.10	.550	.596	.622	.650	.682	.703	.717	.750	.758	.776	.793	.806	.10	
.25	.725	.758	.776	.796	.816	.830	.840	.860	.865	.877	.888	.896	.25	
.50	.967	.978	.983	.989	.994	.998	1.00	1.00	1.01	1.01	1.01	1.01	.50	
.75	1.27	1.25	1.24	1.22	1.21	1.20	1.19	1.17	1.17	1.16	1.15	1.15	.75	
.90	1.60	1.54	1.51	1.48	1.44	1.41	1.40	1.36	1.35	1.33	1.31	1.29	.90	
.95	1.84	1.75	1.70	1.65	1.59	1.56	1.53	1.48	1.47	1.44	1.41	1.39	.95	
.975	2.06	1.94	1.88	1.82	1.74	1.70	1.67	1.60	1.58	1.54	1.51	1.48	.975	
.99	2.35	2.20	2.12	2.03	1.94	1.88	1.84	1.75	1.73	1.68	1.63	1.60	.99	
.995	2.57	2.39	2.29	2.19	2.08	2.01	1.96	1.86	1.83	1.78	1.73	1.69	.995	
.999	3.08	2.83	2.69	2.56	2.41	2.31	2.25	2.11	2.09	2.01	1.93	1.89	.999	
.9995	3.30	3.02	2.87	2.71	2.55	2.45	2.38	2.23	2.19	2.11	2.03	1.98	.9995	

ν_2, DEGREES OF FREEDOM FOR DENOMINATOR

ν_1, DEGREES OF FREEDOM FOR NUMERATOR

	Cum. prop.	1	2	3	4	5	6	7	8	9	10	11	12	Cum. prop.
120	.0005	.0⁴40	.0⁵50	.0⁵51	.016	.031	.049	.067	.087	.105	.123	.140	.156	.0005
	.001	.0³16	.0²10	.0²81	.023	.042	.063	.084	.105	.125	.144	.162	.179	.001
	.005	.0³39	.0²50	.024	.051	.081	.111	.139	.165	.189	.211	.230	.249	.005
	.01	.0³16	.010	.038	.074	.110	.143	.174	.202	.227	.250	.271	.290	.01
	.025	.0²99	.025	.072	.120	.165	.204	.238	.268	.295	.318	.340	.359	.025
	.05	.0²39	.051	.117	.177	.227	.270	.306	.337	.364	.388	.408	.427	.05
	.10	.016	.105	.194	.265	.320	.365	.401	.432	.458	.480	.500	.518	.10
	.25	.102	.288	.405	.481	.534	.574	.606	.631	.652	.670	.685	.699	.25
	.50	.458	.697	.793	.844	.875	.896	.912	.923	.932	.939	.945	.950	.50
	.75	1.34	1.40	1.39	1.37	1.35	1.33	1.31	1.30	1.29	1.28	1.27	1.26	.75
	.90	2.75	2.35	2.13	1.99	1.90	1.82	1.77	1.72	1.68	1.65	1.62	1.60	.90
	.95	3.92	3.07	2.68	2.45	2.29	2.18	2.09	2.02	1.96	1.91	1.87	1.83	.95
	.975	5.15	3.80	3.23	2.89	2.67	2.52	2.39	2.30	2.22	2.16	2.10	2.05	.975
	.99	6.85	4.79	3.95	3.48	3.17	2.96	2.79	2.66	2.56	2.47	2.40	2.34	.99
	.995	8.18	5.54	4.50	3.92	3.55	3.28	3.09	2.93	2.81	2.71	2.62	2.54	.995
	.999	11.4	7.32	5.79	4.95	4.42	4.04	3.77	3.55	3.38	3.24	3.12	3.02	.999
	.9995	12.8	8.10	6.34	5.39	4.79	4.37	4.07	3.82	3.63	3.47	3.34	3.22	.9995
∞	.0005	.0⁶39	.0⁵50	.0⁵51	.016	.032	.050	.069	.088	.108	.127	.144	.161	.0005
	.001	.0⁵16	.0²10	.0²81	.023	.042	.063	.085	.107	.128	.148	.167	.185	.001
	.005	.0³39	.0²50	.024	.052	.082	.113	.141	.168	.193	.216	.236	.256	.005
	.01	.0³16	.010	.038	.074	.111	.145	.177	.206	.232	.256	.278	.298	.01
	.025	.0²98	.025	.072	.121	.166	.206	.241	.272	.300	.325	.347	.367	.025
	.05	.0²39	.051	.117	.178	.229	.273	.310	.342	.369	.394	.417	.436	.05
	.10	.016	.105	.195	.266	.322	.367	.405	.436	.463	.487	.508	.525	.10
	.25	.102	.288	.404	.481	.535	.576	.608	.634	.655	.674	.690	.703	.25
	.50	.455	.693	.789	.839	.870	.891	.907	.918	.927	.934	.939	.945	.50
	.75	1.32	1.39	1.37	1.35	1.33	1.31	1.29	1.28	1.27	1.25	1.24	1.24	.75
	.90	2.71	2.30	2.08	1.94	1.85	1.77	1.72	1.67	1.63	1.60	1.57	1.55	.90
	.95	3.84	3.00	2.60	2.37	2.21	2.10	2.01	1.94	1.88	1.83	1.79	1.75	.95
	.975	5.02	3.69	3.12	2.79	2.57	2.41	2.29	2.19	2.11	2.05	1.99	1.94	.975
	.99	6.63	4.61	3.78	3.32	3.02	2.80	2.64	2.51	2.41	2.32	2.25	2.18	.99
	.995	7.88	5.30	4.28	3.72	3.35	3.09	2.90	2.74	2.62	2.52	2.43	2.36	.995
	.999	10.8	6.91	5.42	4.62	4.10	3.74	3.47	3.27	3.10	2.96	2.84	2.74	.999
	.9995	12.1	7.60	5.91	5.00	4.42	4.02	3.72	3.48	3.30	3.14	3.02	2.90	.9995

ν_2, DEGREES OF FREEDOM FOR DENOMINATOR

For sample sizes larger than, say, 30, a fairly good approximation to the F distribution percentiles can be obtained from

$$\log F_{\alpha}(\nu_1, \nu_2) \approx \left(\frac{a}{\sqrt{h-b}} \right) - cg$$

where $h = 2\nu_1\nu_2/(\nu_1 + \nu_2)$, $g = (\nu_2 - \nu_1)/\nu_1\nu_2$, and a, b, c are functions of α given below:

VALUES OF α

	$\alpha = .50$	.75	.90	.95	.975	.99	.995	.999	.9995
a	0	.5859	1.1131	1.4287	1.7023	2.0206	2.2373	2.6841	2.8580
b	—	.58	.77	.95	1.14	1.40	1.61	2.09	2.30
c	290	.355	.527	.681	.846	1.073	1.250	1.672	1.857

ν₁, DEGREES OF FREEDOM FOR NUMERATOR

Cum. prop.	15	20	24	30	40	50	60	100	120	200	500	∞	Cum. prop.	
.0005	.199	.256	.293	.338	.390	.429	.458	.524	.543	.578	.614	.676	.0005	120
.001	.223	.282	.319	.363	.415	.453	.480	.542	.568	.595	.631	.691	.001	
.005	.297	.356	.393	.434	.484	.520	.545	.605	.623	.661	.702	.733	.005	
.01	.338	.397	.433	.474	.522	.556	.579	.636	.652	.688	.725	.755	.01	
.025	.406	.464	.498	.536	.580	.611	.633	.684	.698	.729	.762	.789	.025	
.05	.473	.527	.559	.594	.634	.661	.682	.727	.740	.767	.785	.819	.05	
.10	.560	.609	.636	.667	.702	.726	.742	.781	.791	.815	.838	.855	.10	
.25	.730	.765	.784	.805	.828	.843	.853	.877	.884	.897	.911	.923	.25	
.50	.961	.972	.978	.983	.989	.992	.994	1.00	1.00	1.00	1.01	1.01	.50	
.75	1.24	1.22	1.21	1.19	1.18	1.17	1.16	1.14	1.13	1.12	1.11	1.10	.75	
.90	1.55	1.48	1.45	1.41	1.37	1.34	1.32	1.27	1.26	1.24	1.21	1.19	.90	
.95	1.75	1.66	1.61	1.55	1.50	1.46	1.43	1.37	1.35	1.32	1.28	1.25	.95	
.975	1.95	1.82	1.76	1.69	1.61	1.56	1.53	1.45	1.43	1.39	1.34	1.31	.975	
.99	2.19	2.03	1.95	1.86	1.76	1.70	1.66	1.56	1.53	1.48	1.42	1.38	.99	
.995	2.37	2.19	2.09	1.98	1.87	1.80	1.75	1.64	1.61	1.54	1.48	1.43	.995	
.999	2.78	2.53	2.40	2.26	2.11	2.02	1.95	1.82	1.76	1.70	1.62	1.54	.999	
.9995	2.96	2.67	2.53	2.38	2.21	2.11	2.01	1.88	1.84	1.75	1.67	1.60	.9995	
.0005	.207	.270	.311	.360	.422	.469	.505	.599	.624	.704	.804	1.00	.0005	∞
.001	.232	.296	.338	.386	.448	.493	.527	.617	.649	.719	.819	1.00	.001	
.005	.307	.372	.412	.460	.518	.559	.592	.671	.699	.762	.843	1.00	.005	
.01	.349	.413	.452	.499	.554	.595	.625	.699	.724	.782	.858	1.00	.01	
.025	.418	.480	.517	.560	.611	.645	.675	.741	.763	.813	.878	1.00	.025	
.05	.484	.543	.577	.617	.663	.694	.720	.781	.797	.840	.896	1.00	.05	
.10	.570	.622	.652	.687	.726	.752	.774	.826	.838	.877	.919	1.00	.10	
.25	.736	.773	.793	.816	.842	.860	.872	.901	.910	.932	.957	1.00	.25	
.50	.956	.967	.972	.978	.983	.987	.989	.993	.994	.997	.999	1.00	.50	
.75	1.22	1.19	1.18	1.16	1.14	1.13	1.12	1.09	1.08	1.07	1.04	1.00	.75	
.90	1.49	1.42	1.38	1.34	1.30	1.26	1.24	1.18	1.17	1.13	1.08	1.00	.90	
.95	1.67	1.57	1.52	1.46	1.39	1.35	1.32	1.24	1.22	1.17	1.11	1.00	.95	
.975	1.83	1.71	1.64	1.57	1.48	1.43	1.39	1.30	1.27	1.21	1.13	1.00	.975	
.99	2.04	1.88	1.79	1.70	1.59	1.52	1.47	1.36	1.32	1.25	1.15	1.00	.99	
.995	2.19	2.00	1.90	1.79	1.67	1.59	1.53	1.40	1.36	1.28	1.17	1.00	.995	
.999	2.51	2.27	2.13	1.99	1.84	1.73	1.66	1.49	1.45	1.34	1.21	1.00	.999	
.9995	2.65	2.37	2.22	2.07	1.91	1.79	1.71	1.53	1.48	1.36	1.22	1.00	.9995	

ν₂, DEGREES OF FREEDOM FOR DENOMINATOR

The values given in this table are abstracted with permission from the following sources:

1. All values for ν₁,ν₂ equal to 50, 100, 200, 500 are from A. Hald, *Statistical Tables and Formulas*, John Wiley & Sons, Inc., New York, 1952.

2. For cumulative proportions .5, .75, .9, .95, .975, .99, .995 most of the values are from M. Merrington and C. M. Thompson, *Biometrika*, vol. 33 (1943), p. 73.

3. For cumulative proportions .999 the values are from C. Colcord and L. S. Deming, *Sankhyā*, vol. 2 (1936), p. 423.

4. For cum. prop. = α < .5 the values are the reciprocals of values for 1 − α (with ν₁ and ν₂ interchanged). The values in Merrington and Thompson and in Colcord and Deming are to five significant figures, and it is hoped (but not expected) that the reciprocals are correct as given. The values in Hald are to three significant figures, and the reciprocals are probably accurate within one to two digits in the third significant figure except for those values very close to unity, where they may be off four to five digits in the third significant figure.

5. Gaps remaining in the table after using the above sources were filled in by interpolation.

$$\alpha = \frac{(\nu_1/\nu_2)^{\frac{1}{2}\nu_1}}{\beta(\frac{1}{2}\nu_1,\frac{1}{2}\nu_2)} \int_{-\infty}^{F\alpha} F^{\frac{1}{2}\nu_1-1}\left(1 + \frac{\nu_1 F}{\nu_2}\right)^{-(\nu_1+\nu_2)/2} dF$$

These subprograms for six probability functions were
written under the direction of Dr. Rolf E. Bargmann and are
part of an interactive statistical package developed under a
THEMIS research grant. They were written for an IBM 360/65
- 2250 console system but can be used without any modification on any system that will accept doubleprecision FORTRAN
subprograms and implicit variable name specifications.
Doubleprecision and implicit statements have not been discussed in this textbook, but can be found in any programming
textbook.

To make the package of subprograms readily available to
students, it is suggested that they be keypunched and stored
in an internal computer library so that they can be called
as though they were built-in FORTRAN functions. If only a
subset of the programs is needed, care must be exercised to
include all routines that are called by the subprograms.

For each of the six distributions, a separate subprogram is available to (1) compute the cumulative probabilities, (2) determine the critical values associated with any
specified probability, (3) find the ordinate.

The six distributions and their functional representation are given below:

(a) the standard normal distribution

$$f(z) = \frac{1}{\sqrt{2\pi}} e^{-\frac{1}{2}z^2} , \quad \infty < z < \infty .$$

(b) the student's t distribution

$$f(t;d) = \frac{\Gamma \frac{d+1}{2}}{\sqrt{d\pi} \; \Gamma\left[\frac{d}{2}\right]} [1 + t^2/d]^{-(d+1)/2} , \quad -\infty < t < \infty, \; d > 0$$

(c) the chi-square distribution

$$f(\chi^2;d) = \frac{1}{2^{d/2} \; \Gamma\left[\frac{d}{2}\right]} (\chi^2)^{(d-2)/2} e^{-\chi^2/2} , \quad 0 \leq \chi^2 \leq \infty, \; d > 0$$

(d) the F distribution

$$f(F;d_1,d_2) = \frac{\Gamma\left[\frac{d_1+d_2}{2}\right]\left[\frac{d_1}{d_2}\right]^{d_1/2} F^{(d_1-2)/2}}{\Gamma\left[\frac{d_1}{2}\right]\Gamma\left[\frac{d_2}{2}\right](1+d_1 F/d_2)^{(d_1+d_2)/2}} \quad , \quad F>0, \ d_1,d_2>0$$

(e) the gamma distribution

$$f(y;\alpha) = \frac{1}{\Gamma(\alpha)} y^{\alpha-1} e^{-y} \quad , \quad 0<y<\infty, \ \alpha>0$$

(f) the beta distribution

$$f(y;\alpha,\beta) = \frac{\Gamma(\alpha+\beta)}{\Gamma(\alpha)\cdot\Gamma(\beta)} y^{\alpha-1} (1-y)^{\beta-1} \quad , \quad 0<y<\infty, \ \alpha>0, \ \beta>0$$

The last two distributions have not been previously defined. Both the t, χ^2, F, gamma, and beta distributions use the gamma function (FUNCTION DLGGM) represented by $\Gamma(\cdot)$ which we have already encountered in Chapter 8 and 10. This doubleprecision function is found at the end of this Appendix.

Each of the six subprograms calculates the cumulative probability, $\Pr(Y \leq Y0)$ for any specified value of Y0. Each routine requires that the value of Y0 and the parameters of the distribution be specified.

(a) the standard normal distribution

$$\text{YORMX}(Y0) = \int_{-\infty}^{Y0} f(z) \; dz$$

(b) the t distribution with d degrees of freedom

$$\text{TTX}(Y0,D) = \int_{-\infty}^{Y0} f(t;d) \; dt$$

(c) the χ^2 distribution with d degrees of freedom

$$\text{CHIX}(Y0,D) = \int_{0}^{Y0} f(\chi^2;d) \; d\chi^2$$

(d) the F distribution with d_1 degrees of freedom in the numerator and d_2 degrees of freedom in the denominator.

$$\text{FFX}(Y0,D1,D2) = \int_{0}^{Y0} f(F;d_1,d_2) \; dF$$

(e) the gamma distribution with parameter α

$$\text{GAMX}(Y0,A) = \int_{0}^{Y0} f(y;\alpha) \; dy$$

(f) the beta distribution with parameters α and β

$$\text{BETAX}(Y0,A,B) = \int_{0}^{Y0} f(y;\alpha,\beta) \; dy$$

```
DOUBLE PRECISION FUNCTION YORMX(DZ)
IMPLICIT  REAL*8(D)
DOUBLE PRECISION YORMX
DPI = .3989422804041433
DX = DABS(DZ)
YORMX = 0.D0
IF(DZ.LT.-18.D0) GO TO 99
YORMX = 1.D0
IF (DZ.GT.9.D0) GO TO 99
IF(DX.GT.3.D0) GO TO 10
DAL = 0.D0
DBL = 1.D0
DAH = DX
DBH = 1.D0
DAN = 0.D0
5    DAN = DAN + 1.D0
DAI = -(2.D0*DAN - 1.D0)*DX*DX
DBI = 4.D0*DAN - 1.D0
DAL = DBI*DAH + DAI*DAL
DBL = DBI*DBH + DAI*DBL
DAI = DX*DX - DAI
DBI = 2.D0 + DBI
DAH = DBI*DAL + DAI+DAH
DBH = DBI*DBL + DAI*DBH
DFA = DAL/DBL
DFB = DAH/DBH
IF(DFB.EQ.0.D0) GO TO 20
IF(DABS((DFB-DFA)/DFA).LE.1.D-14) GO TO 20
GO TO 5
10   DAL = 0.D0
DBL = 1.D0
DAH = 1.D0
DBH = DX
DBI = DX
DAN = 1.D0
DFA = 1.D0/DX
15   DAN = DAN + 1.D0
DAI = DAN - 1.D0
DAC = DBI*DAH + DAI*DAL
DBC = DBI*DBH + DAI*DBL
DFB = DAC/DBC
DAL = DAH
DBL = DBH
DAH = DAC
DBH = DAC
IF(DFB.EQ.0.D0) GO TO 20
IF(DABS((DFB-DFA)/DFB).LE.1.D-14) GO TO 20
DFA = DFB
GO TO 15
20   YORMX = DPI*DFB*DEXP(-DX*DX/2.D0)
IF(DX.LE.3.D0) YORMX = 0.5D0 - YORMX
IF(DZ.GT.0.D0) YORMX = 1.D0 - YORMX
99   RETURN
END
```

```
      DOUBLE PRECISION FUNCTION TTX(DX,DF)
      DOUBLE PRECISION TTX,DX,DF,DS,FFX
      TTX = Ø.DØ
      IF(DF.LE.Ø.DØ) GO TO 99
      DS = DX * DX
      TTX = FFX(DS,1.DØ,DF)
      TTX = Ø.5DØ*(1.DØ - TTX)
      IF(TTX.LT.1.D-5) TTX = Ø.5DØ*FFX(1.DØ/DS,DF,1.DØ)
      IF(DX.GT.Ø.DØ) TTX = 1.DØ - TTX
   99 RETURN
      END
```

```
      DOUBLE PRECISION FUNCTION CHIX(X,DF)
      DOUBLE PRECISION CHIX,X,DF,GAMX
      CHIX = GAMX(Ø.5DØ*X,Ø.5DØ*DF)
      RETURN
      END
```

```
      DOUBLE PRECISION FUNCTION FFX(X,DA,DB)
      DOUBLE PRECISION FFX,Y,DA,DB,BETAX,X,DC
      FFX = Ø.DØ
      IF(DA.LE.Ø.DØ .OR. DB.LE.Ø.DØ) GO TO 9
      DC = DA*X/DB
      Y = DC/(1.DØ + DC)
      FFX = BETAX (Y,Ø.5DØ*DA,Ø.5DØ*DB)
   99 RETURN
      END
```

```
      DOUBLE PRECISION FUNCTION GAMX(DY,DE)
      IMPLICIT REAL*8(D)
      DOUBLE PRECISION GAMX,YORMX
      DY = DX
      DF = DE
      DSUM = Ø.DØ
      IF(DX.GT.Ø.DØ) GO TO 2
      GAMX = Ø.DØ
```

```
         GO TO 99
   2     IF(DF.GT.Ø.DØ) GO TO 4
         GAMX = 1.DØ
         GO TO 99
   4     DT = DF + DLOG(DX)-DX-DLGGM(DF)-DLOG(DABS(DF-DX))
         GAMX = Ø.DØ
         IF(DX.LT.DF.AND.DT.LT.-138.DØ) GO TO 99
         GAMX = 1.DØ
         IF (DX.GE.DF.AND.DT.LT.-39.DØ) GO TO 99
         IF((DF.GE.2ØØ.DØ).AND.(DY.LE.DF)) GO TO 21
         GAMX = 1.DØ
         IF(DX.GE.1.D36) GO TO 99
         IF((DF.GE.2.DØ).AND.(DY.GE.DF+3.DØ*DSORT(DF)))GO TO 4Ø
         DAI = DF
         DDF = DAI*DLOG(DY) - DY - DLGGM(DAI + 1.DØ)
  16     IF(DDF.LE.-8Ø.DØ) GO TO 1Ø
         DFG = DEXP(DDF)
  12     DFH = DFG
         DSUM = DSUM + DFG
         DFG = DFG*DY/(DAI + 1.DØ)
         DAI = DAI + 1.DØ
         IF(DAI.GT.2ØØ.DØ) GO TO 25
         IF(DFG.LT.DFH) GO TO 13
         IF(DAI.GT.2ØØ.DØ) TO TO 25
         GO TO 12
  13     DFH = DFG
         IF(DFG/DSUM.LE.1.D-14) TO TO 15
         GO TO 12
  1Ø     DAI = DAI + 1.DØ
         IF(DAI.GT.2ØØ.DØ) GO TO 25
         DDF = DDF + DLOG(DY/DAI)
         GO TO 16
  21     DH = 2.DØ*DF
  26     DYN = ((DY/DF)**Ø.333333333333333 -1.DØ + 1.DØ/DH)
        1*DSQRT(DH)
         DNMIX = YORMX(DYN)
         GAMX = DNMIX + DSUM
         IF(GAMX.LE.1.D-6Ø)GAMX = Ø.DØ
         IF(GAMX.GE..99999999999999)GAMX = 1.DØ
         GO TO 99
  15     GAMX = DSUM
         IF(GAMX.LE.1.D-6Ø)GAMX = Ø.DØ
         IF(GAMX.GE..99999999999999)GAMX = 1.DØ
         GO TO 99
  25     DH = 9.DØ*DAI
         DF = DAI
         GO TO 26
  4Ø     DFX = DF*DLOG(DY) - DY - DLGGM(DF)
         DAL = Ø.DØ
         DAH = 1.DØ
         DBL = 1.DØ
         DBH = DY
         DBK = 1.DØ
         DBKP = DY
         DK = 1.DØ
  42     DAK = DK - DF
         DAKP = DK
         DAL = DBK*DAH + DAK*DAL
```

```
      DBL = DBK*DBH + DAK*DBL
      DAH = DBKP*DAL + DAKP*DAH
      DBH = DBKP*DBL + DAKP*DBH
      DFA = DAL/DBL
      DFB = DAH/DBH
      IF(DFB.EQ.Ø.DØ) GO TO 45
      IF(DABS((DFA-DFB)/DFB).LE.1.D-13) GO TO 41
      DK = DK + 1.DØ
      GO TO 42
   41 DFX = DFX + DLOG(DFB)
      GAMX = 1.DØ
      IF(DFX.GE.-8Ø.DØ) GAMX = 1.DØ - DEXP(DFX)
      IF(GAMX.LE.1.D-6Ø)GAMX = Ø.DØ
      IF(GAMX.GE.99999999999999)GAMX = 1.DØ
      GO TO 99
   45 GAMX = 1.DØ
   99 RETURN
      END
```

```
      DOUBLE PRECISION FUNCTION BETAX(DX,DY,DN)
      IMPLICIT REAL *8( D,X,B)
      DOUBLE PRECISION GAMX
      NOUT = 6
      DIGEST=.999999999999DØ
      DZ=Ø.DØ
      DE=1.DØ
      DK = DM - DE
      DSUM=DZ
      IF(DM.LE. DZ) GO TO 1
      IF(DN.LE. DZ) GO TO 2
      IF(DX*(DE-DX) .LT. DZ) GO TO 1
      IF(DX.GT.DTEST) GO TO 2
      IF (DX.LT.1.D-42) GO TO 3
      DCUB=DMIN1(2ØØØ.DØ,3Ø.DØ/(DX*(1.DØ-DX)))
      IFLAG=Ø
      IF(DX .LE. DM/(DM+DN') TO TO 1Ø1
      DX=DE-DX
      DHLD=DN
      DN=DM
      DM=DHLD
      DK = DM - DE
      IFLAG=1
  1Ø1 IF(DX.GE.DM/(DM+DN)-1.D-15) TO TO 121
      DT=DLGGM(DM+DN)-DLGGM(DM)-DLGGM(DN)+DM*DLOG(DX)+
     1DN*DLOG(DE-DX)-DLOG(DM*(DE-DX)-DN*DX)
      IF(DT.LT.-138.DØ) GO TO 8
      IF(IFLAG.EQ.1.AND.DT.LT.-39.DØ) TO TO 8
  121 IF(DM.GE.19.DØ) GO TO 7
      IF(DM+DN.GE.DCUB)GO TO 7ØØ
      I=DN+.5DØ
      DNT=I
      IF( DABS(DN-DNT).LT.1.D-7) GO TO 4
```

631

```
      LI= 2Ø.DØ-DM+DE
      DK=DM-DE
      DLGN=DLGGM(DN)
      DNX=DN+DLOG(DE-DX)
      DLDX = DLOG(DX)
      DO 5 I=1,LI
      DK=DK+DE
      IF (I.GE.2) GO TO 3Ø5
      DT=DLGGM(DN+DK)- DLGN-DLGGM(DK+DE)+DK*DLOG(DX)+DNX
      IF(DT.GT. -8Ø.DØ) GO TO 6
      GO TO 5
  3Ø5 DT = DT + DLOG((DN+DK- DE)/DK) + DLDX
      IF (DT.GT.-8Ø.DØ) GO TO 6
    5 CONTINUE
      GO TO 7
    6 DSUM=DEXP(DT)
      DTERM=DSUM
      LI=2Ø.DØ-DK
      DO 9 I=1,LI
      DTERM= (DN+DK)*DX*DTERM/(DK+DE)
      IF(DTERM/DSUM.LT. 1.D-12) GO TO 8
      DSUM=DSUM+DTERM
    9 DK=DK+DE
    7 IF(DK+DN.GT.DCUB)GO TO 7ØØ
      DCFT = DZ
      DSHL = DSUM
      DK = DK + DE
      DSN = DE
      DAL = DZ
      DBL = DE
      DAH = DE
      DBH = DE
      DCFLT = DLGGM(DK+DN) - DLGGM(DK+DE) - DLGGM(DN) +
     1DK*DLOG(DX) + DN * DLOG(DE - DX)
      DCFLU = DCFLT
      KFLAG = 1
      DSUM = DE
      MFLAG = 1
      DO 135 KK = 1,8Ø
      DKK = KK
      DD1 = DK +DKK - DE
      DD2 = DK + 2.DØ*DKK - DE
      DDM1 = -DX*DD1*(DD1+DN)/(DD2*(DD2-DE1))
      DDM2 = DX*DKK*(DN-DKK)/(DD2*(DD2+DE))
      DAL = DAH + DDM1*DAL
      DAH = DAL + DDM2*DAH
      DBL = DBH + DDM1*DBL
      DBH = DBL + DDM2*DBH
      IF(DABS(DBH).LT.1.D-6Ø) GO TO 2Ø1
      IF((DCFLU.LT.-8Ø.DØ).OR.(DCFLU.GT.8Ø.DØ)) GO TO 2Ø1
      IF(KFLAG.EQ.Ø) GO TO 2Ø3
      DCFT = DEXP(DCFLT)*DSN
      KFLAG = Ø
      DSUM = DCFT
      MFLAG = Ø
  2Ø3 DCFU = DSUM
      DSUM = DCFT*DAH/DBH
      IF(DDM2.EQ.DZ) GO TO 2Ø8
```

632

```
      IF (DSUM) 135,2Ø8,2Ø2
2Ø2   IF(DABS(DSUM-DCFU)/DSUM.LT.1.D-12) GO TO 2Ø8
      GO TO 135
2Ø1   IF(DBH.EQ.DZ) GO TO 135
      KFLAG = 1
      DCFLU = DCFLT + DLOG(DABS(DAH)) - DLOG(DABS(DBH)) +
     1DLOG(DSUM)
      DSN = DSIGN(DE,DAH)*DSIGN(DE,DBH)*DSN*DSIGN(DE,DSUM)
135   CONTINUE
2Ø8   IF(MFLAG.EQ.1) DSUM = DZ
      DSUM = DSUM + DSHL
      GO TO 8
7ØØ   DK=DK+DE
      DW1=DN*DX/(DE-DX)
      DU=DK
      IF(DN.GE.DK) GO TO 7Ø7
      DW1=CK*(DE-DX)/DX
      DU=DN
      DSUM=DSUM+DE-GAMX(DW1,DU)
      GO TO 8
7Ø7   DSUM=DSUM+GAMX(DW1,DU)
  8   BETAX=DSUM
      IF(BETAX .LE. 1.D-6Ø) BETAX = Ø.DØ
      IF(BFTAX .GE. .99999999999999) BFTAX = 1.DØ
      IF( IFLAG.EQ. Ø) RETURN
      BETAX=DE-BETAX
      DX=DE-DX
      DM=DN
      DN=DHLD
      RETURN
  4   DN = DNT
      GO TO 7
  1   WRITE(NOUT,1ØØ) DM,DN,DX
1ØØ   FORMAT(///5X, 3ØHERROR IN INPUT PARAMETER BETAX,
     1 9HSET TO Ø. ,/5X,2HM=, D2Ø.8,2HN=,D2Ø.8,2HX=,D2Ø.8)
  3   BETAX=DZ
      IF(BETAX .LE. 1.D-6Ø) BETAX = Ø.DØ
      IF(BETAX .GE. .99999999999999) BETAX = 1.DØ
      RETURN
  2   BETAX=DE
      RETURN
      END
```

Each of the following subprograms determines the value of Y0 which corresponds to a given cumulative probability, P0, where $P0 = Pr(Y \leq Y0)$. The cumulative probability, P0 and the parameters of the distribution must be given.

(a) the standard normal distribution

$$Y0 = YORMP(P0),$$

where $P0 = \int_{-\infty}^{Y0} f(z) \ dz$

(b) the student's t distribution with d degrees of freedom

$$Y0 = TTP(P0,D),$$

where $P0 = \int_{-\infty}^{Y0} f(t; \ d) \ dt$

(c) the chi-square distribution with d degrees of freedom

$$Y0 = CHIP(P0,D)$$

where $P0 = \int_{0}^{Y0} f(\chi^2;d) \ d\chi^2$

(d) the F distribution with d_1 and d_2 degrees of freedom

$$Y0 = FFP(P0,D1,D2)$$

where $P0 = \int_{0}^{Y0} f(F;d_1,d_2) \ dF$

(e) the gamma distribution with parameter α

$$Y0 = GAMP(P0,A)$$

where PO $= \int_0^{Y0} f(y;\alpha)\ dy$

(f) the beta distribution with parameters α and β

$$Y0 = BETAP(PO,A,B)$$

where PO $= \int_0^{Y0} f(y;\alpha,\beta)\ dy$

```
      DOUBLE PRECISION FUNCTION YORMP(DP)
      IMPLICIT REAL*8(D,Y)
      DOUBLE PRECISION YORMZ
      Y=DP
      DOMEG = .9999999999999999D38
      IF(Y)1,1,2
    1 YORMP = -DOMEG
      GO TO 99
    2 IF(Y-1.DØ)3,4,4
    4 YORMP = DOMEG
      GO TO 99
    3 IF(Y-.5DØ)5,6,7
    6 YORMP = Ø.DØ
      GO TO 99
    5 DQ = Y
      GO TO 1Ø
    7 DQ = 1.DØ-Y
   1Ø NCYL = Ø
      DET = DSQRT(-2.DØ*DLOG(DQ))
      DXN = DET-((.Ø1Ø328DØ*DET+.8Ø2853DØ)*DET+2.515517DØ)/
     1(((.ØØ13Ø8DØ*DET+.189269DØ)*DET+1.432788DØ)*DET + 1.DØ)
      IF(Y-.5DØ)12,12,13
   12 DXN = -DXN
   13 DPN = YORMX(DXN)
      DER = DPN -Y
      IF(DABS(DER/Y) - 1.D-12) 14,14,15
   15 NCYL = NCYL +1
      IF(NCYL-2Ø) 17,17,14
   17 DXP = DXN
   29 DZ = YORMZ(DXP)
      IF ( DZ ) 27,27,28
   27 DXP = DXP/2.DØ
      GO TO 29
   28 DXN = DXN-DER/DZ
      GO TO 13
   14 YORMP = DXN
   99 RETURN
      END
```

```
      DOUBLE PRECISION FUNCTION TTP(DP,DF)
      DOUBLE PRECISION TTP,FFP,DP,DF,DP
      DR = 2.DØ*DP - 1.DØ
      IFLAG = 1
      IF(DP.GE.Ø.5DØ) GO TO 2
      DR = - DR
      IFLAG = 2
   2  TTP = DSQRT(FFP(DR,1.DØ,DF))
      GO TO (99,3),IFLAG
   3  TTP = - TTP
  99  RETURN
      END
```

```
      DOUBLE PRECISION FUNCTION CHIP(P,DF)
      DOUBLE PRECISION CHIP,P,DF,GAMP
      CHIP = 2.DØ*GAMP(P,Ø.5DØ*DF)
      RETURN
      END
```

```
      DOUBLE PRECISION FUNCTION FFP(P,DA,DB)
      DOUBLE PRECISION FFP,P,DA,DB,BETAP,DX
      FFP= .99999999999999D38
      DX= BETAP(P,Ø.5DØ*DA,Ø.5DØ*DB)
      IF(DA .LE. Ø.DØ .OR. DX .GE. 1.DØ) GO TO 99
      FFP=DX*DB/(DA*(1.DØ-DX))
  99  RETURN
      END
```

```
      DOUBLE PRECISION FUNCTION GAMP(P,DF)
      DOUBLE PRECISION GAMP,P,DF,Y,ARG,YORMP,XN,RAT,AE,
     1GAMX,PN,XO,ER,GAMZ,Z,OMEG
      Y=P
      OMEG = .999999999999999D38
      IF(DF)1,1,2
   1  GAMP = Ø.
      GO TO 99
   2  IF(P)1,1,3
   3  IF(1.-P)4,4,3Ø
   4  GAMP = OMEG
      GO TO 99
```

```
30    IF(DF-1.)31,31,5
31    ARG = YORMP(P)
      IF(ARG)32,33,33
33    XN=DF+ARG*DSQRT(DF)
      GO TO 8
32    XN = DF/2.
      GO TO 8
 5    ARG = YORMP(P)
      RAT = L./(9.*DF)
      XN=DF*((1.-RAT+ARG*DSQRT(RAT))**3)
      IF(DF-200.) 51,52,52
51    IF(XN) 7,7,8
 7    AE=(0.6931471805599945*(DF-1.)+DLGGM(DF)+DLOG(Y))/DF
      IF(AE+80.)1,1,27
27    XN=DEXP(AE)
 8    NCYCL = 0
18    PN = GAMX(XN,DF)
      XO = XN
      ER = PN-P
      IF (DABS(ER)/Y-1.E-0)9,9,10
10    NCYCL = NCYCL + 1
      IF(NCYCL-10)11,11,9
11    XO = XN-ER/GAMZ(XN,DF)
17    IF(XN-XO)12,9,12
12    IF(XO)13,13,14
13    XO = XN-0.5*(XN-XO)
      GO TO 17
14    XN = XO
      GO TO 18
52    GAMP = XN
      GO TO 99
 9    GAMP=XO
99    RETURN
      END
```

```
      DOUBLE PRECISION FUNCTION BETAP(DP,DM,DN)
      IMPLICIT REAL *8(D,B)
      DIMENSION DARG(4),DFUN(4)
      NOUT = 6
      DZ=0.D0
      DU=1.D0
      IF(DP*(DU-DP)) 95,91,20
20    IF((DM.LE.DZ).OR.(DN.LE.DZ))  GO TO 95
      IF(DM.EQ.DU)  GO TO 90
      IF (DN.EQ.DU)  DO TO 92
      DL = DZ
      DIF = DU/3.D0
      DLX = -DP
      DUX = DU - DP
      JJ = 0
89    JEND = 3
88    JJ = JJ + 1
      DO 80 J=1,JEND
```

```
      DMP = (DU+DL)/2.DØ
      IF((DU-DL).LT.1.D-4Ø) GO TO 1
      IF((DU-DL)/DP.LT.1.D-12).AND.(DL.GT.1.D-12)) GO TO 195
      DO 81 I=1,2
      DARG(I) = DL + (DU-DL)*DIF*DFLOAT(I)
      DFUN(I) = BETAX(DARG(I),DM,DN) - DP
      IF(DFUN(I).EQ.DZ) DMP = DARG(I)
      IF(DFUN(I)) 81,1,82
82    DU = DARG(I)
      DUX = DFUN(I)
      IF(I.EQ.1) TO TO 8Ø
      DL = DARG(1)
      DLX = DFUN(1)
      GO TO 8Ø
81    CONTINUE
      DL = DARG(2)
      DLX = DFUN(2)
8Ø    CONTINUE
      JEND = 2
      DMP =(DU+DL)/2.DØ
      DFD = DUX - DLX
      IF((DFD.LT.1.D-1Ø ).AND.(DFD/DP.LT.1.D-6)) GO TO 1
      DECR = DUX*(DU-DL)/DFD
      DMP = DU - DECR
      IF((DMP - DL).LT.1.D-4Ø)  GO TO 195
      IF(((DMP - DL).LT.1.D-12).AND.(DL.CT.1.D-12))GO TO 195
      DFUN(3) = BETAX(DMP,DM,DN) - DP
      DABF = DABS(DFUN(3))
      DFUNE    = DFUN(3)
      IF((DABF.LT.1.D-1Ø).AND.(DABF/DP.LT.1.D-6))  GO TO 1
      IF (DMP.LT.1.D-4Ø) GO TO 195
      IF(((DU-DMP).LT.1.D-12).AND.(DU.GT.Ø.999999999999DØ))
     1GO TO 195
      IF(DFUN(3)) 83,1,84
83    IF(DECR.LE.Ø.9DØ*(DU-DL)) GO TO 183
      DMPU = DMP
      DMP = 5.DØ*(DMP-DL) + DL
      DFUNE    = BETAX(DMP,DM,DN) - DP
      IF(DFUNE  ) 183,1,4Ø
4Ø    DU = DMP
      DUX = DFUNE
      DL = DMPU
      DLX = DFUN(3)
      IF(JJ-1Ø) 88,89,195
84    IF(DECR.GE.Ø.1DØ*(DU-DL))  GO TO 184
      DMPU = DMP
      DMP = DU - 5.DØ*DECR
      DFUNE    = BETAX(DMP,DM,DN) - DP
      IF(DFUNE  )  41,1,184
41    DU = DMPU
      DUX = DFUN(3)
      DL = DMP
      DLX = DMP
      DLX = DFUNE
      IF(JJ-1Ø) 88,89,195
183   DL = DMP
      DLX = DFUNE
      IF(JJ-1Ø)88,89,195
```

```
184   DU = DMP
      DUX = DFUNE
      IF(JJ-1Ø) 88,89,195
  1   BETAP = DMP
      RETURN
195   DRES = DFUNE + DP
      WRITE(NOUT,196)DP,DM,DN,DMP,DRES
196   FORMAT(1HØ,5X,33HNO CONVERGENCE IN BETAP IN DOUBLE,
     1 9H PRECISION / 1X,9HINPUT P = D18.1Ø,4H M = D18.1Ø =
     2D18.1Ø,9H LAST X = D18.1Ø,13H PRODUCES P = D18.1Ø)
      GO TO 1
 91   DMP = DP
      GO TO 1
 95   DMP = DZ
      WRITE (NOUT,1Ø1) DP,DM,DN
1Ø1   FORMAT (1HØ,28HARGUMENTS FOR BETAP WERE P = D12.4,
     14H M =  D12.4, 4H N = D12.4, 2ØH RESULT HAS BEEN SET
     2 7H TO ZERO )
      GO TO 1
 9Ø   DMP = DU - (DU-DP)**(DU/DN)
      GO TO 1
 92   DMP = DP**(DU/DM)
      GO TO 1
      END
```

The following subprograms determine the ordinate of each probability distribution given the abscissa, Y0, and the parameters of the distribution.

(a) the standard normal distribution

$$YORMZ(Y0) = f(Y0) = \frac{1}{\sqrt{2\pi}} e^{-\frac{1}{2}(Y0)^2}$$

(b) the t distribution with d degrees of freedom

$$TTZ(Y0,D) = f(Y0;d) = \frac{\Gamma\left[\frac{d+1}{2}\right]}{\sqrt{2\pi}\ \Gamma\left[\frac{d}{2}\right]} \left[1 + \frac{(Y0)^2}{d}\right]^{-\frac{d+1}{2}}$$

(c) the χ^2 distribution with d degrees of freedom

$$CHIZ(Y0,D) = f(Y0;d) = \frac{1}{2^{\frac{d}{2}}\ \Gamma\left[\frac{d}{2}\right]} (Y0)^{\frac{d-1}{2}} e^{-\frac{(Y0)^2}{2}}$$

(d) the F distribution with d_1 and d_2 degrees of freedom

$$FFZ(Y0,D1,D2) = f(Y0;d_1,d_2) =$$

$$= \frac{\Gamma\left[\frac{d_1+d_2}{2}\right]\left[\frac{d_1}{d_2}\right]^{\frac{d_1}{2}} (Y0)^{\frac{d_1-2}{2}}}{\Gamma\left[\frac{d_1}{2}\right]\ \Gamma\left[\frac{d_2}{2}\right]\ \left[\frac{1+d_1\cdot(Y0)}{d_2}\right]^{\frac{d_1+d_2}{2}}}$$

(e) the gamma distribution with parameter α

$$\mathrm{GAMZ}(Y0,A) = f(Y0;\alpha) = \frac{1}{\Gamma(\alpha)} \, (Y0)^{\alpha-1} \, e^{-(Y0)}$$

(f) the beta distribution with parameters α and β

$$\mathrm{BETAZ}(Y0,A,B) = f(Y0;\alpha,\beta) = \frac{\Gamma(\alpha+\beta)}{\Gamma(\alpha)\Gamma(\beta)} \, (Y0)^{\alpha-1}(1-Y0)^{\beta-1}$$

D-4a   STANDARD NORMAL ORDINATES

```
    DOUBLE PRECISION FUNCTION YORMZ(X)
    DOUBLE PRECISION X,Y,YORMZ,EY
    Y=X
    EY = -Y*Y/2.DØ
    IF(EY+8Ø.DØ)1,1,2
  1 YORMZ = Ø.DØ
    GO TO 99
  2 YORMZ = .39894228Ø4Ø1432678*DEXP(EY)
 99 RETURN
    END
```

D-4b   t ORDINATES

```
    DOUBLE PRECISION FUNCTION TTZ(DX,DF)
    DOUBLE PRECISION TTZ,DX,DF,DLGGM
    TTZ = Ø.DØ
    IF (DF .LE. Ø.DØ) GO TO 99
    TTZ = DLGGM(Ø.5DØ*(DF+1.DØ))-DLGGM(Ø.5DØ*DF)-Ø.5DØ *
   1DLOG(3.14159265358979*DF)-Ø.5DØ*(DF+1.DØ)*DLOG(1.DØ +
   2DX**2/DF)
    IF (TTZ .GT. -8Ø.DØ) GO TO 3
    TTZ = Ø.DØ
    GO TO 99
  3 TTZ = DEXP(TTZ)
 99 RETURN
    END
```

D-4c χ^2 ORDINATES

```
    DOUBLE PRECISION FUNCTION CHIZ (X,DF)
    DOUBLE PRECISION CHIZ,X,DF,GAMZ
    CHIZ = Ø.5DØ*GAMZ(Ø.5DO*X,Ø.5DØ*DF)
    RETURN
    END
```

```
      DOUBLE PRECISION FUNCTION FFZ(DX,DA,DB)
      DOUBLE PRECISION DX,DA,DB,FFZ,DZ,DC,BETAZ
      FFZ = Ø.DØ
      IF(DA.LE.Ø.DØ .OR. DB.LE.Ø.DØ) GO TO 9
      DC = DA/DB
      DZ = DC*DX/(1.DØ + DC*DX)
      FFZ = DC*(1.DØ-DZ)**2 * BETAZ(DZ,Ø.5DØ*DA,Ø.5DØ*DB)
   99 RETURN
      END
```

```
      DOUBLE PRECISION FUNCTION GAMZ(X,DF)
      DOUBLE PRECISION GAMZ,X,DF,Y,OMEG,F,DLGGM
      Y=X
      OMEG = .999999999999999D38
      IF(Y)1,1,2
    2 IF(DF)5,5,9
    1 IF(DF-1.DØ) 3,4,5
    3 GAMZ = OMEG
      GO TO 99
    4 GAMZ = 1.DØ
      GO TO 99
    5 GAMZ = Ø.DØ
      GO TO 99
    9 F = (DF - 1.DØ)*DLOG(Y) - Y - DLGGM(DF)
      IF(F + 8Ø.DØ) 7,7,8
    7 GAMZ = Ø.DO
      GO TO 99
    8 GAMZ = DEXP(F)
   99 RETURN
      END
```

```
      DOUBLE PRECISION FUNCTION BETAZ (DX,DA,DB)
      IMPLICIT REAL*8(D)
      DOUBLE PRECISION BETAZ
      DOMEG = .999999999999999D38
      BETAZ = Ø.DØ
      IF(DA.LE.Ø.DØ .OR. DB.LE.Ø.DØ) GO TO 9
      IF (DX.GT.Ø.DØ .AND. DX.LT.1.DØ) GO TO 1
      IF(DX.LE.Ø.DØ .AND.DA.GT.1.DØ) GO TO 9
      IF(DX.GE.1.DØ .AND.DB.GT.1.DØ) GO TO 9
      BETAZ = DOMEG
      IF(DX.LE.1.DØ .AND. DA.LT.1.DØ) GO TO 99
      IF(DX.GE.1.DØ .AND. DB.LT.1.DØ) GO TO 99
      BETAZ = 1.DØ
      IF(DX.LE.Ø.DØ .OR. DX.GE.1.DØ) GO TO 99
```

```
      BETAZ = DLGGM(DA+DB) - DLGGM(DA) - DLGGM(DB) +
     1(DA-1.DØ)*DLOG(DX) + (DB-1.DØ)*DLOG(1.DØ-DX)
      IF(BETAZ .GT. -8Ø.DØ .AND. BETAZ .LT. 8Ø.DØ) GO TO 2
      BETAZ = Ø.DØ
      IF(BETAZ .GE. 8Ø.DØ) BETAZ = DOMEG
      GO TO 99
    2 BETAZ = DEXP(BETAZ)
   99 RETURN
      END
```

Although this is the same program as FUNCTION ALGAMA(X), it has a different name since it is a doubleprecision function. It is called by different routines in evaluating the cumulative probabilities, the critical values, and the ordinates.

```
      DOUBLE PRECISION FUNCTION DLGGM(DX)
      IMPLICIT REAL*8(D)
      DY = DX
      DTERM = 1.DØ
      DA = .999999999999999
      DB = 1.ØØØØØØØØØØØØØØ1
      DOMEG = .999999999999999D38
      DLGGM = DOMEG
      IF(DX .LE.Ø.DØ) GO TO 99
      DLGGM = Ø.DØ
      IF(DX.GE.DA .AND. DX.LT.DB) GO TO 99
      IF(DX.GE.DA + 1.DØ .AND. DX.LT.DB + 1.DØ) GO TO 99
    2 IF (DY - 18.DØ)  3,3,4
    3 DTERM = DTERM * DY
      DY = DY + 1.DØ
      GO TO 2
    4 DS = 1.DØ/DY**2
      DZ = .641Ø25641Ø25641ØD-2
      DW = -1.917526917526918 D-3
      DV = .84175Ø84175Ø8418D-3
      DU = -5.95238Ø952380952D-4
      DT = 7.9365Ø79365Ø7937D-4
      DR = -2.7777777777777778D-3
      DQ = 8.333333333333333D-2
      DP = .9189385332Ø46727DØ
      DLGGM = (DY-.5DØ)*DLOG(DY) + DP - DY - DLOG(DTERM) +
     1(((((DZ*DS+DW)*DS+DV)*DS+DU)*DS+DT)*DS+DR)*DS+DQ)/DY
   99 RETURN
      END
```

INDEX

Series, 102
Sequential approach, 339
Set:
  complement of, 147
  element of, 146
  null, 146
Shape parameters, 411
Sigma, 28
Significance level, 238
Simple events, 146
Significant digits, 51
Simple frequency distribution, 10, 28
Simple random sampling:
  confidence limits for
    proportion, 282
  distribution of sample
    proportion, 279
  estimated variance of
    sample mean, 265
  precision compared with
    stratified random sam-
    pling, 544
  sample size needed for
    means, 246, 275
  sample size needed for
    proportions, 282
  variance of sample mean,
    219
  variance of sample
    proportion, 281
Simpson rule, A-5
Simulation, 445
Simultaneous linear equa-
  tion, 573, A-9
Size, determination of:
  for correlation estima-
    tion, 442
  for mean estimation, 246
  for percentage estima-
    tion, 282
  for regression estima-
    tion, 412
  for small samples, 275
  for variance estimation,
    464
  using decision theory
    approach, 374
Slope, 411
Sort routine, A-4
Source deck, 67, 69
Software, 61
Space, 146, 175, 564
Standard:
  moments, 38
  normal distribution, 191
  scores, 49
  Z values, 49, 238

Standard deviation, 33, 41,
  154
Standard deviation of
  errors, 403
Standard deviation of esti-
  mates, 497
States of nature, 299
Statistical inference, 3
Steps in a sample survey,
  542
Stop statement, 89
Stopping rule, 186
Strata, construction of,
  543
Strategy, 298
Stratified random sampling,
  541
Student's distribution, 267
  table of, C-3
Student's t, 265
Subroutines, 71, 223
Summation, notation, 28
Symmetry, 21, 277
  measure of, 33
System cards, 70

T

t distribution, 265, 411,
  476
Test of an hypothesis, 462,
  474
  decision approach, 354
  for proportions, 281
  large sample tests, 236
  small sample tests, 272
Test of significance:
  for correlation, 439
  for goodness of fit, 514
  for regression coeffi-
    cient, 410
  for variance, 462, 474
Third standard moment, 33
Transfer:
  control, 98
  statements, 98, 103
Translator, 62
Transformation, 47, 197
  Z*, 456
Truncation errors, 62
Two-way classification, 488
Type III distribution, 288

U

Unbiased estimator, 536
Uncertainty, 215
Ungrouped data, 8, 28

650

Uniform distribution, 175,
180, 189
Unimodal, 156
Union of sets, 146
Upper class boundary, 12

V

Variables:
  FORTRAN, 80
  random, 151
Variance:
  calculation, short
    method, 41
  definition for mathemat-
    ical model, 154
  estimation of population
    variance, 262
  of observed data, 33
  pooled estimate, 482
Vectors, 308
Venn Diagram, 147

W

Weighted average:
  of average losses, 305
  of risks, 307
Width, class, 15, 48
Write:
  loop, 128, 310
  statement, 95

Y

Y^2 distribution, 277

Z

Z transformation, 47, 197
Z* transformation, 456